Logarithmic Model

Algebraically: A log model has an equation of the form

$$f(x) = a + b \ln x$$

where $b \neq 0$.

Graphically: The graph of a log model has the form of one of the two graphs shown in the figure.

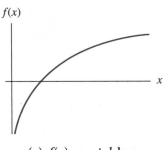

(a) $f(x) = a + b \ln x$
with $b > 0$

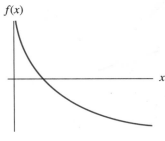

(b) $f(x) = a + b \ln x$
with $b < 0$

Logistic Model

Algebraically: A logistic model has an equation of the form

$$f(x) = \frac{L}{1 + Ae^{-Bx}}$$

We refer to L as the limiting value of the function.

Graphically: The logistic function f increases if B is positive and decreases if B is negative. The graph of a logistic function is bounded by the horizontal axis and the line $y = L$.

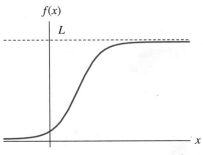

(a) $f(x) = \dfrac{L}{1 + Ae^{-Bx}}$ with $B > 0$

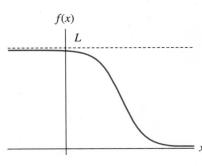

(b) $f(x) = \dfrac{L}{1 + Ae^{-Bx}}$ with $B < 0$

Quadratic Model

Algebraically: A quadratic model has an equation of the form

$$f(x) = ax^2 + bx + c$$

where $a \neq 0$.

Graphically: The graph of a quadratic function is a concave-up parabola if $a > 0$ and is a concave-down parabola if $a < 0$.

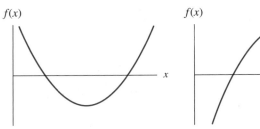

(a) $f(x) = ax^2 + bx + c$ with $a > 0$

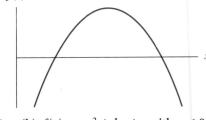

(b) $f(x) = ax^2 + bx + c$ with $a < 0$

Cubic Model

Algebraically: A cubic model has an equation of the form

$$f(x) = ax^3 + bx^2 + cx + d$$

where $a \neq 0$.

Graphically: The graph of a cubic function has one inflection point and no limiting values.

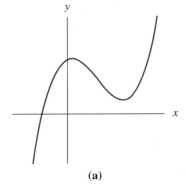

(a)

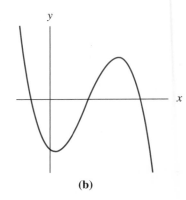

(b)

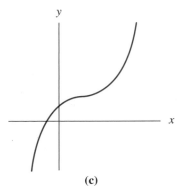

(c)

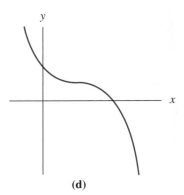

(d)

Calculus
Concepts

Brief Third Edition

BRIEF THIRD EDITION

Calculus Concepts

An Informal Approach to the Mathematics of Change

Donald R. LaTorre
Clemson University

John W. Kenelly
Clemson University

Iris Fetta Reed
Clemson University

Cynthia R. Harris
Reno, Nevada

Laurel L. Carpenter
Charlotte, Michigan

Houghton Mifflin Company
Boston   New York

Publisher: Jack Shira
Sponsoring Editor: Lauren Schultz
Development Editor: David George
Associate Editor: Jennifer King
Editorial Associate: Kasey McGarrigle
Project Editor: Kathleen Deselle
Senior Production/Design Coordinator: Carol Merrigan
Manufacturing Manager: Florence Cadran
Senior Marketing Manager: Danielle Potvin
Marketing Associate: Nicole Mollica

Cover photograph © 2003 Bruce T. Martin Photography

TI-83, TI-86, and TI-89 are registered trademarks of Texas Instruments Incorporated.

Excel, Microsoft, and Windows are either registered trademarks or trademarks of Microsoft Corporation in the United States and/or other countries.

Printed in the U.S.A.

Library of Congress Control Numbers

Student Text: 2003110169
Brief Student Text: 2003115277

International Standard Book Numbers

Student Text: 0-618-40128-8
Instructor's Annotated Edition: 0-618-40129-6
Brief Student Text: 0-618-40127-X
Brief Instructor's Annotated Edition: 0-618-40130-X

3 4 5 6 7 8 9-DOW-08 07 06 05

Contents

2 Ingredients of Change: Nonlinear Models   76

3 Describing Change: Rates

156

4 Determining Change: Derivatives   232

5 Analyzing Change: Applications of Derivatives   296

7 Analyzing Accumulated Change: Integrals in Action

470

To Students

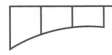

Mastering Concepts

WHAT THIS BOOK IS ABOUT

This book is written to help you understand the inner workings of how things change and to help you build systematic ways to use this understanding in everyday real-life situations that involve change. Indeed, a primary focus of the material is on change, since calculus is the mathematics of change.

Even if you have studied calculus before, this book is probably different from any other mathematics textbook that you have ever used. It is based on three premises:

1. Understanding is as important as the mastery of mathematical manipulations. Algebraic skill and the ability to manipulate expressions must be regularly practiced, or they will fade away. If you understand concepts, you will be able to explain some things in your life forever.

2. Mathematics is present in all sorts of real-life situations. It is not just an abstract subject in textbooks. In real life, mathematics is often messy and not at all like the tidy, neat equations that you were taught to factor and solve. Speaking of equations, where do they come from? Nature seldom whispers an equation into our ears.

3. The new graphics technology in today's calculators and computers is a powerful tool that can help you understand important mathematical connections. Like many tools in various fields, technology frees you from tedious, unproductive work; enables you to engage situations more realistically; and lets you focus on what you do best . . . think and reason.

HOW TO USE THIS BOOK

- Begin by throwing away any preconceived notions that you may have about what calculus is and any notion that you are "not good" in mathematics.

- Make a commitment to learn the material—not just a good intention, but a genuine commitment.

- Study this book. Notice that we said "study," not "read." Reading is a part of study, but study involves much more. You should not only read (and reread) the discussions but work through each example to understand its development.

- Use paper, pencil, and your graphing calculator or computer when you study. These are your basic tools, and you cannot study effectively without them.

- Find a study partner, if at all possible. Each of you will be able to help the other learn. Communicating within mathematics, and about mathematics, is important to your overall development toward understanding mathematics.

- Write your solutions clearly and legibly, being certain to interpret all of your answers with complete sentences using proper grammar. Careful writing will help you sort through your ideas and focus your learning.

- Make every effort not to fall behind. You know the dangers, of course, but we remind you nevertheless.

- Finally, remember that there is no substitute for effective study. You have your most valuable resource with you at all times—your mind. Use it.

Preface

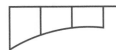

Bridging Concepts

This book presents a fresh, new approach to the concepts of calculus for students in fields such as business, economics, liberal arts, management, and the social and life sciences. It is appropriate for courses generally known as "brief calculus" or "applied calculus."

Our overall goal is to improve learning of basic calculus concepts by involving students with new material in a way that is different from traditional practice. The development of conceptual understanding coupled with a commitment to make calculus meaningful to the student are guiding forces. The material in this book involves many applications of real situations through its data-driven, technology-based modeling approach. It considers the ability to correctly interpret the mathematics of real-life situations of equal importance to the understanding of the concepts of calculus in the context of change. Complete understanding of the concepts is enhanced and emphasized by the continual use of the fourfold viewpoint: numeric, algebraic, verbal, and graphical.

Data-Driven

Many everyday, real-life situations involving change are discrete in nature and manifest themselves through data. Such situations often can be represented by continuous or piecewise continuous mathematical models so that the concepts, methods, and techniques of calculus can be utilized to solve problems. Thus we seek, when appropriate, to make real-life data a starting point for our investigations.

The use of real data and the search for appropriate models also expose students to the reality of uncertainty. We emphasize that sometimes there can be more than one appropriate model and that answers derived from models are only approximations. We believe that exposure to the possibility of more than one correct approach or answer is valuable.

Modeling Approach

We consider modeling to be an important tool and introduce it at the outset. Both linear and nonlinear models of discrete data are used to obtain functional relationships between variables of interest. The functions given by the models are the ones used by students to conduct their investigations of calculus concepts. It is the connection to real-life data that most students feel shows the relevance of the mathematics in this course to their lives and adds reality to the topics studied.

Interpretation Emphasis

This book differs from traditional texts not only in its philosophy but also in its overall focus, level of activities, development of topics, and attention to detail. Interpretation of results is a key feature of this text that allows students to make sense of the mathematical concepts and appreciate the usefulness of those concepts in their future careers and in their lives.

Informal Style

Although we appreciate the formality and precision of mathematics, we also recognize that this alone can deter some students from access to mathematics. Thus we have sought to make our presentations as informal as possible by using nontechnical terminology where appropriate and a conversational style of presentation.

PEDAGOGICAL FEATURES

- *Chapter Opener* Each chapter opens with a real-life situation and several questions about the situation that relate to the key concepts in the chapter. These applications correspond to and reference an activity in the chapter.

- *Concept Objectives* The Concept Objectives appear at the beginning of each chapter and list the goals of the chapter. The objectives are divided into two categories: concepts to be understood and skills to be learned.

- *Concept Inventory* A Concept Inventory listed at the end of each section gives students a brief summary of the major ideas developed in that section.

- *Section Activities* The Activities at the end of each section cement concepts and allow students to explore topics using, for the most part, actual data in a variety of real-world settings. Questions and interpretations pertinent to the data and the concepts are always included in these activities. The activities do not mimic the examples in the chapter discussion and thus require more independent thinking on the part of the students. Possible answers to odd activities are given at the end of the book.

- *Activity Icons* The Activity Icons found in the Instructor's Annotated Edition show the classification of the activities into different categories: Excel, technology, writing, and algebra. Even though these icons can be helpful to instructors, they do not define all that is included in each activity. In particular, some activities could be classified under several different icons.

 An Excel icon marks activities that have been written with the capabilities of spreadsheets in mind.

 The authors suggest that students have access to graphing calculators or computers with appropriate software while working these activities.

 The authors consider *Writing Across the Curriculum* to be important. These activities are designed to encourage students to communicate in written form.

 These activities are designed to draw upon students' knowledge of algebra.

- *Chapter Summary* A Chapter Summary connects the results of the chapter topics and further emphasizes the importance of knowing these results.

- *Concept Check* A check list is included at the end of each chapter summarizing the main concepts and skills taught in the chapter along with sample odd activities corresponding to each item in the list. The sample activities are to help students assess their understanding of the chapter content and identify on which areas to focus their study.

- *Chapter Review Test* A Chapter Review Test at the end of each chapter provides practice with techniques and concepts. Complete answers to the Chapter Review Tests are included in the answer key located at the back of the text.

- *Projects* Projects included after each chapter are intended to be group projects with oral or written presentations. We recognize the importance of helping students develop the ability to work in groups, as well as hone presentation skills. The projects also give students the opportunity to practice the kind of writing that they will likely have to do in their future careers.

CONTENT CHANGES IN THE THIRD EDITION

This new edition contains pedagogical changes intended to improve the presentation and flow of the concepts discussed. It contains many new examples and activities. In addition, many data sets have been updated to include more recent data.

Three important pedagogical and context changes include the addition of the NAVG icon, the incorporation of data appropriate for spreadsheets, and streamlining.

 NAVG Viewing situations from multiple perspectives offers students with different learning styles an opportunity to choose the outlook that best suits their understanding of the concepts. Even though *Calculus Concepts* has always incorporated a fourfold perspective, the use of these viewpoints is emphasized by a NAVG (an acronym for numerical, algebraic, verbal, and graphical) icon that is added occasionally in the text to remind students that mathematics can be seen through these four different perspectives.

 Excel Activities and projects involving real-world data that are appropriate for investigation using Microsoft Excel have been added to many sections in the text. Additionally, one of the two projects at the end of each chapter is designed to use Microsoft Excel. Data tables for these activities, selected answers, and projects are downloadable in spreadsheet format from the *Calculus Concepts* CD-ROM and the web site.

Streamlining Many precalculus topics have either been condensed in the third edition or placed in a supplementary appendix. In many instances, verbiage has been omitted. When discussion is important to developing an idea, a **Concept Development** head marks the dialogue. The increased algebraic rigor of the second edition is optional in the third edition.

 WEB/CD Prerequisite material and extended coverage for some topics is now available in three appendices that are on the CD-ROM and at the *Calculus Concepts* web site:

○ *Appendix A: Trigonometry Basics* This appendix provides a brief overview of traditional trigonometry.

○ *Appendix B: Strengthening the Concepts* This appendix includes supplementary topics that are organized by chapter and section and often contains activities with answer keys. In addition, Appendix B offers more thorough algebraic coverage of topics in some sections for those courses that demand a certain level of algebraic difficulty.

○ *Appendix C: Strengthening the Skills* This appendix provides additional skill-building activities.

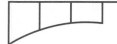

Bridging Technology

TECHNOLOGY FOCUS   ## Graphing Calculators and Spreadsheets

Calculus has traditionally relied upon a high level of algebraic manipulation. However, many nontechnical students are not strong in algebraic skills, and an algebra-based approach tends to overwhelm them and stifle their progress. Today's easy access to technology in the forms of graphing calculators and computers breaks down barriers to learning imposed by the traditional reliance on algebraic methods. It creates new opportunities for learning through graphical and numerical representations. We welcome these opportunities in this book by assuming continual and immediate access to technology.

This text requires that students use graphical representations freely, make numerical calculations routinely, and find functions to fit data. Thus continual and immediate access to technology is essential. Because of their low cost, portability, and ability to personalize the mathematics, the use of graphing calculators or laptop computers with software such as Excel or Maple is appropriate.

Technology Guides

Because it is not the authors' intent that class time be used to teach technology, we provide two Technology Guides for students: a *Graphing Calculator Instruction Guide* containing keystroke information adapted to materials in the text for the TI-83 and TI-86 (available at the web site) models, and an *Excel Instruction Guide* providing the same instruction for Excel spreadsheets. Instruction on using the TI-89 graphing calculator can be found on the companion web site. In the student text, open book icons refer readers to applicable sections within the appropriate technology guide.

It is worth noting that different technologies may give different model coefficients than those given in this book. We used a TI-83 graphing calculator to generate the models in the text and the answer key. Other technologies may use different fit criteria for some models than that used by the TI-83.

Eye on Computers and the Internet

The *Calculus Concepts* **CD-ROM and Web Site** (accessible through **college.hmco.com**) provides an exceptional variety of valuable resources for instructors and students alike. The instructors' CD-ROM includes worksheets, presentation slides, additional projects, data sets categorized by type for use on tests and quizzes, and other resource materials.

The student CD-ROM provides a glossary of terms, skill and drill problems, and graphing calculator keystroke information. In addition, the web icon found in the margins of the textbook directs students to the CD-ROM for further information on **WEB/CD** helpful topics, additional skill practice problems, and links to sites for obtaining data updates.

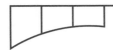

Building Bridges to Better Learning

RESOURCES FOR INSTRUCTORS

In addition to the resources found on the CD-ROM and at the web site, the printed *Instructor's Resource Guide with Complete Solutions* gives practical suggestions for using the text in the manner intended by the authors. It gives suggestions for various ways to adapt the text to your particular class situation. It contains sample syllabi, sample tests, ideas for in-class group work, suggestions for implementing and grading projects, and complete activity solutions.

The *Instructor's Annotated Edition* is the text with margin notes from the authors to instructors. The notes contain explanations of content or approach, teaching ideas, indications of where a topic appears in later chapters, indications of topics that can be easily omitted or streamlined, suggestions for alternate paths through the book, warnings of areas of likely difficulty for students based on the authors' years of experience teaching with *Calculus Concepts*, and references to topics in the *Instructor's Resource Guide* that may be helpful.

Also included in the *Instructor's Annotated Edition* are icons that show the classification of each section's activities into different solution categories. In-line headers have been added so that instructors can easily identify activities that would be of interest to their students.

LEARNING RESOURCES FOR STUDENTS

1. The *Calculus Concepts Video Series* contains chapter-by-chapter lectures by a master teacher. The video series can be used by students who miss a class or by students who think they would benefit from seeing another teacher explain a particular topic. These videos can also be used as training tools for graduate teaching assistants.

2. The *Student Solutions Manual* contains complete solutions to the odd activities.

3. The *Graphing Calculator Instruction Guide* contains keystroke information adapted to material in the text for the TI-83 and TI-86 models. Instruction on using the TI-89 graphing calculator can be found on the companion web site.

4. An *Excel Instruction Guide* provides basic instruction on this spreadsheet program.

These two Technology Guides contain step-by-step solutions to examples in the text and are referenced in this book by a supplements icon.

5. The *Calculus Concepts* Web Site (accessible through **college.hmco.com**) contains extra practice problems, help with algebra, links to updated data needed for certain activities, a glossary of terms, practice quizzes, and other assistance. The web **WEB/CD** icon in the text directs you to the book-specific web site when appropriate.

Acknowledgments

We gratefully acknowledge the many teachers and students who have used this book in its preliminary, first, and second editions and who have given us feedback and suggestions for improvement. In particular, we thank the following reviewers whose many thoughtful comments and valuable suggestions guided the preparation of this third edition revision.

Patricia Beaulieu—*University of Louisiana at Lafayette*
Joseph Cieply—*Elmhurst College*
Mark Clark—*Palomar College*
John Haverhals—*Bradley University*
Michael Helinger—*Clinton Community College*
Larry Johnson—*Metropolitan State College of Denver*
Christine Joy—*University of Oklahoma*
Greg Klein—*Texas A&M University*
Biyue Liu—*Monmouth University*
Diana McCoy—*Florida International University*
Lawrence Merbach—*North Dakota State College of Science*
Carl Mueller—*Georgia Southwestern State University*
Patrick O'Brien—*Mesa Community College*
David Platt—*Mesa Community College*
Richard Porter—*Northeastern University*
Carol Rardin—*Central Wyoming College*
Helen Read—*University of Vermont*
Bernd Rossa—*Xavier University*
Charlotte Simmons—*University of Central Oklahoma*
Donna Stein—*University of Tennessee, Knoxville*
Paul Vaz—*Arizona State University*

We especially acknowledge the help of

Sherry Biggers—*Clemson University*
Jennifer LaVare—*Clemson University*
Kristin Miller Bunnell—*Clemson University*

Special thanks to Sudhir Goel for his careful work in checking the text and answer key for accuracy. The authors express their sincere appreciation to Charlie Hartford, who first believed in this book, and to Lauren Schultz, Dave George, Jennifer King, Marika Hoe, Kathy Deselle, and their associates at Houghton Mifflin Company for all their work in bringing this third edition into print.

Heartfelt thanks to our husbands, Jim Reed and Dean Carpenter, without whose encouragement and support this edition would not have been possible. Thanks also to Jessica, Travis, Lydia, and Carl, whose cooperation was much appreciated.

Calculus Concepts 3/e Promotes Student

Technology Focus

This graphing calculator-dependent text creates new opportunities for learning through graphical and numerical representations. Rather than overwhelm students with an algebraic approach, this text uses technology to support students as they learn difficult calculus concepts. Spreadsheet options are offered to expose business students to applications they will encounter in their future careers.

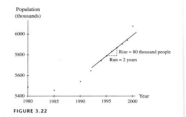

FIGURE 3.22

If you picture a smooth curve through these data points, then the slope of the secant line in Figure 3.22 should be close to the slope of the curve at 1996.

Although *symmetric difference quotient* is a cumbersome name, each word in the phrase is descriptive and should help you remember how to calculate a symmetric difference quotient. We use the term *symmetric* because we choose a close point on either side of our point of interest and the same distance away. (This technique should be used with care if the data are not equally spaced around the point at which we desire to estimate the rate of change). We use the term *difference* because we find the difference in the inputs and outputs of the two points and the term *quotient* because we divide to find the slope that is used to estimate the rate of change.

Again, consider the population data in Table 3.4 and the task of using a symmetric difference quotient to estimate the instantaneous rate of change in 1994. Note that the points on either side of 1994 are not equal distances away from 1994. In order to use a symmetric difference quotient, you would have to calculate the slope of the secant line using the years 1992 and 1996:

$$\text{Slope} = \frac{5835 - 5649}{1996 - 1992} = \frac{186 \text{ thousand people}}{4 \text{ years}} = 46.5 \text{ thousand people per year}$$

To summarize, symmetric difference quotients are useful if you want a fast estimate, if you cannot construct a model for a given data set, or if you need to estimate a rate of change that cannot be calculated from a given model.

> **Using Symmetric Difference Quotients**
>
> A symmetric difference quotient may be used to approximate an instantaneous rate of change if
>
> ○ You have a set of data but do not wish to take the time to find a model for the data.
>
> ○ You have a set of data that cannot be modeled by one of the functions we have studied or whose model does not have a tangent line at the point at which you need to know the instantaneous rate of change.

Instructor Note
Functions similar to those graphed in Figures 3.21a, c, and d occur in many real-life modeling situations in this text. Graphs like the one shown in Figure 3.21b are seen in chemistry with titration data.

The graph in Figure 3.21d has a break at *P* and, therefore, is not continuous at *P*. This graph is similar to the Indiana population graph, and the slope does not exist at the break in the function. The slope does exist at all other points on the graph.

The graph in Figure 3.21d illustrates a general rule relating continuity and rates of change.

> If a function is not continuous at a point, then the instantaneous rate of change does not exist at that point.

3.2.3

If you keep in mind the relationships among instantaneous rates of change, slopes of tangent lines, slopes of secant lines, and local linearity, then you should have little difficulty determining the times when the instantaneous rate of change does not exist.

Recall from our discussion in Chapter 1 that calculus is applied to continuous portions of functions. As we have just seen, instantaneous rates of change do not exist where a function is not continuous. Modeling is a tool by which we transform discrete data into a continuous function. Discrete data are useful for finding change, percentage change, and average rates of change, but a continuous or piecewise continuous function is necessary in order to find instantaneous rates of change.

Approximating Instantaneous Rates of Change

Instructor Note
Although it is possible to omit a discussion of symmetric difference quotients, many activities throughout the text ask students to use this technique.

Concept Development: Symmetric Difference Quotient Although a continuous or piecewise continuous model is necessary to calculate an instantaneous rate of change, it is possible to use a **symmetric difference quotient** to *estimate* a rate of change from a set of data without taking the time to construct a model. Quick approximations are sometimes as helpful as a more involved calculation.

TABLE 3.4

Year	1980	1985	1990	1992	1994	1995	1996	1997	1998	1999	2000
Population (thousands)	5490	5459	5544	5649	5746	5792	5835	5872	5908	5943	6080

(Source: *Statistical Abstract*, 2001.)

Consider the data in Table 3.4, which show the population of Indiana between 1980 and 2000. To estimate how quickly the population was changing in 1996 using only the data, simply find the slope of the line between the points on either side of 1996.

$$\frac{5872 - 5792}{1997 - 1995} = \frac{80 \text{ thousand people}}{2 \text{ years}} = 40,000 \text{ people per year}$$

In 1996, the population of Indiana was growing at a rate of approximately 40,000 people per year.

A scatter plot of the data and the secant line whose slope is our symmetric difference quotient are shown in Figure 3.22.

Multiple Representations

A compass icon indicates when students can see the mathematics through four different perspectives: numerically, algebraically, verbally, and graphically. This feature supports students of various learning styles—particularly those who are visual learners.

Success and Motivates Students

Section Activities

Each section ends with activities that allow students to investigate topics using real-world data. These exercises require students to think more critically about a particular concept. To further engage the students, the chapter openers include applications that correspond to and reference activities in the chapter.

3.4 Numerically Finding Slopes 213

○ *Slopes of piecewise functions*
○ *The derivative does not exist at a point if the function is not continuous at that point or if the function is continuous but not smooth, that is, if the limiting value of slopes of nearby secant lines from the left does not equal the limiting value of slopes of secant lines from the right.*

3.4 Activities

1. **a.** Sketch a line tangent to the graph of $y = 2^x$ at the point corresponding to $x = 2$, and estimate its slope.

$y = 2^x$

b. Use the equation $y = 2^x$ to estimate numerically the slope of the line tangent to the graph at $x = 2$.

2. **a.** Sketch a line tangent to the graph of $y = -x^2 + 4x$ at the point corresponding to $x = 3$, and estimate its slope.

$y = -x^2 + 4x$

b. Use the equation $y = -x^2 + 4x$ to estimate numerically the slope of the line tangent to the graph at $x = 3$.

3. **a.** Numerically estimate the limit of slopes of secant lines on the graph of $g(x) = x^3 + 2x^2 + 1$ between the point corresponding to $x = 2$ and close points to the left of $x = 2$.

b. Numerically estimate the limit of slopes of secant lines on the graph of $h(x) = x^2 + 16x$ between the point corresponding to $x = 2$ and close points to the right of $x = 2$.

c. Combine the functions g and h to form the piecewise continuous function

$$f(x) = \begin{cases} x^3 + 2x^2 + 1 & \text{when } x \le 2 \\ x^2 + 16x & \text{when } x > 2 \end{cases}$$

Is the graph of f continuous at $x = 2$? Is the graph of f smooth at $x = 2$?

d. Is it possible to sketch a line tangent to the graph of f at $x = 2$? If so, what is its slope? If not, why not?

4. **a.** What is the slope of $f(t) = 4t + 15$ at $t = 3$?

b. Numerically estimate the limit of slopes of secant lines on the graph of $h(t) = 8(1.5^t)$ between the point corresponding to $t = 3$ and close points to the right of $t = 3$.

c. Combine the functions f and h to form the piecewise continuous function

$$g(t) = \begin{cases} 4t + 15 & \text{when } t \le 3 \\ 8(1.5^t) & \text{when } t > 3 \end{cases}$$

Is the graph of g continuous at $t = 3$? Is the graph of g smooth at $t = 3$?

d. Is it possible to sketch a line tangent to the graph of g at $t = 3$? If so, what is its slope? If not, why not?

5. **AIDS** The graph depicts the number of AIDS cases diagnosed between 1994 and 2000. The equation for the graph is $\text{Cases} = \dfrac{36.631}{1 + 0.051e^{1.123965x}} + 42$ thousand where x is the number of years since 1994.

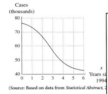

Cases (thousands)

Years since 1994

(Source: Based on data from *Statistical Abstract,* 2

Instructor Note You may wish to specify a desired accuracy in the numerical estimates of tangent line slopes in these activities. Of prepared to accept any reasonable answer.

---

216 CHAPTER 3 Describing Change: Rates

14. Discuss the advantages and disadvantages of finding rates of change graphically and numerically. Include in your discussion a brief description of when each method might be appropriate to use.

15. **Price** Cattle prices (for choice 450-pound steer calves) from April 1998 through December 1998 can be modeled by the equation

$$p(m) = \begin{cases} 0.3172m^3 - 2.0820m^2 \\ \quad -1.7895m + 98.6398 & \text{when} \\ \quad \text{dollars per 100 pounds} & 0 \le m \le 5 \\ 1.669m + 68.904 \text{ dollars} & \text{when} \\ \quad \text{per 100 pounds} & 0 < m \le 8 \end{cases}$$

where m is the number of months since April 1998.
(Source: Based on data from *USDA: Cattle and Beef Industry Statistics,* March 1999.)

a. What is the limiting value of slopes of secant lines on the graph of p from the left of $m = 5$?

b. What is the slope of the portion of the graph of p to the right of $m = 5$?

c. What do your answers to parts a and b tell you about the derivative of p at $m = 5$?

d. Estimate the rate of change of cattle prices in September of 1998.

16. **Jails** The capacity of jails in a southwestern state has been increasing since 1990. The average daily population of one of the jails can be modeled by

$$j(t) = \begin{cases} 8.10t^3 - 55.53t^2 + 128.8t + 626.8 \\ \quad \text{inmates} & \text{when} \le t \le 5 \\ 18.8t + 800.6 \text{ inmates} & \text{when } t > 5 \end{cases}$$

where t is the number of years since 1990.
(Source: Based on data from Washoe County Jail, Reno, Nevada.)

a. What is the limiting value of slopes of secant lines on the graph of j from the left of $t = 5$?

b. What is the slope of the portion of the graph of j to the right of $t = 5$?

c. What do your answers to parts a and b tell you about the derivative of j at $t = 5$?

d. Estimate the rate of change of jail population in 1995.

17. **Profit** Let $P(x) = 1.02^x$ Canadian dollars be the profit from the sale of x mountain bikes. On

November 25, 2002, P Canadian dollars were worth $C(P) = \dfrac{P}{1.5786}$ American dollars. Assume that this conversion applies today.

a. Write a function for profit in American dollars from the sale of x mountain bikes.

b. What is the profit in Canadian and in American dollars from the sale of 400 mountain bikes?

c. How quickly is profit (in American dollars) changing when 400 mountain bikes are sold?

18. **Profit** Refer to the functions P and C in Activity 17.

a. Write a function giving average profit per mountain bike for the sale of x mountain bikes in Canadian dollars.

b. Write a function for average profit in American dollars.

c. How quickly is average profit (in American dollars) changing when 400 mountain bikes are sold?

19. **Social Security** The table for this activity (located on the *Calculus Concepts* CD-ROM and web site) gives the number of retired workers and survivors receiving OASI benefits. Consider the year and widow/widower beneficiaries columns in this table.

a. Divide the data into two subsets that have the 1996 data point in common. Find an equation to fit each set of data, and combine the two to form a piecewise continuous model W with input x years after 1975.

b. Use the model in part a and at least five carefully chosen close points from each direction to estimate numerically the rate of change of W at $x = 23$. Interpret this result.

c. What is the limiting value of the slopes of secant lines of the graph of W from the left of $x = 21$? What is the slope of the portion of the graph of W to the right of $x = 21$?

d. What do your answers to part c tell you about the derivative of W at $x = 21$?

e. Estimate the rate of change of the number of widow/widower beneficiaries in 1996.

20. **Population** The table for this activity (located on the *Calculus Concepts* CD-ROM and web site) gives the U.S. population for ages 0 through 85 by gender between 1990 and 1999 with projections from 2000 through 2020.

Modeling Approach

Students use real data and graphing technology to build their own models and interpret results. This approach emulates what students will have to do in their careers.

Calculus Concepts 3/e Provides a Variety

Spreadsheet and Graphing Calculator Usage

Spreadsheet and graphing calculator usage is integrated throughout the text.

Sections in the manuals are referenced in the text by a supplements icon next to the particular example discussed in the technology guide.

The natural logarithm function is not defined for negative input values or for an input of zero. But as x approaches 0 from the right ($x \rightarrow 0^+$), the outputs of the natural log function decrease without bound. The tilt of the function appears to become vertical. That is, the slope is increasing without bound as x approaches 0 from the right. Also, as x increases without bound, $\ln x$ increases without bound, but more and more slowly. The slope never becomes zero. Thus, as $x \rightarrow \infty$, the derivative function approaches 0. (See Figures 4.38a and b.)

FIGURE 4.38

(a)   **(b)**

Numerically investigating the slope (to three decimal places) for a few input values (to three decimal places) once again helps to establish magnitude. (See Table 4.10.) Unlike the case for the exponential function, whose derivative at a certain input value is dependent on the corresponding output value, these derivatives are dependent on the input values. Note that each derivative value is the reciprocal of the input value—that is, it is 1 divided by the input value. The derivative of $y = \ln x$ at $x = 2$ is $\frac{1}{2}$, the derivative at $x = 4$ is $\frac{1}{4}$, the derivative at $x = 10$ is $\frac{1}{10}$, and so on.

TABLE 4.10

x	Derivative of $y = \ln x$
$\frac{1}{2}$	2.000
1	1.000
2	0.500
4	0.250
10	0.100

Derivative of ln x

If $y = \ln x$, then $\frac{dy}{dx} = \frac{1}{x}$ for positive x values.

Our next example illustrates the use of all of the derivative rules presented in this section and the previous section.

EXAMPLE 2   *Using Simple Derivative Rules*

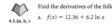

4.3.2a, b, c

Find the derivatives of the following functions:

a. $f(x) = 12.36 + 6.2 \ln x$ b. $g(t) = 4e^t + 19$

Project 3.2 Doubling Time

EXCEL

Setting

Doubling time is defined as the time it takes for an investment to double. Doubling time is calculated by using the compound interest formula $A = P\left(1 + \frac{r}{n}\right)^{nt}$ or the continuously compounded interest formula $A = Pe^{rt}$. An approximation to doubling time can be found by dividing 72 by 100r. This approximating technique is known as the **Rule of 72.**

Dr. C. G. Bilkins, a nationally known financial guru, has been criticized for giving false information about doubling time and the Rule of 72 in seminars. Your team has been hired to provide mathematically correct information for Dr. Bilkins to use.

Tasks

1. Construct a table of doubling times for interest rates from 2% to 20% (in increments of 0.25%) when interest is compounded annually, semiannually, quarterly, monthly, weekly, and daily. Construct a table of doubling-time approximations for interest rates of 2% through 20% when using the Rule of 72. Devise similar rules for 71, 70, and 69. Then construct tables for these rules. Examine the tables and determine the best approximating rule for interest compounded semiannually, quarterly, monthly, weekly, and daily. Justify your choices.

   For each interest compounding listed above, compare percent errors when using the Rule of 72 and when using the rule you choose. Percent error is $\frac{\text{estimate} - \text{true value}}{\text{true value}}$ 100%. Comment on when the rules overestimate, when they underestimate, and which is preferable.

2. Dr. Bilkins is interested in knowing how sensitive doubling time is to changes in interest rates. Estimate rates of change of doubling times at 2%, 8%, 14%, and 20% when interest is compounded quarterly. Interpret your answers in a way that would be meaningful to Dr. Bilkins.

Reporting

1. Prepare a written report for Dr. Bilkins in which you discuss your results in Tasks 1 and 2. Be sure to discuss whether Dr. Bilkins should continue to present the Rule of 72 or present other rules that depend on the number of times interest is compounded.

2. Prepare a summary document for Dr. Bilkins. It should include (a) a brief summary of how to estimate doubling time using an approximation rule and (b) a statement about the error involved in using the approximation. Also include a brief statement summarizing the sensitivity of doubling time to fluctuations in interest rates. Include the document in your written report.

3. (Optional) Prepare a brief (15-minute) presentation of your study. You will be presenting it to Dr. Bilkins. Your presentation should be only a summary, but you need to be prepared to answer any technical questions that may arise.

Instructor Note
If you are not using Excel and wish to assign this project to your students, scale down the tables to be created in Task 1 to increments of 2% and omit weekly compounding.

231

Projects

Instructors have the option to assign projects to their students. These projects allow students to strengthen their writing and presentation skills to use in their future careers and learn how to communicate mathematically.

# of Teaching and Technology Options

22. Phone Bill The graph shows the average monthly cellular phone bill since 1987. The slope of the curve at point *A* is −6.23.

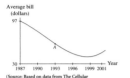

(Source: Based on data from The Cellular Telecommunications and Internet Association.)

a. What should be the units on the slope at point *A*?

b. How rapidly was the dollar amount growing in 1993?

c. What is the slope of the tangent line at point *A*?

d. What is the instantaneous rate of change of the amount at point *A*?

23. Growth Rate The growth of a pea seedling as a function of time can be modeled by two quadratic functions as shown. The slopes at the labeled points are (in ascending order) −4.2, 1.3, and 5.9.

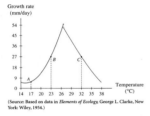

(Source: Based on data in *Elements of Ecology*, George L. Clarke, New York: Wiley, 1954.)

a. Match the slopes with the points, *A*, *B*, and *C*.

b. What are the units on the slopes for each of these points?

c. How quickly is the growth rate changing with respect to temperature at 23°C?

d. What is the slope of the tangent line at 32°C?

e. What is the instantaneous rate of change of the growth rate of pea seedlings at 17°C?

24. Beetles The graph shows the survival rate (percentage surviving) of three stages in the development of a flour beetle (egg, pupa, and larva) as a function of the relative humidity.

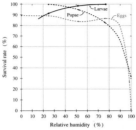

(Source: Chapman, *Animal Ecology.* New York: McGraw-Hill, 1931.)

In parts *a* through *g*, fill in each of the following blanks with the appropriate stage (eggs, pupae, or larvae).

a. At 60% relative humidity, the instantaneous rate of change of the survival rate of _____ is approximately zero.

b. An increase in relative humidity improves the survival rate of _____ and reduces the survival rate of _____.

c. At 97% relative humidity, the survival rate of _____ is declining faster than that of _____.

d. Any tangent lines drawn on the survival curve for _____ will have negative slope.

e. Any tangent lines drawn on the survival curve for _____ will have positive slope.

f. At 30% relative humidity, the survival rates for _____ and _____ are changing at approximately the same rate.

Diverse Exercise Sets

New headings in the exercise sets indicate the type of problem that follows. These exercises include writing, applications, technology, and explorations. The headings make it easy for instructors to customize assignments to the diverse needs and interests of their students. Instructors also have a wide variety of exercise formats to choose from when creating homework assignments.

Activities Key

At the beginning of each exercise set there is an activities key, which labels the types of exercices that can be found in the activities set. This feature allows professors to quickly see the type of exercises they are including in their assignments. The technology icons are more prominent in this edition, making it easier for instructors to assign the appropriate exercises. The Activities Key is not shown in the student text in order to encourage students to determine independently the best method to solve the problem.

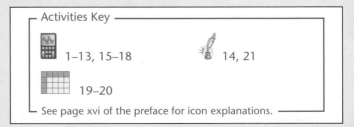

Calculus Concepts 3/e Offers Instructor

Instructor Resources

HM ClassPrep CD-ROM with HMTesting
(0-618-40137-7)

This package includes both HM ClassPrep and HMTesting. It allows instructors to access both lecture aids and algorithmic testing software in one place.

Instructor's Resource Guide with Solutions
(0-618-40134-2)

This guide provides practical suggestions, sample syllabi, sample tests, group work ideas, and the solutions to all the activities to help instructors plan their lessons and homework assignments.

Instructor's Annotated Edition
Full Version: 0-618-40129-6
Brief Version: 0-618-40130-X

Unique for this course, the IAE includes margin notes that contain comments from the authors to instructors. The notes are based on the authors' years of experience teaching with *Calculus Concepts* and include explanations of content or approach, teaching ideas, warnings of areas that may present difficulty to students, and references to topics in the *Instructor's Resource Guide* that may be helpful.

Instructional Videos/DVDs
DVD: 0-618-40135-0
Videos: 0-618-40136-9

The Video/DVD series provides additional explanation of concepts, sample problems, and applications to help students review essential topics. The comprehensive video coverage, organized by textbook section, can be used by students who miss a class, by students who would like to see another presentation, or by graduate teaching assistants as an instructional tool.

and Student Support Packages

Student Resources

HM mathSpace Student CD-ROM
(0-618-40138-5)

WEB/CD This CD-ROM contains a glossary, tutorial practice problems, and graphing calculator keystroke information. Students who need additional help can practice difficult calculus concepts using this CD-ROM.

Student Companion Web Site

WEB/CD Web icons found in the textbook margins direct students to further information on helpful topics found on the web, such as additional practice problems and links to sites for obtaining data updates.

Technology Guides

There are two technology guides available: a **Graphing Calculator Instruction Guide** (ISBN: 0-618-40132-6) and an **Excel Instruction Guide** (ISBN: 0-618-40131-8). These manuals show students how to solve certain examples in the text. They include instructions for the TI-83 and TI-86 calculators as well as for Excel. Instruction for using the TI-89 is provided on the companion web site.

SMARTHINKING

Houghton Mifflin and *SMARTHINKING*™ provide students with online, text-specific tutoring when they need it. SMARTHINKING password cards can be packaged with this text. To learn more about this online tutoring service, instructors should contact their local Houghton Mifflin sales representative.

Student Solutions Manual
(0-618-40133-4)

Students can see the complete solution worked out in the Student Solutions Manual for all the odd-numbered problems.

Calculus
Concepts

Brief Third Edition

CHAPTER

1

Ingredients of Change: Functions and Linear Models

The primary goal of this book is to help you understand the two fundamental concepts of calculus—the derivative and the integral—in the context of the mathematics of change. This first chapter is therefore devoted to a study of the key ingredients of change: functions and mathematical models. Functions provide the basis for analyzing the mathematics of change because they enable us to describe relationships between variable quantities.

This chapter introduces you to the process of building mathematical models. Because many of the models we use are formed from data, it is necessary to understand which values of the variables make sense in applied situations. We also use functions to explore the calculus concepts of the infinitely large and the infinitesimally small.

Concept Objectives

This chapter will help you understand the concepts of

- ○ Mathematical model
- ○ Function
- ○ Discrete versus continuous
- ○ Function composition
- ○ Inverse function
- ○ Limit at a point and end behavior
- ○ Continuity
- ○ Linear function, slope, and rate of change

and you will learn to

- ○ Set up and appropriately label a model
- ○ Find and interpret inputs and outputs of functions
- ○ Interpret graphs
- ○ Find composite functions
- ○ Work with piecewise continuous functions
- ○ Estimate limits graphically and numerically
- ○ Determine end behavior
- ○ Use technology to fit a linear function to data
- ○ Determine the rate of change of a linear model

Jose Luis Pelaez, Inc./CORBIS

Concept Application

Chlorofluorocarbons (CPCs) released into the atmosphere are believed to thin the ozone layer. A worldwide protocol calling for the phasing out of all CFC production was ratified in 1987. Here are some questions about the release of CFCs into the atmosphere that can be answered mathematically by using functions and calculus:

○ What was the amount of CFCs released into the atmosphere in 1975? in 1995?

○ At what rate was the release of CFCs declining between 1974 and 1980?

○ On the basis of data accumulated since 1987, in what year will there no longer be any CFCs released into the atmosphere?

This chapter gives you the tools that are necessary to answer such questions. The information needed to answer these questions is found in Activity 27 of Section 1.4.

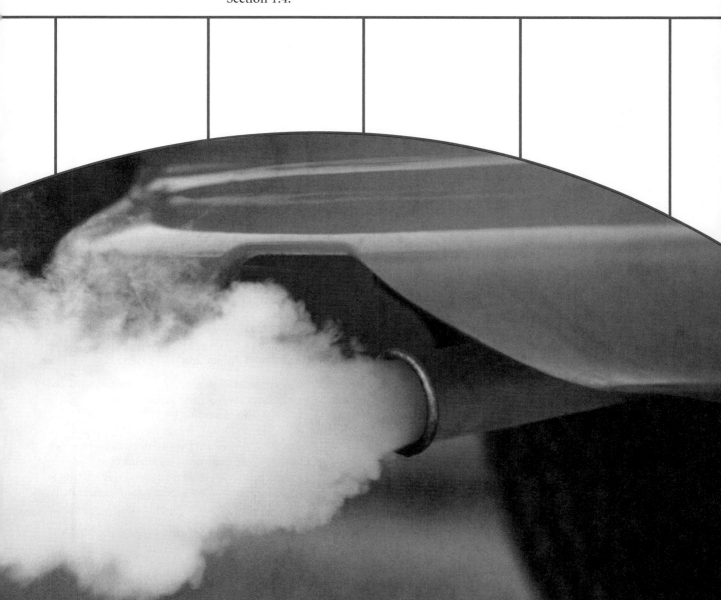

1.1   Models, Functions, and Graphs

Calculus is the study of change—how things change and how quickly they change. We begin our study of calculus by considering how we describe change. Let us start with something that many Americans desire to change: their weight. Suppose Joe weighs 165 pounds and begins to diet, losing 2 pounds a week. We can describe how Joe's weight changes in several ways:

With numerical data, such as Table 1.1.
With a graph, such as Figure 1.1.

Even though this text does not always illustrate each of the four views side by side, in certain places we have made a point to show all four representations. When this occurs, we highlight the fourfold view with the symbol . This "NAVG" symbol should serve as an occasional reminder that you may find it helpful to consider the four representations for each formula you encounter.

TABLE 1.1

Weeks after diet starts	Weight (pounds)
0	165
1	163
2	161
3	159
4	157
5	155

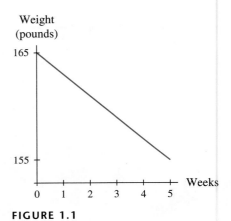

FIGURE 1.1

With words:

> $W(t)$ is Joe's weight (in pounds) t weeks after he begins to diet if he weighs 165 pounds at the start and loses 2 pounds each week.

With an equation:

> $W(t) = 165 - 2t$ pounds
>
> where $W(t)$ is Joe's weight after t weeks of dieting

Most of the mathematical formulas considered in this text can be viewed from each of four perspectives: numerical, algebraic, verbal, and graphical. Each of these representations adds a different facet to our understanding of the formula and what it represents.

Considering a formula from each of the four perspectives is much like observing an island from four different directions. A sailor examining an island from the west may see aspects of the coastline that would be hidden from a sailor who is looking at the island from the north. Similarly, we may notice an aspect of a formula in its graphical form that is less obvious in its algebraic form, or vice versa.

Mathematical Models

Although all four representations are helpful, only the equation and the graph enable us to apply calculus concepts to a particular situation in order to study change. The process of translating a real-world problem into a usable mathematical equation is called **mathematical modeling,** and the equation (with the variables described in the problem context) is referred to as a **model.** We use models to describe numerical data or verbal information. Then we use the model to answer a variety of questions about the situation, interpreting our answers and stating them in numerical and verbal form. This process of translating a real-life situation into a usable equation, applying mathematics to the equation to answer questions about the situation, and translating the answers back into the real-world situation is the focal point of this text.

EXAMPLE 1   *Model Construction and Use*

Tunnel   A civil engineer is planning to dig a tunnel through a mountain as shown in Figure 1.2. (Note that Figure 1.2 is not drawn to scale.) The tunnel will begin 575 feet above sea level and will be constructed with a constant upward slope of 5%; that is, the tunnel will rise vertically 5 feet for every 100 feet of horizontal distance. Table 1.2 shows the amount of vertical rise for several horizontal distances.

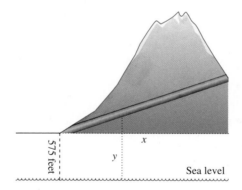

FIGURE 1.2

TABLE 1.2

Horizontal increase (feet)	Vertical increase (feet)
100	5
200	10
300	15
400	20
500	25

a. Use the verbal description and the table of data to write a model for the elevation above sea level of the tunnel in terms of the horizontal distance from where the tunnel begins at the base of the mountain.

b. Find the elevation of the tunnel at a horizontal distance of 2500 feet from the starting point.

c. If the tunnel exits the mountain at a horizontal distance of 7000 feet from where it began, what is the elevation of the tunnel when it emerges from the mountain?

d. If the tunnel will cost $120 per foot to construct, what will be the cost (to the nearest thousand dollars) of building the tunnel?

Solution

a. The elevation of the tunnel begins at 575 feet and rises 5 feet for every 100 horizontal feet. We write this as

    Elevation of tunnel = 575 + 5 feet for every 100 horizontal feet

Using y as the elevation and x as the horizontal feet in hundreds, we convert this statement to the equation

$$y = 575 + 5x \text{ feet above sea level}$$

where x is the horizontal distance from the starting point in hundreds of feet. This equation (with the variable descriptions) is a model for the elevation of the tunnel.

b. Any of the following three methods can be used to determine the elevation. *Algebraically:* Substitute $x = 25$ in the equation to obtain 700 feet above sea level. *Graphically:* Draw a graph of the equation $y = 575 + 5x$ with x-values between 0 and 50 and suitable y-values. Read the y-value from the graph that corresponds to $x = 25$. It is $y = 700$. *Numerically:* Use the fact that the tunnel rises 25 feet for each 500 feet of horizontal increase to extend Table 1.2 as shown in Table 1.3.

TABLE 1.3

Horizontal increase (feet)	500	1000	1500	2000	2500
Vertical increase (feet)	25	50	75	100	125

Because the elevation at the starting point is 575 feet, the height of the tunnel 2500 feet from the starting point is $125 + 575 = 700$ feet above sea level.

c. Using any of the methods described in part b, we find that the elevation of the tunnel when it emerges from the mountain is 925 feet.

d. Using a right triangle (see Figure 1.3) and the Pythagorean Theorem, we find that

$$\text{Length of tunnel} = d = \sqrt{7000^2 + 350^2} = \sqrt{49{,}122{,}500} \approx 7009 \text{ feet}$$

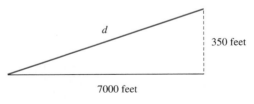

d

350 feet

7000 feet

FIGURE 1.3

The cost of construction is (7009 feet)($120 per foot), or approximately $841,000. ●

Example 1 serves as a preview to our use of mathematical models in calculus. It uses the available data to produce a mathematical equation (model) that describes the relationship between the variable quantities of interest (elevation of the tunnel above sea level and horizontal distance from where the tunnel enters the mountain). It then uses the model to answer associated questions.

Functions

How Joe's weight changes during the course of his diet and how the tunnel's height above sea level changes as the horizontal distance changes are examples of *functions*. The four representations (data, words, graph, and equation) all describe the same function. Informally, a function is a description of how one thing, called the **output** (Joe's weight or the tunnel's elevation), changes as something else, called the **input** (the number of weeks on the diet or the horizontal distance), changes. To understand functions thoroughly, however, we need a more precise definition.

Think of Joe's weight function as a *rule* that tells you what he weighs when you know how long he has been dieting. We visualize the relationship between weight and time spent dieting with the **input/output diagram** shown in Figure 1.4. A rule is a function if each input produces exactly one output. If any particular input produces more than one output, then the rule is not a function.

Your ability to understand and identify input and output will be essential in your ability to understand and interpret derivatives (how quickly output changes as input changes) later on.

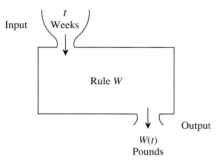

FIGURE 1.4

> A **function** is a rule that assigns exactly one output to each input.

To verify that the rule for Joe's weight is a function, we must ask, "Can Joe have more than one weight after a certain number of weeks on his diet?" Assuming that Joe's weight is measured at a fixed time on the same scale once a week, there will be only one weight that corresponds to each week. Thus Joe's weight is a function of the number of weeks he has been on his diet.

You probably recall from previous math courses that the standard terms for the set of inputs and the set of outputs of a function are *domain* and *range*, respectively. Other terms for input and output include *independent variable* and *dependent variable* and *controlled variable* and *observed variable*. In this book, however, we use the terms *input* and *output*.

In the table representation of Joe's weight function, the set of inputs is {0, 1, 2, 3, 4, 5} and the set of outputs is {165, 163, 161, 159, 157, 155}. In the graph representing Joe's weight, the set of inputs is all real numbers (not just integers) between 0 and 5 ($0 \leq$ weeks ≤ 5) and the set of outputs is all real numbers between 155 and 165 ($155 \leq$ weight ≤ 165). We call this graph continuous because it can be drawn without lifting the writing instrument from the page. We will give a more precise definition of *continuous* in Section 1.3.

EXAMPLE 2   *Identifying Functions, Inputs, and Outputs*

Grades   Suppose $Q(s)$ is the grade (out of 30 points) that student s scored on the first quiz in a course attended by the five students listed in Table 1.4.

TABLE 1.4

Student, s	J. DeCarlo	S. Dyers	J. Lykin	E. Mills	G. Schmeltzer
Grade, $Q(s)$	28	12	25	21	25

a. Identify the set of inputs and the set of outputs.

b. Is Q a function of s?

Solution

a. The set of inputs is simply the set of students, and the set of outputs is {12, 21, 25, 28}.

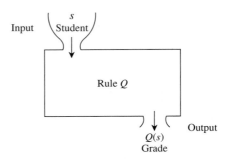

FIGURE 1.5

b. Q is a function of s because each student (input) corresponds to only one quiz grade (output). Note that Q is a function even though two students have the same grade. The input/output diagram for Q is shown in Figure 1.5. ●

Determining Outputs

In the example of Joe's diet, t is the number of weeks on the diet (the input) and W is the name of the rule. The notation for the output is $W(t)$. The t is inside the parentheses to remind us that t is the input, and the W is outside to remind us that it is the rule that gives the output. Thus $W(4) = 157$ means that when Joe has been on his diet for 4 weeks (the input), he weighs 157 pounds (the output).

The way to find the output that corresponds to a known input depends on how the function is represented. In a table, you simply locate the desired input in the input row or column. The output is the corresponding entry in the adjacent row or column. For example, the output corresponding to the input S. Dyers in Table 1.4 is 12, and we write $Q(\text{S. Dyers}) = 12$.

In a function represented by a graph, the input is traditionally on the horizontal axis. Locate the desired value of the input on the horizontal axis, move directly up (or down) until you reach the graph, and then move left (or right) until you encounter the vertical axis. (You will find a see-through ruler helpful for improved accuracy.)

The value at that point on the vertical axis is the output. In Figure 1.6, the graph of average faculty salaries at a private liberal arts college, the output is approximately $56,000 when the input is 2003.

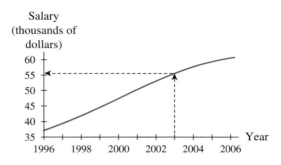

FIGURE 1.6

Finally, if a function is represented by a formula, you simply substitute the value of the input everywhere that the variable appears in the formula and calculate the result. To use the formula $W(t) = 165 - 2t$ to find Joe's weight after 3 weeks, substitute 3 for t in the formula: $W(3) = 165 - 2(3) = 165 - 6 = 159$ pounds.

Units of Measure

In understanding how functions and graphs model the real world, it is important that you understand the **units of measure** of the input and output of functions. In the weight loss example, the unit of measure of the input is *weeks* and the unit of measure of the output is *pounds*. The units of measure in Figure 1.6 can be read from the graph. The unit of measure of the output is *thousands of dollars*, and the unit of measure of the input is *years*. Note that the unit of measure is a word or short phrase telling *how* the variable is measured, not an entire description telling what the variable represents. For example, it would be incorrect to say that the unit of measure of the output in Figure 1.6 is "average faculty salary in thousands of dollars." This is a description of the output variable, not the unit of output.

EXAMPLE 3   *Using Function Representations*

1.1.1a–j

This symbol indicates that instructions specific to this example for using your calculator or computer are given in a technology supplement.

Land Value   The value of a certain piece of property between 1985 and 2005 is given by the equation

$$v(t) = 3.622(1.093^t) \text{ thousand dollars}$$

where t is the number of years since the end of 1985.

A graph and some output values of this function are given in Figure 1.7 and Table 1.5, respectively.

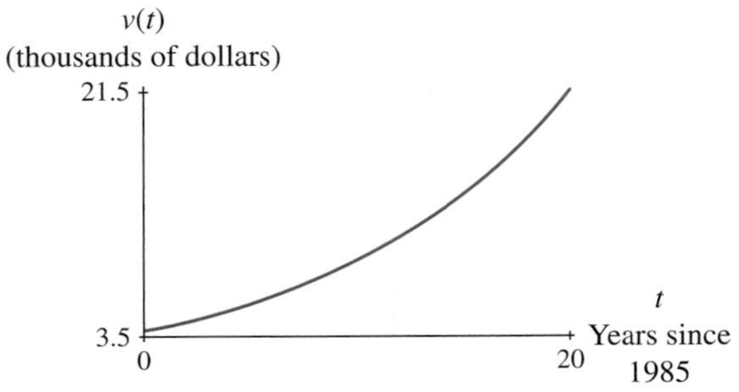

$v(t)$
(thousands of dollars)

FIGURE 1.7

TABLE 1.5

Year	Value (thousand dollars)
1985	5.622
1990	5.650
1995	8.814
2000	13.748

The situation modeled by v can be described as follows: The value of a piece of property worth \$3,622 at the end of 1985 increased by 9.3% each year since 1985.

a. Describe the input variable. What are its units?

b. Describe the output variable. What are its units?

c. What was the land value in 2000?

d. When did the land value reach \$20,000?

Solution

a. The variable t is the number of years since the end of 1985. Its units are years.

b. The output $v(t)$ is the value of a piece of property. Its units are thousand dollars.

c. The input $t = 15$ corresponds to 2000, so the value of the land in 2000 was

$$v(15) = 3.622(1.093^{15}) \approx \$13,748$$

WEB/CD

Strengthening the Concepts: Solving Exponential Equations Algebraically

d. In this question we know the output, and we need to find the corresponding input. This requires us to solve for t in the equation $20 = 3.622(1.093^t)$. You can either solve the equation algebraically (using logarithms) or use technology to solve it. In either case, you should find that $t \approx 19.2$. Note that because t is defined as the number of years since the *end* of 1985, $t = 19$ corresponds to the end of 2004, so $t = 19.2$ corresponds to early in the year 2005. Thus the land reached a value of \$20,000 in 2005. ●

It is important to be able to describe the behavior of a function or graph. We use the words *increasing, decreasing,* and *constant* to describe the output behavior of a graph or function. If a graph is rising as you move from left to right along the horizontal axis, then it is said to be **increasing.** A graph is said to be **decreasing** if it is falling as you move from left to right along the horizontal axis. The portion of a graph that neither rises nor falls is called **constant.**

Graphically Recognizing Functions

It is easy to determine whether a graph represents a function. Examine the two graphs in Figure 1.8. The top graph does not describe y as a function of x because

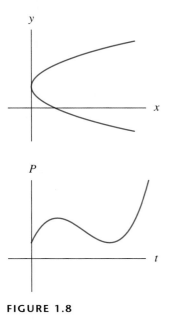

FIGURE 1.8

each positive x-value produces two different y-values. The bottom graph does show P as a function of t because every input produces only one output. The quickest way to find out whether a graph represents a function is to apply the **Vertical Line Test.**

Vertical Line Test

Suppose that a graph has inputs located along the horizontal axis and outputs located along the vertical axis. If at any input you can draw a vertical line that crosses the graph in two or more places, then the graph does not represent a function.

Figure 1.9 shows a graph that fails the Vertical Line Test.

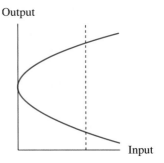

FIGURE 1.9   This graph is not a function.

One way to determine whether a formula represents a function is to graph the equation and then apply the Vertical Line Test.

EXAMPLE 4   *Determining When a Graph Represents a Function*

Sales   A model for the quarterly sales of a corporation over a 2-year period is

$$q(t) = 0.07t^2 - 0.65t + 2.38 \text{ million dollars}$$

sold during the tth quarter.

a.   Graph the equation. Label the graph with the appropriate input and output units.

b.   Explain why the graph represents a function.

c.   Estimate the input over which the output is increasing.

d.   Estimate the input over which the output is decreasing.

Solution

a. The graph of q appears in Figure 1.10. The input units are *quarters* and the output units are *million dollars*.

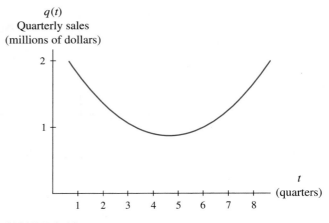

FIGURE 1.10

b. The graph in Figure 1.10 represents a function because any vertical line drawn through the graph crosses the graph at only one point.

c. The graph is increasing during the 5th through 8th quarters.

d. The graph is decreasing during the 1st through 4th quarters.   ●

Modeling in Business

Modeling in the business world often involves terms that are easily understood. Examples of such business terms are *fixed costs* (also called *overhead*), *variable costs, total cost, revenue, profit,* and *break-even point.* To help you understand these terms, consider an enterprising college student who opens a dog-grooming business. An initial capital investment is needed to purchase equipment, buy a business license, set up a lease agreement for a shop, and so on.

Once the business officially opens, the owner sets aside money at the beginning of each week to pay rent, utilities, salaries, and the like. These costs remain the same regardless of the number of dogs that are groomed during the week and are called **fixed costs.** Other costs change with the number of customers. These **variable costs** include expenditures such as advertising, supplies, laundry and custodial fees, and overtime. The fixed costs plus the variable costs give the **total cost** of doing business.

> Total cost = fixed costs + variable costs

When the commodity produced (dogs groomed, in this example) can be measured in units, then the total production cost divided by the number of units produced is known as the **average cost.**

$$\text{Average cost} = \frac{\text{total cost}}{\text{number of units produced}}$$

Businesses receive **revenue,** which is, in general, the quantity of a commodity sold times the price of that commodity. In the pet-grooming business, revenue is the number of dogs groomed times the grooming price. **Profit** is revenue minus total cost.

$$\text{Profit} = \textbf{revenue} - \text{total cost}$$

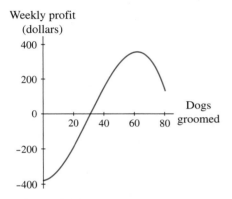

Weekly profit (dollars)

FIGURE 1.11

The graph in Figure 1.11 shows the weekly profit for this business as a function of the number of dogs groomed. The places where a graph crosses (or touches) the horizontal and vertical axes are called **intercepts.** The intercepts often have interpretations in the context of the situation that the graph represents. The horizontal axis intercept in Figure 1.11 is approximately 31, and the vertical axis intercept is about ‾380. Therefore, from this graph we can conclude that the fixed costs each week are approximately $380 (the cost associated with grooming no dogs). Profit is negative until 31 dogs are groomed. This point on the graph is called the **break-even point,** the point at which revenue equals total cost so that profit is zero.

The **break-even point** occurs when revenue equals total cost and profit equals 0.

Profit appears to peak at approximately $350. This occurs when about 62 dogs are groomed each week. Thus 62 is the optimal number of dogs for this business to groom each week. If more than 62 dogs were groomed each week, profit would actually decline, perhaps because of the need to pay overtime or hire another employee. Cost, profit, revenue, and break-even points are fundamental concepts that are important for business and non-business students to understand. We will refer to them regularly throughout this text.

In this section we have considered models, functions, and graphs. These three concepts form the foundation for the remainder of this text. Calculus is the study of change. Because change is most precisely described by functions, it is vital that you have a thorough understanding of functions. The remainder of this chapter and the next chapter are devoted to a further exploration of functions.

1.1   Concept Inventory

- ○ *Mathematical modeling*
- ○ *Function*
- ○ *Inputs and outputs*
- ○ *Units of measure*
- ○ *Reading and interpreting graphs*
- ○ *Graph terms: continuous, increasing, decreasing, constant, axes intercepts*
- ○ *Vertical Line Test*
- ○ *Business terms: fixed costs, variable costs, average cost, total cost, revenue, profit, break-even point*

1.1   Activities

For each of the rules in Activities 1 through 4, specify the input and output, the input and output variables, and the input and output units. Determine whether each rule is a function.

1. $R(w)$ = the first-class domestic postal rate (in cents) of a letter weighing w ounces

2. $H(a)$ = your height in inches when you were age a years old

3. $A(m)$ = the amount (in dollars) you pay for lunch on the mth day of any week

4. $C(m)$ = the amount of credit (in dollars) that Citibank Visa will allow a 20-year-old with a yearly income of m dollars

Determine whether the tables in Activities 5 through 8 represent functions. Assume that the input is in the left column.

5.

Age	Percent using computers at work
18–24	37.1
25–29	52.5
30–39	53.3
40–49	54.9
50–59	50.7
over 59	32.6

(Source: *Statistical Abstract*, 2001.)

6.

Military rank (4 years service)	Basic monthly pay in 2002 (dollars)
Second Lieutenant	2639
First Lieutenant	3170
Captain	3422
Major	3928
Lt. Colonel	4440
Colonel	5177
Brigadier General	6418
Major General	7571

(Source: Defense Finance and Accounting Service.)

7.

Person's height	Person's weight (pounds)
5'3"	139
6'1"	196
5'4"	115
6'0"	203
5'10"	165
5'3"	127
5'8"	154
6'0"	189
5'6"	143

8.

Year	Total wagers in Nevada on the Super Bowl (millions)
1996	70.9
1997	70.9
1998	77.3
1999	76.0
2000	71.0
2001	67.7
2002	71.5

(Source: www.vegasinsider.com.)

9. Which of the following graphs represent functions? (The input axis is horizontal.)

a.
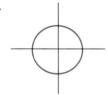

b.

c.
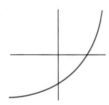

10. Which of the following graphs represent functions? (The input axis is horizontal.)

a.
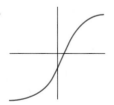

b.
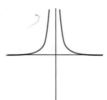

c.
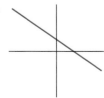

11. Prices $P(m)$ is the median sale price (in thousands of dollars) of existing one-family homes in metropolitan area m in 2000. Write the following statements in function notation.
(Source: *Statistical Abstract*, 2001.)

a. The median sale price in Honolulu was $295,000.

b. In Providence, RI, the median sale price was $137,800.

c. $170,100 was the median sale price in Portland, OR.

12. Darkness Hours $H(d)$ is the number of hours of darkness in Anchorage, Alaska, on the dth day of the year. Write the following statements in function notation.

a. On the 121st day of the year Anchorage has 7.5 hours of darkness.

b. The duration of darkness in Anchorage on the 361st day of the year is 18.5 hours.

c. In Anchorage there are only 4.5 hours of darkness on the 181st day of the year.

13. Exports $E(t)$ is the value of cotton exports in millions of dollars in year t. Write sentences interpreting the following mathematical statements.
(Source: *Statistical Abstract*, 1994.)

a. $E(1988) = 1975$

b. $E = 1999$ when $t = 1992$

14. Population $P(r)$ is the percentage of U.S. residents in the year 2000 who were of origin r. Write sentences interpreting the following mathematical statements.
(Source: *Statistical Abstract*, 2001.)

a. $P(\text{Hispanic}) = 11.8$

b. $P = 4.1$ when $r = $ Asian and Pacific Islander

15. Sales A music store offers one free CD with the purchase of four CDs priced at $18 each (tax included).

a. Construct a graph showing the cost of buying x CDs. Show input values from 0 to 10.

b. What is the cost of 6 CDs?

c. How many CDs could you buy if you had $36?

d. How many CDs could you buy if you had $100?

e. What is the average price of each CD if you buy 3 CDs? 6 CDs?

16. **Sales**   A fraternity is selling T-shirts on the day of a football game. The shirts sell for $8 each.

   a. Complete the table.

Number of shirts sold	Revenue (dollars)
1	
2	
3	
4	
5	
6	

   b. Construct a revenue graph by plotting the points in the table.

   c. How many T-shirts can be purchased with $25?

   d. If an 8% sales tax were added, how many T-shirts could be purchased with $25?

17. **Payment**   You are interested in buying a used car and in financing it for 60 months at 10% interest. As a special promotion, the dealer is offering to finance with no down payment. The graph below shows the value of the car as a function of the amount of the monthly payment.

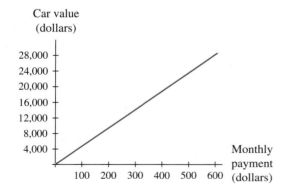

Car value (dollars)

   a. Estimate the value of the car you can buy if your monthly payment is $200.

   b. Estimate the monthly payment for a car that costs $16,000.

   c. Estimate the amount by which your monthly payment will increase if you buy a $20,000 car rather than a $15,000 car.

   d. How would the graph change if the interest rate were 12.5% instead of 10%?

18. **Payment**   You have decided to purchase a car for $18,750. You have 20% of the purchase price to use as a down payment, and the purchase will be financed at 10% interest. The graph below shows the monthly loan payment in terms of the number of months over which the loan is financed.

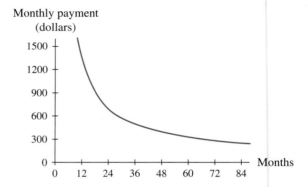

Monthly payment (dollars)

   a. What is the actual amount being financed?

   b. Estimate your monthly payment if you finance the purchase over 36 months.

   c. Estimate the number of months you will have to pay on the loan if you can afford to pay only $300 a month.

   d. If you decrease the time financed from 48 to 36 months, by how much will your payment increase?

   e. How would the graph change if you were buying a $20,000 car?

19. **Social Security**   The accompanying graph shows the percentage by which Social Security checks increased as a result of cost-of-living adjustments every year from 1996 through 2002.

   a. What was the cost-of-living increase in 1997?

   b. When was the cost-of-living increase the greatest? Estimate the cost-of-living increase in that year.

   c. When was the cost-of-living increase 2.1%?

   d. Did Social Security benefits increase or decrease between 2001 and 2002? Explain.

Social Security
cost-of-living
increase
(percent)

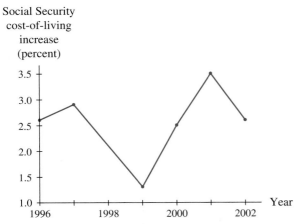

(Source: Based on data from Social Security Administration.)

20. Wages The accompanying graph compares the minimum wage with its adjusted value in constant 2000 dollars.

Minimum wage
(dollars)

(Source: Based on data from Bureau of Labor Statistics.)

a. Estimate the minimum wage in 1980 and the adjusted minimum wage in that same year.

b. Over what intervals did the real minimum wage remain constant?

c. During the years when the minimum wage did not increase, what happened to the buying power of the minimum wage?

d. In what years was the adjusted minimum wage above $5.75?

e. Estimate the values at which the graphs intercept the vertical axis. Interpret these values.

21. Weight A baby weighing 7 pounds at birth loses 7% of her weight in the 3 days after birth and then, over the next 4 days, returns to her birth weight. During the next month, she steadily gains 0.5 pound per week. Sketch a graph of the baby's weight from birth to 4 weeks. Accurately label both axes.

22. Growth The graph depicts a certain girl's height plotted according to her age.

Height

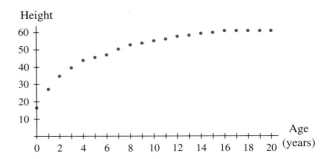

a. What are the units along the vertical axis?

b. How long was the girl when she was born?

c. Approximately how much did she grow during the first year?

d. Over what time interval does the graph appear to remain constant?

e. At approximately what age did she reach her full height? How tall did she become?

f. Did the girl grow faster during her first 3 years or during the last 3 years before she attained her full height?

g. Would you expect the graph to increase or decrease after age 20? Why?

23. Snow Depth Donner Pass in the Sierra Nevada mountain range of northern California is the most important transmontane route between Reno and San Francisco. Donner Pass is named after George and Jacob Donner, who in 1846 attempted to lead a party of more than 80 immigrants through the pass to the Sacramento Valley. They were unable to make it through the pass before they became snowbound. A little over half the party survived the winter and finished their journey.

The accompanying graph represents the depth of the snow cover at Donner Memorial State Park from December 1993 through February 1994.

a. How deep was the snow on December 1, 1993?

b. For approximately how many days after December 2 did the snow remain at the same depth?

Snow
(inches)

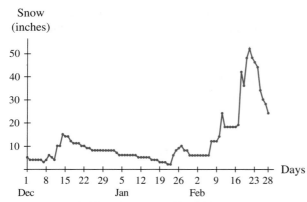

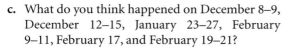

(Source: Compiled from data in *Monthly Climatological Data: California*, vol. 97, no. 12 [December 1993] and vol. 98, nos. 1 and 2 [January–February 1994].)

Monthly profit
(dollars)

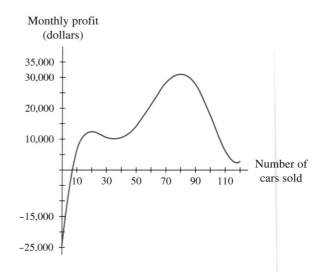

c. What do you think happened on December 8–9, December 12–15, January 23–27, February 9–11, February 17, and February 19–21?

d. Why might there have been an immediate decrease in the depth of the snow cover after each one of the peaks?

e. On what day was the snow deepest? How deep (in feet and inches) was the snow at that time?

f. What event may have taken place to cause the rapid decrease of the graph after the last peak?

g. During what day(s) did the most snowfall probably occur?

24. **Profit**   The accompanying graph depicts the average monthly profit for Slim's Used Car Sales based on data Slim gathered from his five franchises over the last year. Slim needs your help in interpreting this graph.

a. Slim pays rent on the lot where the cars to be sold are displayed, and he pays himself and his salespeople a fixed salary at the beginning of each month. There is also the cost of electricity, water and sewer, health insurance, and so on. These costs remain the same regardless of the number of cars sold during the month. Estimate Slim's fixed costs at the beginning of each month.

b. Slim also has variable costs, such as the prices he pays for the cars. Other variable costs include paying someone to wash and maintain the cars on the lot, minor repairs to the cars, and so on. Slim receives revenue from the customer for each

car he sells. Where is Slim's total cost more than his revenue?

c. What is Slim's monthly break-even point?

d. Between what numbers of cars sold is Slim's profit increasing?

e. The monthly profit peaks at two places. Estimate the number of cars sold and the monthly profit generated from the sale of those cars at these two points.

f. Give plausible reasons why Slim's profit sometimes decreases.

g. On the basis of the information presented in the graph, how many cars should Slim try to sell each month to maximize his monthly profit?

25. **Medicine**   A patient is instructed by her doctor to take one pill containing 500 milligrams of a drug. Assume that her body uses 20% of the original amount of the drug daily. Let y equal the number of milligrams of the drug remaining in the patient's body after x days.

a. Complete the table below to describe numerically the elimination of the drug from the patient's body.

x	0	1	2	3	4	5
y	500	400				

b. Find a model that describes the elimination of the drug from the patient's body.

c. What are the smallest and largest values of x that make sense in this problem? What are the smallest and largest values of y that make sense in this problem? Graph your model.

d. Calculate the x- and y-intercepts for the graph of your model. Interpret these values in the context of this problem.

e. For which values of x is y decreasing?

f. Use a graph of the model to estimate how much of the drug is left in the patient's body after 3.5 days. Use your model to find the exact value.

g. Use a graph of the model to estimate when the concentration of the drug will be 60 milligrams. Use your model to find the exact value.

26. **Medicine**   Repeat Activity 25, assuming that each day the patient's body uses 20% of the amount of drug that was in her body *the previous day.*

For Activities 27 through 30, find the output of the **WEB/CD** function corresponding to each input value given.

Skills

27. $t = 3.2s + 6; s = 5, s = 10$

28. $y = 7x + 3; x = 20, x = -2$

29. $R(w) = 39.4(1.998^w); w = 3, w = 0$

30. $S(t) = \dfrac{120}{1 + 3.5e^{-2t}}; t = 10, t = 2$

 WEB/CD This symbol indicates that additional activities are available in Appendix C: Strengthening the Skills on the *Calculus Concepts* CD-ROM **Skills** and web site.

For Activities 31 through 34, find the input of the **WEB/CD** function corresponding to each output value given.

Skills

31. $R(w) = 39.4(1.998^w); R(w) = 78.8,$ $R(w) = 394$

32. $y = 7x + 3; y = 24, y = 13.5$

33. $Q(x) = 0.32x^3 - 7.9x^2 + 100x - 15;$ $Q(x) = 515, Q(x) = 33.045$

34. $S(t) = \dfrac{120}{1 + 3.5e^{-2t}}; S(t) = 60, S(t) = 90$

For each of the functions in Activities 35 through 38, **WEB/CD** determine whether an input or an output value is given, and find its corresponding output or input.

Skills

35. $A = 3200e^{0.492t}; t = 15$

36. $g(x) = 49x^2 + 32x - 134; g(x) = 5086$

37. $y = (39.4)(\ln 1.998)(1.998^x); y = 2.97$

38. $p = \dfrac{100}{1 + 25e^{-0.37m}}; p = 95$

39. Discuss the differences between an equation and a function. In your discussion, describe how to identify functions graphically and numerically.

40. Discuss the advantages of each of the four perspectives from which a mathematical model can be viewed.

1.2   Constructed Functions

In Section 1.1, we examined what a function is and how functions can be represented. Now we consider constructing more complicated functions.

Combining Functions

Concept Development: Function Addition and Multiplication   We can combine two or more functions to create a new function by adding or subtracting. Consider functions S and C that give sales and costs for a corporation from 1996 through 2001. The equation $S(t) = 3.570(1.105^t)$ gives the sales in millions of dollars t years after 1996, and the equation $C(t) = -39.2t^2 + 540.1t + 1061.0$ gives the costs in thousands of dollars t years after 1996. Input/output diagrams for S and C are shown in Figure 1.12.

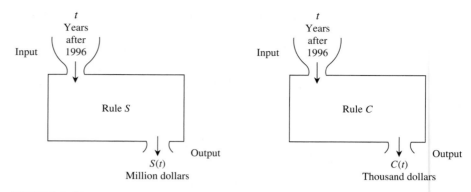

FIGURE 1.12

1.2.1 a–c

By subtracting costs from sales, we can get a new function representing profit. Before doing so, however, we need to ask ourselves two questions: "Is the set of inputs the same for both functions?" and "Are the outputs given in the same units?" The answer to the first question is yes, but the answer to the second is no. However, we can easily change one of the functions to achieve the same output units.

To construct an equation P for the corporation's profit between 1996 and 2001, we must first multiply C by 0.001 to convert the output units from thousands of dollars to millions of dollars. Figure 1.13 shows the input/output diagram for the adjusted cost function.

The profit can now be found by subtracting the adjusted cost function from the sales function. The input/output diagram for the profit function is shown in Figure 1.14.

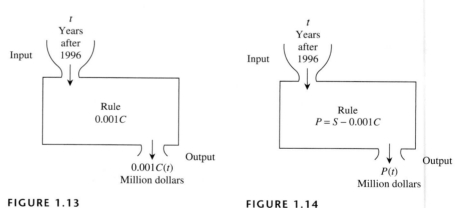

FIGURE 1.13   **FIGURE 1.14**

The profit function can be written as

$$P(t) = S(t) - 0.001C(t)$$
$$= 3.570(1.105^t) - 0.001 \left(-39.2t^2 + 540.1t + 1061.0\right) \text{ million dollars}$$

where t represents the number of years since 1996.

Further, we can now use P to estimate the profit in given year. For instance, the profit in 1998 was $P(2) \approx 2.375$ million dollars.

Product functions are created by multiplying two functions. Again, this makes sense only if both functions have the same set of inputs and if the output units are compatible so that, when multiplied, they give a meaningful result.

Suppose that $M(x)$ is the selling price (in dollars) of a gallon of milk on the xth day of last month, and $G(x)$ is the number of gallons of milk sold. The input/output diagrams are shown in Figure 1.15.

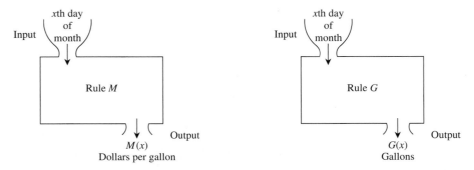

FIGURE 1.15

The inputs of both functions are the same, and when the outputs are multiplied, with units ($ per gallon)(gallons), the result is the sales in dollars on the xth day of last month. The input/output diagram for the new function $T = M \cdot G$ is given in Figure 1.16.

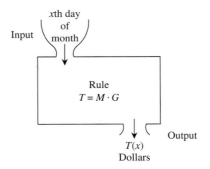

FIGURE 1.16

$M(x) = 0.007x + 1.492$ dollars per gallon and $G(x) = 31 - 6.332(0.921^x)$ gallons are functions that model the selling price and the number of gallons of milk sold on the xth day of last month. Thus milk sales can be modeled by the function

$$T(x) = [M(x)] \cdot [G(x)]$$
$$= [0.007x + 1.492] \cdot [31 - 6.332(0.921^x)] \text{ dollars}$$

on the xth day of last month.

Concept Development: Function Composition   Another way in which we combine functions is called **function composition.** Consider Tables 1.6 and 1.7, which give the altitude of an airplane as a function of time and give air temperature as a function of altitude, respectively.

TABLE 1.6

t = time into flight (minutes)	$F(t)$ = feet above sea level
0	4500
1	7500
2	13,000
3	19,000
4	26,000
5	28,000
6	30,000

TABLE 1.7

F = feet above sea level	$A(F)$ = air temperature (degrees Fahrenheit)
4500	72
7500	17
13,000	-34
19,000	-55
26,000	-62
28,000	-63
30,000	-64

It is possible to combine these two data tables into one table showing air temperature as a function of time. This process requires using the output from Table 1.6 as the input for Table 1.7. The only restriction is that both the numerical values and the units in the output of one function match those in the input of the other function. Figure 1.17 illustrates the output of one function being used as the input of the next function. The resulting new function is shown in Table 1.8.

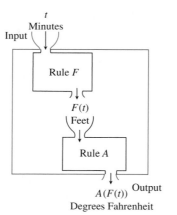

FIGURE 1.17

TABLE 1.8

t = time into flight (minutes)	$A(F)$ = air temperature (degrees Fahrenheit)
0	72
1	17
2	-34
3	-55
4	-62
5	-63
6	-64

Because we used the output of Table 1.6, $F(t)$, as the input for Table 1.7, the new function output is commonly written as $A(F(t))$. The A outside the parentheses is to remind us that the output is air temperature, and putting $F(t)$ inside the parentheses reminds us that altitude from Table 1.6 is the input. It is common to refer to A as the *outside function* and to F as the *inside function*. The mathematical symbol for the composition of an inside function F and an outside function A is $A \circ F$, so

$$(A \circ F)(t) = A(F(t))$$

1.2.2a, b

The altitude data can be modeled as $F(t) = -222.22t^3 + 1755.95t^2 + 1680.56t + 4416.67$ feet above sea level, where t is the time into the flight in minutes. The air temperature can be modeled as $A(F) = 277.897(0.99984^F) - 66$ degrees Fahrenheit, where F is the number of feet above sea level. The composition of these two functions is

$$(A \circ F)(t) = A(F(t))$$
$$= 277.897[0.99984^{(-222.22t^3 + 1755.95t^2 + 1680.56t + 4416.67)}] - 66°F$$

where t is the time in minutes into the flight.

We formalize the process of function composition as follows: Given two functions f and g, we can form their composition if the outputs from one of them, say f, can be used as inputs to the other function, g. In terms of inputs and outputs, we have those shown in Figure 1.18. In this case, we can replace the portion of the diagram within the teal box by forming the composite functions $g \circ f$ whose input/output diagram is shown in Figure 1.19.

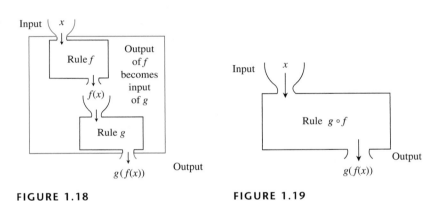

FIGURE 1.18   **FIGURE 1.19**

Note that the inputs to the composite function $g \circ f$ are the inputs to the inside function f and that the outputs are from the outside function g.

EXAMPLE 1   *Finding a Composite Function*

Lake Contamination   Consider the word descriptions of the following two functions and their input/output diagrams shown in Figure 1.20.

$C(p)$ = parts per million of contamination in a lake when the population of the surrounding community is p people

$p(t)$ = the population in thousands of people of the lakeside community in year t

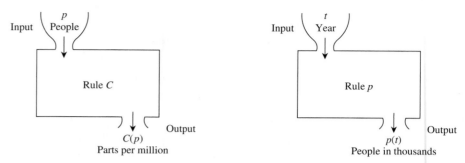

FIGURE 1.20

a. Draw an input/output diagram for the composition function that gives the contamination in a lake as a function of time.

b. The contamination in the lake can be modeled as $C(p) = \sqrt{p}$ parts per million when the community's population is p people. The population of the community can be modeled by $p(t) = 0.4t^2 + 2.5$ thousand people t years after 1980. Write the function that gives the lake contamination as a function of time.

Solution

a. Note that the output of the second function is almost the same as the input of the first function. If we multiply $p(t)$ by 1000 to convert the output from thousands of people to people, we have the function $P(t) = 1000p(t)$ people in year t. Now we can compose the two functions C and P to create the new function $C \circ P$ whose input/output diagram is shown in Figure 1.21.

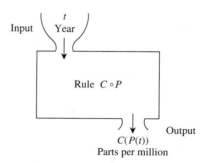

FIGURE 1.21

b. To convert the output of the function p to people, multiply by 1000 to obtain the function $P(t) = 1000(0.4t^2 + 2.5)$ people t years after 1980. Thus the contamination in the lake t years after 1980 can be modeled as

$$(C \circ P)(t) = C(P(t)) = \sqrt{1000(0.4t^2 + 2.5)} \text{ parts per million} \quad \bullet$$

Piecewise Continuous Functions

Another way in which we will construct functions is putting two or more functions together to create a **piecewise continuous function.** Suppose you invest $100 at a 7% annual percentage rate. We call the initial amount invested the *principal.* Because the interest is compounded annually, the amount in your account can be represented by the piecewise continuous function

$$B(t) = \begin{cases} \$100 & \text{when } 0 \leq t < 1 \\ \$107 & \text{when } 1 \leq t < 2 \\ \$114.49 & \text{when } 2 \leq t < 3 \\ \$122.50 & \text{when } 3 \leq t < 4 \\ \$131.08 & \text{when } 4 \leq t < 5 \\ \$140.26 & \text{when } t = 5 \end{cases}$$

t years after the $100 was invested. A graph of the function is shown in Figure 1.22.

We call this equation a piecewise continuous function because it is made up of portions of six different continuous functions and is defined for all values of *t* between 0 and 5. The values of *t* where the formula changes (here *t* = 1, 2, 3, 4, 5) are called the **break points** of the function. Piecewise continuous functions will be especially helpful in modeling situations in which a quantity demonstrates a distinct and sudden change in behavior. Such a situation is illustrated in Example 2.

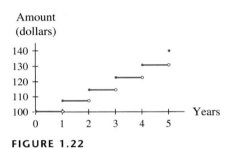

Amount (dollars)

FIGURE 1.22

EXAMPLE 2   *Graphing and Interpreting Piecewise Functions*

1.2.3

Population   The population of West Virginia from 1985 through 1999 is given for certain years in Table 1.9.

TABLE 1.9

Year	Population (thousands)	Year	Population (thousands)
1985	1907	1993	1816
1987	1858	1995	1821
1989	1807	1997	1816
1990	1793	1999	1807
1991	1798		

(Source: *Statistical Abstract,* 1998 and 2001.)

These data can be modeled by

$$P(t) = \begin{cases} -23.373t + 3892.220 \text{ thousand people} & \text{when } 85 \leq t < 90 \\ -1.013t^2 + 193.164t - 7387.836 \text{ thousand people} & \text{when } 90 \leq t \leq 99 \end{cases}$$

where *t* is the number of years since 1900. The graph of *P* is shown in Figure 1.23.

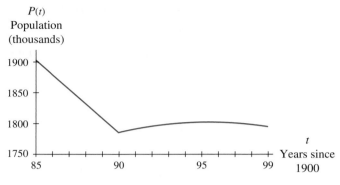

FIGURE 1.23

a. According to the model, what was the population of West Virginia in 1987? in 1998?

b. Find $P(85)$ and $P(90)$. Interpret your answers.

WEB/CD

Strengthening
the Concepts:
Graphing
Piecewise
Functions

Solution

a. To determine the population in 1987, we substitute 87 for t in the top function.

$$P(87) = -23.373(87) + 3892.220 \approx 1859 \text{ thousand people}$$

To determine the population in 1992, we use the bottom function.

$$P(98) = -1.013(98)^2 + 193.164(98) - 7387.836 \approx 1813 \text{ thousand people}$$

b. $P(85) = -23.373(85) + 3892.220 \approx 1906$ thousand people. In 1985 the population of West Virginia was approximately 1,906,000 people.

$$P(90) = -1.013(90)^2 + 193.164(90) - 7387.836 \approx 1791 \text{ thousand people}.$$

In 1990 the population of West Virginia was approximately 1,791,000 people. ●

Inverse Functions

Given a function, we can sometimes create a new function by reversing the input and output of the original function. We call this new function an **inverse function.** To illustrate, we begin by examining data (Table 1.10) showing new-home prices in the United States as a function of the year.

TABLE 1.10

Year y	1970	1980	1990	1995	2000
Average sale price of a new home $P(y)$	$23,400	$64,600	$122,900	$133,900	$169,000

(Source: *Statistical Abstract*, 1998 and 2000.)

Figure 1.24 shows an input/output diagram for this function. If we swap the inputs and outputs, we get

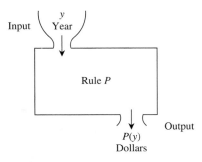

FIGURE 1.24

TABLE 1.11

Average sale price of a new home	Year
$23,400	1970
$64,600	1980
$122,900	1990
$133,900	1995
$169,000	2000

Each average price (input) corresponds to only one year (output), so Table 1.11 represents a function, the inverse of the original function. Figure 1.25 shows an input/output diagram for the inverse function.

When an equation or a graph represents a function, we know that for each function input there corresponds exactly one output. To determine if a function has an inverse, we ask the question "For each function output, does the inverse relationship define a unique input?" That is, in order for a function to have an inverse, it must be one-to-one. A function f is called a **one-to-one function** if for any two different input values of the function, say a and b, we have $f(a) \neq f(b)$. For instance, the function $y = x^2$ is not a one-to-one function (and therefore does not have an inverse function) because when $x = 2$ and $x = -2$, $y = 4$. A graphical test that can be used to determine if a function is one-to-one is the **Horizontal Line Test**.

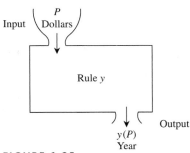

FIGURE 1.25

Horizontal Line Test

Suppose that a graph has inputs located along the horizontal axis and outputs located along the vertical axis. If at any output you can draw a horizontal line that crosses the graph in two or more places, the function is not a one-to-one function.

Figure 1.26 shows a graph that passes the Horizontal Line Test and is a one-to-one function. The graph shown in Figure 1.27 fails the Horizontal Line Test and therefore does not have an inverse function.

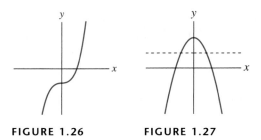

FIGURE 1.26   **FIGURE 1.27**

EXAMPLE 3   *Graphically Finding an Inverse Function*

Office Space   The graph in Figure 1.28 shows the percentage of vacant office space in Houston from 1993 through 2000. Does the function represented by this graph have an inverse function?

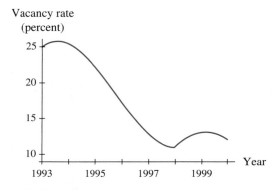

FIGURE 1.28
(Source: *Statistical Abstract*, 2001.)

Solution   The graph in Figure 1.29 shows that the vacancy rate function does not pass the Horizontal Line Test. It is not a one-to-one function and therefore does not have an inverse function.

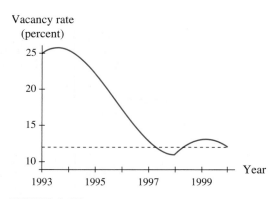

FIGURE 1.29

Finding an algebraic expression for an inverse function requires that you solve for the input variable of a function in terms of the output variable. Sometimes this is simple, sometimes it is difficult, and sometimes it is impossible. You may recall from previous math courses that when a function f is used as the input for its inverse function g, the result is simply the input variable. That is, when a function and its inverse are combined to create a composite function, the composition has the result of "undoing" the effects of the two functions.

Composition Property of Inverse Functions

If f and g are inverse functions, then

$$f(g(x)) = (f \circ g)(x) = x \qquad \text{and} \qquad g(f(x)) = (g \circ f)(x) = x$$

At this point we remind you of a few examples of inverse functions where b is a real number.

Examples of Inverse Function Pairs

$f(x)$	$g(x)$	$f(g(x))$	$g(f(x))$
$x + b$	$x - b$	$(x - b) + b = x$	$(x + b) - b = x$
bx	$\dfrac{x}{b}, b \neq 0$	$b\left(\dfrac{x}{b}\right) = x$	$\dfrac{bx}{x} = x$
$\dfrac{b}{x}$	$\dfrac{b}{x}$	$\dfrac{b}{\frac{b}{x}} = x$	$\dfrac{b}{\frac{b}{x}} = x$

For $x > 0$ and positive integer n,

x^2	$\sqrt{x}$	$\left(\sqrt{x}\right)^2 = x$	$\sqrt{x^2} = x$
x^n	$\sqrt[n]{x}$	$\left(\sqrt[n]{x}\right)^n = x$	$\sqrt[n]{x^n} = x$
e^x	$\ln x$	$e^{\ln x} = x$	$\ln(e^x) = x$

The pair $f(x) = e^x$ and $g(x) = \ln x$ (read *natural logarithm* of x) shows up in life sciences and finance and is discussed in Section 2.2.

Calculus is the mathematics of change that occurs in continuous portions of functions. As we embark on our study of calculus, we will consider modeling primarily as a tool for developing continuous or piecewise continuous functions. Once we can model data with a function, we can use calculus to analyze that model—especially to determine rates of change and identify maxima, minima, and inflection points.

1.2   Concept Inventory

- ○ *Adding and subtracting functions*
- ○ *Multiplying and dividing functions*
- ○ *Function composition*
- ○ *Piecewise continuous functions*
- ○ *Inverse functions*
- ○ *One-to-one functions*

1.2   Activities

1. **Transplants** The number of kidney and liver transplants in the United States between 1992 and 1996 can be modeled by

$K(x) = 9088.859 + 1697.717 \ln x$ kidney transplants
$L(x) = 2424.764 + 915.025 \ln x$ liver transplants

where x is the number of years since 1990.
(**Source: Based on data from** *Statistical Abstract*, **1998.**)

   a. Construct a model for the number of kidney and liver transplants between 1992 and 1996.

   b. Use the model to estimate the number of kidney and liver transplants in 1995.

2. **Earnings** The salary of one of Compaq Computer Corporation's senior vice presidents from 1996 through 1998 can be modeled by

$$S(t) = 69{,}375t + 380{,}208 \text{ dollars}$$

t years after 1996. His other, nonsalary compensations during the same period can be modeled by

$$C(t) = -31.67t^2 + 137.15t + 233.5 \text{ thousand dollars}$$

t years after 1996. In addition, each year he received an average bonus of $650,000.
(**Source: Based on data from Compaq's 1999 Notice of Annual Meeting.**)

   a. Construct a model for this VP's total yearly salary package, including nonsalary compensation and bonuses.

   b. Estimate the VP's 1997 total salary package.

3. **Milk Consumption** Per capita milk consumption in the United States between 1980 and 1999 can be modeled by

$$M(t) = -0.2190t + 45.2325 \text{ gallons per person per year}$$

and consumption of whole milk for the same period can be modeled by

$$W(t) = 0.01685t^2 - 3.48778t + 187.95962 \text{ gallons per person per year}$$

In both models, t is the number of years since 1900. Use the models to construct a model giving the per capita consumption of milk other than whole milk and estimate this per capita consumption for the year 2000.
(**Source: Based on data from** *Statistical Abstract*, **2001.**)

4. **Profit** State lotteries have gained in popularity over the past 20 years. The cost of the lotteries to the states (prizes and expenses) between 1980 and 2000 can be modeled by

$$C(x) = 142.773x^2 + 136.719x + 1448.336 \text{ million dollars}$$

where x is the number of years after 1980. The profit realized by the states as a result of the lotteries can be modeled by

$$P(x) = -3.226x^3 + 80.111x^2 + 203.746x + 1007.542 \text{ million dollars}$$

where x is the number of years after 1980. Find an equation for the revenue the states earned as a result of lotteries, and estimate the total lottery revenue in the year 2001.
(**Source: Based on data from** *Statistical Abstract*, **2001.**)

5. **Farming** The percentage of Iowa corn farmers in two communities who had heard about, and who had planted, hybrid seed corn t years after 1924 can be modeled as follows:

Percentage hearing:

$$h(t) = \frac{100}{1 + 128.0427e^{-0.7211264t}} \text{ percent}$$

Percentage planting:

$$p(t) = \frac{100}{1 + 913.7241e^{-0.607482t}} \text{ percent}$$

(Source: Based on data from Ryan and Gross, "The Diffusion of Hybrid Seed Corn in Two Iowa Communities," *Rural Sociology*, March 1943.)

a. Write an equation for the percentage of Iowa corn farmers who had heard about, but who had not yet planted, hybrid seed corn *t* years after 1924.

b. Estimate what percentage of Iowa corn farmers had heard about but not yet planted the hybrid seed corn in 1929.

6. **Credit Cards** The total amount of credit card debt from 1998 and projected through 2005 can be expressed by the function

$$D(y) = 42.3793y + 219.5172 \text{ billion dollars}$$

y years after 1990. The number of cardholders during that same time interval can be expressed by the function

$$H(y) = 1.7y + 140.3 \text{ million cardholders}$$

(Source: Based on data from *Statistical Abstract*, 2000 and 2001.)

a. Construct a function that expresses the average credit card debt per cardholder.

b. Estimate the average debt of a cardholder in 2005.

7. **Births** The number of births during the 1980s to women who were 35 years of age or older can be modeled as

$$n(x) = -0.034x^3 + 1.331x^2 + 9.913x + 164.447$$
$$\text{thousand births } x \text{ years after 1980}$$

The ratio of cesarean-section deliveries performed on women in the same age bracket during the same time period can be modeled as

$$p(x) = -0.183x^2 + 2.891x + 20.215 \text{ deliveries}$$
$$\text{per 1000 live births } x \text{ years after 1980}$$

Write an expression for the number of cesarean-section deliveries performed on women 35 years of age or older during the 1980s.

(Source: Based on data from *Statistical Abstract*, 1992.)

8. **Natural Gas Trade** The following table shows two functions of the year *t*: *I* is the projected amount of natural gas imports in quadrillion Btu,

and *E* is the projected natural gas exports in trillion Btu. Show how you could combine the two functions to create a third function giving net trade of natural gas in year *t*. (Net trade is negative when imports exceed exports.)

t	2005	2010	2015	2020
I(t)	4.91	5.61	6.71	6.58
E(t)	330	430	530	630

9. **Debit Cards** The table below shows two functions of the year *y*: *D* is the total number of debit card transactions, and *P* is the number of point-of-sale transactions. Show how you could combine the two functions to create a third function showing the percentage of debit card transactions that were conducted at the point of sale in year *y*.

Year	D(y) (millions)	P(y) (millions)
1994	9078	624
1995	10,464	775
1996	11,780	1096
1997	12,580	1600
1998	13,160	2000
1999	13,316	2428

(Source: *Statistical Abstract*, 2000.)

10. **Gas Prices** The table below represents two functions of the year *t* using 1983 as the base year: *R* is the average price in dollars of a gallon of regular unleaded gasoline, and *P* is the purchasing power of the dollar as measured by consumer prices. Indicate how to combine the two functions to create a third function showing the price of gasoline in constant 1983 dollars.

Year	R(t)	P(t)
1990	1.22	0.77
1992	1.19	0.71
1994	1.17	0.68
1996	1.29	0.64
1997	1.29	0.62
1998	1.19	0.61
1999	1.19	0.60

(Sources: Energy Information Administration; *Statistical Abstract*, 1998; and www.Economagic com.)

Determine whether the pairs of functions in Activities 11 through 14 can be combined by function composition. If so, give function notation for the new function, and draw and label its input/output diagram.

11. $P(c)$ is the profit in dollars generated by the sale of c computer chips.

$C(t)$ is the number of computer chips a manufacturer has produced after t hours of production.

12. $C(t)$ is the number of cats in the United States at the end of year t.

$D(c)$ is the number of dogs in the United States at the end of year c.

13. $C(t)$ is the average number of customers in a restaurant on a Saturday night t hours after 4 P.M.

$P(c)$ is the average amount in tips in dollars generated by c customers.

14. $R(x)$ is the revenue in deutsche marks from the sale of x soccer uniforms.

$D(r)$ is the dollar value of r deutsche marks.

In Activities 15 through 20, rewrite each pair of functions as one composite function.

15. $f(t) = 3e^t$ $t(p) = 4p^2$

16. $h(p) = \dfrac{4}{p}$ $p(t) = 1 + 3e^{-0.5t}$

17. $g(x) = \sqrt{7x^2 + 5x - 2}$ $x(w) = 4e^w$

18. $s(q) = -14 + 7455 \ln q$ $q(p) = 6.439(0.07^p)$

19. $g(t) = 3$ $t(m) = 4m + 17$

20. $c(x) = 3x^2 - 2x + 5$ $x(t) = 4 - 6t$

21. Water Ski Sales   Yearly sales of water skis in the United States (excluding Alaska and Hawaii) between 1985 and 1992 can be modeled by

$$S(x) = \begin{cases} 12.1x - 905.4 \text{ million dollars} \\ \qquad \text{when } 85 \le x \le 88 \\ -14.8x + 1414.9 \text{ million dollars} \\ \qquad \text{when } 89 \le x \le 92 \end{cases}$$

where x is the number of years since 1900.
(Source: Based on data from *Statistical Abstract*, 1994.)

a. Find $S(85)$, $S(88)$, $S(89)$, and $S(92)$.

b. Use the answers to part a to sketch a graph of S.

c. Is S a function of x? Explain.

22. Population   The population of North Dakota between 1985 and 1999 can be modeled by

$$P(t) = \begin{cases} -7.35t + 676.3 \text{ thousand people} \\ \qquad \text{when } 0 \le t \le 6 \\ -0.084t^3 + 2.027t^2 - 13.969t + 662.667 \\ \qquad \text{thousand people} \qquad \text{when } 6 < t \le 14 \end{cases}$$

where t is the number of years since 1985.
(Source: Based on data from *Statistical Abstract*, 1998 and 2001.)

a. Find $P(0)$, $P(4)$, $P(9)$, and $P(14)$.

b. According to the model, what was the population in 1991?

c. Use the answers to parts a and b to sketch a graph of P.

d. Is P a function of t? Explain.

23. Shipping Charges   A mail-order company charges a percentage of the amount of each order for shipping. For orders of up to $20, the charge is 20% of the order amount. For orders of greater than $20 up to $40, the charge is 18%. For orders of greater than $40 up to $75, the charge is 15%. For orders of greater than $75, the charge is 12%.

a. Why would it benefit a company to assess the shipping charges described here?

b. What will shipping charges be on orders of $17.50? $37.95? $75.00? $75.01?

c. Write a formula for the shipping charge for an order of x dollars.

d. Sketch a graph of the formula in part c.

e. Which of the following representations of the shipping charge function do you believe it would be best for the company to put in its catalog? (i) a word description, (ii) a formula as in part c, (iii) a graph, or (iv) some other representation (specify)? Explain the reasons for your choice.

24. Sales   A music club sells CDs for $14.95. When you buy five CDs, the sixth one is free.

a. Sketch a graph of the cost to buy x CDs for values of x between 1 and 12.

b. Write a formula for the graph you sketched in part a.

Swap the inputs and outputs for each of the functions in Activities 25 through 28. Draw an input/output diagram for each new rule. Then write out the new rule in words. Which of the new rules are inverse functions?

25. $R(w) =$ the first-class domestic postal rate (in cents) of a letter weighing w ounces

26. $B(t) =$ the amount in an investment account (in dollars) after t years, assuming no withdrawals are made during the t years

27. $B(x) =$ the number of students in this class whose birthday is on the xth day of the year, assuming it is not a leap year

28. $D(r) =$ the number of years it takes for an investment to double if the annual percentage rate (APR) is $r\%$

Reverse the inputs and outputs in the function tables in Activities 29 through 32, and determine whether the results are inverse functions.

29.

Age	Percent using computers at work
18–24	37.1
25–29	52.5
30–39	53.3
40–49	54.9
50–59	50.7
over 59	32.6

(Source: *Statistical Abstract*, 2001.)

30.

Military rank (4 years service)	Basic monthly pay in 2002 (dollars)
Second Lieutenant	2639
First Lieutenant	3170
Captain	3422
Major	3928
Lt. Colonel	4440
Colonel	5177
Brigadier General	6418
Major General	7571

(Source: Defense Financial and Accounting Service.)

31.

Person's height	Person's weight (pounds)
5'3"	139
6'1"	196
5'4"	115
6'0"	203
5'10"	165
5'3"	127
5'8"	154
6'0"	189
5'6"	143

32.

Year	Total wagers in Nevada on the Super Bowl (millions of dollars)
1996	70.9
1997	70.9
1998	77.3
1999	76.0
2000	71.0
2001	67.7
2002	71.5

(Source: www.vegasinsider.com.)

33. Complaints The number of health insurance complaints received in 2001 by a state's board of health can be modeled by

$$C(m) = 319.6m + 801 \text{ complaints}$$

in the mth month of 2001.

a. In what month were 2399 complaints filed?

b. How many complaints were filed in the first quarter of the year 2001?

c. Does C have an inverse? Explain.

34. Livestock The number of livestock that can be supported by a crop of rye grass can be modeled by

$$L(r) = \frac{9.52(15,000)}{r} \text{ animals per hectare}$$

where r is the annual energy requirement of the animal in megajoules per animal.

(Source: R. S. Loomis and D. J. Connor, *Crop Ecology: Productivity and Management in Agricultural Systems*. Cambridge, Eng.: Cambridge University Press, 1992.)

a. Does L have an inverse? Explain.

b. If milking cows require 64,000 megajoules per cow, how many cows can this crop of rye grass support?

c. If the crop is supporting 7 animals per hectare, what is the energy requirement of the animals?

35. Labor Force   As tabulated in December of each year between 1991 and 2000, the number of women in accounting, auditing, and bookkeeping jobs in the United States can be modeled by the function

$$w(t) = -0.24t^3 + 5.3t^2 - 19.9t + 321.67$$
$$\text{thousand women}$$

where t is the number of years after 1990.
(Source: Based on data from U.S. Bureau of Labor Statistics.)

a. Construct a graph of w.

b. Is w a one-to-one function? Explain your reasoning.

c. Does w have an inverse function? Explain.

36. Earnings   The average hourly earnings of U.S. production workers in the accounting, auditing, and bookkeeping industry between 1991 and 2001 can be modeled by the function

$$E(x) = 0.0116x^2 + 0.309x + 12.05 \text{ dollars}$$

where x is the number of years since 1990. Basing your answer on a graph of the function, determine if E is a one-to-one function and has an inverse.
(Source: Based on data from U.S. Bureau of Labor Statistics.)

37. Why is it important to understand the units of measure of input and output for a given function? How can labeling units help in function construction?

38. Describe the types of input/output units compatibility necessary for each of the following types of function construction: addition, multiplication, division, and composition.

1.3   Functions, Limits, and Continuity

The basic difference that sets calculus apart from algebra is that in calculus, the behavior of the output of a function is considered for inputs that can become infinitely large or for input intervals that are infinitesimally small. In order to consider a function in either the telescopic or microscopic sense, calculus uses the concepts of limits and continuity. We briefly discuss these concepts in this section in order to set the stage for their use as they appear throughout this text.

Limits and the Infinitely Large

As we study functions in calculus, we pay attention to how the output of a function behaves as the input becomes larger and larger in the positive direction (increases without bound) or the magnitude of the input becomes larger and larger in the negative direction (decreases without bound). This behavior is called the **end behavior** of the function. End behavior analysis is similar to looking at a function through an infinitely powerful telescope. We will use end behavior analysis in Chapter 2 and in later chapters as we study the topics of limits of sums and integration. The outputs of many functions become infinitely large as the magnitude of the inputs become larger and larger.

We say that the output of a function is **increasing without bound** if the output continues to increase in height infinitely. If the output of a function continues to decrease infinitely, we say that the output is **decreasing without bound.** This end behavior is seen in the linear and polynomial functions that we will discuss in Sections 1.4 and 2.4. (See Figures 1.30 and 1.31.) However, there are many functions with outputs that do not grow infinitely large as the inputs become arbitrarily large.

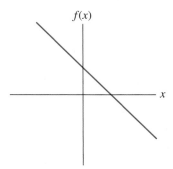

FIGURE 1.30 $f(x)$ decreases without bound as x becomes arbitrarily large in the positive direction.

$f(x)$ increases without bound as x becomes arbitrarily large in the negative direction.

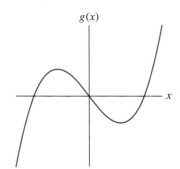

FIGURE 1.31 $g(x)$ increases without bound as x approaches positive infinity.

$g(x)$ decreases without bound as x approaches negative infinity.

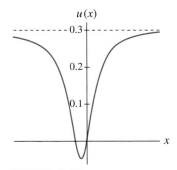

FIGURE 1.32

For an example of end behavior, consider the graph of the function shown in Figure 1.32. As x becomes arbitrarily large in the positive direction (moving from $x = 0$ to the right without bound along the graph of the function), the function outputs $u(x)$ increase while becoming closer and closer to 0.3. As the magnitude of x becomes larger and larger in the negative direction (moving from $x = 0$ to the left without bound on the function graph), again the function outputs become closer and closer to 0.3. We express these two facts mathematically by writing

$$\lim_{x\to\infty} u(x) = 0.3 \quad \text{and} \quad \lim_{x\to-\infty} u(x) = 0.3$$

or with the combined statement

$$\lim_{x\to\pm\infty} u(x) = 0.3$$

When a function approaches a number k as the input increases or decreases without bound, we call the horizontal line that has equation $y = k$ a **horizontal asymptote.** The function depicted in Figure 1.32 has a horizontal asymptote, $y = 0.3$. We say that the output of the function has a *limiting value* of $y = 0.3$.

In addition to estimating end behavior from a graph, it is also possible to estimate the end behavior of a function numerically by evaluating the function at increasingly large values of the input variable. This process is illustrated in Example 1.

EXAMPLE 1   *Numerically Estimating End Behavior*

1.3.1a, b

An equation for the function depicted in Figure 1.32 is $u(x) = \dfrac{3x^2 + x}{10x^2 + 3x + 2}$.
Numerically estimate $\lim\limits_{x \to \infty} u(x)$ and $\lim\limits_{x \to -\infty} u(x)$.

Solution   To estimate the positive end behavior of u numerically, choose increasingly large positive values of x, as shown in Table 1.12.

TABLE 1.12

$x \to \infty$	10	100	1000	10,000
$u(x)$	0.300388	0.3000094	0.3000099	0.3000009$\overline{9}$

It appears that as x approaches positive infinity, the function outputs approach 0.3. Next, consider the negative end behavior of u by choosing negative values of x with increasingly large magnitudes, as shown in Table 1.13.

TABLE 1.13

$x \to -\infty$	−10	−100	−1000	−10,000
$u(x)$	0.298354	0.299894	0.299990	0.299999

It also appears that the limiting value of the outputs is 0.3.   ●

$j(x)$

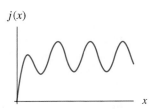

$\lim\limits_{x \to \infty} j(x)$ does not exist.

FIGURE 1.33

Keep in mind that a function may oscillate and not approach any specific output value as x approaches infinity. Such is the case with the function j in Figure 1.33. In this case, we say that the limit of j as x approaches infinity *does not exist*. The sine and cosine functions are good examples of such functions.

Limits and the Infinitesimally Small

In calculus we are also interested in the behavior of the output of a function as the input of that function gets closer and closer to a certain value. This type of behavior analysis is similar to looking at the function through a microscope and increasing the power of the magnification so as to zoom in on a very small portion of that function. We will use this microscopic approach when we consider instantaneous rates of change of functions in Chapters 3 and 4 as well as when we consider limits of sums in Chapter 6.

Concept Development: Limit at a Point   Consider $r(t) = \dfrac{t^2 - 16}{t - 4}$. The graph of r appears to be a line, but as we zoom in on the graph near $t = 4$, a hole in the graph becomes evident. (See Figure 1.34).

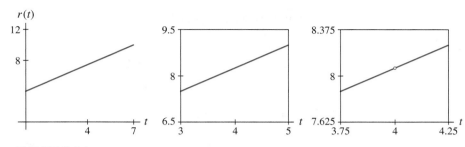

FIGURE 1.34

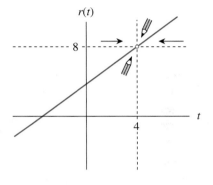

FIGURE 1.35

To graphically estimate the behavior of r as t gets closer and closer to 4, place the tips of two pencils on the graph in Figure 1.35 on either side of $t = 4$ and, keeping the pencils on the graph, move them toward each other. The pencil tips should move toward the hole in the function. Indeed, as the input becomes closer and closer to 4 from either direction, the graph becomes closer and closer to an output of 8. The fact that $r(t)$ is never actually 8 does not affect the limit, because when determining limits, we are interested in where the function is headed, not in whether it ever actually arrives there.

The output $r(t)$ is not actually 8 at $t = 4$ because the output is not defined for $t = 4$. However, we can say that as the input t approaches (gets closer and closer to) 4, the output $r(t)$ approaches 8. Mathematically, we say "as t approaches 4, the limiting value of r is 8". Thus, we write $\lim_{t \to 4} r(t) = 8$.

It is also possible to estimate this *limit* numerically. To do so, we evaluate the function at values increasingly close to, and on either side of, $t = 4$. Table 1.14 shows $r(t)$ values as t approaches 4 from the left (this is denoted $t \to 4^-$), and Table 1.15 shows $r(t)$ values as t approaches 4 from the right (this is denoted $t \to 4^+$).

TABLE 1.14

$t \to 4^-$	$r(t)$
3.8	7.8
3.9	7.9
3.99	7.99
3.999	7.999

TABLE 1.15

$t \to 4^+$	$r(t)$
4.2	8.2
4.1	8.1
4.01	8.01
4.001	8.001

Caution: Do not confuse the symbols used to indicate approaching an input value from the right or from the left with the statement "$x \to \pm c$" or "$x \to \pm \infty$." A plus or minus sign placed *in front of* a number or the infinity symbol indicates whether the number or infinity is positive or negative, whereas a plus or minus sign written *after* a number ($x \to c^+$ or $x \to c^-$) indicates a specific direction from which that number is to be approached.

In Table 1.14 it appears that as t approaches 4 from the left, the output is becoming closer and closer to 8. We symbolize this idea by writing

$$\lim_{t \to 4^-} r(t) = 8$$

This is read as "the limit of $r(t)$ as t approaches 4 from the left is 8." On the basis of Table 1.15, we make a similar statement: "The limit of $r(t)$ as t approaches 4 from the right is 8," which we write as

$$\lim_{t \to 4^+} r(t) = 8$$

Note that the input values near 4 in the tables were arbitrarily chosen. You should arrive at the same conclusion if you choose other values increasingly close to $t = 4$. Because the limits from the left and right of 4 are the same, we conclude that $\lim_{t \to 4} r = 8$; that is, the limit of $r(t)$ as t approaches 4 is 8.

Another function behavior possible is the outputs increase or decrease without bound as the inputs approach a specific value from the left or from the right. Consider Example 2.

EXAMPLE 2   *Numerically Estimating a Limit at a Point*

1.3.2a–b

For $u(x) = \dfrac{3x}{9x + 2}$, numerically estimate $\lim_{x \to -2/9} u(x)$.

Solution   Note that $u(x)$ is not defined at $x = \dfrac{-2}{9}$ because the denominator is zero when $x = \dfrac{-2}{9}$. Consider values of x increasingly close to $\dfrac{-2}{9} \approx -0.222222$ from both the left and the right as shown in Table 1.16.

TABLE 1.16

$x \to \dfrac{-2^-}{9}$	$u(x)$	$x \to \dfrac{-2^+}{9}$	$u(x)$
-0.23	9.9	-0.21	-5.7
-0.223	95.6	-0.219	-22.7
-0.2223	952.7	-0.2219	-229.6
-0.22223	9524.1	-0.22219	-2298.5

It appears that the output values of u are becoming increasingly large as x moves closer and closer to $\dfrac{-2}{9}$ from the left. Also, the output values of u are negative, and their magnitudes seem to become increasingly large as x moves closer and closer to $\dfrac{-2}{9}$ from the right. On the basis of this numerical investigation, we conclude that the limits as x approaches $\dfrac{-2}{9}$ from the left and right do not exist. To indicate that $u(x)$ increases without bound as x approaches $\dfrac{-2}{9}$ from the left, we use the notation $\lim_{x \to -2/9^-} u(x) \to \infty$. We also write $\lim_{x \to -2/9^+} u(x) \to -\infty$ to indicate that $u(x)$ decreases without bound as x approaches $\dfrac{-2}{9}$ from the right.

Keep in mind that this numerical method gives only an estimate. Figure 1.36 shows a graph of the function u.

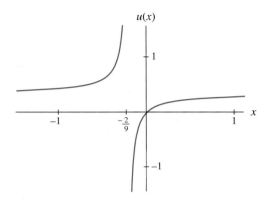

FIGURE 1.36

The graph seems to support our previous estimates that $\lim\limits_{x \to -2/9^-} u(x) \to \infty$ and that $\lim\limits_{x \to -2/9^+} u(x) \to -\infty$. We use the numerical technique illustrated in Example 2 to evaluate the behavior of special constructed functions in Chapter 3. ●

Continuity

Before applying the ideas of calculus when looking at a function in an infinitely large or infinitesimally small manner, we must be guaranteed that the function is continuous (that is, has no holes or breaks) over the interval on which we wish to evaluate the function. The functions r and u already seen in this section are not continuous functions. Because calculus involves the study of continuous portions of functions, it is important to have a complete understanding of what makes a function *continuous* or discontinuous.

Let us first consider some functions that are not continuous and the types of situations that lead to a function's being discontinuous at a point. Consider the four functions shown in Figure 1.37. We observe that a function is not continuous at a point where a hole in the graph occurs (see f), the output becomes infinitely large (see g), or a break or jump occurs (see h and j).

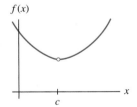
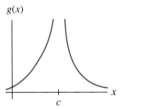
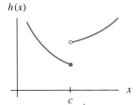
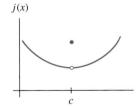

FIGURE 1.37

Intuitively, we say that a function is **continuous** if the function can be drawn without lifting the writing instrument from the page. That is, a function is continuous on an (open) interval if the output of the function is defined at every point on the interval and there are no breaks or jumps in the function output.

WEB/CD

**Strengthening
the Concepts:
Continuity
Defined**

Continuity

A function is continuous on an (open) interval if the output of the function is defined at every point on the interval and there are no breaks or jumps in the function output.

Example 3 investigates continuity in more detail.

EXAMPLE 3   *Determining Continuity for a Piecewise Continuous Function*

Population   Suppose that a small nation is ruled by a ruthless dictator who wishes to rid a part of his country of all people of a certain ethnic background. For 10 years he practices subtle ethnic cleansing, virtually unnoticed by the world. After 10 years he suddenly begins such violent purging that many of the people flee the country or are killed in a matter of a few weeks. The extreme violence ends, but over the next year some of the people continue to seek refuge in neighboring countries. At the end of the 11th year of his reign, the dictator is assassinated and a democratic government is established. Immediately the refugees flood back into the country. The graph in Figure 1.38 shows the approximate size of the population of the affected region of the country for the 20 years after the dictator's rule began.

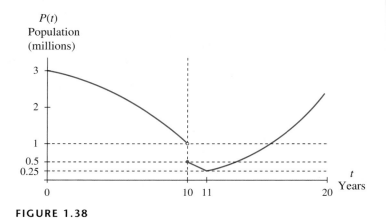

FIGURE 1.38

a. Find and interpret $P(10)$. Explain why $\lim_{t \to 10} P(t)$ does not exist.

b. Find $P(11)$ and $\lim_{t \to 11} P(t)$. What do these values indicate about the population when $t = 11$?

c. Discuss the continuity of the function P.

Solution

a. The graph indicates that $P(10) = 0.5$ million people. The population was 0.5 million people 10 years after the dictator began the "cleansing." The limit of $P(t)$ does not exist at $t = 10$ because the output jumps from 1 million people to 0.5 million people at this input.

b. The graph indicates that $P(11) = 0.25$ million people. Graphical estimates of the function output as t approaches 11 from the left and from the right show that
$$\lim_{t \to 11^-} P(t) = 0.25 \text{ million people and } \lim_{t \to 11^+} P(t) = 0.25 \text{ million people. Thus,}$$
$$\lim_{t \to 11} P(t) = 0.25 \text{ million people. Both the limit as } t \text{ approaches 11 and the}$$
function value at $t = 11$ indicate that the population 11 years after the cleansing began was 0.25 million people.

c. The function P appears to be defined in three parts: the first part on $0 \le t < 10$, the second part on $10 \le t \le 11$, and the third part on $t \ge 11$. Because $P(t)$ exists at $t = 11$ and there are no jumps or breaks at that input, P is continuous at $t = 11$. However, because there is a jump at $t = 10$, P is not continuous at $t = 10$. Thus, the function is continuous for t between 0 and 20 except at $t = 10$.   ●

Continuity and Function Interpretation

In order to apply mathematical functions and calculus in real-world situations, it is important to understand when a function is continuous and how it is to be interpreted. There are times when we use a continuous function, represented by a continuous graph, in a situation that is not actually continuous. Perhaps the best example of this is interest earned on an account.

Consider again the investment of $100 at a 7% annual percentage rate that was given in the discussion of piecewise continuous functions in Section 1.2. Table 1.17 shows the account balance for the first 5 years after you make the investment.

TABLE 1.17

Year	0	1	2	3	4	5
Balance (dollars)	100.00	107.00	114.49	122.50	131.08	140.26

Instead of using a piecewise function to model the account balance, it is common to use a continuous function. In this case, interest is compounded annually, so the following familiar *interest formula* applies:

$$A = P(1 + r)^t$$

where A is the dollar amount accumulated after t years when P dollars are invested at an interest rate of $100r\%$ compounded annually. Thus, the formula for the amount in your account t years after you made the investment is $A(t) = 100(1.07^t)$ dollars. Its graph is shown in Figure 1.39.

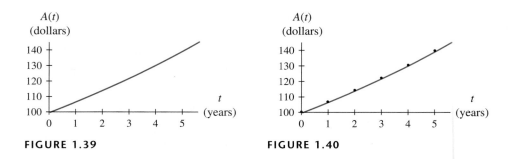

FIGURE 1.39   **FIGURE 1.40**

The six black points (without the curve) in Figure 1.40 constitute a graph of the data in Table 1.17 that is known as a **scatter plot.** The information given by scatter plots and tables of numerical data is considered to be *discrete.* That is, the information appears only for given points. However, the information given by the function A and the graph in Figure 1.39 are continuous. Although the equation and graph in Figure 1.39 are continuous, they make sense only for integer values of $t \geq 0$ in the context of annually compounded interest.

Suppose that 6 months after you invest the $100, you have to withdraw the money. According to the equation, when $t = 0.5$ year, the amount in the account should be $100(1.07^{0.5}) \approx \$103.44$. However, because interest is compounded once a year (at the end of the year), your money has earned no interest after 6 months, and the amount in the account is only $100. The only points on the continuous graph that have meaning in this investment situation are those corresponding to positive integer inputs, some of which are shown in black in Figure 1.40. Because A, which is a continuous function, has meaning in this context only at distinct points, we say that A is a **continuous function with discrete interpretation.**

As the previous discussion illustrates, when you are working with a continuous function with discrete interpretation, it is important to consider carefully what is happening to both the input and the output. You must take special care not simply to round input values in these situations without thinking about the context of the question you are answering and the nature of the input and output of the model.

Examples of situations that can be represented by continuous functions that have restrictions on the input include interest that is not compounded continuously and anything that is a yearly total, such as sales, profit, revenue, number of live births, number of overseas telephone calls, and yearly high price of gasoline.

Functions That Are Continuous without Restriction

There are situations that are not technically continuous but for which we use a continuous model that is not discretely interpreted. For example, consider a function representing the population of a nation over a year. It is not the case that people are born and die continuously. That is, if you were to consider zooming in on this function to observe the population over an hour or day, you would notice jumps in the function. However, because the "jumps" in the graph are insignificant in context over an extended period of time, we use a continuous function, interpreted as a continuous function, to model population.

In situations that are truly continuous (such as temperature over time) and in situations that are technically discrete but that we treat as continuous (such as population growth over time and cumulative totals), we don't restrict the inputs (over some interval) that can be used in a continuous model for the situation. When any value in a certain interval can be used as input, we call the model **continuous without restriction** or simply **continuous.**

The models used throughout this text are either continuous or piecewise continuous. However, not every modeling situation has a continuous interpretation, so you need to consider what inputs make sense in context. Ask yourself the question: "For this model, do inputs of any value within some interval make sense in context?" If the answer is "yes," then we say that the model is continuous without restriction on that interval. If the answer is "no," then the model has restrictions on its input and must be interpreted carefully.

EXAMPLE 4   *Classifying Functions*

Water Evaporation   Consider the following two situations:

a.   A dish containing 5 liters of water is set out to evaporate in a lab. The amount that evaporates depends on the temperature and humidity in the air surrounding the dish. At the end of each day, the amount of water evaporated from the dish during that day is calculated. The data is converted to a continuous model for the amount of water evaporated each day as a function of the day.

b.   The same dish containing 5 liters of water is set out to evaporate in a lab. At the end of each day, the amount of water *remaining* in the dish is recorded. The data is converted to a continuous model for the amount of water remaining in the dish as a function of the day.

Solution   Can both continuous models be used without restriction? In other words, will any input correspond to an output that is meaningful in the situation? One way to determine the answer to these questions is to consider whether or not the "counter" is being reset every day.

a.   In this situation, we are measuring the amount of water that evaporates during one day. So, at the beginning of each day there is no water evaporated. An input corresponding to times other than ends of days may produce an output, but that output has no interpretation in the context.

To understand why, suppose the humidity rises steadily during the week so that the amount of water evaporated each day decreases according to the model $E(d) = 0.4 - 0.05d$ liters at the end of day d. On day 1, $E(1) = 0.35$ liter evaporated, and on day 2, $E(2) = 0.3$ liter evaporated. Is $E(1.5) = 0.325$ liter meaningful in this context? If $d = 1$ corresponds to the end of the first day, then $d = 1.5$ corresponds to the middle of the second day. Does it make sense that in the middle of the second day, 0.325 liter had evaporated? No. The daily amount evaporated begins at 0 liters at the beginning of day 2 and ends at 0.3 liter at the end of day 2. It is impossible for 0.325 liter to have evaporated in the middle of the day.

The key is that the evaporation counter is reset to zero each day. A continuous model of this situation must be discretely interpreted. (A common mistake is to reason that the water is evaporating in a continuous manner, so any function related to the amount of water will be continuous without restriction. It is the way in which we are measuring that demands a discrete interpretation.)

b. In this situation, we are measuring the amount remaining in the dish at the end of each day. That means that at the beginning of each day, the amount of water in the dish is not zero. Suppose that the amount remaining in the dish is modeled by $R(d) = 5 - 0.3d$ liters at the end of day d. On day 1, $E(1) = 4.7$ liters remain, and on day 2, $E(2) = 4.4$ liters remain. Is $E(1.5) = 4.55$ liters meaningful in this context? Is it reasonable that in the middle of the second day, 4.55 liters remain in the dish? Yes. This function is continuous without restriction. ●

A Financial Illustration of Continuity, Interpretation, and a Limit

As mentioned earlier, it is important to understand how to use a continuous function that models a discrete situation. The world of finance yields a classic example of such a situation. We have already mentioned the interest formula

WEB/CD

Strengthening the Concepts: APR and APY

$$A = P(1 + r)^t$$

where A is the dollar amount accumulated after t years when P dollars are invested at an interest rate of $100r\%$ compounded annually. (Note that r is a decimal number.) Often interest is compounded more than once a year. In this case the general **compound interest formula** applies.

Compound Interest Formula

The amount accumulated in an account after t years when P dollars are invested at an annual interest rate of $100r\%$ compounded n times a year is

$$A = P\left(1 + \frac{r}{n}\right)^{nt} \text{ dollars}$$

Here $\left(\frac{r}{n}\right)100\%$ is the interest rate that is applied to the balance in the account at the end of each compounding period, and nt is the total number of compounding periods in t years.

Consider $1000 invested at a 5% annual interest rate compounded monthly. The equation for the accumulated amount $A(t)$ after t years is

$$A(t) = 1000\left(1 + \frac{0.05}{12}\right)^{12t} \text{ dollars}$$

This equation can be rewritten in the form

$$A(t) \approx 1000(1.004166667^{12})^t \approx 1000(1.05116^t) \text{ dollars}$$

Even though the compound interest formula is a continuous function, it has a discrete interpretation because the amount changes only at the actual times of compounding. For instance, we can use the monthly compounding function

$A(t) = 1000(1.05116^t)$ dollars after t years to find the amount of the investment at the end of the 3rd month of the 6th year by calculating $A(6.25) \approx \$1365.95$. (This value was calculated using the unrounded value of the base.) However, it would be incorrect to use $t = 6.2$ to calculate the amount in the account on the 14th day of the 3rd month of the 6th year, because interest is compounded monthly, not daily.

There is one well-known interest formula that has continuous interpretation, the **continuously compounded interest** formula.

Continuously Compounded Interest Formula

The amount accumulated in an account after t years when P dollars are invested at an annual interest rate of $100r\%$ compounded continuously is

$$A = Pe^{rt} \text{ dollars}$$

This formula is used to model situations where interest is considered to be compounded continuously. It arises as a direct result of considering what happens to the compound interest formula as compoundings occur more and more frequently. Considering a 1-year period, the more often compounding occurs, the larger n, the number of compoundings, becomes. In order to consider compoundings to occur continuously, we must consider what happens as n becomes infinitely large. Numerically, it can be shown that $\lim_{n \to \infty} \left(1 + \frac{1}{n}\right)^n = e \approx 2.71828182846$. Using algebra it can be shown that $\lim_{n \to \infty} P\left(1 + \frac{r}{n}\right)^{nt} = Pe^{rt}$. We will explore these limits numerically in the activities.

As we will see in subsequent chapters, limits and continuity are the cornerstone of calculus. Your ability to understand limits intuitively as well as to interpret the continuity of functions will greatly enhance your ability to understand the calculus concepts of differentiation and integration.

1.3 Concept Inventory

- Increasing and decreasing without bound
- End behavior
- Horizontal asymptote
- Numerically and graphically estimating end behavior
- Left and right limits
- Numerically and graphically estimating a limit at a point
- Continuity
- Continuous functions with discrete interpretation
- Continuous functions without restriction
- Compound interest applications

1.3 Activities

Use the graphs shown in Activities 1 through 4 to estimate the limits given.

1. a. $\lim_{x\to\infty} h(x)$ **b.** $\lim_{x\to-\infty} h(x)$ **c.** $\lim_{x\to 2} h(x)$

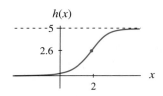

2. a. $\lim_{x\to\infty} s(x)$ **b.** $\lim_{x\to-\infty} s(x)$

c. $\lim_{x\to 0} s(x)$

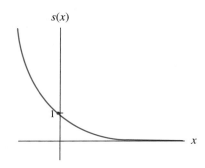

3. a. $\lim_{t\to 1^-} m(t)$ **b.** $\lim_{t\to 1^+} m(t)$

c. $\lim_{t\to 1} m(t)$ **d.** $\lim_{t\to\infty} m(t)$

e. $\lim_{t\to-\infty} m(t)$

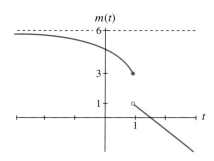

4. a. $\lim_{p\to-2} t(p)$ **b.** $\lim_{p\to\pm\infty} t(p)$

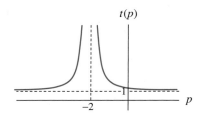

5. Consider the function

$$g(t) = \begin{cases} 2^t - 4 & \text{when } t < 2 \\ 0 & \text{when } 2 \le t < 4 \\ \dfrac{t^2 - 16}{t - 4} & \text{when } t > 4 \end{cases}$$

a. Are there any horizontal asymptotes for this function? If so, give the equation(s) of the asymptote(s). If not, explain why not.

b. Discuss whether the function is continuous for the following inputs:

i. $x = -5$ **ii.** $x = 2$

iii. $x = 3$ **iv.** $x = 4$

6. Consider the function

$$f(x) = \begin{cases} \dfrac{6x}{x^2 - 6x} & \text{when } x < 6 \\ x - 7 & \text{when } x \ge 6 \end{cases}$$

a. Are there any horizontal asymptotes for this function? If so, give the equation(s) of the asymptote(s). If not, explain why not.

b. Discuss whether the function is continuous for the following inputs:

i. $x = 2$ **ii.** $x = 6$ **iii.** $x = 10$

Numerically estimate the limits given in the Activities 7 through 10.

7. $\lim_{x\to-1/3} \dfrac{x^3 + 6x}{3x + 1}$ **8.** $\lim_{x\to 3} \dfrac{1}{(x - 3)^2}$

9. $\lim_{t\to\infty} \dfrac{14}{1 + 7e^{0.5t}}$ **10.** $\lim_{q\to-\infty} (-2^q + 6)$

11. If f is a continuous function, is it possible for there to be an input value of f for which the limit does not exist? Explain.

12. **Chemical Spill**   The cost of cleaning up 100*x*% of a small chemical spill from a mountain stream is given by

$$C(x) = \frac{5}{1 + 5e^{-0.4x}} \text{ thousand dollars}$$

   a. What is the cost of removing 50% of the chemical spill?

   b. Find $\lim_{x \to \infty} C(x)$. Does this limit have meaning in the context of this problem? Explain.

   c. What is the cost of cleaning up all of the spill?

13. **Interest**   Suppose that you invest $1 for 1 year at a rate of 100% compounded *n* times during the year.

   a. Use the compound interest formula to develop a function for the amount at the end of 1 year given *n*, the number of compoundings.

   b. Fill in the values of *n* on the following table.

Compounding	*n*	Amount
yearly		
semiannually		
quarterly		
monthly		
weekly		
daily		
every hour		
every minute		
every second		

   c. Use the formula from part *a* to fill in the amount column of the table. (Round values to two decimal places.)

   d. According to the table, what is the limit of amount as *n* grows larger and larger?

   e. Write the limit found in part *d* in mathematical notation.

14. **Interest**   Suppose that you invest $1 for 1 year at a rate of 100*r*% compounded *n* times during the year.

   a. Use the compound interest formula to develop a function for the amount at the end of 1 year. (Hint: *n* and *r* will both appear as variables in the formula.)

   b. Suppose $r = 0.1$ and create a table that shows *n* and the amount at the end of 1 year for the following compounding options: yearly, semiannually, quarterly, monthly, weekly, daily, every hour, every minute, and every second. (Round the amount to four decimal places.)

   c. According to your table, what is the limit of the amount as the number of compoundings becomes larger and larger?

   d. Use the information for this account and the continuous compounding interest formula to calculate the amount at the end of 1 year. How does your answer here compare to your estimate in part *c*?

   e. Repeat parts *b* through *d* for $r = 0.5$.

15. **Drug Concentration**   Ultram is a non-narcotic pain reliever. Suppose a person takes two 50-milligram tablets of Ultram. A model for the amount in her blood *t* hours after the initial dose is

$$U(t) = 100(e^{-1.3t} - e^{-9.5t}) \text{ milligrams}$$

   a. Examine a graph of *U*, and describe the concentration of Ultram in this person's blood.

   b. Estimate $\lim_{t \to \infty} U(t)$.

   c. When do you feel it is safe for this person to take another dose of Ultram?

16. **Population**   The average annual world population growth rate between 1985 and 2000 and the projected growth rate through 2045 can be modeled by

$$P(x) = 1.724e^{-0.02x} \text{ percent}$$

   x years after 1985.

   a. What is the end behavior of *P*? Interpret it in context.

   b. Discuss how your answer to part *a* is related to the term *horizontal asymptote*.

If a continuous function is used to model the function descriptions in Activities 17 through 20, can it be used without restriction or must it be discretely interpreted?

17. $P(m)$ is the amount, in millions of dollars, of property value loss resulting from an earthquake of magnitude *m* centered in a particular major metropolitan area.

18. $H(x)$ is the number of hits at a certain web site on the xth day of 2001.

19. $Q(t)$ is your body temperature in degrees Celsius t hours after you take 500 mg of Tylenol.

20. $C(r)$ is the number of children who register for summer day camp when the registration fee is set at r dollars.

21. **Rainfall**   The table shows the predicted amount of rainfall as a function of the wind speed of a tropical storm.

Wind speed (mph)	10	8	5	3
Predicted rainfall (inches)	10	15	23	33

   a. Draw a scatter plot of these data.

   b. Use your scatter plot and the Vertical Line Test to verify that the table represents a function.

   c. Can you determine, using the given information, the rainfall predicted when the wind speed is 6 mph? If so, find the amount. If not, explain why not.

   d. Is this function representation discrete or continuous? Give reasons for your answer.

22. **Rainfall**   The predicted amount of rainfall for a tropical storm can be modeled by the equation

$$R(v) = 54.1(0.847^v) \text{ inches}$$

where v is the wind speed (in miles per hour) of the storm.

   a. Describe the input and output of the function R.

   b. State the units of measure of the input and output.

   c. Is R an increasing, decreasing, or constant function for wind speeds between 3mph and10 mph?

   d. Can you determine, using the given information, the rainfall predicted when the wind speed is 6 mph? If so, find the amount. If not, explain why not.

   e. Is this model discrete, continuous without restriction, or continuous with discrete interpretation? Give reasons for your answer.

   f. According to this function, at what wind speed will there be 28 inches of rain from the storm?

23. **Marketing**   The average daily profit resulting from telephone marketing calls by an employee of a certain company can be modeled by

$$P(c) = 0.043c^3 - 3.129c^2 + 71.133c - 315.524 \text{ dollars}$$

when c calls are made daily.

   a. Find $P(25)$ and interpret the answer.

   b. Find the value of c between 0 and 25 for which $P(c) = 180$, and interpret the answer.

   c. Graph this function for 0 to 30 calls per day. Use the graph to estimate when the greatest daily profit occurs.

24. **Cable Rates**   The table lists the average basic cable rate in dollars per month for selected years from 1970 through 2000.

Year	Average basic cable rate (dollars per month)
1970	5.50
1975	6.50
1980	7.69
1985	9.73
1990	16.78
1995	23.07
2000	30.08

(Source: *Statistical Abstract*, 1998 and 2001.)

   a. Draw a scatter plot of these data.

   b. Use the scatter plot and the Vertical Line Test to verify that the table represents a function.

   c. Use A to represent the output, average basic cable rate in dollars per month, and y to represent the input in years since 1900. Write "In 1970 the average basic cable rate was $5.50 per month" using function notation.

   d. Explain the meaning of the notation $A(90.5)$. Does $A(90.5)$ make sense in the context of this problem?

   e. Find the value of y for which $A(y) = 23.07$. Interpret your answer.

25. **Cable Subscribers**   On the basis of data recorded between 1975 and 2000, the number of cable subscribers in the United States as a function of the average basic cable rate can be modeled as

$S(r) = -56.1105 + 37.201 \ln r$ million subscribers

when the average basic rate is r dollars per month.
(Source: Based on data from *Statistical Abstract*, 2001.)

a. Describe the input and output of the function S.

b. State the units of measure of the input and output.

c. Should this continuous model be discretely interpreted? Why or why not?

d. Use the model to estimate the number of subscribers when the basic monthly cable rate was $15.

e. According to the model, what is the total revenue from the basic monthly rate when there are 10 million subscribers?

26. **Kidney Transplants**   The number of yearly kidney transplants performed in the United States between 1992 and 1996 can be modeled by the equation

$$K(t) = 9088.859 + 1697.717 \ln t \text{ transplants}$$

t years after 1990.
(Source: Based on data from *Statistical Abstract*, 1998.)

a. Describe the input and output of the function K.

b. State the units of measure of the input and output.

c. Should this continuous model be discretely interpreted? Why or why not?

d. Use the model to estimate the number of transplants in 1995.

e. Find $K(2.5)$ and interpret the answer.

f. According to the model, when did the number of yearly transplants first exceed 13,000?

27. **Stock Price**   The price of Microsoft stock on October 10, 2002, can be modeled by

$$P(h) = 0.368h + 43.99 \text{ dollars per share}$$

h hours after the market opened.

a. Describe the input and output of the function P.

b. State the units of measure of the input and output.

c. Should this continuous model be discretely interpreted? Why or why not?

d. What was the price of this stock after 3 hours of trading?

e. Did the price reach $47 per share during this particular day? (The market opens at 9:30 A.M. and closes at 4:00 P.M.)

28. **Customer Base**   The customer base in the late 1990s for Amazon.com can be modeled by

$$A(t) = 2.3(4.652^t) \text{ million customers}$$

t years after 1998.
(Source: Based on data from "Amazon.com takes loss for quarter," *Reno Gazette–Journal*, July 23, 1999, p. 4B.)

a. Describe the input and output of the function A.

b. State the units of measure of the input and output.

c. Should this continuous model be discretely interpreted? Why or why not?

d. Estimate the size of the customer base at the end of 2000.

e. When was the customer base 9 million?

29. **Flex Schedules**   The percentage of workers of age a years with flex schedules can be estimated using the function

$$P(a) = 0.000498a^3 - 0.0686a^2 + 3.044a - 14.952 \text{ percent}$$

where a is between 16 and 65.
(Source: Based on data from *Statistical Abstract*, 1998.)

a. Find $P(20)$ and interpret your answer.

b. State the units of measure of the input and output.

c. Are there any restrictions on the input of this model? Explain.

30. **Temperature**   Air temperature at high altitudes can be estimated using the function

$$T(a) = 278(0.852^a) - 66°F$$

where a is the number (in thousands) of feet above sea level.

a. Find $T(20)$ and interpret your answer.

b. State the units of measure of the input and output.

c. Are there any restrictions on the input of this model? Explain.

31. Discuss the possible limitations of using a technology-generated graph to determine the continuity and end behavior of a function.

1.4 Linear Functions and Models

Having explored the concept of a function earlier in this chapter, we now turn our attention to several specific types of functions that will be helpful as we seek to describe real-life situations with mathematical models. Our goal here and in the next chapter is to give you an understanding of the behavior underlying certain functions in order to help you determine what function is appropriate in a particular modeling situation.

Representations of a Linear Model

We begin with the simplest of all functions: the linear function. A **linear function** is one that repeatedly and at even intervals adds the same value to the output. The output values form the pattern of a line when graphed; thus we use the term *linear* to describe the data. For example, consider a newspaper delivery team that makes weekly deliveries of newspapers and devotes their Saturday mornings to selling new subscriptions. The function describing the number of customers can be represented four ways:

With words:

The team starts with 80 customers, and 5 new customers are added each week.

With data (see Table 1.18):

TABLE 1.18

Weeks	0	1	2	3	4	5	6	7
Number of customers	80	85	90	95	100	105	110	115

With graphs (see Figure 1.41):

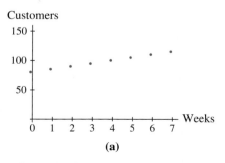

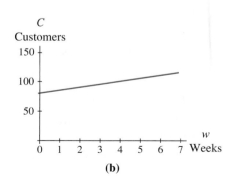

FIGURE 1.41

With an equation:

$C = 5w + 80$ customers, where w stands for the number of weeks since the team began the subscription drive.

Each of these four representations are indicative of a linear model.

The Parameters of a Linear Model: Slope and Intercept

A linear equation is determined by two parameters: a starting value and the amount of the incremental change. All linear functions appear algebraically as

$$f(x) = ax + b$$

where a is the incremental change per unit input and b is the starting value.*

Linear functions are graphed as lines where a is the *slope* (a measure of the line's steepness) and b is the *vertical axis intercept* (that is, the output value at which the line crosses the vertical axis). The slope of a graph is of primary importance in our study of calculus. As we shall see in Chapter 3, the slope of most functions is determined using calculus; however, the slope of a line can be calculated more simply.

The directed horizontal distance from one point on a graph to another is called the *run*, and the corresponding directed vertical distance is called the *rise*. The quotient of the rise divided by the run is the **slope** of the line connecting the two points. Consider again the newspaper subscription graph in Figure 1.41. We have chosen the points on the graph that correspond to 0 and 7 weeks. The rise and run corresponding to those two points are shown in Figure 1.42.

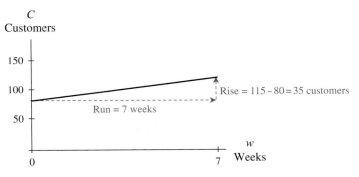

FIGURE 1.42

The slope is calculated as $\dfrac{\text{rise}}{\text{run}} = \dfrac{35 \text{ customers}}{7 \text{ weeks}} = 5$ customers per week. (This slope value will be the same regardless of which two points are chosen for the calculation.) The steeper a line, the greater the magnitude of the value of the slope. Lines that fall rather than rise have negative slope. Unlike the slope, the steepness of a line does not depend on whether it rises or falls.

* $y = a + bx$ is an alternative form of the linear equation. This form is commonly used by statisticians.

The slope of a graph of a function at a particular point is a measure of how quickly the output is changing as the input changes at that point. We call this measure the **rate of change** of the function at that point and discuss it more thoroughly in Chapters 3 and 4. Because linear functions are characterized by constant incremental change, their slope (rate of change) is constant at all points. Other types of functions have a different slope at every point; that is, their rate of change is not constant.

EXAMPLE 1 *Calculating Slopes*

Sales The resale value of a used car is represented graphically in Figure 1.43.

a. Locate and interpret the horizontal- and vertical-axis intercepts.

b. Calculate the slope of the graph.

c. Does the slope depend on the direction in which it was calculated?

d. Interpret the slope.

e. Write a linear model for the graph.

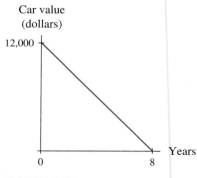

FIGURE 1.43

Solution

a. Recall that the places where the line crosses the horizontal and vertical axes are called **intercepts.** The vertical axis intercept is $12,000 and corresponds to the value of the car when it was purchased. The horizontal axis intercept is 8 years and indicates the first year in which the car had essentially no value.

b. To travel from the horizontal axis intercept 8 to the vertical axis intercept 12,000 on the graph requires that you move 12,000 units up ($+\$12,000$ = rise) and 8 units to the left (-8 years = run). See Figure 1.44. The quotient of the rise divided by the run is $\frac{\$12,000}{-8 \text{ years}} = -\$1500$ per year.

c. If you traveled from the vertical axis intercept 12,000 to the horizontal axis intercept 8, the run would be positive and the rise negative. The quotient of the rise divided by the run would be $\frac{-\$12,000}{8 \text{ years}} = -\$1500$ per year. Note that the quotient $\frac{\text{rise}}{\text{run}}$ is the same either way you compute it.

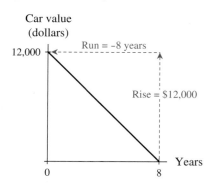

FIGURE 1.44

d. The quotient $\frac{\text{rise}}{\text{run}}$ is equal to the slope of the linear model and is the rate of change of the value of the car. We say that the car depreciated at a rate of $1500 per year.

The equation of the line that represents the value of the car has slope -1500 and starting value 12,000. If we let the variable t represent the number of years since the car was purchased, the equation for the value of the car is

Value = -1500t + 12,000 dollars   ●

Although the slope of a particular linear model never changes, the graph of the model may look different when the horizontal or vertical scale is changed. You should therefore always use the same horizontal and vertical views when comparing two different graphs. For example, the two graphs in Figure 1.45 show energy production and energy consumption in the United States from 1975 through 1980.

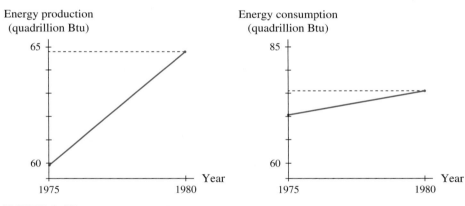

FIGURE 1.45
(Source: Based on data from *Statistical Abstract,* 1994.)

At first glance, it appears that the graph of energy production is steeper than that of energy consumption. In other words, it appears that the amount of energy being produced is increasing at a faster rate than that of the amount of energy being consumed. However, by carefully looking at the graphs, we notice that the vertical scales are different. Calculating slopes using these different scales shows that energy production was increasing by approximately 0.98 quadrillion Btu per year from 1978 through 1980 and that energy consumption was increasing by 1.08 quadrillion Btu per year during that same time period. That is, the energy consumption graph is actually the steeper graph. Appearances can deceive, so be careful when comparing graphs with differing scales.

We summarize our discussion of linear equations as follows:

Linear Model

Verbally: A linear model is one that has a constant rate of change.

Algebraically: A linear model has an equation of the form

$$f(x) = ax + b$$

where a and b are constants.

Graphically: The graph of a linear function is a line (see Figure 1.46.) The parameter a in the equation is the constant rate of change of the output and is the slope of the line. The parameter b is the vertical axis intercept.

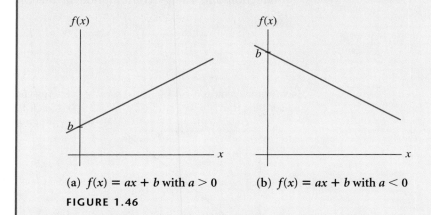

(a) $f(x) = ax + b$ with $a > 0$ (b) $f(x) = ax + b$ with $a < 0$

FIGURE 1.46

Finding a Linear Model from Data

Suppose you are sole proprietor of a small business that has seen no growth in sales for the last several years. You have noticed, however, that your federal taxes have increased as shown in Table 1.19.

TABLE 1.19

Year	1999	2000	2001	2002	2003	2004
Tax	$2532	$3073	$3614	$4155	$4696	$5237

Upon close examination of the data, you note that taxes increased by the same amount every year.

1.4.1a–d

The changes in successive output, which are called **first differences,** are constant. This constant increase is the incremental change where the increment is one year. Because the incremental change is constant, we know that the data represent a linear increase in taxes. Because the data show the tax amount each year, this incremental change is the rate of change of the tax amount and the slope of the underlying linear model. If the data had shown taxes every other year, the first differences would be constant, indicating a linear pattern, but the value of the first differences would be twice the slope. Slope values are always expressed per unit increase in input.

On the basis of the calculation of first differences, we make the following equivalent statements:

Taxes increased by $541 per year.
The slope of the linear function described by the data is $541 per year.
The rate of change of the tax amount is $541 per year.

If we consider 1999 to be our starting point, then the starting tax value is $2532. We express the tax function verbally: The tax amount began in 1999 at $2532 and increased each year by $541. We express the function algebraically as

Tax = $541t + 2532$ dollars

where t represents the number of years since 1999. Such models are important because they often enable us to analyze the results of change. With certain assumptions, they may even allow us to make cautious predictions about the short-term future. For example, to predict the tax amount owed in 2007, we substitute $t = 8$ into the tax model.

Tax = $541(8) + 2532 = \$6860$

Admittedly, in this instance, an equation is not necessary to make such a prediction, but there are many situations in which it is difficult to proceed without a model.

It is important to understand that when we use mathematical models to make predictions about the short-term future, *we are assuming that future events will follow the same pattern as past events.* This assumption may or may not be true. That does not mean that such predictions are useless, *only that they must be viewed with extreme caution.*

A Word of Caution

When you use a model to predict output values for input values that are within the interval of input data used to obtain the model, you are using a process called **interpolation.** Predicting output values for input values that are outside the interval of the input data is called **extrapolation.** Because you do not know what happens outside the range of given data, **estimates obtained by extrapolation must be viewed with caution and may result in misleading predictions.**

We have already noted that the rate of change of the tax amount is $541 per year, which is how much taxes increase during a 1-year period. Thinking about rate of change in this way helps us answer questions such as

* If you pay taxes twice a year, how much will taxes increase each time?

$$(\$541 \text{ per year})\left(\tfrac{1}{2} \text{ year}\right) = \$270.50$$

- How much will taxes increase each time if you pay taxes quarterly?

$$(\$541 \text{ per year})\left(\tfrac{1}{4} \text{ year}\right) = \$135.25$$

- How much will taxes increase during the next 3 years?

$$(\$541 \text{ per year}) (3 \text{ years}) = \$1623$$

EXAMPLE 2   *Writing a Linear Model*

TABLE 1.20

Years	Companies (percent)
5	50
6	47
7	44
8	41
9	38
10	35

(Source: Cognetics, Cambridge, Mass., 1998.)

Business Survival   Table 1.20 shows the percent of U.S. companies that are still in business after a given number of years in operation.

a.   Find the constant rate of change in the percent of businesses surviving.

b.   Find a model for the percent of companies still in business as a function of years of operation. Graph the model and give a verbal description of it.

Solution

a.   Calculating the first differences of the data in Table 1.20 indicates that the percent of businesses surviving decreased by 3 percentage points per year. We could also say that the rate of change of the survival percentage is −3 percentage points per year.

> *Caution:* Note that this result is not the same as a rate of change of −3% per year. When the output of a function is a percent, units on the rate of change must be expressed in terms of percentage points per input unit. More will be said about this in Chapter 3.

b.   The data reveal that the percent begins with 50% and that each year after the first table value, the percent declines by 3 percentage points. We write this mathematically as

$$P(t) = -3t + 50 \text{ percent of businesses}$$

t years after the fifth year in operation. Figure 1.47 depicts the graph of P.

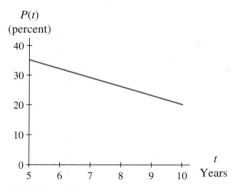

FIGURE 1.47

The function $P(t) = -3t + 50$, its graph, and the data in Table 1.20 all show that the percent of U.S. companies that are still in business declines by 3 percentage points for each year after 5 years of operation and that at the end of 5 years 50 percent of those companies are still in business.   ●

The preceding examples illustrate methods of finding a linear model of the form $f(x) = ax + b$ for data points that fall on a line. However, real-life data values are seldom perfectly linear. For instance, the tax data we considered earlier are not likely to occur in real-life situations because tax rates and revenues of most businesses change from year to year. Consider the following modification to the tax data:

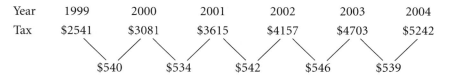

Year	1999	2000	2001	2002	2003	2004
Tax	$2541	$3081	$3615	$4157	$4703	$5242

$540 $534 $542 $546 $539

The first differences are not constant but are "nearly constant." A linear model may be used if first differences are close to constant. Be sure to calculate first differences only for data that are evenly spaced.

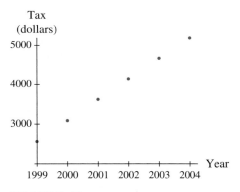

FIGURE 1.48

1.4.2a–d

An examination of the scatter plot in Figure 1.48 reinforces our earlier observation, from first differences, that the data are close to being linear. How do we get a linear model in this situation? Use the linear regression routine that is built into your calculator or computer to find a linear equation that fits the data. You should find the equation to be

Tax1 $= 540.371t - 1{,}077{,}663.581$ dollars

where t is the year. Overdrawing the line of best fit on the scatter plot confirms our observation that a linear model is appropriate.

Aligning Data

When the input data are years, it is often desirable to modify how the years are numbered to reduce the number of digits you have to enter, as well as to reduce the magnitude of some of the coefficients in the model equation. We refer to the process of renumbering data as **aligning** the data. For example, if we renumber the years in the tax data so that 1999 is year 0, 2000 is year 1, and so on, we obtain the data shown in Table 1.21.

1.4.3a–e

TABLE 1.21

Aligned year	0	1	2	3	4	5
Tax	$2541	$3081	$3615	$4157	$4703	$5242

The aligned years are described as "the number of years since 1999." Using a calculator or computer to obtain a linear model for this aligned data, we have

$$\text{Tax}2 = 540.371t + 2538.905 \text{ dollars}$$

where t is the number of years since 1999. Note that this equation and the equation for the unaligned data have the same slope (rate of change) but differ in the vertical axis intercept. That is, the parameter a in the equation $f(x) = ax + b$ is unchanged but the b value differs. Aligning the data has the effect of shifting the data (usually to the left). This does not change the slope of the line, but it does change where the line crosses the vertical axis.

When using a model obtained from aligned data, it is important that you keep in mind at all times the description of the input variable. If we are willing to assume that the rate of change remains constant at about $540.37 per year, then we can estimate the tax in 2006 by using a value of $t = 2006$ in the first model or a value of $t = 7$ in the second model (the number of years that 2006 is from 1999).

$$\text{Tax}1 = 540.371(2006) - 1{,}077{,}663.58 \text{ dollars} \approx \$6322$$
$$\text{Tax}2 = 540.371(7) + 2538.905 \text{ dollars} \approx \$6322$$

These models give the same output values. They are equivalent. Note that the model with the lower-numbered input (Tax2) has the smaller intercept value. It also has the advantage that the constant term ($2538.905) is the approximate beginning amount shown in the table.

To summarize, when the input is years, it is desirable to align data to make the data entry faster and to reduce the magnitude of some of the coefficients. In general, the smaller the magnitude of the input data, the smaller the magnitude of the coefficients in the function that models the data.

EXAMPLE 3   *Aligning Data to Find Linear Models*

Hepatitis A   Table 1.22 shows the incidents of hepatitis A per 100,000 people in Clark County, Nevada, in the mid-1990s.

TABLE 1.22

Year	1993	1994	1995	1996	1997
Incidents per 100,000	7.1	12.8	20.2	24.4	32.3

(Source: Nevada Health Division.)

a. Use a calculator or computer to find three linear models for these data by aligning the years in three different ways.

b. Use each model in part *a* to estimate the incidents of hepatitis A in 1999.

c. Is the calculation in part *b* interpolation or extrapolation?

Solution

a. Three possible models are

$$H(t) = 6.2t - 12{,}349.64 \text{ incidents per 100,000 in year } t$$
$$J(t) = 6.2t - 569.64 \text{ incidents per 100,000 } t \text{ years after 1900}$$
$$K(t) = 6.2t + 6.96 \text{ incidents per 100,000 } t \text{ years after 1993}$$

Note that in each function, the slope value is the same and the intercept value is different. Also note that the smaller the input values, the smaller the magnitude of the intercept value. The third model has the advantage that the vertical intercept is close to the first value shown in the table.

b. To estimate the incidents of hepatitis A in 1999, we substitute 1999 for *t* in $H(t)$, 99 for *t* in $J(t)$, and 6 for *t* in $K(t)$.

$$H(1999) \approx 44.2 \text{ cases per 100,000 people}$$
$$J(99) \approx 44.2 \text{ cases per 100,000 people}$$
$$K(6) \approx 44.2 \text{ cases per 100,000 people}$$

We estimate that in 1999, there were approximately 44.2 cases of hepatitis A per 100,000 people in Clark County, Nevada.

c. The estimate in part *b* is an extrapolation because 1999 is outside the year values given in the data. This estimate may or may not be valid. If health officials wanted to predict accurately the number of hepatitis A cases in a particular year, they would take many factors other than past incidence into account. ●

What Is "Best Fit"?

When you use a calculator or computer to fit a linear equation to data, you should always use the routine that produces the line of best fit. How do we decide that a particular line best fits the data? After all, there are many lines that can be drawn through the data. Figure 1.49 (on next page) shows a line fit to some data points.

A visual indication of how well the line fits the data is the extent to which the data points deviate from the line. We determine a numerical measure of this visual observation in the following way. Calculate the amount by which the line misses the data; that is, calculate the vertical distance (the *deviation*) of each data point from the line. These deviations are simply the lengths of the vertical segments shown in Figure 1.50 (on next page). Each deviation measures the *error* between the associated data point and the line.

$$\text{Deviation} = \text{error} = y_{\text{data}} - y_{\text{line}}$$

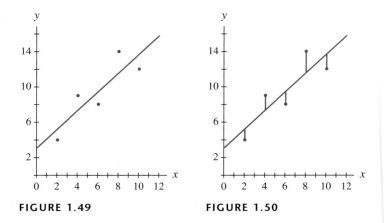

FIGURE 1.49   **FIGURE 1.50**

To obtain a single numerical measure, we square the errors and sum:

SSE = the sum of squared errors

$$= (\text{first error})^2 + (\text{second error})^2 + \cdots + (\text{last error})^2$$

This number, denoted by **SSE (sum of squared errors),** is an overall measure of how well the line fits the data.

1.4.4

The key idea to the correct use of SSE is this: Smaller values of SSE arise from lines where the errors are small, and larger values of SSE arise from lines where the errors are large. Thus a common strategy for choosing the best-fitting line is to choose the line for which the sum of squared errors (SSE) is as small as possible. Such a line is designated as the line of "best fit," and the procedure for choosing it is called the **method of least squares.**

Although it is possible to use calculus to find a line of best fit according to the method of least squares, we will use the routine in a calculator or computer whenever we wish to fit a line to data. In the next chapter, we will look at other functions that can be used to fit data. The method of least squares is the accepted method of finding a best-fitting line, but there are other methods of fitting different kinds of functions to data in which SSE does not play a role. Even though it is possible to compute SSE for each of these functions, SSE should *not* be used to compare models of different types. Such comparison is not a valid statistical procedure.

Numerical Considerations: Reporting Answers

When dealing with numerical results, it is important to understand their accuracy and how precise the results need to be. For instance, imagine that you have a bank account with a balance of \$18,532.71 paying interest at a rate of $17\frac{1}{4}\%$ compounded annually. At the end of the year, the bank calculates your interest to be $(18,532.71)(0.1725) = \$3196.892475$. Would you expect your next bank statement to record the new balance as \$21,729.60248? Obviously, when reporting numerical results dealing with monetary amounts, we do not consider partial pennies. The bank reports the balance as \$21,729.60. The required precision is only two decimal places.

> You should always round numerical results in a way that *makes sense* in the context of the problem.

TABLE 1.23

Year	Net sales (millions of dollars)
1998	115.6
1999	80.6
2000	45.7

If a situation allows us to report answers to the nearest cent, the value $170.533333333 million is not appropriate because this answer indicates a value of $170,533,333.333 with the final digit representing a fraction of a cent.

Generally, results that represent people or objects should be rounded to the nearest whole number. Results that represent money usually should be rounded to the nearest cent or, in some cases, to the nearest dollar. Consider, however, an international company that reports net sales as shown in Table 1.23.

A linear model for these data is $y = -34.95x + 69,945.683$ million dollars, where x is the year. If we wished to estimate net sales in 2001, we would substitute $x = 2001$ in the model to obtain $y = \$10.73333333$ million. Should we report the answer as $10,733,333.33 or $10,733,333? Neither! When we are dealing with numerical results, our answer can be only as accurate as the least accurate output data. In this case, the answer would have to be reported as $10.7 million or $10,700,000.

> You should round numerical results to the same accuracy as the given output data.

Keep in mind that a number by itself is likely to be absolutely worthless. For example, it would not make sense for an international company to publish in its annual report that net sales were 10.7. This could mean 10.7 dollars, 10.7 thousand Euros, 10.7 million Yen, and so on. The label makes a big difference in our understanding of the number.

> A number is useless without a label that clearly indicates the units involved.

Numerical Considerations: Calculating Answers

Although the correct rounding and reporting of numerical results is important, it is even more important to calculate the results correctly. Because you must sometimes round your answers, it is tempting to think that you can round during the calculation process. Don't! Never round a number unless it is the final answer that you are reporting. Rounding during the calculation process may lead to serious errors. Your calculator or computer is capable of working with many digits. Keep them all while you are still working toward a final result.

When you use your calculator or computer to fit a function to data, it finds the parameters in the equation to many digits. Although the text shows rounded coefficients and it may be acceptable to your instructor for you to round the coefficients when reporting a model, make sure that you use all of the digits while working with the model. This helps reduce the possibility of round-off error.

For example, suppose that your calculator or computer generates the following equation for a set of data showing weekly profit for an airline for a certain route as a function of the ticket price.

$$\text{Weekly profit} = -0.00374285714285x^2 + 2.5528571428572x - 52.71428571429$$
$$\text{thousand dollars}$$

where x is the ticket price in dollars. In the answer key, you would see the model reported to three decimal places:

$$\text{Weekly profit} = -0.004x^2 + 2.553x - 52.714 \text{ thousand dollars}$$

However, if you used the rounded model to calculate weekly profit, your answers would be incorrect because of round-off error. Table 1.24 shows the inconsistencies between the rounded and unrounded models.

TABLE 1.24

Ticket Price	Profit from rounded model	Profit from unrounded model
$200	$298 thousand	$308 thousand
$400	$328 thousand	$370 thousand
$600	$39 thousand	$132 thousand

> Never use a rounded model to calculate and never round intermediate answers during the calculation process.

We feel it is important to print the fewest digits possible under the conditions that (a) the rounded model visually fits the data, and (b) the rounded model gives values fairly close to answers obtained with the full model. In particular, if there is a difference between the results from the full model and those from the rounded model, that difference should appear only in the last digit for which we can claim accuracy.

The number of decimal places shown in models in the text will vary. However, whenever we calculate with a model, we use all the digits available at that point in the text. If the data are given, the unrounded model will be used in calculations. When a model found earlier is used in a later section and the data are not repeated, all calculations will be done with the rounded model given in that later section. For convenience and consistency, the answer key will report all models with three decimal places in coefficients and six decimal places in exponents.

The Four Elements of a Model

A model is an equation describing the relationship between an output variable and input variable together with their defining statements. The first element of a model is the equation itself. As previously noted, equations obtained from data using technology usually have coefficients with many digits. When using the model, do not round the coefficients, but when reporting the model, it may be acceptable to your instructor if you round the parameters to three or four decimal places as we have done throughout the text.

The second element of a model is a description of what the output variable represents. Look back to the models we obtained for the small business tax. Tax is *what* is measured, and dollars is a label telling *how* it is measured. Always label a model equation with output units to tell how the output is measured.

The third element of a model is the description of the input variable. If all that had been written for the tax model was Tax2 = 540.37143t + 2538.905 dollars, you might have erroneously predicted that your federal taxes in 2006 would be 540.37143(2006) + 2538.905 = $1,086,524. However, there is a statement with the equation that reads "where t is the number of years since 1999." Because 2006 is the 7th year since 1999, you should correctly predict your federal taxes in 2006 to be 540.37143(7) + 2538.905 = $6322.

The final element of a model is a description of the valid input interval. The tax model was obtained from data between 1999 and 2004. It is important, when giving a model that will be used without the data from which it was obtained, that you indicate the interval of input values over which the model is valid. This is the only way that other people who use the model will know whether they are interpolating or extrapolating. Here is the convention we will adopt: When a model is presented without the data from which it was obtained, we will indicate the range of input data used. For example, we would report the tax model by saying, "The tax between 1999 and 2004 can be modeled as Tax = 540.37143t + 2538.905 dollars t years after 1999." This statement incorporates the four elements of a model.

> When both the data and a model are present, we do not always repeat the interval information. This is especially true in activities where data are given. The answer key does not always restate the interval.

There are four important elements of every model:

1. An equation

2. A label denoting the units on the output

3. A description (including units) of what the input variable represents

4. An indication of the interval of input values over which the model is valid. This information should be given whenever the model is presented or used apart from the data from which it was obtained.

As you proceed through this course, take the time to record your models completely and to report your answers accurately. If you practice proper reporting while you work through the activities on linear models, you will begin to develop a habit that will help you throughout your study of calculus in this text. In the next chapter we will consider models other than the linear model. These reporting guidelines hold true for them as well.

1.4   Concept Inventory

○ *Algebraic form of a linear function:*
  $f(x) = ax + b$

○ *Rate of change (slope) of a linear model is constant*

○ *Calculating slope as* $\frac{rise}{run}$

○ *Interpreting slope*

○ *First differences*

○ *Aligning input data*

○ *Interpolating and extrapolating*

○ *The method of least squares*

○ *Calculation guidelines and rounding rules*

○ *Labeling units on answers*

○ *Four elements of a model*

1.4   Activities

1. **Profit**   The accompanying graph shows a corporation's profit in millions of dollars over a period of time.

   a. Estimate the slope of the graph, and write a sentence explaining the meaning of the slope in this context.

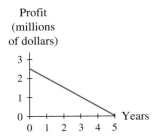

   b. What is the rate of change of the corporation's profit during this time period?

   c. Identify the horizontal and vertical axis intercepts, and explain their significance to this corporation.

2. **Temperature**   The air temperature in a certain location from 8 A.M. to 3 P.M. is shown in the accompanying graph.

   a. Estimate the slope of the graph, and explain its meaning in the context of temperature.

   b. How fast is the temperature rising between 8 A.M. and 3 P.M.?

   c. What would be the meaning of a horizontal axis intercept for this temperature graph?

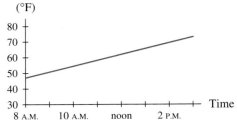

3. **Organ Donors**   The number of organ donors in the United States between 1988 and 1996 can be modeled by

   $$D(t) = 382.5t + 5909 \text{ donors}$$

   t years after 1988.
   (Source: Based on data from United Network for Organ Sharing.)

   a. According to the model, what is the rate of change of the number of organ donors?

   b. Sketch a graph of the model. What is the slope of the graph?

   c. Find the vertical axis intercept, and explain its significance in context.

4. **Bankruptcy**   Total Chapter 12 bankruptcy filings between 1996 and 2000 can be modeled by

   $$B(t) = -83.9t + 1063 \text{ filings}$$

   t years after 1996.
   (Source: Based on data from Administrative Office of U.S. Courts.)

   a. What is the rate of change of the number of Chapter 12 bankruptcy filings?

   b. Sketch a graph of the model. What is the slope of the graph?

   c. Find the vertical axis intercept, and explain its significance in context.

5. **Population**   The population of a certain country can be modeled by $P(t) = 6.98t + 12.39$ million. Estimate the population of that country in 2007.

6. **Medicine**   The absorption of a certain drug into the body can be modeled by $A(x) = -0.076x + 0.3$ grams. Find the amount of this drug that has been absorbed 3 hours after the drug was taken.

7. **Revenue**   The revenue for International Game Technology was $744.0 million in 1997 and $824.1 million in 1998. Assume that revenue was increasing at a constant rate.
   (Source: International Game Technology Annual Report, 1998.)

   a. Find the rate of change of revenue.

   b. By how much did revenue increase each quarter of 1998?

   c. Assuming that the rate of increase remains constant, complete the following table.

Year	Revenue (millions of dollars)
1997	
1998	
1999	
2000	

   d. Find an equation for revenue in terms of the year.

8. **Car Value** Suppose you bought a Mazda Miata in 1999 for $24,000. In 2001 it was worth $18,200. Assume that the rate at which the car depreciates is constant.

   a. Find the rate of change of the value of the car.

   b. Complete the following table:

Year	Value
1999	
2001	
2002	
2003	
2004	

   c. Find an equation for the value in terms of the year.

   d. How much will the value of the car change during a 1-month period? Round your answer to the nearest dollar.

9. **Sales** A house sells for $73,000 at the end of 1990 and for $97,500 at the end of 2002.

   a. If the market value increased linearly from 1990 through 2002, what was the rate of change of the market value?

   b. If the linear increase continues, what will the market value be in 2005?

   c. In what year might you expect the market value to be $75,000? $100,000?

   d. Find a model for the market value. What does your model estimate for the market value in 1999? Do you believe this estimation? What assumption was made when you created the model? Is this assumption necessarily true?

10. **Births** Twenty-eight percent of the births in 1990 were to unmarried women. The percent in 1995 was 32.
   (Source: *Statistical Abstract,* 1998.)

   a. Find the rate of change of the percentage of births, assuming that it is constant.

   b. Estimate the percentage of births to unmarried women in 1996.

11. **Credit** Consumer credit in the United States was $838.6 billion in 1993 and $1235.8 billion in 1997. Assume that consumer credit increases at a constant rate.
   (Source: *Statistical Abstract,* 1998.)

   a. Find the rate of increase.

   b. On the basis of the rate of increase, estimate consumer credit in 2000.

   c. Is the assumption that consumer credit increases at a constant rate valid? Explain.

   d. Assuming a constant rate of change, when will consumer credit reach $2 trillion?

   e. Do you expect consumer credit to reach $2 trillion sooner or later than your answer in part *d* indicates? Why?

12. **Tuition** The 2002–2003 tuition changes for students attending a certain university in the southeast are given in the table below. A full-time student is a student who enrolls in 12 or more credit hours.

Schedule of academic charges	Resident	Nonresident
Full-time academic fee	$2737	$6286
Part-time academic fee (per credit hour)	238	534
Auditing academic fee (per credit hour)	119	267
Graduate assistant fee	780	780

   (Source: www.Clemson.edu.)

   Find formulas for the amount of tuition that each of the following students pays when she or he registers for *x* credit hours.

   a. Resident student

   b. Nonresident student

   c. Graduate assistant

13. **Pets** To promote Be Kind to Animals Week, the American Humane Association ran a notice in the *Anderson Independent-Mail* in the first week of May 1993. The notice read, "Every hour in the United States, more than 2000 dogs and 3500 cats are born. . . . Add these animals to an existing pet population of 54 million dogs and 56 million cats and the total exceeds one billion!" For the questions that follow, consider only dog and cat births, not deaths.

   a. What is the rate of change of the number of dogs in the United States?

   b. What is the rate of change of the number of cats in the United States?

c. Find an equation for the number of dogs t days after the beginning of 1993.

d. Find an equation for the number of cats t days after the beginning of 1993.

e. How many total animals (cats and dogs) do you estimate there were at the end of 1993? Does this agree with the prediction in the article?

14. Break Even You and several of your friends decide to mass-produce "I love calculus and you should too!" T-shirts. Each shirt will cost you $2.50 to produce. Additional expenses include the rental of a downtown building for a flat fee of $675 per month, utilities estimated at $100 each month, and leased equipment costing $150 per month. You will be able to sell the T-shirts at the premium price of $14.50 because they will be in such great demand.

a. Give the equations for monthly revenue and monthly cost as functions of the number of T-shirts sold.

b. How many shirts do you have to sell each month to break even? Explain how you obtained your answer.

15. Population In 1999, the U.S. Bureau of the Census estimated that the world's population had reached 6 billion people. A newspaper article about the population stated that "Despite a gradual slowing of the overall rate of growth, the world population is still increasing by 78 million people a year. . . . [T]he number of humans on the planet could double again to 12 billion by 2050 if the current growth rate continues."
(Source: "Population of world ready to hit 6 billion," *Chicago Tribune*, October 21, 1999, p. A1.)

a. According to the article, what is the rate of change of the world's population?

b. Find a linear model for the population of the world, assuming that the population was 6 billion at the beginning of 2000.

c. According to the model in part b, when will the world's population be 12 billion? Does this agree with the estimate given in the article?

d. What assumption did you make when you made the prediction in part c? What can you conclude about the assumptions made by the Census Bureau in making their estimate of the year in which the population would reach 12 billion?

16. Cost A jewelry supplier charges a $55 ordering fee and $2.50 per pair of earrings regardless of the order size.

a. If a retailer orders 100 pairs of earrings, what is the cost?

b. What is the average cost per pair of earrings when 100 pairs are ordered?

c. If the retailer marks up the earrings 300% of the average cost, at what price will the retailer sell the 100 pairs of earrings? (Sale price = cost + 300% of cost)

d. Find an equation for cost as a function of the number bought.

e. Find an equation for the average cost per pair as a function of the number bought.

f. Find an equation for the sale price based on a 300% markup.

g. If 100 pairs of earrings are purchased, how many must the retailer sell at a 300% markup to recover the cost of the earrings (break-even point)?

17. ATM Fee In 1996 the average ATM transaction fee for U.S. banks was $0.97. In 1999 the average fee was $1.37.
(Source: U.S. Public Interest Research Group.)

a. Assuming that ATM fees increased linearly between 1996 and 1999, calculate the rate of change in the average fee.

b. By aligning the years differently, find three linear models for the amount of the average ATM fee between 1996 and 1999.

c. Use your models to estimate the average fee in 1998 and 2003.

d. Are the estimates in part c interpolations or extrapolations?

18. Heating Oil The following table gives the number of gallons of oil in a tank used for heating an apartment complex t days after January 1 when the tank was filled.

t	Oil (gallons)
0	30,000
1	29,600
2	29,200
3	28,800
4	28,400

a. What is the rate of change of the amount of oil?

b. How much oil can be expected to be used during any particular week in January?

c. Predict the amount of oil in the tank on January 30. What assumptions are you making when you make predictions about the amount of oil? Are the assumptions valid? Discuss.

d. Find and graph an equation for the amount of oil in the tank.

e. When should the tank be refilled?

19. Explain the difference between using a model for interpolation and using it for extrapolation. Does interpolation always give an accurate picture of what is happening in the real world? Does extrapolation? Why or why not?

20. **Taxes**   The table below shows the amount of motor vehicle taxes collected by a certain county between 1994 and 1998.

Year	1994	1995	1996	1997	1998
Amount collected (millions of dollars)	108	122	137	155	169

(Source: Nevada Legislative Counsel Bureau.)

a. Calculate the first differences of the collected amounts. Would you classify the differences as constant, nearly constant, or neither?

b. Find three models for the data using three different alignments.

c. Estimate the amount collected in 2000.

d. According to your models, when will the amount collected be $300,000,000? Under what conditions will your estimate be accurate?

21. **Enrollment**   The table below shows the enrollment for a university in the southeast from 1965 through 1969.

Year	1965	1966	1967	1968	1969
Students	5024	5540	6057	6525	7028

a. Find three linear models for these data by aligning the years differently.

b. Use all three models to estimate the enrollment in 1970.

c. The actual enrollment in 1970 was 8038 students. How far off was your estimate? Do you

consider the error to be significant or insignificant? Explain your reasoning.

d. Would it be wise to use these models to predict the enrollment in the year 2000? Explain.

22. **Union Pressure**   In order to put pressure on a company to negotiate a contract that is more favorable to employees, a workers' union may order a slowdown on labor. Consider, for instance, a slowdown of laborers at ports along the U.S. Pacific seaboard. The accompanying table shows the volume (in cartons) of cargo that is unloaded each day by one team of cargo handlers during a work slowdown when cargo handlers reduce their efficiency by x%.

Cartons	108	90	60	42	24
Percent slow-down	10	25	50	65	80

a. Find a linear model for the volume of cargo handled by a team during an x% slowdown.

b. Use the model to estimate the volume handled if no slowdown occurs.

c. What are the vertical-axis and horizontal-axis intercepts for your model? Discuss in context the information given by the intercepts.

d. What is the slope for your model? Interpret the slope in context.

23. **Price**   The price of tickets to a college's home football games is given in the following table.

Year	Price (dollars)	Year	Price (dollars)
1981	10	1989	16
1983	12	1991	18
1985	13	1994	20
1987	15	1999	25

(Source: Clemson University Athletic Department.)

a. Find a linear model for the price of tickets.

b. Use the model to find the ticket price in 1984, 1992, and 2002. (Round to the nearest dollar.) Were you interpolating or extrapolating to find these prices?

c. In 2002, tickets cost $26. Was your prediction accurate?

d. Suppose you graduate, get married, and have a child who attends that college. Predict the price you will have to pay to attend a football game with your child during his or her freshman year. Describe in detail how you arrived at your answer.

e. Repeat part *d* assuming that you have a grandchild who attends the same college.

f. What assumptions are you making when you use the linear model to make predictions? Are these assumptions reasonable over long periods of time? Explain.

g. What conclusions can you draw about the validity of extrapolating from a model?

24. **Cost**   You are an employee in the summer at a souvenir shop. The souvenir shop owner wants to purchase 650 printed sweatshirts from a company. The catalog contains a table of costs and directions to call the company for costs for orders more than 350. The catalog costs are shown in the table. The shop owner, who has tried unsuccessfully for a week to contact the company, asks you to estimate the cost for 650 shirts.

Number purchased	Total cost (dollars)	Number purchased	Total cost (dollars)
50	250	250	700
100	375	300	825
150	500	350	950
200	600		

a. Find a linear model to fit the data.

b. Use the model to predict the cost for 650 shirts. Note that all costs in the table are integer multiples of $25.

c. Determine the average cost per shirt for 650 shirts.

d. The shop owner is preparing a newspaper advertisement to be published in a week. If the standard markup is 700%, what should the advertised price be?

e. How many of the 650 shirts will need to be sold at the price determined in part *d* in order to pay for the cost of all 650 shirts (break-even point)?

25. **Funding**   The percentage of funding for public elementary and secondary education provided by the federal government during the 1980s is given in the following table.

School year	% Funding	School year	% Funding
1981–82	7.4	1986–87	6.4
1982–83	7.1	1987–88	6.3
1983–84	6.8	1988–89	6.2
1984–85	6.6	1989–90	6.1
1985–86	6.7		

(Source: *Statistical Abstract*, 1992 and 1994.)

a. Find a linear model for the percentage of funding.

b. What is the rate of change of the model you found?

c. On the basis of the model, estimate the percentage of funding provided by the federal government in the 1993–94 school year.

d. Judging by the model, determine in what school year federal government funding will first be less than 5% of educational funding.

26. **Postal Rates**   The table below shows year 2003 first-class domestic postage for mail up to 9 ounces.

Weight not exceeding	Postage
1 oz	$0.37
2 oz	0.60
3 oz	0.83
4 oz	1.06
5 oz	1.29
6 oz	1.52
7 oz	1.75
8 oz	1.98
9 oz	2.21

(Source: U.S. Postal Service.)

a. Observe a scatter plot of these data. Determine visually whether a linear model is appropriate.

b. Verify your observations in part *a* by calculating first differences in the postal rates.

c. Find a formula for the postage in terms of weight. Be specific about what the variables represent.

27. **Emissions**   Chlorofluorocarbons (CFCs) released into the atmosphere are believed to thin the ozone layer. In the 1970s, the United States banned the use of CFCs in aerosols, and the atmospheric release of CFCs declined. The decline was temporary, how-

ever, because of increased CFC use in other countries. In 1987, the Montreal Protocol, calling for a phasing out of all CFC production, was ratified. The following table shows atmospheric release of CFC-12, one of the two most prominent CFCs, from 1974 through 1992.

Year	CFC-12 atmospheric release (millions of kilograms)
1974	418.6
1976	390.4
1978	341.3
1980	332.5
1982	337.4
1984	359.4
1986	376.5
1988	392.8
1990	310.5
1992	255.3

(Source: Ronald Bailey, ed., *The True State of the Planet.* New York: The Free Press, for the Competitive Enterprise Institute, 1995.)

a. Look at a scatter plot of the data. Does the behavior of the data seem to support the foregoing description of the atmospheric release of CFCs since the 1970s? Explain.

b. Use three linear models to create a piecewise model for CFC-12 atmospheric release.

c. Sketch a graph of the data and the three-piece model.

d. Use your model to answer the questions in the chapter opener on page 3.

28. Population   The population of West Virginia from 1985 through 2000 is shown in the following table.

Year	Population (thousands)	Year	Population (thousands)
1985	1907	1993	1816
1987	1858	1995	1821
1989	1807	1997	1816
1990	1793	1999	1807
1991	1798	2000	1807

(Source: *Statistical Abstract,* 1998, U.S. Census Bureau.)

a. Observe a scatter plot of the data. What two years are the dividing points that should be used to create a piecewise continuous linear function?

b. Divide the data in the years you determined in part *a*. Include the dividing points in each data set. Find linear models for each set of data, and write the function in correct piecewise continuous function notation. Compare your model with the one given in Example 2 in Section 1.2.

c. Use your model to estimate the population in 2002. Compare your estimate with the actual population of 1,802,000.

29. Population   The population of North Dakota between 1985 and 1996 is shown in the following table.

Year	Population (thousands)	Year	Population (thousands)
1985	677	1993	635
1987	661	1994	640
1989	646	1995	641
1991	633	1996	643

(Source: *Statistical Abstract,* 1998.)

a. Observe a scatter plot of the data. What year is the dividing point that should be used to create a piecewise continuous function?

b. Divide the data in the year you determined in part *a*. Include the dividing point in both data sets. Find linear models to fit each set of data, and write the function in correct piecewise continuous function notation.

c. Use your model to estimate the population in 1997. Compare your estimate to the actual population of 641,000.

30. Girl Scouts   Membership in the Girl Scouts of the U.S.A. in selected years from 1970 through 1992 is shown in the table on the next page.

a. Look at a scatter plot of the data. Describe the behavior of the data from 1970 through 1980 and from 1985 through 1992.

b. Find linear models for the 1970–1980 data and for the 1985–1992 data. Write the two models as one piecewise model.

c. Use the models to estimate the membership in 1973 and in 1993. Categorize your estimates as interpolation or extrapolation.

Year	Membership (thousands)	Year	Membership (thousands)
1970	3922	1988	3052
1975	3234	1989	3166
1980	2784	1990	3269
1985	2802	1991	3383
1986	2917	1992	3510
1987	2947		

(Source: *Statistical Abstract*, 1994.)

31. Mile Run   The table for this activity (located on the *Calculus Concepts* CD-ROM and web site) shows the world record from 1913 through 2002 for the men's 1-mile race. Note: The 1999 record was valid through the end of 2002.

a. In what year was the world's record for the mile run first under 4 minutes?

b. Convert the data in the time columns to seconds and construct a scatter plot of the data. Does the overall pattern seem to be linear?

c. Between which two consecutive world records did the greatest change occur?

d. Fit a linear function to the data. Discuss the fit of the function to the data.

e. Use the model in part *d* to predict when the 3.5-minute mile will be broken. Discuss the reasonableness of this prediction.

32. In your own words describe the necessity for each of the four elements of a model.

33. Compare and contrast the terms *slope, steepness,* and *rate of change* as used to describe linear models.

SUMMARY

Mathematical Modeling and Functions

Mathematical modeling is the process by which we construct a mathematical framework to represent a real-life situation. In this book we often use mathematical modeling to mean fitting a line or curve to data. The resulting equation, together with output label, input description, and interval description, which we refer to as the mathematical model, provides a representation of the underlying relationship between the variable quantities of interest. A function is a description of how one thing (output) changes as something else (input) changes. We encounter functions represented in four ways: tables of data, graphs, word descriptions, and equations.

Constructed Functions

There are several ways to create new functions by combining two or more other functions whose input and output units are compatible. The basic construction techniques are function addition, multiplication, com-

position, and inversion. In each of these constructions, knowing the input and output units of the functions is the key to understanding how to combine the functions. Also, piecewise continuous functions can be constructed by defining the function differently over various input intervals.

Limits and Continuity

The idea of a limiting value of a function is a fundamental theme of calculus that can be intuitively understood to be the behavior of the outputs of a function as the inputs of a function become infinitesimally close to a specific value. Limits can also be used to describe the end behavior of a function as the magnitude of the inputs becomes infinitely large.

Another fundamental concept of calculus is the idea of continuity. We define a function to be continuous on an interval if for every value in that interval the output exists and there are no breaks or jumps in the outputs. It is important that when using a continuous function to model real-world data, we understand when the context places restrictions on the set of inputs of the function.

Linear Functions and Models

A linear function models a constant rate of change. Its underlying equation is that of a line: $y = ax + b$ where the parameter a is called the slope of the line and is calculated as $\frac{\text{rise}}{\text{run}}$.

Because the slope of a line is a measure of its rate of increase or decrease, the slope is also known as the rate of change for the linear model. The parameter b appearing in the linear model $y = ax + b$ is simply the output of the model when the input is zero.

The Role of Technology

In order to construct mathematical models from data, we must use appropriate tools. Normally, these tools are graphing calculators or personal computers. You should clearly understand that our use of technology will simply be as a tool in the service of mathematics and that no tool is a substitute for clear, effective thinking. Technology carries only the graphical and numerical computational burden. You will have to perform the mathematical analyses, interpret the results, make the appropriate decisions, and then communicate your conclusions in a clear and understandable manner.

CONCEPT CHECK

Can you

- Identify functions?
- Correctly use function notation?
- Correctly interpret graphs?
- Correctly apply business terms?
- Accurately work with functions?
- Combine two functions?
- Identify when a function has an inverse?
- Determine end behavior?
- Estimate limits graphically?
- Estimate limits numerically?
- Determine continuity?
- Correctly work with input restrictions on continuous functions?
- Interpret the parameters of a linear model?
- Construct and work with a linear model?
- Construct and work with a piecewise continuous model?

To practice, try

Section 1.1	Activities 1, 5, 9
Section 1.1	Activities 11, 13
Section 1.1	Activity 19
Section 1.1	Activity 24
Section 1.1	Activities 29, 31
Section 1.2	Activities 3, 7, 11
Section 1.2	Activities 25, 35
Section 1.3	Activity 5
Section 1.3	Activity 3
Section 1.3	Activities 7, 9
Section 1.3	Activities 6, 11
Section 1.3	Activities 25, 29
Section 1.4	Activity 3
Section 1.4	Activity 21
Section 1.4	Activity 29

REVIEW TEST

1. **Wetlands**   The graph shows the yearly gain of wetlands in the United States between 1987 and 1994.

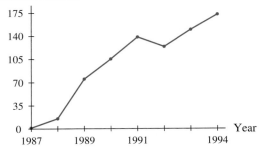

Gain in wetlands
(thousands of acres)

(Source: Ronald Bailey, ed., *The True State of the Planet.* New York: The Free Press, for the Competitive Enterprise Institute, 1995.)

   a. Did the number of acres of wetlands in the United States increase or decrease between 1991 and 1992? Explain.

   b. Estimate the slope of the portion of the graph between 1989 and 1991. Interpret your answer.

2. **Sales**   Let $T(d)$ be the number of tickets sold by the Majestic Theater box office for the Broadway musical *Phantom of the Opera* on the dth day of 2000.

   a. Draw and label an input/output diagram for T.

   b. Is T a function of d? Why or why not?

   c. When the inputs and outputs of T are reversed, is the result an inverse function? Why or why not?

3. **Mortality**   Of the three top causes of death (heart disease, cancer, and stroke), cancer is the only one for which the death rate is increasing. According to the National Cancer Institute, the percent of the U.S. population dying from cancer in 1973 was 18% and in 1995 was 23%. Assume that the percent of the population dying from cancer between 1973 and 1995 increased at a constant rate.

   a. Find the rate of increase of the percent of deaths due to cancer between 1973 and 1995.

   b. On the basis of the increase found in part a, estimate the percent of cancer deaths in 2002. Under what conditions is this estimate valid?

4. Consider the function

$$f(x) = \begin{cases} 2x + 5 & \text{when } x < 1 \\ -2x + 9 & \text{when } 1 < x < 2 \\ \dfrac{3x - 1}{x + 2} & \text{when } x \geq 2 \end{cases}$$

   a. Determine the following limits:

   i. $\lim\limits_{x \to 1} f(x)$ **ii.** $\lim\limits_{x \to 2} f(x)$ **iii.** $\lim\limits_{x \to \infty} f(x)$

   b. Is f continuous at $x = 1$ and $x = 2$? Explain why or why not.

5. **Cropland**   In *The True State of the Planet*, the statement is made that "the amount of arable and permanent cropland worldwide has been increasing at a slow but relatively steady rate over the past two decades." Selected data between 1970 and 1990 reported in *The True State of the Planet* are shown in the following table.
   (Source: Ronald Bailey, ed., *The True State of the Planet.* New York: The Free Press, for the Competitive Enterprise Institute, 1995.)

Year	Cropland (millions of square kilometers)
1970	13.77
1975	13.94
1980	14.17
1985	14.31
1990	14.44

a. Find a linear model for the data.

b. Is the model in part *a* discrete, continuous with discrete interpretation, or continuous without restriction?

c. What is the rate of change of your model in part *a*? Write a sentence interpreting the rate of change.

d. Do the data and your model support the statement quoted above? Explain.

e. According to your model, what was the amount of arable and permanent cropland in 1995?

Project 1.1 Tuition Fees

Setting

Nearly all students pursuing a college degree are classified as either part-time or full-time. Many colleges and universities charge tuition for part-time students according to the number of credit hours they are taking. They charge full-time students a set tuition regardless of the number of credits they take, except possibly in an overload situation.

Tasks

1. Find the tuition charges at your college or university. Some schools have different tuition rates for different classifications of students (for instance, residents may be charged a different tuition than nonresidents). Pick one classification to model. Write a piecewise linear model for tuition as a function of the number of credit hours a student takes. Consider as part of the college tuition charges such as matriculation and activity fees that all students must pay.

2. Use your model to generate a table of tuition charges for your school. Compare your table with the published tuition charges. Are there any discrepancies? If so, explain why they might have occurred.

3. Even though you would not pay tuition if you registered for 0 credit hours, your model may have an interpretation at this point. What is that interpretation?

Reporting

Write a letter containing your findings, addressed to the Committee on Tuition. You should explicitly define to whom the model applies and what the variables in the model represent. Include an explanation of how to use the model, a statement about which input values are valid and which are invalid for use in the model, and a discussion of the preceding tasks.

Project 1.2 United States Population

Setting

The table (located on the *Calculus Concepts* CD-ROM and web site) gives the U.S. population for ages 0 through 85 by gender between 1990 and 1999 with projections from 2000 through 2020.

Tasks

1. Analyze the data for the male population by examining a scatter plot and discussing its shape.

2. Find the first differences in the male populations and discuss what the information tells you. Compare your findings to your observations in Task 1.

3. Fit a linear model to the data. Plot a graph of the model against a scatter plot of the data, and discuss the fit.

4. Calculate the deviations and squared errors as well as the sum of squared errors for your model and the data.

5. Repeat Tasks 1 through 4 for the female population data.

Reporting

Write a report presenting your findings. Discuss your preliminary analysis of the population data sets. Compare your model results to your preliminary analysis, and discuss what the deviations and sum of squared errors indicate about the fit of the models. Attach copies of your spreadsheets and graphs as addenda to the report.

Ingredients of Change: Nonlinear Models

Although linear functions and models are among the most frequently occurring ones in nonscience settings, nonlinear models apply in a variety of situations.

This chapter begins with a study of exponential and logarithmic functions and includes a section on their applications. We consider situations in which exponential growth is restricted in some way. Such situations often can be modeled by a logistic function. The final functions we consider are quadratic and cubic, along with their applications to real-world situations.

We conclude the chapter with a summary of the steps taken in examining a set of data and determining which model is best suited to describe the data and answer questions about changes in the underlying context.

Concept Objectives

This chapter will help you understand the concepts of

- ○ Exponential models and percentage change
- ○ Logarithmic models
- ○ Compound interest, present value, and future value
- ○ Exponential growth and decay
- ○ Logistic models
- ○ Quadratic and cubic models
- ○ End behavior
- ○ Concavity and inflection points

and you will learn to

- ○ Find one of six equations to fit data
- ○ Use models to answer many questions
- ○ Answer financial investment and debt questions
- ○ Work with exponential growth and decay models
- ○ Vertically shift output to better fit an exponential or logistic function
- ○ Horizontally shift input to better fit a logarithmic function
- ○ Choose an appropriate model on the basis of a scatter plot and how the model will be used

Stone/Getty Images

Concept Application

The national gender ratio is the number of males per 100 females in the United States. Factors such as immigration, war, and advances in health care affect the gender ratio. The Census Bureau keeps track of gender ratio data. By examining the behavior of the data, we can choose an appropriate function to model the gender ratio. The model then can be used to answer questions about the gender ratio such as

○ For what age is the gender ratio 100?
○ For what age are there twice as many women as men?
○ What is the expected future trend in the gender ratio?

Answers to questions such as these are important to government agencies, health care providers, retailers, and others. The information needed to answer these questions is found in Activities 27 and 29 of Section 2.4.

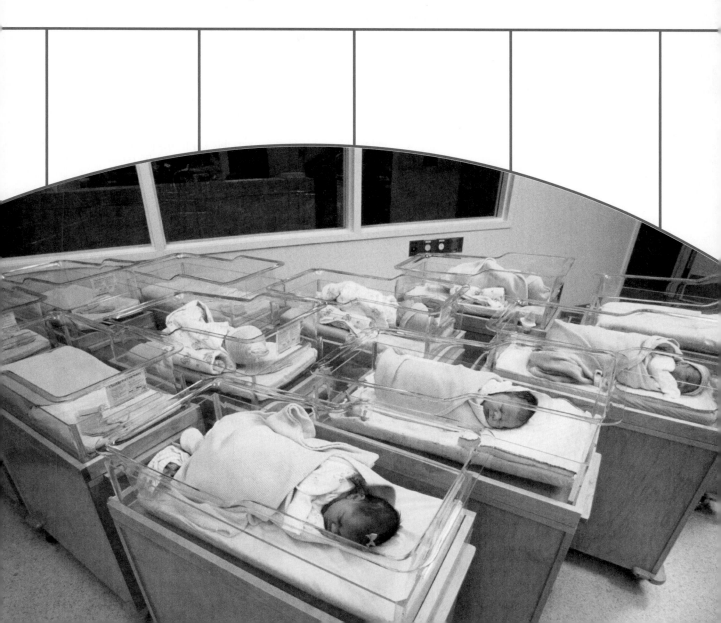

2.1   Exponential Functions and Models

In Chapter 1 we studied linear functions whose output resulted from the repeated *addition* of a constant at regular intervals. We now turn to a function whose output is the result of repeated *multiplication* by a constant at regular intervals.

For example, in the case of a bacterial culture that starts with 10,000 cells and doubles every hour, the current size of the culture is determined by repeated multiplication. The size of the culture at the end of each of the first four hours is shown in Table 2.1.

TABLE 2.1

Hour	0	1	2	3	4
Cells	10,000	20,000	40,000	80,000	160,000

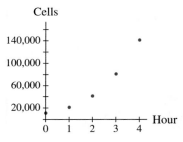

FIGURE 2.1

When we examine a scatter plot of the culture size over the first 10 hours we see that the points certainly do not fall on a line. (See Figure 2.1.) The culture grows more rapidly in the later hours. We note that the starting culture size is 10,000 cells, and the size each hour is twice the prior size. The repeated multiplication by 1.05 makes the year variable t appear as an exponent on 1.05; that is, the year value counts the number of times that 1.05 has been used as a multiplier. An equation for the bacterial culture size is

$$P(t) = 10,000(2^t) \text{ cells}$$

t hours after the culture was first counted. Because the variable t appears in the exponent, we call the equation an **exponential equation** and refer to the growth as **exponential growth.**

Exponential equations also arise when we see a *decreasing* amount of the original substance, such as in the study of radioactive material. For example, if 400 grams of a radioactive substance decays by 2% per day, then the amount of the substance remaining each day is 98% of the previous day's amount. The amount of the radioactive substance after x days of decay is given by the equation

$$f(x) = 400(0.98^x) \text{ grams}$$

In this case, the multiplier is between 0 and 1. Such a situation, wherein an amount diminishes by a constant multiplier, is referred to as **exponential decay.**

Any change in a quantity that results from repeated multiplication generates an exponential function. In general, if we start with an amount a and multiply by a constant positive factor b each year, then the quantity that we have at the end of x years is given by the exponential equation

$$f(x) = ab^x$$

defining f as a function of x.

Percentage Change and Exponential Models

Linear functions exhibit a constant rate of change, but exponential functions exhibit a constant **percentage change.** Percentage change occurs when the amount of growth or decay is determined by the current size. For instance, when a new product is introduced into the economic marketplace, word-of-mouth advertising often takes place. Each satisfied customer immediately tells another potential customer about the exciting new product. For that reason, exponential models are often used to analyze a product's rapid sales growth.

For the exponential equation $f(x) = ab^x$ we determine the constant percentage change by calculating $(b - 1)100\%$. For example, for a bank account whose balance is modeled by $B = 100(1.05^t)$ dollars t years after the initial investment is made, the constant percentage change of the balance is

WEB/CD

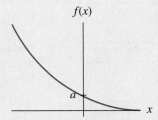

Strengthening the Concepts: Simplifying Exponential Expressions

$$(1.05 - 1)100\% = (0.05)100\% = 5\%$$

That is, each year the balance of the account increases by 5%. Likewise, for the bacterial growth model $P(t) = 10,000(2^t)$ cells, the constant percentage change is $(2 - 1)100\% = 100\%$; that is, the bacterial population increases by 100% each hour. Finally, for the remaining amount of radioactive substance given by $f(x) = 400(0.98^x)$ grams after x days of decay, the constant percentage change is $(0.98 - 1)100\% = (-0.02)100\% = -2\%$. Thus the amount decreases by 2% each day.

An alternative form of the exponential model $f(x) = ab^x$ is $f(x) = ae^{kx}$. It is easy to convert from one form to the other using $k = \ln b$ and $b = e^k$.

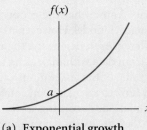

Exponential Model

Verbally: An exponential model has a constant percentage change.

Algebraically: An exponential model has an equation of the form

$$f(x) = ab^x$$

where $a \neq 0$ and $b > 0$. The percentage change is $(b - 1)100\%$, and the parameter a is the output corresponding to an input of zero.

Graphically: An exponential model graph has the form of one of the two graphs in Figure 2.2.

(a) **Exponential growth**
$f(x) = ab^x$ with $b > 1$

(b) **Exponential decay**
$f(x) = ab^x$ with $0 < b < 1$

FIGURE 2.2

Note that the equation in the box defines a function f with input x. From now on, when we refer to an exponential equation or function, we will consider the equation to be of the form $f(x) = ab^x$. In the case of either exponential growth or exponential decay (as long as $a \geq 0$), the exponential function $f(x) = ab^x$ is concave up $\underline{}$ or $\diagdown\underline{}$. It's outputs approach zero in one direction and increase without bound in the opposite direction.

EXAMPLE 1   *Using Percentage Change to Write an Exponential Model*

Dr. Laura   In the late 1990s, Dr. Laura Schlessinger hosted the fastest-growing national radio program in history.[*] The program debuted in July of 1994 with 20 affiliate stations, and during the 1990s the number of stations grew by 86.4% per year.

a.   Why is an exponential model appropriate for the number of affiliate stations?

b.   Find a model for the number of affiliate stations.

c.   How many affiliate stations carried the "Dr. Laura" program in July of 1999?

Solution

a.   An exponential model is appropriate because the percentage change is constant.

b.   Because the percentage change is 86.4% per year, the constant multiplier is $1 + 0.864 = 1.864$. The number of affiliate stations can be modeled by the equation

$$L(x) = 20(1.864^x) \text{ stations}$$

x years after July of 1994.

c.   According to the model, the number of affiliate stations in July of 1999 was $L(5) = 20(1.864^5) \approx 450$ stations.   ●

TABLE 2.2

Year	Population
1994	7290
1995	6707
1996	6170
1997	5677
1998	5223
1999	4805
2000	4420
2001	4067
2002	3741
2003	3442

Percentage Differences from Data

Often we do not know what the proper repeated multiplier should be in order to write an exponential model. However, if we are given data, sometimes we can look at percentage differences to determine the repeated multiplier. **Percentage differences** are calculated from data with increasing input values by dividing each first difference by the output value of the lesser input value and multiplying by 100.

We look at how to calculate percentage differences by considering a small town's dwindling population. According to the town's records, the population data from 1994 through 2003 are as shown in Table 2.2. A scatter plot of the data is shown in Figure 2.3.

[*] "Dr. Laura celebrates five years," *Reno Gazette–Journal*, July 27, 1999, p. 2D.

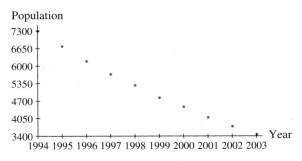

FIGURE 2.3

Because the inputs are evenly spaced, examine the output data in greater detail by calculating first differences:

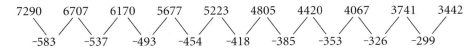

2.1.1a, b

Concept Development: Percentage Differences As indicated by the shape of the scatter plot in Figure 2.3, the first differences show us that the change in population from year to year is not constant. However, if we calculate the percentages that these yearly changes represent, we notice a pattern. From 1994 to 1995, the population decreased by 583 people. This represents an 8% (approximate) decrease from the 1994 population of 7290 people.

The percent symbol (%) means to divide by 100. For example, $8\% = \frac{8}{100} = 0.08$. When changing a decimal number into a percent, rewrite the number using "divide by 100." For example, $4.39 = \frac{439}{100}$. When you remove "divide by 100," insert "%." That is, $\frac{439}{100} = 439\%$.

$$\frac{^-583 \text{ people}}{7290 \text{ people}} \approx -0.07997 \approx -8\%$$

From 1995 to 1996, the population decreased by 537 people. This also represents about an 8% decrease from the previous year's (1995) population.

$$\frac{^-537 \text{ people}}{6707 \text{ people}} \approx -0.08007 \approx -8\%$$

In fact, every year the population decreased by approximately 8%.

$$1996 \text{ to } 1997: \frac{^-493}{6170} \approx -0.07990 \approx -8\%$$

$$1997 \text{ to } 1998: \frac{^-454}{5677} \approx -0.07997 \approx -8\%$$

$$1998 \text{ to } 1999: \frac{^-418}{5223} \approx -0.08003 \approx -8\%$$

$$1999 \text{ to } 2000: \frac{^-385}{4805} \approx -0.08012 \approx -8\%$$

$$2000 \text{ to } 2001: \frac{^-353}{4420} \approx -0.07986 \approx -8\%$$

$$\text{2001 to 2002: } \frac{-326}{4067} \approx -0.08016 \approx -8\%$$

$$\text{2002 to 2003: } \frac{-299}{3741} \approx -0.07993 \approx -8\%$$

The percentage differences here are constant. However, this is often not the case. When considering a set of data, percentage differences in the output data vary from point to point unless every data point falls on the exponential equation that is fit to the data. In that case and when the input values are 1 unit apart, the constant percentage differences equal the percentage change, $(b - 1)100\%$. When this situation arises, we use the terms *percentage change* and *percentage differences* interchangeably.

Finding Exponential Models

Again consider a small town's population, shown numerically in Table 2.2 and graphically in Figure 2.3. We found that the percentage change in output is -8%, so each year the population is approximately 92% of what it was the previous year. In other words, each year the population is 0.92 times the previous year's population.

Let us now develop an algebraic description of the declining population. The equation for the population decay must show the 1994 population of 7290 people repeatedly multiplied by 0.92. Thus an exponential equation that models the population data is

$$P(x) = 7290(0.92^x) \text{ people}$$

where x is the number of years since 1994.

2.1.2

We can also use technology to determine an equation for the data. Enter the data points (aligning the years with $x = 0$ in 1994, $x = 1$ in 1995, $x = 2$ in 1996, etc.), and have your calculator or computer fit an exponential equation. You should obtain the equation

$$T(x) = 7290.25032(0.91999^x) \text{ people}$$

where x is the number of years since 1994.

Why is this equation not identical to the one we developed previously? Remember that as we computed percentage differences for each year, we rounded, so the percentage decline was not exactly 8% each year. This rounding contributed to the difference between the models. Note that the technology-generated equation indicates an 8.001% decline.

If we assume that the population continues to decline by the same percentage (approximately 8%), we can estimate that during 2004 (that is, from the end of 2003 through the end of 2004), the town's population was 92% of the population in 2003. Thus the town's population in 2004 will be approximately

$$3442(0.92) \approx 3167 \text{ people}$$

This estimation can also be computed using the unrounded equation for population with input $x = 2004 - 1994 = 10$:

$$T(10) = 7290.25032(0.91999^{10}) \approx 3167 \text{ people}$$

This estimate from the model agrees with the one calculated using only the data.

EXAMPLE 2   *Finding an Exponential Model from Data*

Tire Sales   Table 2.3 shows sales data for a tire manufacturer.

TABLE 2.3

Year	Tire sales (millions of dollars)
1976	23
1980	38.4
1984	64
1988	107
1992	179
1996	299
2000	499
2004	833

a. What are the percentage differences in the tire sales data?

b. Use the percentage change to estimate tire sales in 2008.

c. Use technology to find an exponential equation for tire sales. Write the model for tire sales. What is the percentage change in this equation?

d. Graph the equation in part *c* together with a scatter plot of the data. Describe the end behavior of the function as time increases.

e. Use the model in part *c* to estimate tire sales in 2008.

Solution

a. Calculating percentage differences, we see that every 4 years, sales have increased by approximately 67%.

$$1976 \text{ to } 1980: \frac{15.4}{23} \approx 0.670 \approx 67\% \qquad 1992 \text{ to } 1996: \frac{120}{179} \approx 0.670 \approx 67\%$$

$$1980 \text{ to } 1986: \frac{25.6}{38.4} \approx 0.667 \approx 67\% \qquad 1996 \text{ to } 2000: \frac{200}{299} \approx 0.669 \approx 67\%$$

$$1986 \text{ to } 1988: \frac{43}{64} \approx 0.672 \approx 67\% \qquad 2000 \text{ to } 2004: \frac{334}{499} \approx 0.669 \approx 67\%$$

$$1988 \text{ to } 1992: \frac{72}{107} \approx 0.673 \approx 67\%$$

The percentage differences are all approximately 67%. Thus 67% is the approximate percentage change of the tire sales every 4 years.

b. If the percentage growth remains constant, then sales in 2008 will increase 67% over the 2004 sales.

$$\begin{matrix} \text{Amount of increase} \\ \text{(rounded to the} \\ \text{nearest million)} \end{matrix} = 67\% \text{ of 2004 sales} = (0.67)(\$833) \approx \$558 \text{ million}$$

Estimation for 2008:

2004 sales + amount of increase = 833 + 558 = $1391 million

c. Entering the tire sales data points into a calculator or computer (aligning the years with $t = 0$ in 1976, $t = 4$ in 1980, $t = 8$ in 1984, etc.) and having the calculator or computer fit an exponential equation produces the model

$$S(t) = 22.98235(1.13683^t) \text{ million dollars}$$

t years after 1976.

Note that the percentage change in this exponential function is calculated as $(1.13683 - 1)100\% \approx 14\%$. Were you expecting to see 67%? What happened?

Recall that in part *a* we calculated the percentage growth to be 67% *every 4 years*. The percentage growth in the model is the *annual* percentage growth; that is, it is a growth of 14% *per year*. If we multiply 1.13683 (the repeated multiplier in the function *S*) by itself 4 times, we obtain 1.67: $1.13683^4 \approx 1.67$ representing a 4-year growth of 67%.

d. A graph of the equation and the tire sales data is shown in Figure 2.4.

As the years increase, the output *S(t)* increases without bound. We say $\lim\limits_{t \to \infty} S(t) \to \infty$.

e. Using the exponential model *S*, we estimate the sales for 2008 as $S(32) \approx \$1392$ million. ●

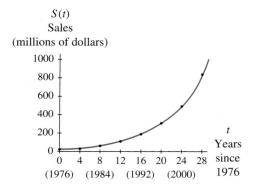

S(t)
Sales
(millions of dollars)

t
Years
since
1976

FIGURE 2.4

Note the slight difference between the answers to part *b* and part *e* in Example 2. Two factors are involved in this difference. First, when calculating $1391 million in part *b*, we used a rounded percentage change: 67% every 4 years instead of the percentage change in the technology-generated model. Second, the result $1391 million was calculated from a data point (the 2004 sales) instead of from the sales generated by the model (part *e*).

Although both of these predictions are valid, for consistency we adopt the following rule of thumb:

> Once an equation has been fitted to data, we will use the equation to answer questions rather than using data points or rounded estimates.

Doubling Time and Half-Life

One property of exponential models is that when the quantity being modeled either doubles (during exponential growth) or halves (during exponential decay), it does so over a constant interval. For instance, an initial investment of $1000 that doubles every 14 years can be modeled by the exponential function

$$I(t) = 1000(1.050^t) \text{ dollars}$$

t years after the initial investment. The investment will be worth $2000 by $t = 14$ years, and it will double again to be worth $4000 by $t = 28$ years. We say that 14 years is the *doubling time* of the investment. In general, the **doubling time** in exponential growth is the amount of time it takes for the output to double.

Similarly, the **half-life** in exponential decay is the amount of time it takes for the output to decrease by half. Exponential decay occurs when drugs are taken into the body. Once the drug is absorbed, its level in the blood stream is at a peak. Each time the blood passes through the kidneys, liver, spleen, and other organs, a percentage of the drug is removed. Pharmaceutical companies report half-life information in terms of either elimination half-life or plasma concentration half-life. *Elimination half-life* is the time it takes for half of the drug to be removed from the body. *Plasma half-life* is the amount of time it takes for the concentration of the drug in the plasma to reach half of its peak concentration. Plasma concentration is usually measured in micrograms per milliliter (μg/mL). When given information on the amount of a substance at a given time and information on the *doubling time* or *half-life,* we can use exponential growth or decay information to write an exponential model. Example 3 illustrates this process.

EXAMPLE 3   *Writing an Exponential Decay Model*

Medicine   Dilantin* is a drug used to control epileptic seizures. On Monday a patient takes a 300-mg Dilantin tablet at 4 P.M. Eight hours later the Dilantin reaches its peak plasma concentration of 15 μg/mL. The average plasma half-life of Dilantin is 22 hours.

a. Write a model for the concentration of Dilantin in the patient's plasma as a function of the time after peak concentration is reached.

b. What is the concentration of Dilantin in this person's plasma at 8 A.M. on Tuesday?

c. The minimum desired concentration of Dilantin is 10 μg/mL. Will the concentration dip below that level before the patient takes a second tablet on Tuesday at 4 P.M.?

Solution

a. The input of the model is the time after the peak concentration that occurs at midnight on Monday, and the initial amount we consider is the peak concentration of 15 μg/mL. After 22 hours, the concentration will be half its peak, or 7.5 μg/mL. Using the two data points $(0, 15)$ and $(22, 7.5)$, we obtain the model

$$C(h) = 15(0.96898^h) \ \mu\text{g/mL}$$

h hours after midnight Monday.

b. Eight hours after midnight, the plasma concentration is

$$C(8) = 15(0.96898^8) \approx 11.7 \ \mu\text{g/mL}$$

c. To determine when the plasma concentration reaches 10 μg/mL, solve the equation $10 = 15(0.96898^h)$ using technology or algebra. You should obtain $h \approx 12.9$ hours. It takes approximately 12.9 hours after the peak concentration at midnight for the concentration to reach 10 μg/mL. This corresponds to approximately 1 P.M. on Tuesday. Thus the concentration will be below 10 μg/mL when the patient takes another tablet at 4 P.M. on Tuesday.   ●

*Based on information obtained from **www.parke-davis.com.** Accessed 6/11/00.

Numerical Considerations

When you use a calculator or computer to fit an exponential equation to data, it is important to align the input values so that they are small in magnitude. If you fail to align the input values appropriately, the technological numerical computation routine may return an invalid result because of improper scaling, numerical overflow, or round-off errors. For instance, if you use $x = 1994, 1995, 1996$ instead of $x = 0, 1, 2$ in the small town population example, the model returned is likely to be $y = ab^x$, where $a \approx 1.187 \cdot 10^{76}$. This huge value for a could cause problems computationally.

It is also important that the input values be aligned in the tire sales example to small numbers such as $t = 0, 4, 8, \ldots$ instead of as $t = 1976, 1980, 1984, \ldots$. If you use the actual year as input, then the equation that is returned is likely to be $y = ab^t$, where $a \approx 0$. It is also possible that an overflow error will result and no equation will be returned.

As you work with exponential models, keep in mind the following principle:

Aligning Exponential Data

When using technology to find the equation for an exponential model, align the input data using reasonably small values to avoid numerical computation errors.

Graphically, this aligning of input values does not change the nature of the exponential function. Aligning input values simply shifts the graph of the function to the left as illustrated in Figure 2.5.

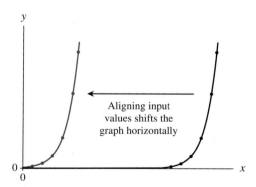

FIGURE 2.5   Original data and function in black. Aligned data and function in teal.

Exponential functions with their constant percentage change are common in finance, health and life sciences, social sciences, and many other areas. As we continue our exploration of calculus concepts, you will become increasingly familiar with these functions.

2.1 Concept Inventory

- ○ *Exponential function: $f(x) = ab^x$ (characterized by constant percentage change)*
- ○ *Exponential growth and decay*
- ○ *Calculating percentage change*
- ○ *Finding percentage differences*
- ○ *Doubling time and half-life*
- ○ *Importance of aligning data*

2.1 Activities

For Activities 1 through 8, match each graph with its equation.

1. $f(x) = 2(1.3^x)$

$f(x) = 2(0.7^x)$

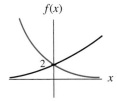

2. $f(x) = 2(1.3^x)$

$f(x) = -2(1.3^x)$

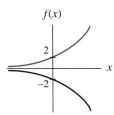

3. $f(x) = 3(1.2^x)$

$f(x) = 3(1.4^x)$

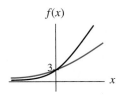

4. $f(x) = 2(0.8^x)$

$f(x) = 2(0.6^x)$

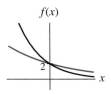

For Activities 5 through 8, indicate whether the function describes exponential growth or decay, and give the constant percentage change.

5. $f(x) = 72(1.05^x)$

6. $K(r) = 33(0.92^r)$

7. $y(x) = 16.2(0.87^x)$

8. $A(t) = 128.57(1.035^t)$

For Activities 9 and 10, find the constant percentage change, and interpret it in context.

9. Bacteria After h hours, the number of bacteria in a Petri dish during a certain experiment can be modeled as $B(h) = 100(0.61^h)$ thousand bacteria.

10. Membership The membership of a popular club can be modeled by $M(x) = 12(2.5^x)$ members by the end of the xth year after its organization.

11. Imports The U.S. Energy Information Administration projected that imports of petroleum products in 2005 would be 4.81 quadrillion Btu and would increase by 5.47% per year through 2020. (Source: *Annual Energy Outlook,* 2001.)

   a. Find a model for projected petroleum product imports between 2005 and 2020.

   b. According to the model, when will imports exceed 10 quadrillion Btu?

   c. Describe the end behavior of the model as time increases.

12. Assets In 1994, Charles Schwab & Co. had approximately \$135 billion of assets in customer accounts, and throughout the late 1990s that value grew by approximately 39% per year.

   a. Find a model for Schwab's customer account assets in the late 1990s.

   b. Use the model to estimate Schwab's customer account assets in 2000.

13. Cost A New York City subway token cost \$1.50 in 1997. It has been estimated that the cost will rise by 7.46% per year through 2010.

a. Find a model for New York City subway token prices from 1997 through 2010.

b. What does the model predict as the 2010 price?

14. **Emissions**   In 1975 the EPA emissions standard for cars was 3.1 grams of nitrogen oxide per mile of driving. Assume that the EPA standard decreased by 9.3% per year between 1975 and 2000.

   a. Find a model for the emission standard between 1975 and 2000.

   b. Estimate the EPA standard in 2000.

15. **Social Security**   According to the Social Security Advisory Board, the number of workers per beneficiary of the Social Security program was 3.3 in 1996 and is projected to decline at a rate of 1.46% per year through 2030.

   a. Find a model for the number of workers per beneficiary from 1996 through 2030.

   b. What does the model predict the number of workers per beneficiary will be in 2030? How will this number affect your life?

16. **Sales**   When a company ceases to advertise and promote one of its products, sales often decrease exponentially, provided that other market conditions remain constant. At the time that publicity was discontinued for a newly released popular animated film, sales were 520,000 videotapes per month. One month later, videotape sales had fallen to 210,000 tapes per month.

   a. What was the monthly percentage decline in sales?

   b. Assuming that sales decreased exponentially, give the equation for sales as a function of the number of months since promotion ended.

   c. What will sales be 3 months after the promotion ends? 12 months after?

   d. Walt Disney has been known to produce videotapes of some of its films for a limited time before completely stopping production. Do you believe that this is a wise business decision? Explain.

   e. What would be necessary to ensure that the exponential decrease in sales of a product did not occur?

17. **Population**   Carefully read the accompanying newspaper article (from the *Chicago Tribune*,

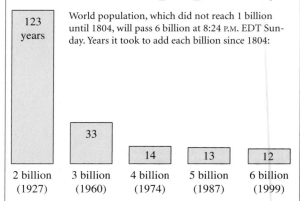

6,000,000,000 people today

| 123 years | World population, which did not reach 1 billion until 1804, will pass 6 billion at 8:24 P.M. EDT Sunday. Years it took to add each billion since 1804: |

2 billion (1927)   3 billion (1960)   4 billion (1974)   5 billion (1987)   6 billion (1999)

33   14   13   12

Population of world ready to hit 6 billion

LOS ANGELES TIMES
WASHINGTON—

Call this Y6B: The year of 6 billion, a milestone the world's population is expected to reach this weekend.

The birth of the planet's 6 billionth inhabitant, projected by the U.S. Census Bureau, also will mark another historic first: The world's population has doubled in less than 40 years.

Despite a gradual slowing of the overall rate of growth, the world population is still increasing by 78 million people a year. That's the equivalent of adding a city nearly the size of San Francisco every three days, or the combined populations of France, Greece and Sweden every year,

according to a coalition of environmental and population groups.

"It took all of human history for the world's population to reach 1 billion in 1804, but little more than 150 years to reach 3 billion in 1960. Now, not quite 40 years later, we are twice that number," said Amy Coen, president of Population Action International.

Even with a decelerating growth rate, the number of humans on the planet could double again to 12 billion by 2050 if the current growth rate continues, the coalition projects.

(Source: "6,000,000,000 People Today," from *USA Today*. Copyright © 1999, *USA Today*. Reprinted with permission.)

October 21, 1999), and answer the questions that follow.

a. Use the data shown in the figure accompanying the article to find an exponential model for world population from 1927 through 1999. Examine a scatter plot of the data with a graph of the exponential function. Discuss how well the function fits the data.

b. The article gives a current rate of change of population. On the basis of that rate of change and the 1999 population, find a linear model for world population.

c. Rewrite the equation in part *b* so that its input matches that of the function in part *a*. For example, if the input in part *a* is years since 1900 and the input in part *b* is years since 1999, then use the relationship

$$\frac{\text{Linear equation}}{\text{input}} = \frac{\text{exponential equation}}{\text{input}} - 99$$

to transform the linear equation to one whose input matches that in the exponential equation.

d. Graph the linear function together with the scatter plot and exponential function from part *a*. Discuss what you observe about the two graphs.

e. Look at the graph in part *d* from 1900 through the year 2050. (Be sure to adjust the vertical axis appropriately.) Comment on the two models and their use in estimating the population from 1900 through 2000 and predicting the population beyond 2000.

f. Use the two models to estimate the population in 2000 and 2050. What assumptions do you make about population growth when you predict the population using these two models?

g. Compare the predictions from part *f* for 2050 with the prediction given in the article. What does this tell you about the rate-of-change assumptions the U.S. Census Bureau used in making their 2050 prediction?

18. **Milk Storage** According to the back of a milk carton sold by Model Dairy, the number of days that milk will keep when stored at various temperatures is as shown in the table.

Temperature (degrees Fahrenheit)	Days
30	24
38	10
45	5
50	2
60	1
70	0.5

a. Find an exponential model for the data.

b. If a refrigerator is adjusted to 40°F from 37°F, how much sooner will milk spoil when stored in this refrigerator?

19. **Waste** The Environmental Protection Agency reports the total amount of municipal sold waste (MSW) recycled between 1960 and 2000 as shown in the table.

Year	Total MSW recycling (million tons)
1960	5.6
1970	8.0
1980	14.5
1990	33.2
2000	69.9

(Source: *Municipal Solid Waste in the United States: 2000 Facts and Figures Executive Summary.*)

a. Find an exponential model for the data.

b. According to the model in part *a*, what was the yearly percentage growth in recycled MSW from 1960 through 2000?

20. **Computing Power** During the last three decades, computing power has grown enormously. The accompanying table gives the number of transistors (in millions) in Intel processor chips.

a. Find an exponential model for the data.

b. According to the model found in part *a*, what is the annual percentage increase in the number of transistors used in an Intel computer processor chip?

c. Is it reasonable to expect this exponential growth rate to continue beyond the year 2003? Explain your reasoning.

Processor	Year	Number of transistors (millions)
4004	1971	0.0023
80286	1982	0.134
386DX	1986	0.275
486DX	1989	1.2
Pentium®	1993	3.1
Pentium® Pro	1995	5.5
Pentium® II	1997	7.5
Celeron	1999	19.0
Pentium IV	2000	42.0
Pentium IV	2002	55.0

(Source: www.Intel.com. Accessed 9/23/02.)

21. Farms   The number of U.S. farms with milk cows has been declining. See the table.

Year	Farms (thousands)
1980	334
1985	269
1990	193
1995	140
2000	105

(Source: *Statistical Abstract*, 1998 and 2001.)

a. Find an exponential model for this data. What is the percentage change indicated by the model?

b. Express the end behavior of the model as time increases. Does this end behavior reflect what you believe will happen in the future to the number of farms with milk cows?

c. Give at least two reasons for the decline in the number of farms with milk cows.

22. Bottled Water Consumption   The per capita consumption of bottled water in the United States has increased dramatically in the past 20 years. The accompanying table shows selected years and the per capita bottled water consumption in those years.

a. Find linear and exponential models for the data. Graph the equations for these models on a scatter plot of the data. Which model do you think better describes the per capita bottled water consumption?

Year	Bottled water consumption (gallons per person per year)
1980	2.4
1985	4.5
1990	8.0
1995	11.6
1999	18.1

(Source: *Statistical Abstract*, 2001.)

b. Give the rate of change of the linear model and the percentage change of the exponential model.

c. Use the two models to estimate bottled water consumption in 2000.

d. According to each model, when will per capita bottled water consumption exceed 25 gallons per person per year?

23. Decay   Carbon-14, ^{14}C, has a half-life of approximately 5580 years. If a sample of an artifact contains 0.027 gram of ^{14}C, how long ago did it contain 0.05 gram of ^{14}C?

24. Decay   An abandoned building is found to contain radioactive radon gas. Thirty hours later, 80% of the initial amount of gas is still present.

a. Find the half-life of this radon gas.

b. Give a model for the amount of radon gas present after t hours.

c. What is the limit of the function in part b as time approaches infinity? Interpret this answer in the context of the radon gas.

25. Half-Life   The elimination half-life of a certain type of penicillin is 30 minutes.

a. Write a model for the amount of this penicillin left in a person's body if the initial dose is 250 mg.

b. If it is safe to take another dose of this penicillin once the amount in the body is less than 1 mg, when should another dose be taken?

26. Medicine   If a person takes Digoxin, a heart stimulant, the concentration in the person's blood stream t hours after the Digoxin reaches its peak concentration is

$$D(t) = D_0 e^{-0.0198t} \; \mu g/mL$$

where D_0 is the peak concentration in micrograms per milliliter. Find the half-life of Digoxin.

27. Doubling Time How long would it take an investment to double under each of the following conditions? (Hint: Exponential formulas for compound interest problems can be found in Section 1.3.)

  a. Interest is 6.3% compounded monthly.

  b. Interest is 8% compounded continuously.

  c. Interest is 6.85% compounded quarterly.

28. Mile Run The table for this activity (located on the *Calculus Concepts* CD-ROM and web site) shows the world record from 1913 through 2002 for the men's 1-mile race.

  a. What do you predict to be the end behavior of these data as time increases?

  b. Convert the time data from minutes and seconds to seconds and examine a scatter plot of the data.

  c. Find an exponential model for the data. Does the end behavior of the equation fit the end behavior suggested by the context?

  d. Use the model in part *c* to predict when the 3.5-minute mile will be broken.

29. Why does it make sense to talk of *doubling time* and *half-life* for exponential models but not for linear models?

30. Is it possible to have data that exhibit a concave-up scatter plot but do not imply an exponential model? Explain.

2.2 Logarithmic Functions and Models

In Sections 1.4 and 2.1 we examined two functions that can be fit to data in order to model a real-world situation. In this section we introduce a third model.

Logarithmic Models

Altitude and air pressure are intrinsically related. Altimeters are instruments that determine altitude by measuring air pressure. For an altimeter, the input is air pressure, and the output is altitude. Table 2.4 shows altimeter data, and Figure 2.6 shows a scatter plot of the data.

TABLE 2.4

Air pressure (inches of mercury)	Altitude (thousands of feet)
13.76	20
5.56	40
2.14	60
0.82	80
0.33	100

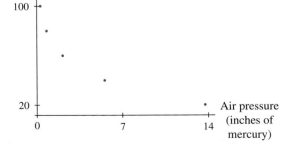

FIGURE 2.6

Although the scatter plot has the basic declining, concave-up appearance of an exponential function, an exponential function does not fit these data well. However, there is another function, called a **logarithmic function** or **log function,** that exhibits similar behavior. The log equation used by most technologies is

$$f(x) = a + b \ln x$$

This equation defines a function f with input x. The b-term in this equation determines whether the function increases or decreases and how rapidly the increase or decrease occurs. The a-term determines the vertical shift of the function. From now on, when we refer to a log equation or function, we will consider the equation to be of the form $f(x) = a + b \ln x$.

The end behavior of log functions is particularly important in determining when they are appropriate to use to fit data. An increasing log function increases without bound as the input increases, and a decreasing log function decreases without bound as the input increases. Log functions are not defined for negative or zero input, and as the input approaches zero from the right, the function either increases or decreases without bound.

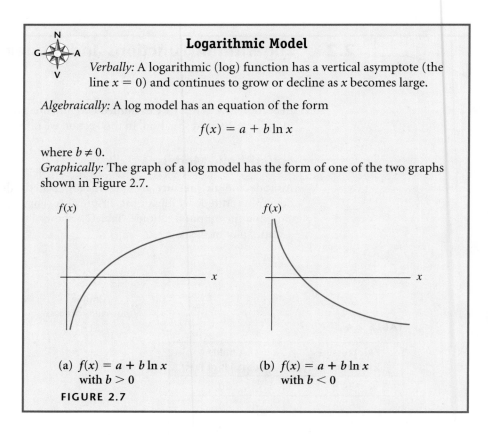

Logarithmic Model

Verbally: A logarithmic (log) function has a vertical asymptote (the line $x = 0$) and continues to grow or decline as x becomes large.

Algebraically: A log model has an equation of the form

$$f(x) = a + b \ln x$$

where $b \neq 0$.

Graphically: The graph of a log model has the form of one of the two graphs shown in Figure 2.7.

(a) $f(x) = a + b \ln x$
    with $b > 0$

(b) $f(x) = a + b \ln x$
    with $b < 0$

FIGURE 2.7

Again consider the graphs in Figure 2.7. Note that log functions exhibit increasingly smaller changes in output for constant changes in input. This slow growth or decline characterizes log functions. The characteristic that makes log functions different from any other functions we will study is that their growth or decline becomes

2.2.1

increasingly slower but their output never approaches a horizontal limiting value, as do the outputs declining exponential functions and, as we will see later, logistic functions.

Returning to the altitude example, we use technology to find the log equation that fits the data in Table 2.5.

$$A(p) = 76.174 - 21.331 \ln p \text{ thousands of feet above sea level}$$

where p is the air pressure in inches of mercury. Figure 2.8 shows a graph of the function A overdrawn on a scatter plot of the data in Table 2.5.

TABLE 2.5

Air pressure (inches of mercury)	Altitude (thousands of feet)
13.76	20
5.56	40
2.14	60
0.82	80
0.33	100

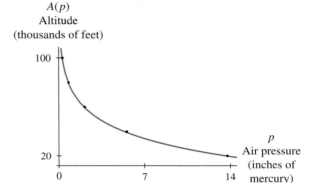

FIGURE 2.8

EXAMPLE 1   *Finding a Log Model*

Investment   An international investment fund manager models bond rates of countries as a tool when making investment decisions. The manager uses the data in Table 2.6 to create a yield curve for Germany, where long-term bond rates are higher than short-term rates.

TABLE 2.6

Time to maturity (years)	German bond rate (percent)	Time to maturity (years)	German bond rate (percent)
1	3.60	6	4.65
2	4.10	7	4.75
3	4.25	8	4.80
4	4.40	9	4.90
5	4.50	10	4.95

(Source: *Investment Digest,* VALIC, vol. 12, no. 1, 1998.)

a.   Sketch a scatter plot of the data.

b.   Find a log model for the data.

c. This investment manager estimates a 5.25% rate for 20-year bonds and a 5.50% rate for 30-year bonds. How closely does the log model match these estimates?

Solution

a.

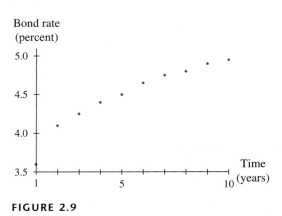

FIGURE 2.9

The scatter plot in Figure 2.9 suggests the slow growth modeled by a log function.

b. A log model for the data is

$$R(t) = 3.6296 + 0.5696 \ln t \text{ percent}$$

for a maturity time of t years.

c. The model in part b predicts the following rates for 20- and 30-year bonds:

$$R(20) \approx 5.34\% \text{ and } R(30) \approx 5.57\%$$

These predictions are slightly higher than the estimates made by the fund manager.   ●

Aligning Log Data

Recall that it is convenient to align yearly data to small input values when finding a linear model and necessary when finding an exponential model so that the coefficients in the equation will not be unnecessarily large values and to avoid round-off error. In both the linear and exponential cases, aligning input does not affect how well the model fits the data. It simply causes a graph of the model to be shifted horizontally closer to or farther from the origin.

However, because log models have the property that the output approaches negative or positive infinity as the inputs approach 0 from the right, differently aligned data results in better- or worse-fitting models. When aligning the input data for a log model, it is important to remember that the log function is not defined for negative or 0 input. If you align the data so that the first input value is 0, your calculator or computer will return an error message when you attempt to find a log equation.

> ### Aligning Logarithmic Data
>
> When you align input data before fitting an equation for a log model, all aligned input values must be numbers greater than zero.

Another property of the log function to consider when aligning input data is the function's behavior as the inputs approach 0 from the right. The steepness of change in this region affects the shape of the function and thus affects how the function fits the data. Example 2 illustrates aligning input data in order to achieve a better-fitting log model.

EXAMPLE 2   *Horizontally Shifting Data*

Voter Turnout   A community in central Indiana has been funding a local campaign to increase voter turnout in local elections. The campaign began in 2000. The percentages of all eligible voters voting in the yearly local elections for years between 1999 and 2004 are shown in Table 2.7.

TABLE 2.7

Year	1999	2000	2001	2002	2003	2004
Voter turnout (%)	2.0	4.1	5.3	6.2	6.8	7.4

a. Examine a scatter plot of the data. What characteristics of the data indicate that a log model may be appropriate?

b. Find a log model for the data with the number of years since 1900 as the input. Examine a graph of the model on a scatter plot of the data, and comment on the fit.

c. Find a log model for the data with the number of years since 1998 as the input. How does the fit of this model compare to that of the one in part *b*?

d. Why did we not use the number of years since 1999 or 2000 as the input?

Solution

a.

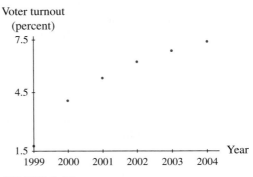

FIGURE 2.10

The scatter plot in Figure 2.10 shows an increasing, concave-down pattern. Because there is no indication that the data will decline, a log model is more appropriate than a quadratic model.

b. $V(t) = 477.789 + 104.567 \ln t$ percent turnout when t is the number of years since 1900.

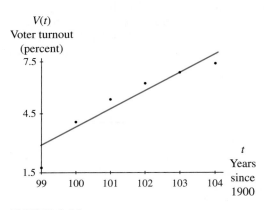

FIGURE 2.11   **FIGURE 2.12**

A graph of the function V (shown in Figure 2.11) appears to be linear compared with the curved data.

c. $T(x) = 2 + 3 \ln x$ percent turnout when x is the number of years since 1998. A graph of this function (shown in Figure 2.12) appears to be an excellent fit to the data.

d. If we align the input as years since 1999, then the first input value is 0. If we align the input as years since 2000, the first input value is -1. Zero and negative numbers are not valid inputs for a log function. ●

It is important to realize that there is no one correct way to shift data either horizontally or vertically. The more experience you have fitting equations to data, the better equipped you will be to make decisions about shifting.

An Important Inverse Relationship

2.2.2

We should not leave a discussion of log and exponential functions and how they can be used to describe change in the real world without mentioning that these two functions are inverses. If we have data whose input/output relationship can be modeled by an exponential function, then the inverse (output/input) relationship can be modeled by a log function, and vice versa. This relationship in its simplest form can be stated:

If $f(x) = \ln x$ and $g(x) = e^x$, then $f(g(x)) = \ln(e^x) = x$ and $g(f(x)) = e^{\ln x} = x$ as long as x is positive.

The inverse relationships for the model equations are more involved but still exist:

If $f(x) = a + b \ln x$ with $b \neq 0$, then $f^{-1}(x) = AB^x$ where $A = e^{-\frac{a}{b}}$ and $B = e^{\frac{1}{b}}$.

If $f(x) = ab^x$, then for $x > 0$, $b > 0$, and $b \neq 1$, $f^{-1}(x) = A + B \ln x$ where

$$A = \frac{-\ln a}{\ln b} \quad \text{and} \quad B = \frac{1}{\ln b}$$

Log functions with their slowing but not leveling end behavior are common in health and life sciences as well as social sciences. Their importance leads us to include them in our discussion. As we continue our exploration of calculus concepts, you will become increasingly familiar with both log and exponential functions.

2.2   Concept Inventory

○ *Log function:* $f(x) = a + b \ln x$

○ *End behavior of log and exponential functions*

○ *Shifting data horizontally*

○ *Log and exponential functions as inverses*

2.2   Activities

For Activities 1 through 4, match each graph with its equation.

1. $f(x) = 2 \ln x$
   $f(x) = -2 \ln x$

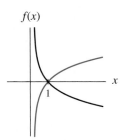

2. $f(x) = 3 + \ln x$
   $f(x) = \ln x$

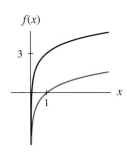

3. $f(x) = 2 \ln x$
   $f(x) = 4 \ln x$

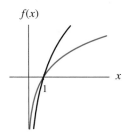

4. $f(x) = -4 \ln x$
   $f(x) = -2 \ln x$

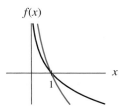

5. Contaminant   The American Association of Pediatrics has stated that lead poisoning is the greatest health risk to children in the United States. Because of past use of leaded gasoline, the concentration of lead in soil can be described in terms of how close the soil is to a heavily traveled road. The accompanying table shows some distances and the corresponding lead concentrations in parts per million.

Distance from road (meters)	Lead concentration (ppm)
5	90
10	60
15	40
20	32

(Source: Estimated from information in "Lead in the Inner City," *American Scientist*, January–February 1999, pp. 62–73.)

a. Find a log model for these data.

b. An apartment complex has a dirt play area located 12 meters from a road. Estimate the lead concentration in the soil of the play area.

c. Find an exponential model for the data. Compare this model to the log model found in part *a*. Which of the two models better displays the end behavior suggested by the context?

6. **Weight**   The body weight of mice used in a drug experiment is recorded by the researcher. The data are given in the accompanying table.

Age beyond 2 weeks (weeks)	Weight (grams)
1	11
3	20
5	23
7	26
9	27

(Source: Estimated from information given in "Letters to Nature," *Nature*, vol. 381, May 30, 1996, p. 417.)

a. Find a log model for the data.

b. Use limit notation to discuss the end behavior of the function.

c. Does the end behavior fit the context? What does this tell you about using your model to extrapolate?

d. Estimate the weight of the mice when they are 4 weeks old.

7. **Bond Rates**   The fund manager in Example 1 on page 93 also models the yield curve for New Zealand, in which long-term bond rates are lower than short-term rates. He uses the data given in the accompanying table.

Time to maturity (years)	New Zealand bond rate (%)
0.25	9.40
2	7.90
3	7.65
4	7.50
6	7.30
10	7.10

a. Find a log model for the bond rate data.

b. The fund manager estimates 15-year rates at 7.00%. How close does your model come to this estimate?

8. **Medicine**   The concentration of a drug in the blood stream increases the longer the drug is taken on a daily basis. The accompanying table gives estimated concentrations (in micrograms per milliliter) of the drug piroxicam taken in 20-mg doses once a day.

Days	Concentration (µg/mL)	Days	Concentration (µg/mL)
1	1.5	11	6.5
3	3.2	13	6.9
5	4.5	15	7.3
7	5.5	17	7.5
9	6.2		

a. Find a log model for the data.

b. Express the end behavior of the equation using limits.

c. Does the end behavior of the equation fit the end behavior suggested by the context?

d. Estimate the concentration of the drug after 2 days of piroxicam doses.

9. **Salaries**   The table shows the average salaries of public elementary and secondary school teachers from 1980 through 1997.

Year	Average salary	Year	Average salary
1980	$15,970	1995	$36,685
1985	$23,500	1996	$37,716
1990	$31,367	1997	$38,562
1994	$35,737		

(Source: *Statistical Abstract*, 1998.)

a. Find a log model for the average salary as a function of the number of years since 1970.

b. Use the log model to estimate the average salary in 1993. What does the log model predict for the average salary in 2000? In which of these two estimates can you be more confident and why?

c. According to the model, when will the average salary reach $40,000?

10. **Cable TV** The table shows the projected number of homes with a cable system offering Internet access.

Year	Homes (millions)	Year	Homes (millions)
1998	19.1	2002	51.8
1999	29.0	2003	57.3
2000	39.0	2004	62.8
2001	45.4	2005	67.4

(Source: Paul Kagan Associates, Inc., *Cable TV Technology*.)

a. Find a log model for the number of homes as a function of the number of years since 1990.

b. According to the model in part *a,* when will the number of homes reach 75 million?

c. What end behavior is suggested by the model equation as time increases? Write this end behavior using limit notation.

11. **Peaches** The average yearly consumption of peaches per person based on that person's yearly family income when the price of peaches is $1.50 per pound is given in the table.

Yearly income (tens of thousands of dollars)	Consumption of peaches (pounds per person per year)
1	5.0
2	6.4
3	7.2
4	7.8
5	8.2
6	8.6

a. Explain why the data is neither linear nor exponential.

b. Find a log function to fit the data.

c. Use the log model to estimate consumption for a person in a family with yearly income of $35,000.

12. **Height** The length of an average girl *t* months after birth is shown in the accompanying table.

a. Explain why the data is neither linear nor exponential.

Age (months)	4	10	18	27	35
Length (centimeters)	62	72	81	89	95

(Source: Based on information from the Fels Longitudinal Study, Wright State University School of Medicine, Yellow Springs, Ohio.)

b. Find a log function to fit the data.

c. Use the log model to estimate the length of an average 3-year-old girl.

d. Does this model adequately describe the length for very young girls (that is, near birth)? Explain.

13. **CDs** The values of manufacturer shipments of compact disc albums between 1993 and 1997 are as shown in the table.

Year	CD shipments (millions of dollars)
1993	6511
1994	8465
1995	9377
1996	9935
1997	9915

(Source: *Statistical Abstract*, 1998.)

a. Find a log model for the value of shipments as a function of the number of years since 1900. Comment on how well the equation fits the data.

b. Realign the input as the number of years since 1992, and find a new log model for the data. How does the fit of this equation compare to that of the equation in part *a*?

c. Realign the input as the number of years since 1993 plus 0.5, and find another log model for the data. How well does this equation fit compared with those in parts *a* and *b*?

d. What is the behavior of the function in part *c* beyond 1997? Does the data suggest this same behavior?

e. Why can we not align the input data as the number of years since 1993 if we wish to use a log model?

14. **CPI** The consumer price index (CPI) values for refuse collection between 1990 and 1997 are shown in the table (1982–1984 = 100) on the next page.

Year	CPI	Year	CPI
1990	171.2	1994	231.4
1991	189.2	1995	241.2
1992	207.3	1996	246.0
1993	220.5	1997	250.5

(Source: *Statistical Abstract*, 1998.)

a. Align the input data as the number of years since 1900, and find a log model for the data. How well does this equation fit the data?

b. Realign the input data as the number of years since 1988. How well does this equation fit the data?

c. What does this model predict will happen to CPI values beyond 1997? Do you believe that this model accurately describes future CPI for refuse collection?

15. **Solution pH** The pH of a solution, measured on a scale from 0 to 14, is a measure of how acidic or how alkaline that solution is. The pH is a function of the concentration of the hydronium ion, H_3O^+. The accompanying table shows the H_3O^+ concentration and associated pH for several solutions.

Solution	H_3O^+ concentration (moles per liter)	pH
Cow's milk	$3.981 \cdot 10^{-7}$	6.4
Distilled water	$1.000 \cdot 10^{-7}$	7.0
Human blood	$3.981 \cdot 10^{-8}$	7.4
Lake Ontario water	$1.259 \cdot 10^{-8}$	7.9
Seawater	$5.012 \cdot 10^{-9}$	8.3

a. Find a log model for pH as a function of the H_3O^+ concentration.

b. What is the pH of orange juice with H_3O^+ concentration $1.585 \cdot 10^{-3}$?

c. Black coffee has a pH of 5.0. What is its concentration of H_3O^+?

d. A pH of 7 is neutral, a pH less than 7 indicates an acidic solution, and a pH greater than 7 shows an alkaline solution. Beer has H_3O^+ concentration $3.162 \cdot 10^{-5}$. Is beer acidic or alkaline?

16. **Height** The length of an average boy t months after birth is given in the accompanying table.

Age (months)	0	3	6	24	35
Length (centimeters)	51	61	68	88	96

(Source: Based on information from the Fels Longitudinal Study, Wright State University School of Medicine, Yellow Springs, Ohio.)

a. Examine a scatter plot of this data. Explain why shifting the data may be appropriate.

b. Align the input data by adding 3 months, and find a log function to fit this aligned data.

c. Align the input data by adding 1 year, and find a log function to fit these aligned data.

d. Which log model (from part b or part c) appears to be better? Explain.

17. **Cable Subscribers** On the basis of data recorded between 1975 and 2000, the number of cable subscribers in the United States as a function of the average monthly basic cable rate can be modeled as

$$S(r) = -56.1105 + 37.201 \ln r \text{ million subscribers}$$

when the average basic rate is r dollars per month. (Source: Based on data from *Statistical Abstract*, 2001.)

a. Use the model to estimate the number of subscribers (in millions) for the following monthly basic rates: $15, $30, $45.

b. Switch the input/output values for the three points found in part a. Use the three inverted points to find an inverse function R with input s.

18. **Milk Storage** According to the back of a milk carton sold by Model Dairy, the number of days that milk will keep when stored at various temperatures is shown in the table.

Temperature (degrees Fahrenheit)	Days
30	24
38	10
45	5
50	2
60	1
70	0.5

a. Find a function that fits the data and gives the minimum temperature T at which the milk must be stored in order to be preserved for d days.

b. How cold must the temperature be in order to keep milk for one week?

19. Altimeter The altimeter discussion in this section included the following table:

Air pressure (inches of mercury)	Altitude (thousands of feet)
13.76	20
5.56	40
2.14	60
0.82	80
0.33	100

a. Does it make sense to reverse inputs and outputs for these data? Why?

b. Find a function that fits the reversed data.

20. In essay form, comment on the importance of scale in the appearance of the graphs of exponential and log models and conditions under which the graphs of exponential and log functions appear to be linear.

21. Using the idea of limits, describe the end behavior of the exponential and log models. Explain how this end behavior can help in identifying which of these two functions to fit to a data set.

2.3   Logistic Functions and Models

Exponential Growth with Constraints

Although exponential models are common and useful, it is sometimes unrealistic to believe that exponential growth can continue forever. In many situations, there are forces that ultimately limit the growth. Here is a situation that may seem familiar to you.

You and a friend are shopping in a music store and find a new compact disc that you are certain will become a hit. You each buy the CD and rush back to campus to begin spreading the news. As word spreads, the total number of CDs sold begins to grow exponentially as shown in Figure 2.13.

However, this trend cannot continue indefinitely. Eventually, the word will spread to people who have already bought the CD or to people who have no interest in it, and the rate of increase in total sales begins to decline (Figure 2.14). In fact, because there is only a limited number of people who will ever be interested in buying the CD, total sales ultimately must level off. The graph representing the total sales of the CD as a function of time is a combination of rapid exponential growth followed by a slower increase and ultimate leveling off. See Figure 2.15.

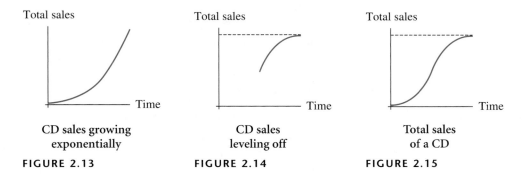

CD sales growing exponentially

FIGURE 2.13

CD sales leveling off

FIGURE 2.14

Total sales of a CD

FIGURE 2.15

S-shaped behavior such as this is common in marketing situations, the spread of disease, the spread of information, the adoption of new technology, and the growth of certain populations. A mathematical function with such an S-shaped curve is called a **logistic function.** Its equation is of the form

$$f(x) = \frac{L}{1 + Ae^{-Bx}}$$

From now on, when referring to a logistic equation or function, we will consider an equation of this form. The number L appearing in the numerator of a logistic equation determines a horizontal asymptote $y = L$ for a graph of the function f.

Be aware that some technologies do not have a built-in logistic regression routine. See the *Excel Instruction Guide* to obtain a logistic curve-fitting procedure. Also consult the Note for Excel Users on page 104.

Excel 2.3.0

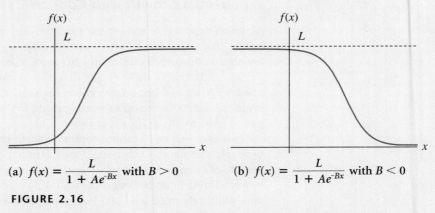

Logistic Model

Algebraically: A logistic model has an equation of the form

$$f(x) = \frac{L}{1 + Ae^{-Bx}}$$

We refer to L as the limiting value of the function.

Graphically: The logistic function f increases if B is positive and decreases if B is negative. The graph of a logistic function is bounded by the horizontal axis and the line $y = L$.

(a) $f(x) = \dfrac{L}{1 + Ae^{-Bx}}$ with $B > 0$ (b) $f(x) = \dfrac{L}{1 + Ae^{-Bx}}$ with $B < 0$

FIGURE 2.16

The graph of a positive, increasing logistic function (as in Figure 2.16a) is trapped between the horizontal axis ($y = 0$) and the horizontal asymptote $y = L$. For an increasing logistic function f, where $f(x) = \frac{L}{1 + Ae^{-Bx}}$ with $B > 0$,

$$\lim_{x \to -\infty} f(x) = 0 \qquad \text{and} \qquad \lim_{x \to \infty} f(x) = L$$

Similarly, for a decreasing logistic function f, where $f(x) = \frac{L}{1 + Ae^{-Bx}}$ with $B < 0$ (as shown in Figure 2.16b) the end behavior can be described by the limit statements

$$\lim_{x \to -\infty} f(x) = L \qquad \text{and} \qquad \lim_{x \to \infty} f(x) = 0$$

In social science and life science applications, the limiting value is often called the *carrying capacity* or the *saturation level.* In other applications it is sometimes referred to as the *leveling-off value.*

Thus any logistic function has two horizontal asymptotes. We refer to $y = 0$ as the *lower asymptote* and to $y = L$ as the *upper asymptote*. We also refer to L as the *limiting value* of the function.

Finding Logistic Models

2.3.1

Consider an office with 10 networked computers. One computer has a virus that will once every hour randomly attempt to infect another computer on the network. Assuming that there are only 10 computers that can get the virus, as time increases the number of infected computers can approach, but never exceed, 10. Consider the following representations (Table 2.8 and Figure 2.17) of the function describing the spread of the computer virus.

TABLE 2.8

Time	Total number of infected computers
0 (before 8 A.M.)	1
1 (8 to 9 A.M.)	2
2 (9 to 10 A.M.)	4
3 (10 to 11 A.M.)	6
4 (11 A.M. to noon)	8
5 (noon to 1 P.M.)	9

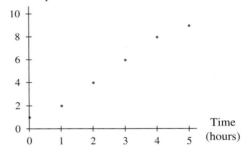

Total number of infected computers

FIGURE 2.17

2.3.2a, b

A logistic equation* calculated by a calculator or computer using a best-fit technique and the data in Table 2.8 is

$$I(t) = \frac{9.98632}{1 + 9.33812e^{-0.89511t}} \text{ computers infected}$$

t hours after 8 A.M. Note that the leveling-off value here is $L = 9.98632$, which is approximately, but not exactly, 10 computers.

A graph of the function I is shown in Figure 2.18. The dotted, horizontal line portrays the upper asymptote, $L \approx 10$ rounded from 9.98632. Note that the curvature on the left side of this graph is **concave up** ⌣ whereas the curvature on the right side is **concave down** ⌢. The point on the graph at which the concavity changes is called the **inflection point.** The inflection point on the logistic graph modeling the spread of the computer virus (Figure 2.18) is marked with a black dot. In some situations, inflection points have very important interpretations. We later use calculus to find and help interpret these special points.

---

*Different calculators and computer software may use slightly different methods for calculating a "best-fit" equation. This results in slightly different coefficients and exponents in the models. Recall that our goal in fitting curves to data is to develop models so that we can use calculus to study the general behavior of the data. Slightly different equations will not significantly affect our study of the general behavior of the data. We report logistic models with equations obtained from a TI-83 calculator.

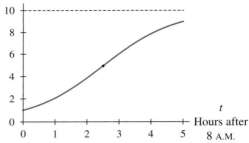

FIGURE 2.18

When you use a calculator or a computer program to find an equation and construct the graph of a logistic model for a set of data, be sure to align the input data. The same numerical computation problems that we discussed for exponential models can occur with logistic models because of the exponential term in the denominator.

EXCEL   **Note for Excel Users**

The technique for finding a best-fit logistic function is explained in detail in the technology supplement. It is important to note that the resulting function depends on the number of iterations and initial guesses for the parameters L, A, and B. Results comparable to those given in this text can be found with the following steps.

Step 1: Using a scatter plot of the data, estimate the location of the inflection point (x_0, y_0).

Step 2: Use the largest output value as an initial estimate of L. You should also use the relationship $L \approx 2y_0$ to check your inflection point estimate. Revise Step 1 if necessary.

Step 3: For increasing logistic functions, use these initial estimates: $A \approx e^{x_0}$, $B \approx 1$. For decreasing logistic functions, use $A \approx e^{-x_0}$, $B \approx -1$.

Step 4: Access the Excel SOLVER and reset these options: maximum number of iterations = 500 and tolerance = 1%. Run the SOLVER. If A is extremely large in the solution, revise Step 3 by using $A \approx e^{0.1x_0}$, $B \approx 0.1$ for increasing logistic functions and $A \approx e^{-0.1x_0}$, $B \approx -0.1$ for decreasing logistic functions. Again use the SOLVER. Running the SOLVER and the current estimates once more may produce a better solution.

Step 5: Check the fit of the logistic function by drawing it on a scatter plot of the data.

EXAMPLE 1   *Finding a Logistic Model*

Bacteria   Table 2.9 shows the number of bacteria counted in a biology experiment.

TABLE 2.9

Day	Bacteria	Day	Bacteria
1	4	6	439
2	12	7	648
3	25	8	748
4	58	9	769
5	230		

Find a logistic model that fits the data. What is the end behavior of the model as time increases?

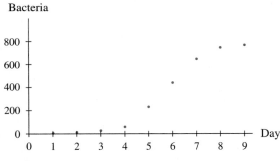

Bacteria

FIGURE 2.19

Solution   A scatter plot of the data (Figure 2.19) suggests that a logistic model is appropriate.

A possible logistic model is

$$B(x) = \frac{786.70445}{1 + 1464.70161e^{-1.26110x}} \text{ bacteria}$$

in day x.

The limiting value of this model is $L = 786.70445$ or, in context, approximately 787 bacteria. The graph of B on a scatter plot appears to fit fairly well (see Figure 2.20). The dotted line in Figure 2.20 denotes the upper asymptote. As time increases, the number of bacteria approaches 787.

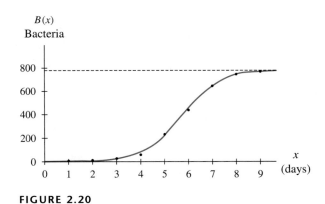

$B(x)$
Bacteria

FIGURE 2.20

The preceding logistic curves begin near zero and then increase toward a limiting value L. As previously mentioned, there are also situations where the curve begins near its limiting value and then decreases toward zero.   ●

EXAMPLE 2   *Aligning Input and Determining End Behavior*

Height   Of a group of 200 college men surveyed, the number who were taller than a given number of inches is recorded in Table 2.10.

TABLE 2.10

Inches	Number of men	Inches	Number of men	Inches	Number of men
65	198	69	139	73	26
66	195	70	101	74	11
67	184	71	71	75	4
68	167	72	43	76	2

Find an appropriate model for the data. What is the end behavior of the model as height increases?

Solution   A scatter plot of the data is shown in Figure 2.21.

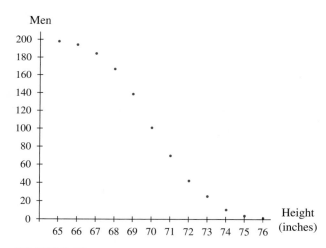

FIGURE 2.21

We choose to align the input data by subtracting 65 from each value. A logistic model for the aligned data is

$$M(x) = \frac{205.66709}{1 + 0.02998e^{0.69528x}} \text{ men out of 200 men surveyed}$$

who are taller than $x + 65$ inches. The two horizontal asymptotes of a graph of this function are $y = 0$ and $y \approx 205.7$. Because the function decreases, the end behavior as height increases is given by the lower asymptote; that is, as the heights increase, the number of men taller than a particular height approaches zero.   ●

Recall that a logistic function is of the form $f(x) = \frac{L}{1 + Ae^{-Bx}}$. What happened to the negative sign before the B in Example 2? In that case, $B = -0.69528$. When the formula was written, the negatives canceled.

$$\text{Number of men} = \frac{205.66709}{1 + 0.02998e^{-(-0.69528)x}} = \frac{205.66709}{1 + 0.02998e^{0.69528x}}$$

This will always be the case for a logistic function that decreases from its upper asymptote.

It is likely that a logistic model will have a limiting value that is lower or higher than the one indicated by the context. For instance, in Example 2 only 200 men were surveyed, but the limiting value is $L \approx 206$ men. This does not mean that the model is invalid, but it does indicate that care should be taken when extrapolating from logistic models.

Shifting Logistic and Exponential Data

We have seen two functions whose end behavior is to approach a number in one or both directions. The output of an exponential function approaches zero as the input becomes increasingly positive or negative, and the output of a logistic function approaches zero in one direction and a nonzero number in the other direction. These functions are useful, but they have limitations. Neither the logistic nor the exponential model in its basic form allows for data that approach a non-zero lower limit. Consider, for example, the data shown in Table 2.11. A scatter plot of the data (Figure 2.22) suggests an exponential function.

TABLE 2.11

x	f(x)
0	101
2	104
4	116
6	164
8	356

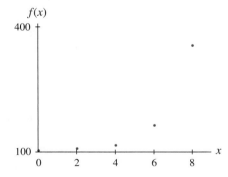

FIGURE 2.22

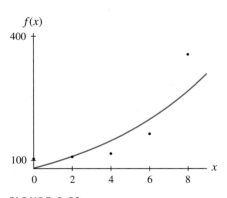

FIGURE 2.23

When we find an exponential equation to fit the data, we obtain the function $f(x) = 81.6597(1.1604^x)$. A graph of this function on a scatter plot of the data is shown in Figure 2.23. The equation is a poor fit for the data and is not what we expect to see.

The problem is that the output data are not approaching zero to the left. They are shifted up by 100, so they approach 100. However, the exponential regression routine built into most technologies does not allow for a vertical shift. We can compensate for this limitation by shifting the data ourselves, modeling them, and adjusting the model accordingly. We modify the data in Table 2.11 by subtracting 100 from all the output values, but we do not change the input values. See Table 2.12.

TABLE 2.12

x	0	2	4	6	8
f(x) − 100	1	4	16	64	256

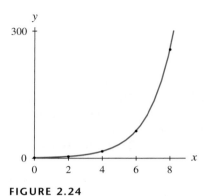

FIGURE 2.24

A scatter plot of the shifted data appears exactly like that in Figure 2.22 except that the vertical axis values are different. An exponential function that fits these adjusted data is $y = 2^x$. A graph of this function on a scatter plot of the shifted data is shown in Figure 2.24.

The fit is perfect, because the data are perfectly exponential. We now must compensate for the shifted data by reversing our shift. We originally subtracted 100, so we now reverse the process by adding 100 to the function for the shifted data. Our final equation for the data in Table 2.11 is

$$g(x) = 2^x + 100$$

Vertically shifting data may be appropriate when a scatter plot suggests a curve with a horizontal limiting value, but the technology-generated equation for that curve is a poor fit to the data and does not reflect the observed behavior of the data. Example 1 gives another situation in which vertical shifting is necessary. Some technologies use functions in different forms from those we present. For example, some technologies use a logistic function of the form $f(x) = \frac{L}{1 + Ae^{-Bx}} + D$. The parameter D is the vertical shift parameter. If the technology you use produces a logistic function of this form, then there is no need for you to shift logistic data in order to obtain a better fit. You need to know your technology and the functions it uses well enough to know whether the discussion of shifting logistic data that is presented in Example 1 applies to you.

EXAMPLE 3   *Vertically Shifting Data*

2.3.3

Investment Clubs   Table 2.13 shows the number of investment clubs in existence during the 1990s. A scatter plot of the data is shown in Figure 2.25.

TABLE 2.13

Year	Number of clubs	Year	Number of clubs
1990	7085	1995	16,054
1991	7360	1996	25,409
1992	8267	1997	31,828
1993	10,033	1998	36,500
1994	12,429		

(Source: National Association of Investors Corp.)

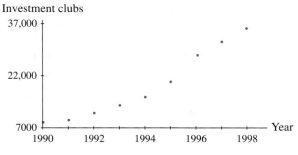

FIGURE 2.25

Find an appropriate model for the data, and use it to estimate the number of investment clubs in 2000.

Solution

The scatter plot and context suggest a logistic curve because the data exhibit an obvious inflection point sometime in 1995 and suggest a horizontal asymptote at $y = 0$. Market saturation in the context of the data suggests a possible upper horizontal asymptote. We align the input values to years since 1990 in order to avoid computational error. When we fit a logistic equation to the aligned data, we obtain

$$N(t) = \frac{244{,}096.932}{1 + 48.227e^{-0.2721t}} \text{ investment clubs}$$

t years after 1990. This model should look suspicious to you, because the limiting value of 244,097 is much greater than the one suggested by the data. A graph of the function on the scatter plot is shown in Figure 2.26.

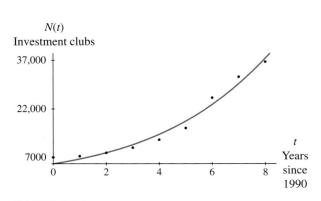

FIGURE 2.26

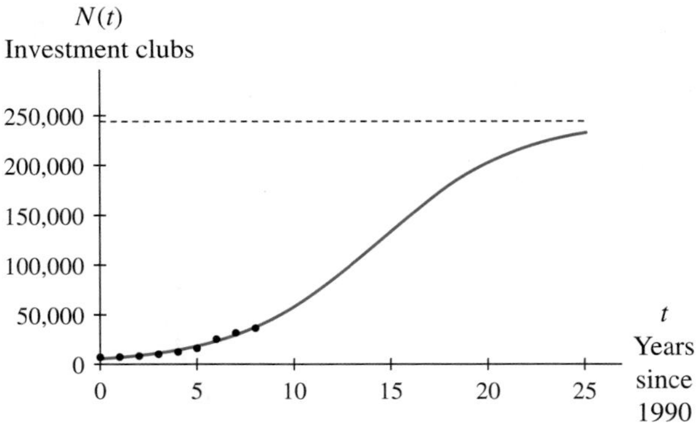

FIGURE 2.27

Although at first glance it might seem that this function is a reasonable fit, look at it more closely. The logistic graph does not exhibit an inflection point that coincides with the inflection point suggested by the data, and the graph does not level off where we expect it to. A view of the graph from a different perspective (see Figure 2.27) shows that the data are modeled by only the concave-up portion of this logistic function.

The underlying problem is that the logistic function N approaches zero on the left, but the data do not approach zero on the left. They appear to approach a value closer to 7000. To compensate for this difference, we shift the data down, closer to zero. We choose to do so by subtracting 7000 from the output data. This value is arbitrarily chosen, and other values will produce similar results. Table 2.14 shows the shifted data. The scatter plot will look exactly like the one in Figure 2.25 except that the vertical axis labels will be different.

TABLE 2.14

Aligned year	Number of clubs − 7000
0	85
1	360
2	1267
3	3033
4	5429
5	9054
6	18,409
7	24,828
8	29,500

When we fit a logistic equation to this shifted data, we obtain the model

$$f(t) = \frac{34{,}280.654}{1 + 207.434e^{-0.8991t}} \text{ investment clubs above 7000}$$

t years after 1990. To compensate for the fact that this model is for the number of clubs above 7000, we simply add 7000 to the output $f(t)$ to obtain a model for the data.

$$I(t) = \frac{34{,}280.654}{1 + 207.434e^{-0.8991t}} + 7000 \text{ investment clubs}$$

t years after 1990. A graph of this function is shown in Figure 2.28.

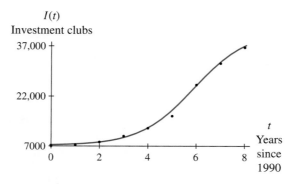

FIGURE 2.28

This model has an inflection point that better matches that indicated by the data. The horizontal asymptotes of the function I are $y = 7000$ (lower) and $y = 34{,}280.654 +$

7000 = 41,280.654 (upper). The limiting value is 41,280.654. According to the model, the number of investment clubs will level off at approximately 41,281.

Using the two models to estimate the number of clubs in 2000, we see that the difference in the estimates is significant:

$$N(10) \approx 58,481 \text{ clubs} \quad \text{and} \quad I(10) \approx 40,418 \text{ clubs}$$

The estimate from the second model is probably more accurate, provided that the logistic pattern continued into 2000. ●

Exponential and logistic models should be shifted vertically by realigning output data *only* when there is reason to believe that the data approach a different limiting position than the input axis; or, for the case of the logistic model, when it is obvious that there is an upper limiting value and the shift will significantly improve end-behavior fit.

Don't forget to use the unrounded equations to calculate output.

2.3   Concept Inventory

○ *Logistic function:* $f(x) = \dfrac{L}{1 + Ae^{-Bx}}$
   (exponential in nature with two horizontal asymptotes)
○ *Equations of horizontal asymptotes*
○ *Concave up and concave down*
○ *Inflection point*
○ *Limiting value*
○ *Importance of aligning data*
○ *Shifting data vertically: exponential and logistic functions*
○ *Asymptotes for shifted models*

2.3   Activities

Identify the scatter plots in Activities 1 through 6 as linear, exponential, logarithmic, logistic, or none of these. If you identify the scatter plot as none of these, give reasons.

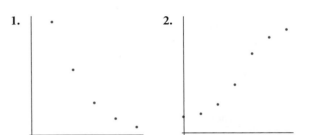

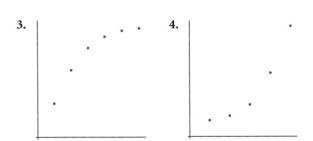

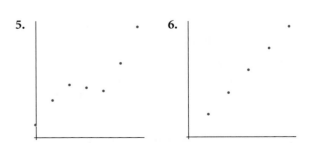

For Activities 7 through 10, indicate whether the function is an increasing or a decreasing logistic function. Also identify the limiting value of the function and write the equations of the two horizontal asymptotes.

7. $f(x) = \dfrac{100}{1 + 9.8e^{-0.98x}}$

8. $A(t) = \dfrac{1925}{1 + 321.1e^{1.86t}}$

9. $h(g) = \dfrac{39.2}{1 + 0.8e^{0.325g}}$

10. $k(x) = \dfrac{16.5}{1 + 1.86e^{-0.43x}}$

11. Postage The following table gives the number of European, North American, and South American countries that issued postage stamps from 1840 through 1880.

Year	Total number of countries
1840	1
1845	3
1850	9
1855	16
1860	24
1865	30
1870	34
1875	36
1880	37

(Source: "The Curve of Cultural Diffusion," *American Sociological Review,* August 1936, pp. 547–556.)

a. Find a logistic model for the data, and discuss how well the equation fits.

b. Sketch the graph of the equation in part *a,* and mark where the curve is concave up and where it is concave down. Label the approximate location of the inflection point.

12. P.T.A. The accompanying table gives the total number of states associated with the national P.T.A. organization from 1895 through 1931.

Year	Total number of states	Year	Total number of states
1895	1	1915	30
1899	3	1919	38
1903	7	1923	43
1907	15	1927	47
1911	23	1931	48

(Source: Hamblin, Jacobsen, and Miller, *A Mathematical Theory of Social Change.* New York: John Wiley & Sons, 1973.)

a. Find a logistic model for the data.

b. What is the maximum number of states that could have joined the PTA by 1931? How does this number compare to the limiting value given by the model in part *a?*

13. Patents The approximate numbers of patents for plow sulkies beginning in 1865 are given in the accompanying table.

Year	Cumulative number of plow sulky patents
1871	200
1877	340
1883	980
1889	1800
1895	2200
1901	2400
1907	2500
1913	2550
1919	2620
1925	2700

(Source: Hamblin, Jacobsen, and Miller, *A Mathematical Theory of Social Change.* New York: John Wiley & Sons, 1973.)

a. What is a *plow sulky,* and to what modern farm implement was it a precursor?

b. Find a logistic model for the data.

c. Discuss why it is logical that the total number of patents for a new invention would increase according to a logistic equation.

14. Flu In the fall of 1918, an influenza epidemic hit the U.S. Navy. It spread to the Army, to American civilians, and ultimately to the world. It is estimated that 20 million people had died from the epidemic by 1920. Of these, 550,000 were Americans—over 10 times the number of World War I battle deaths. The accompanying table gives the total numbers of Navy, Army, and civilian deaths that resulted from the epidemic in 1918.

| Week ending | Total deaths | | Total civilian deaths in 45 major cities |
	Navy	Army	
August 31	2		
September 7	13	40	
September 14	56	76	68
September 21	292	174	517
September 28	1172	1146	1970
October 5	1823	3590	6528
October 12	2338	9760	17,914
October 19	2670	15,319	37,853
October 26	2820	17,943	58,659
November 2	2919	19,126	73,477
November 9	2990	20,034	81,919
November 16	3047	20,553	86,957
November 23	3104	20,865	90,449
November 30	3137	21,184	93,641

(Source: A.W. Crosby, Jr., *Epidemic and Peace 1918*. Westport, Conn.: Greenwood Press, 1976.)

a. Find logistic equations to fit each set of data. In each case, graph the equation on the data.

b. Write models for each of the data sets.

c. Compare the limiting values of the models in part *b* with the highest data values in the table. How many more people do the models indicate died from the epidemic after November 30? Do you believe that the limiting values indicated by the models accurately reflect the ultimate number of deaths? Explain.

15. **Epidemic**   In 1949 the United States experienced the second worst polio epidemic in its history. (The worst was in 1952.) The accompanying table gives the cumulative number of polio cases diagnosed on a monthly basis.

a. Observe a scatter plot of the data from January through June. Describe the concavity and estimate the location of the inflection point indicated by the scatter plot. Does this portion of the data appear to be logistic?

Month	Total number of polio cases
January	494
February	759
March	1016
April	1215
May	1619
June	2964
July	8489
August	22,377
September	32,618
October	38,153
November	41,462
December	42,375

(Source: *Twelfth Annual Report,* National Foundation for Infantile Paralysis, 1949.)

b. Observe a scatter plot of the entire data set given. Does the entire data set appear to be logistic?

c. Find a logistic model for the data.

d. Observe the model in part *c* graphed on a scatter plot of the data between January and June. Discuss how well the graph fits this portion of the data.

16. **Visitors**   The total numbers of visitors to an amusement park that stays open all year are given in the accompanying table.

Month	Cumulative number of visitors by the end of the month (thousands)
January	25
February	54
March	118
April	250
May	500
June	898
July	1440
August	1921
September	2169
October	2339
November	2395
December	2423

a. Find a logistic model for the data.

b. The park owners have been considering closing the park from October 15 through March 15 each year. How many visitors will they potentially miss by this closure?

17. **Chemical Reaction** A chemical reaction begins when a certain mixture of chemicals reaches 95°C. The reaction activity is measured in Units (U) per 100 microliters ($100\mu L$) of the mixture. Measurements during the first 18 minutes after the mixture reaches 95°C are listed in the accompanying table.

Time (minutes)	Activity (U/100μL)	Time (minutes)	Activity (U/100μL)
0	0.10	10	1.40
2	0.10	12	1.55
4	0.25	14	1.75
6	0.60	16	1.90
8	1.00	18	1.95

(Source: David E. Birch et al., "Simplified Hot Start PCR," *Nature*, vol. 381, May 30, 1996, p. 445.)

a. Examine a scatter plot of the data. Estimate the limiting value. Estimate at what time the inflection point occurs.

b. Find a logistic model for the data. What is the limiting value for this logistic function?

c. Use the model to estimate by how much the reaction activity increased between 7 and 11 minutes.

18. **Stolen Bases** San Francisco Giants legend Willie Mays's cumulative number of stolen bases between 1951 and 1963 are as shown in the accompanying table.

a. Find a logistic model for the data. Comment on how well the logistic equation fits the data.

b. What is the interpretation of first differences in this context?

c. Use the model in part *a* to estimate the number of bases that Mays stole in 1964. Compare this estimate with 19, the actual number of stolen bases in 1964.

Year	Cumulative stolen bases
1951	7
1952	11
1953	11
1954	19
1955	43
1956	83
1957	121
1958	152
1959	179
1960	204
1961	222
1962	240
1963	248

19. **Population** A 1998 United Nations population study reported the world population between 1804 and 1987 and projected the population through 2071. These populations are as shown in the accompanying table.

Year	Population (billions)	Year	Population (billions)
1804	1	1999	6
1927	2	2011	7
1960	3	2025	8
1974	4	2041	9
1987	5	2071	10

a. Find a logistic model for world population. Discuss how well the equation fits the data.

b. Do you consider the model appropriate to use in predicting long-term world population behavior? According to the model, what will ultimately happen to world population?

c. Do you believe the model should be used to estimate the world population in 1850? in 1990? Explain.

20. **Surcharges** The percentages of all banks that levied surcharges on transactions at automated teller machines between 1996 and 2001 are as shown in the accompanying table.

a. Examine a scatter plot of the data. Discuss the curvature suggested by the scatter plot.

Year	1996	1997	1998	1999	2001
ATM surcharges (percent)	15	45	71	93	94

(Source: U.S. Public Interest Research Group National Survey.)

    b. Would a logistic model be appropriate in this situation? Why or why not?

    c. Find a logistic model for the data. Express the limiting value of the function using a limit statement. What does the limiting value tell you about using the model to extrapolate?

21. Social Security   The table for this activity (located on the *Calculus Concepts* CD-ROM and web site) gives the number of mother and/or father OASI beneficiaries.

    a. Construct a scatter plot of the data in the year and number of mother and/or father beneficiaries columns. Give the approximate location of the inflection point.

    b. Construct a column in the table for the number of mother and/or father beneficiaries over 190 thousand. Find a logistic model for the number of mother and/or father beneficiaries over 190 thousand as a function of the number of years after 1975.

    c. Add 190 thousand to the equation in part *a* to convert it into a model for the number of mother and/or father OASI beneficiaries. Use this model to estimate the number of mother and/or father beneficiaries in 2002.

22. Subscribers   The table for this activity (located on the *Calculus Concepts* CD-ROM and web site) lists the average number of basic cable TV subscribers between 1975 and 2000.

    a. Construct a column for the number of years after 1975 and another column for the average number of basic cable subscribers in millions.

    b. Construct a scatter plot using the data in the columns in part *a*. Estimate the location of the inflection point.

    c. Construct another column in the table that gives the average number of basic cable subscribers over 9.7 million. Find a logistic model for the number of subscribers over 9.7 million as a function of the number of years after 1975.

    d. Add 9.7 million to the equation in part *a* to convert it into a model for the average number of subscribers in millions. Discuss the fit of this equation to the data in the columns you constructed in part *a*.

23. Credit Cards   Consumer credit card debt, in billions of dollars, between 1990 and 2000 was as shown in the accompanying table.

Year	Debt (billions of dollars)	Year	Debt (billions of dollars)
1990	172	1996	410
1992	200	1998	460
1994	280	2000	473

(Source: Consumer Federation of America.)

    a. Observe a scatter plot of the data. Describe the concavity suggested by the data. In what year do you estimate the concavity changed? Which equations could be used to model these data?

    b. Find a logistic model for the data. Comment on how well the equation fits the data.

    c. Shift the data down by 165, and find a logistic model for the shifted data. Comment on how well the equation fits the shifted data. Compare the fit of this equation with that of the logistic equation in part *b*.

    d. On the basis of the model in part *c*, write a model for credit card debt.

    e. What are the two horizontal asymptotes for the function in part *d*? Compare these with the horizontal asymptotes for the original logistic function in part *b*.

24. Traffic   A regional transportation commission estimates that the daily traffic counts for a certain intersection will be as shown in the following table:

Year	1999	2010	2020	2030
Traffic count (vehicles)	33,700	47,600	66,200	74,000

(Source: Reno/Sparks, Nev., Regional Transportation Commission for McCarren Blvd. East at Prater Way.)

    a. Align the input data as the number of years since 2000. Observe a scatter plot of the data. Estimate the inflection point and horizontal asymptote values suggested by the scatter plot.

b. Find a logistic model for the data. How well does it match the behavior indicated by the scatter plot? What is the limiting value of this logistic function?

c. Shift the output data down by 33,000. Find a logistic model for the shifted data. How well does this new equation fit the shifted data?

d. Use the model in part c to write a new model for traffic counts. What are the two horizontal asymptotes of this new function? What does this model assume about the population of the surrounding metropolitan area?

e. According to the model in part d, what will the daily traffic counts be in 2005? In what year will traffic counts reach 60,000 vehicles per day?

25. **Population** The population of Iowa for selected years between 1987 and 1997 is given in the table.

Year	Population (thousands)	Year	Population (thousands)
1987	2767	1993	2820
1989	2771	1995	2841
1991	2791	1997	2852

(Source: *Statistical Abstract*, 1998.)

a. Align the input as the number of years since 1980, and shift the output data down by 2700. Observe a scatter plot of the aligned and shifted data. Why is a logistic model appropriate?

b. Find a logistic model for the aligned and shifted data. Discuss how well the equation fits these data.

c. Subtract 60 from the shifted output data in part a, and find a logistic model for these new data. How well does this equation fit these data?

d. Using the model in part c, write a model for the population of Iowa from 1987 through 1997. What are the two horizontal asymptotes of this logistic function?

26. **Postage** The data in the accompanying table show how postal rates increased through the year 2000 for first-class letters weighing up to 1 ounce.

Year of rate increase	Rate (cents)
1958	4
1963	5
1968	6
1971	8
1974	10
1975	13
1978	15
1981	20
1985	22
1988	25
1991	29
1995	32
1999	33
2000	34

a. Find a logistic model for the postal rate data. How well does the equation fit the data? What is the limiting value of this function?

b. Shift the postal rates down by 3, and find a logistic model for the shifted data. Does this equation fit the shifted data better than the equation in part a fits the original data?

c. Use the model in part b to write a model for postal rates. What are the two horizontal asymptotes of this function? Do you believe the model accurately describes what will happen to postal rates in the future?

27. **Birthweight** The accompanying table shows the percentage of babies born in 2000 with low birthweight (less than 5 pounds, 8 ounces) as a function of the number of pounds the mother gained during pregnancy.

a. Find an exponential model for the data. Examine a graph of the equation on a scatter plot.

b. Shift the percentages down by 5, and find an exponential model for the shifted data. Examine a graph of this model on a scatter plot of the shifted data. Does this equation appear to fit the shifted data better than the equation in part a fits the original data?

Weight gain of mother (pounds)	Low birthweight babies (percent)
18	10.4
23	8.0
28	6.5
33	5.5
38	5.2
43	5.2

(Source: *National Vital Statistics Reports*, vol. 50, no. 5, February 12, 2002.)

c. Using the model in part *b*, write a model for low birthweight babies. Use the model to estimate the percentage of babies born with low birthweights to mothers who gain only 14 pounds during pregnancy.

28. Rearing Children The table below gives, for selected years, the percentage of children living with their grandparents.

Year	1970	1980	1993	1997	2000
Percent	3.2	3.6	5.0	5.5	6.3

(Source: U.S. Bureau of the Census.)

a. Find an exponential model for the data. Examine the graph of the function on a scatter plot of the data. How well does the equation appear to fit the data?

b. Shift the percentages down by 3, and find an exponential model for the shifted data. How does the fit of this equation compare with the fit of the equation in part *a*? What are the limits of this shifted function as the input increases and decreases without bound?

c. Using the model in part *b*, write a model for the percentage of children living with their grandparents. According to this model, when will the output reach 8%? Do you believe the model accurately reflects the future trend of this percentage?

29. Rearing Children The numbers of children living with their grandparents during selected years between 1970 and 1997 are as shown in the table.

Year	1970	1980	1992	1997
Number (millions)	2.2	2.3	3.3	3.9

(Source: U.S. Bureau of the Census.)

a. Find an exponential model for these data. How well does the equation appear to fit the data?

b. By vertically shifting the data, find a better exponential model for the number of children living with their grandparents.

30. Describe the graph of a logistic function using the words *concavity, inflection,* and *increasing/decreasing.*

31. Using the idea of limits, describe the end behavior of the logistic model, and explain how this end behavior differs from that of the exponential and log models.

2.4 Polynomial Functions and Models

Polynomial functions and models have been used extensively throughout the history of mathematics. Their successful use stems from both their presence in certain natural phenomena and their relatively simple application. Prior to the availability of inexpensive computing technology, ease of use made polynomials the most widely applied models because they were often the only ones that could be calculated with pencil-and-paper techniques. However, we no longer have to deal with this restriction, and we call on polynomials only when they are appropriate.

Even though higher-degree polynomials are useful in some situations, we limit our discussion of polynomial functions and models to the linear, quadratic, and cubic cases.

Quadratic Modeling

A large roofing company in Miami keeps track of the number of roofing jobs it completes each month. The data from January through June follow.

Month	January	February	March	April	May	June
Number of Jobs	117	140	224	368	575	842

The first differences and percentage differences are not close to being constant, so we conclude that some model other than a linear model or an exponential model is appropriate in this case. Note, however, that the differences between the first differences are nearly constant. We call these **second differences.**

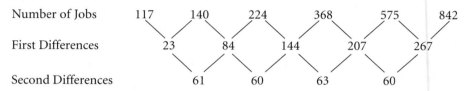

When first differences are constant, the data can be modeled by a linear equation. When second differences are constant, the data can be modeled by the **quadratic function** $f(x) = ax^2 + bx + c$ as long as $a \neq 0$. A quadratic equation of this form defines a function f with input x. From now on, when we refer to a quadratic equation or function, we will consider an equation of the form $f(x) = ax^2 + bx + c$.

The graph of a quadratic function is a **parabola.** By examining a scatter plot of the roofing company data with January aligned as month 1, we observe a plot (see Figure 2.29) that suggests a portion of the familiar parabolic shape.

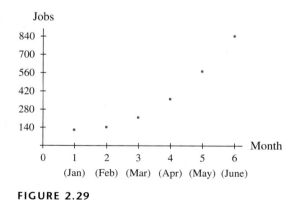

FIGURE 2.29

For this illustration, the rate of change in roofing jobs measures how rapidly the number of jobs is increasing (or possibly decreasing if the trend changes) each month. Unlike the case for a linear function, the rate of change for a quadratic function is *not* constant. We later discuss in detail the rate of change of a quadratic function.

2.4.1a, b

How do we obtain a quadratic model in this situation? After noting that the second differences are nearly, but not quite, constant and observing that the scatter plot suggests that the data are close to quadratic, we use technology to obtain a quadratic equation that fits the data. You should find the quadratic model to be

$$J(x) = 30.571x^2 - 69.029x + 155.6 \text{ jobs}$$

where $x = 1$ in January, $x = 2$ in February, and so on. Note in Figure 2.30 that the function J provides an excellent fit to the data.

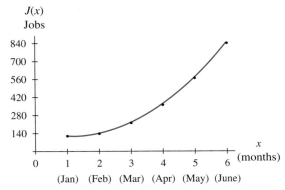

FIGURE 2.30 Quadratic model

If we are willing to assume that this quadratic function continues to model the number of roofing jobs for the next 3 months, how many jobs would we predict the company will have in August? Substituting $x = 8$ into the quadratic equation yields

$$J(8) = 30.571(8)^2 - 69.029(8) + 155.6 = 1559.943 \text{ jobs}$$

Recall from the numerical guidelines established in Section 1.4 that output obtained from a model can be only as accurate as the output data from which the model was obtained. Because the roofing job data were reported as integers, we must report the August prediction as 1559 jobs.

EXAMPLE 1 *Finding a Quadratic Model*

Birthweight The percentage of low birthweight babies born before 37 weeks gestation as a function of the amount of weight gained by the mother is given in Table 2.15.

TABLE 2.15

Weight gain of mother (pounds)	18	23	28	33	38	43
Percentage of babies born before 37 weeks weighing less than 5 pounds 8 ounces	48.2	42.5	38.6	36.5	35.4	35.7

(Source: *National Vital Statistics Report*, vol. 50, no. 5, February 12, 2002.)

a. Find a quadratic model for the data.

b. Compare the minimum of the parabola with the minimum of the data.

Solution

a. A scatter plot of the data suggests a concave-up shape with a minimum around 38 pounds. See Figure 2.31.

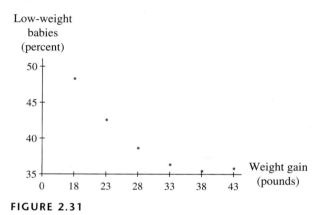

FIGURE 2.31

A quadratic model for the data is

$$P(g) = 0.0294g^2 - 2.286g + 79.685 \text{ percent}$$

when the mother's weight gain is g pounds.

b. Looking at the quadratic function graphed on the scatter plot (see Figure 2.32), we observe that the minimum of the parabola is slightly to the right and below the minimum data point. This means that the model slightly underestimates the minimum percentage and estimates that the minimum occurs slightly after it does in the data table.

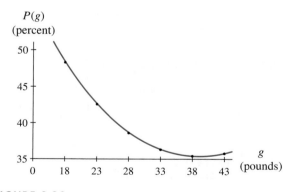

FIGURE 2.32

In the foregoing examples, you have seen data sets that appear to be quadratic. That is, they may be reasonably modeled by a quadratic equation. In each case, the equation had a positive coefficient before the squared term. This value is called the **leading coefficient** because it is usually the first one that you write down in the equation $f(x) = ax^2 + bx + c$. Also, in each case, the graph of the equation appeared to be part of a parabola opening upward. Remember that we call such curvature *concave up*.

Table 2.16 gives the population of Cleveland, Ohio, from 1900 through 1980.

TABLE 2.16

Year	1900	1910	1920	1930	1940
Population	381,768	560,663	796,841	900,429	878,336
Year	1950	1960	1970	1980	
Population	914,808	876,050	750,879	573,822	

(Source: *Statistical Abstract*, 1998.)

A scatter plot of the data suggests a parabola opening downward (or *concave down*). The scatter plot and the graph of a quadratic equation fitted to the data are shown in Figure 2.33.

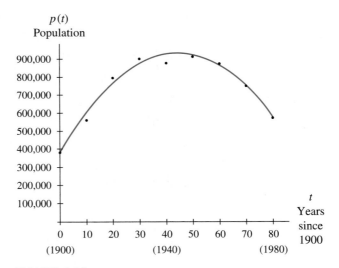

FIGURE 2.33

Would you expect the leading coefficient of the equation to be positive or negative? The population of Cleveland is given by

$$p(t) = -279.995t^2 + 24,919.057t + 374,959.903 \text{ people}$$

where t is the number of years since 1900. When the graph of a quadratic function is concave up, its leading coefficient is positive; when the graph of a quadratic function is concave down, its leading coefficient is negative. Curvature will be important in later discussions that involve concepts of calculus.

We summarize quadratic models as follows:

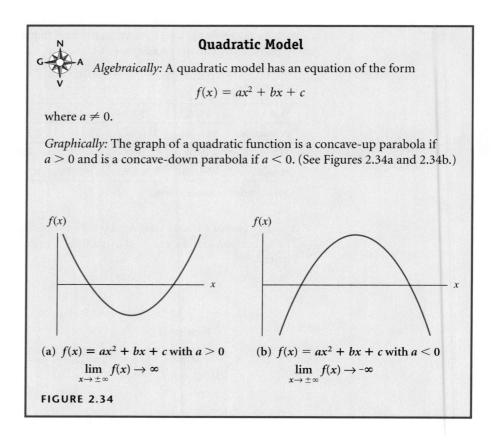

Quadratic Model

Algebraically: A quadratic model has an equation of the form

$$f(x) = ax^2 + bx + c$$

where $a \neq 0$.

Graphically: The graph of a quadratic function is a concave-up parabola if $a > 0$ and is a concave-down parabola if $a < 0$. (See Figures 2.34a and 2.34b.)

(a) $f(x) = ax^2 + bx + c$ with $a > 0$
$$\lim_{x \to \pm\infty} f(x) \to \infty$$

(b) $f(x) = ax^2 + bx + c$ with $a < 0$
$$\lim_{x \to \pm\infty} f(x) \to -\infty$$

FIGURE 2.34

Quadratic or Exponential?

Data sets that exhibit an obvious maximum or minimum are more easily identified as quadratic than data sets without a maximum or minimum. Sometimes all that is indicated by a scatter plot is the left side or right side of a parabola.

You may have noticed that the scatter plot of the roofing job data (Figure 2.29) looks as if an exponential curve might fit, even though the percentage differences are not constant. If you found an exponential model, it would be

$$H(x) = 68.3985(1.5177^x) \text{ jobs}$$

where $x = 1$ in January, $x = 2$ in February, and so on. A graph of the exponential function H on the scatter plot is shown in Figure 2.35. The exponential equation and the quadratic equation (Figure 2.30) both appear to fit the roofing job data well.

In cases in which two equations appear to fit a set of data equally well, you need to consider carefully which one you would prefer to use. (See the hints for model selection in Section 2.5.) Often, however, it is obvious that one model is more appropriate than the other. This is the case in Example 2.

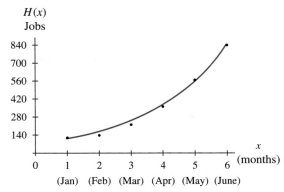

FIGURE 2.35 **Exponential model**

EXAMPLE 2 *Distinguishing Between Quadratic and Exponential Models*

Population Table 2.17 shows the population of the contiguous states of the United States for selected years between 1790 and 1930.

TABLE 2.17

Year	Population (millions)	Year	Population (millions)
1790	3.929	1870	39.818
1810	7.240	1890	62.948
1830	12.866	1910	91.972
1850	23.192	1930	122.775

(Source: *Statistical Abstract,* 1998.)

Find an appropriate model for the data.

Solution An examination of the scatter plot shows an increasing, concave-up shape (see Figure 2.36).

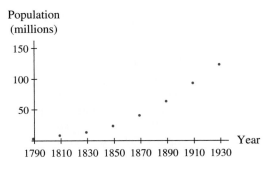

FIGURE 2.36

At first, it seems logical to try an exponential model for the data. An exponential model is

$$E(t) = 4.57753(1.02538^t) \text{ million people}$$

where t is the number of years since 1790. However, this equation does not seem to fit the data very well, as Figure 2.37 shows.

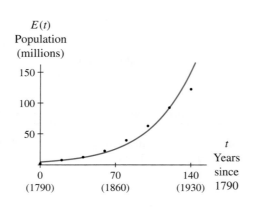

FIGURE 2.37   **Exponential model**

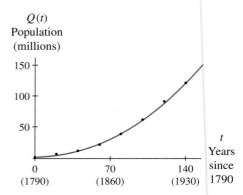

FIGURE 2.38   **Quadratic model**

Another possibility is a quadratic model. The right half of a parabola that opens upward could fit the scatter plot. A quadratic model is

$$Q(t) = 0.00660t^2 - 0.07757t + 4.80892 \text{ million people}$$

where t is the number of years since 1790. When graphed on the scatter plot, this equation appears to be a very good fit (see Figure 2.38). It is the more appropriate model for the population of the contiguous United States using the given data. ●

Cubic Modeling

We saw that when the first differences of a set of evenly spaced data are constant, the data can be modeled perfectly by the linear equation $y = ax + b$. Likewise, when the second differences of evenly spaced input data are constant, the data can be modeled perfectly by the quadratic equation $y = ax^2 + bx + c$. It is possible for the third differences to be constant. In this case, the data can be modeled perfectly by a cubic equation. A cubic equation of the form $f(x) = ax^3 + bx^2 + cx + d$ is a function with input x. From now on, when we refer to a cubic equation or function, we consider the equation to be of this form.

Because in the real world we are extremely unlikely to encounter data that are perfectly cubic, we will not look at third differences. Instead, we will examine a scatter plot of the data to see whether a cubic model may be appropriate. Figure 2.39 shows the graphs of some cubic equations.

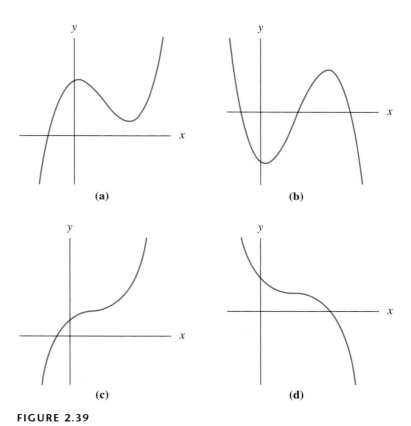

FIGURE 2.39

The four graphs in Figure 2.39 are typical of all cubic equations. That is, every cubic function $f(x) = ax^3 + bx^2 + cx + d$ with $a \neq 0$ has a graph that resembles one of these four. Figures 2.39a and 2.39c correspond to equations in which $a > 0$, and Figures 2.39b and 2.39d are graphs of equations in which $a < 0$. For a cubic equation $f(x) = ax^3 + bx^2 + cx + d$ with $a \neq 0$, the end behavior shows that $f(x)$ increases without bound in one direction and decreases without bound in the opposite direction.

Figure 2.40 shows scatter plots of data sets that could be modeled by cubic equations.

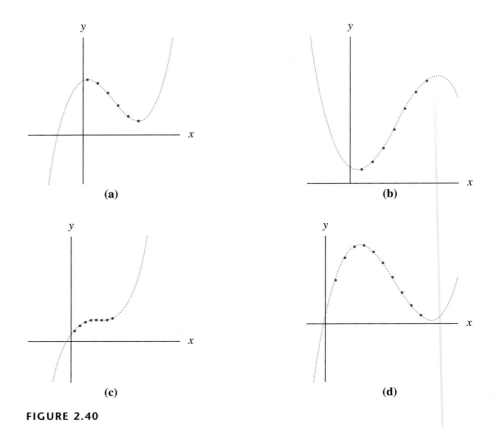

FIGURE 2.40

You may have already noticed that in every cubic function, the curvature of the graph changes once from concave down to concave up, or vice versa. As we noted with the graph of a logistic curve, the point on the graph at which concavity changes is called the *inflection point*. All cubic functions have one inflection point. The approximate location of the inflection point in each of the graphs in Figure 2.41 is marked with a dot. In Chapter 5, we see how calculus can be used to determine the exact location of the inflection point of a cubic function.

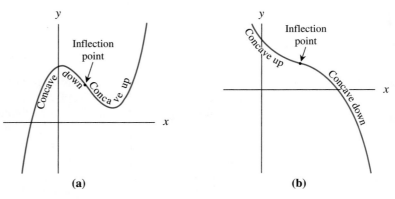

FIGURE 2.41

It is often the case that a portion of a cubic function appears to fit extremely well a set of data that can be adequately modeled with a quadratic function. In an effort to keep things as simple as possible, we adopt the following convention:

> If the scatter plot of a set of data fails to exhibit an inflection point, then it is not appropriate to fit a cubic equation to the data.

We must be extremely cautious when using cubic models to extrapolate. For the data sets whose scatter plots are shown in Figures 2.40a and 2.40c, the functions indicated by the dotted curves appear to follow the trend of the data. However, in Figure 2.40b, it would be possible for additional data to level off (as in a logistic model), whereas the cubic function takes a downward turn. Also, additional data in Figure 2.40d might continue to get closer to the *x*-axis, whereas the cubic function that is fitted to the available data begins to rise.

Cubic Model

Algebraically: A cubic model has an equation of the form

$$f(x) = ax^3 + bx^2 + cx + d$$

where $a \neq 0$.

Graphically: The graph of a cubic function has one inflection point and no limiting values. (See Figure 2.39b.)

EXAMPLE 3 *Finding a Cubic Model*

2.4.2

Gas Price The average price in dollars per 1000 cubic feet of natural gas for residential use in the United States for selected years from 1980 through 2000 is given in Table 2.18.

TABLE 2.18

Year	Price (dollars)	Year	Price (dollars)
1980	3.68	1995	6.06
1982	5.17	1998	6.82
1985	6.12	2000	7.71
1990	5.80		

(Source: *Statistical Abstract*, 1992 and 2001.)

a. Find an appropriate model for the data. Would it be wise to use this model to predict future natural gas prices?

b. Use the model in part *a* to estimate the price in 1993.

c. According to the model, when did the price of 1000 cubic feet of natural gas first exceed $6.00?

Solution

a. An examination of a scatter plot shows that a cubic model is appropriate. Note that the scatter plot shown in Figure 2.42 appears to be mostly concave down between 1980 and 1990 but then is concave up between 1990 and 2000. That is, there appears to be an inflection point (a change of concavity) near 1990.
A cubic model for the price of natural gas is

$$P(x) = 0.00276x^3 - 0.0853x^2 + 0.8034x + 3.776 \text{ dollars}$$

where *x* is the number of years since 1980. A graph of the equation over the scatter plot is shown in Figure 2.43.
Note that the graph is increasing to the right of about 1993. Natural gas prices will probably not continue to rise indefinitely as the cubic function does, so it is unwise to use the model to predict future prices of natural gas. Additional data should be obtained to see the pattern past 2000.

b. In 1993 the average price of 1000 cubic feet of natural gas was $P(13) \approx \$5.87$.

c. To determine when the average price first exceeded $6.00, solve the equation $P(x) = 6$. In Figure 2.44, a dotted line is drawn at $P(x) = \$6.00$. The line intersects the graph at the three places, so there are three solutions to the equation $P(x) = 6$. The solutions are $x \approx 4.96, 10.58,$ and 15.36.

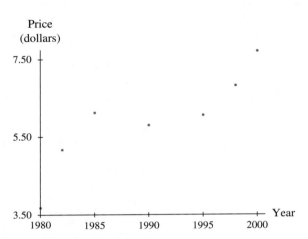

FIGURE 2.42

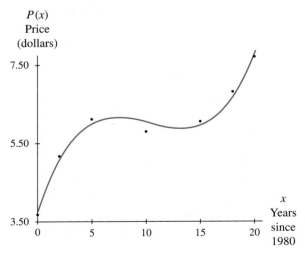

FIGURE 2.43

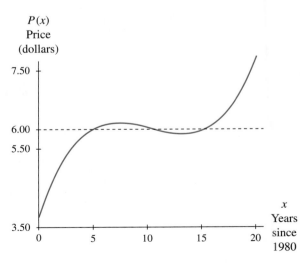

FIGURE 2.44

We seek the smallest solution ($x \approx 4.96$), which corresponds to the first time the average price is $6.00. However, because the data represent yearly averages, we must interpret the model discretely. In 1984 the average price was less than $6.00, and in 1985 the average price was slightly more than $6.00. Therefore, the average price first exceeded $6.00 in 1985. ●

We can also model the natural gas prices in Example 3 with a cubic equation by renumbering the years so that x is the number of years since 1900:

$$\text{Price} = 0.00276x^3 - 0.7473x^2 + 67.4043x - 2018.3896 \text{ dollars}$$

Although the two models have different inputs, they yield the same results. Notice that the coefficient on x^3 is the same in both models. This will be the case for any alignment.

2.4 Concept Inventory

○ *Second differences*

○ *Quadratic function:* $f(x) = ax^2 + bx + c$ *(constant second differences, no change in concavity)*

○ *Parabola*

○ *Cubic function:* $f(x) = ax^3 + bx^2 + cx + d$ *(one change in concavity)*

○ *Inflection point*

2.4 Activities

Identify the curves in Activities 1–6 as concave up or concave down. In each case, indicate the portion of the horizontal axis over which the part of the curve that is shown is increasing or decreasing.

1.

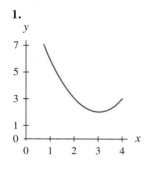

2.

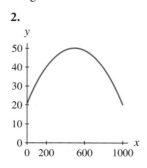

3.

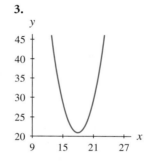

4.

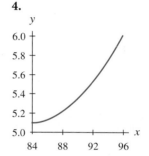

5.

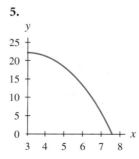

6.

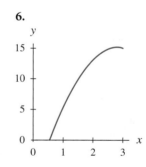

7. Missile Height During a training mission in the South Pacific, a Tomahawk cruise missile misfires. It goes over the side of the ship and hits the water. Suppose the data showing the height of the missile are collected via telemetry as shown in the accompanying table.

a. Without graphing, show that the data in the table are quadratic.

b. Without finding an equation, complete the accompanying table by filling in the missing values.

c. Find a quadratic model for the complete data.

Seconds from launch	Feet above water
0	128
0.5	140
1	144
1.5	140
2	128
2.5	108
3	80
3.5	
4	

d. Use the model to determine when the missile hits the water.

8. **Sales**   A discount store has calculated from past sales data the weekly revenue resulting from the sale of blenders. These amounts are shown in the accompanying table.

Number of blenders sold	Revenue (dollars)
4	114
7	247
10	356
13	418
16	451
19	446
22	412

a. Why does selling 19 blenders result in less revenue than selling 16 blenders?

b. Examine a scatter plot of the data in the table, and find a quadratic model for the data.

c. What is the revenue when 17 blenders are sold? 18 blenders?

d. How could the data in the table and the answers in part *c* be beneficial to the manager of the store?

9. **Jobs**   Suppose that the roofing company discussed in this section operates in Michigan instead of Miami.

a. Using the quadratic model

$$J(x) = 30.571x^2 - 69.029x + 155.6 \text{ jobs}$$

where $x = 1$ in January, $x = 2$ in February, and so on, determine the number of jobs in December.

b. Do you feel this answer is reasonable? Why or why not?

c. Would you answer to part *b* be different if the roofing company operated in your home state?

10. **Profit**   The accompanying table gives the price in dollars of a round-trip ticket from Denver to Chicago on a certain airline and the corresponding monthly profit, in millions of dollars, for that airline.

Ticket price (dollars)	Profit (millions of dollars)
200	3.08
250	3.52
300	3.76
350	3.82
400	3.70
450	3.38

a. Is a quadratic model appropriate for the data? Explain.

b. Find a quadratic model for the data.

c. As the ticket price increases, the airline should collect more money. How can it be that when the ticket price reaches a certain amount, profit decreases?

d. At what ticket price will the airline begin to post a negative profit (a net loss)?

11. **Marriage Age**   As listed in *The 1999 World Almanac,* the median age (in years) at first marriage of females in the United States is shown in the accompanying table.

Year	1960	1970	1980	1990
Age	20.3	20.8	22.0	23.9

a. Refute or defend the following statement: "The data are perfectly quadratic."

b. Without finding an equation, estimate the age that corresponds to the year 2000.

c. Find a quadratic model for the data.

d. Use the model to estimate the age in the year 2000. Is it the same as your answer to part *b*?

WEB/CD **e.** Find the most current data available for the median age at first marriage of women. Compare the model value with the actual value.

Source

f. When you add the new data to the table, is a quadratic model still appropriate? Explain.

12. **Profit** The monthly profit *P* (in dollars) from the sale of *x* mobile homes at a dealership is given by the accompanying table.

x	*P* (dollars)	*x*	*P* (dollars)
7	43,700	14	63,000
8	48,000	15	63,750
11	57,750	19	61,750
13	61,750	22	55,000

Examine a scatter plot of the data. Find a quadratic model to fit the data, and graph it on the scatter plot. Does it appear that the quadratic function is a good fit?

13. **Cost** A factory makes 7-millimeter aluminum ball bearings. Company analysts have determined how much it costs to make certain numbers of ball bearings in a single run. These costs are shown in the accompanying table.

Number of ball bearings	Cost (dollars)
500	3.10
1000	4.25
3000	8.95
4500	12.29
7000	18.45

a. Should the data be modeled by using a linear or a quadratic equation? What is the model?

b. How much is the overhead for a single run?

c. How much will it cost to make 5000 ball bearings?

d. Ball bearings are made in sets of 100. If the company is planning to make 5000 ball bearings in a run, how much more would it cost them to make one extra set of 100 in that same run? What term is used in economics to express this idea?

e. Ball bearings are sold in cases of 500. Rewrite the cost equation in terms of the number of cases of ball bearings.

14. **Sales** The factory discussed in Activity 13 sells ball bearings in cases of 500. It charges the prices shown in the table.

Number of cases	Price (dollars)
1	78.47
10	168.85
20	263.83
100	817.88
200	999.96

a. Find a quadratic model for the data.

b. According to the model, how much revenue is made on 10 cases of ball bearings?

c. Using the cost model found earlier, $C(u) = 1.17627u + 1.87986$ where *u* is the number of cases of ball bearings, and the revenue model you just found, write an expression to represent the profit made on *u* cases of ball bearings.

d. How much profit does the factory make on the sale of 500, 1000, and 10,000 ball bearings, respectively?

e. How many ball bearings must be sold to make a profit of $800?

f. How many ball bearings must the factory sell at one time to a single customer before it no longer makes a profit from that sale?

15. **Veggies** The per capita utilization of commercially produced fresh vegetables in the United States from 1980 through 2000 is as shown in the table.

Year	Vegetable consumption (pounds per person)
1980	149.1
1985	156.0
1990	167.1
1995	179.1
2000	201.7

(Sources: *Statistical Abstract*, 2001; and **www.ers.usda.gov.** Accessed 9/25/02.)

a. Find a quadratic model for the data, and examine the equation graphed on a scatter plot of the data.

b. Do you believe the equation in part *a* is a good fit? Explain.

c. The per capita consumption in 2001 had not yet been tabulated when the data in the table was published. What does the quadratic model give as the per capita consumption in 2001? Do you believe that this estimate is reliable?

d. According to your model, in what year will consumption exceed 225 pounds per person?

WEB/CD e. Do you believe this model gives a good or a poor representation of what will happen? Support your answer with more current data.

Source

16. **Mortality** The accompanying table lists the death rates (number of deaths per thousand people whose age is *x*) in 1998 for the United States.

x	Death rate (deaths per thousand people)
40	2.0
45	3.0
50	4.4
55	6.7
60	10.7
65	16.5

(Source: *Statistical Abstract*, 2001.)

a. Find a quadratic model for the data in the table. Discuss how well the equation fits the data.

b. Use the model to complete the following table:

Age	Model prediction	Actual rate
51		4.7
52		5.1
53		5.6
57		8.1
59		9.7
63		14.1
70		25.5
75		38.0
80		59.2

c. What can you conclude about using a model to make predictions?

17. **Lead Paint** Lead was banned as an ingredient in most paints in 1978, although it is still used in some specialty paints. Lead usage in paints from 1940 through 1980 is as shown in the accompanying table.

Year	Lead usage (thousands of tons)
1940	70
1950	35
1960	10
1970	5
1980	0.01

(Source: Estimated from information in "Lead in the Inner Cities," *American Scientist*, January–February 1999, pp. 62–73.)

a. Examine a scatter plot of the data. Find quadratic and exponential models for lead usage. Comment on how well each equation fits the data.

b. What is the end behavior suggested by the data? Which function in part *a* has the end behavior suggested by the context?

c. Which model would be more appropriate to use to estimate the lead usage in 1955? Use that model to estimate the usage in 1955.

18. From a magazine, newspaper, or some other source, collect data that you feel may be modeled by one of the models we have studied. Find a model, and discuss why you chose the model and how well it fits the data. Do you feel that predictions based on this model are likely to be realistic? Why or why not?

For the graphs in Activities 19 through 24, describe the curvature by indicating the portions of the displayed horizontal axis over which each curve is concave down or concave up. Mark the approximate location of the inflection point on each curve.

19.

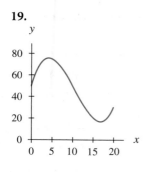

20.

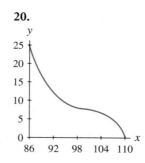

21.

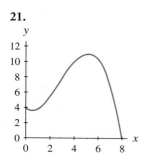

22.

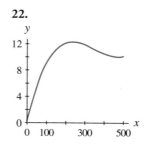

23.

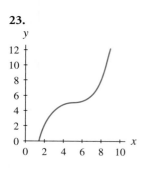

24.

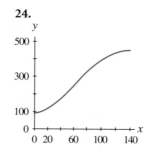

25. Education The following table shows the amounts spent on reducing sizes of first-grade through third-grade public school classes in a certain state.

Year	Amount (millions)
1990	$3
1992	$31
1994	$37
1996	$42
1998	$66

(Source: Nevada Department of Education.)

a. Examine a scatter plot of the data. How does a scatter plot indicate that a cubic model might be appropriate?

b. Find a cubic model for the amount spent on class size reduction.

c. Use the model to estimate the amounts in 1993 and 1999. In which of these estimates can you have more confidence?

d. Compare the estimates from part *c* with the actual amounts of $34 million spent in 1993 and $99 million spent in 1999. Does this comparison support the statements you made in part *c* concerning the reliability of the estimates?

26. Births The numbers of live births in the United States for selected years between 1950 and 2000 to women 45 years of age and older are as given in the accompanying table.

Year	Births	Year	Births
1950	5322	1980	1200
1960	5182	1990	1638
1970	3146	2000	4604

(**Source: www.infoplease.com.** Accessed 9/24/02.)

a. Examine a scatter plot of the data in the table. Find a cubic model for the data, and graph the equation on the scatter plot. Discuss how well the equation fits the data.

b. What trend does the model indicate before 1940 and beyond 2000? Do you believe the trends are valid? Explain.

c. Use the model to estimate the number of live births in 1940. The actual number was 7558. How close is the model estimate?

d. Do you believe your model would be an accurate predictor of the number of live births to women 45 years of age and older for the current year? Why or why not?

27. Gender Ratio The table on the next page shows the number of males per 100 females in the United States calculated using census data. This number is referred to as the *gender ratio*.

Year	Males per 100 females
1900	104.6
1910	106.2
1920	104.1
1930	102.6
1940	100.8
1950	98.7
1960	97.1
1970	94.8
1980	94.5
1990	95.1
2000	96.3

(Source: U.S. Bureau of the Census.)

a. Examine a scatter plot of the data. What behavior of the data suggests that a cubic model is appropriate?

b. Find a cubic model for the gender ratio. What trend does the model indicate for the years beyond 2000? Do you believe this is a good predictor of future gender ratios? Why or why not?

c. What factors might have contributed to changes in the gender ratio in this country?

28. **Cost**   The following table shows a manufacturer's total cost (in hundreds of dollars) to produce from 1 to 33 forklifts per week.

Weekly production	Total cost (hundreds of dollars)
1	18.5
5	80
9	125
13	160
17	185
21	210
25	225
29	245
33	280

a. Examine a scatter plot of the data in the table. What characteristics of the scatter plot indicate that a cubic model would be appropriate?

b. Find a cubic model for total manufacturing cost.

c. What does the model predict as the cost to produce 23 forklifts per week? 35 forklifts per week?

d. Convert the cubic equation in part b for total cost to one for average cost.

e. Graph the average cost function.

f. How could the average cost graph help a production manager make production decisions?

29. **Gender Ratio**   The gender ratio in the United States is different for different age groups. The following table shows gender ratios corresponding to age that were calculated from the 2000 census data.

Age	Males per 100 females
under 1	105.0
10	105.2
20	104.7
30	102.8
40	99.7
50	96.6
60	92.0
70	81.9
80	63.1
90	37.6
100 and over	24.9

(Source: U.S. Bureau of the Census.)

a. According to the data, for what age are the numbers of males and females equal?

b. Find cubic and logistic models for the gender ratio as a function of age. Compare the fit of the two equations. Which equation do you believe is better for modeling these data?

c. Use the model you chose in part b to find the age for which there are twice as many women as men. What does this information tell you about death rates of men and women?

30. **Births**   The accompanying table gives the Cesarean delivery rate per 100 live births in the United States from 1990 through 2000.

a. Examine a scatter plot of the data in the table, and discuss its curvature. Should the data be

Year	Cesarean delivery rate (per 100 live births)
1990	22.7
1992	22.3
1994	21.2
1996	20.7
1998	21.2
2000	22.9

(Source: *National Vital Statistics Report,* vol. 50, no. 5, February 12, 2000.)

modeled by a quadratic or a cubic equation? What is the model?

b. What does the model estimate as the Cesarean delivery rate in the United States in 1989 and 1999? Compare the model estimates to the actual rates of 22.8 births per 100 in 1989 and 22.0 births per 100 in 1999.

c. Discuss why you feel the model you chose would or would not be a good estimator of the Cesarean delivery rate in the United States in the year 2005.

WEB/CD d. Find more recent data on Cesarean delivery rates, and add the values you found to the table. Is the model you chose appropriate for the expanded data set? Explain.

Source

31. Earnings   The net earnings of Gap Inc. from 1992 through 2001 is shown in the accompanying table.

Year	Net earnings (millions of dollars)
1992	211
1993	258
1994	320
1995	354
1996	453
1997	534
1998	825
1999	1127
2000	877
2001	−8

(Source: Gap Inc., *Annual Report 2001.*)

a. Observe a scatter plot of the data in the table, and explain why a piecewise model is needed to describe this data set.

b. Divide the data into two subsets, each sharing a common data point. Find an equation for each subset of data, and combine the equations to form a piecewise continuous model.

c. Use your model to estimate the net earnings of Gap Inc. in 1991 and in 2002.

d. List as many reasons as you can for the change in the net earnings after 1999.

e. What behavior does your model indicate for Gap Inc.'s earnings beyond 2001?

WEB/CD f. Find more recent data on Gap's earnings, and add it to the table. Is the model you chose appropriate for the expanded data set? Explain.

Source

32. Hospital Stay   Consider the average hospital stays for selected years between 1993 and 2000 as shown in the following table.

Year	Average stay (days)	Year	Average stay (days)
1993	6.0	1997	5.1
1994	5.7	1998	5.1
1995	5.4	1999	5
1996	5.2	2000	4.9

(Source: U.S. National Center for Health Statistics.)

a. Observe a scatter plot of the data. Comment on the curvature suggested by the scatter plot.

b. Find a cubic model for average hospital stay. Does the curvature of a graph of the function match that of the scatter plot?

c. Divide the data into two sets: from 1993 through 1998 and from 1998 through 2000. Include the 1998 data point in both sets. Find models for each set of data, and use the models to write a piecewise continuous model for average hospital stay. How well does this piecewise function match the curvature of the data?

d. Use the model in part *c* to estimate the average hospital stay in 2002.

e. List as many factors as you can that contribute to changes in the average hospital stay.

f. What current issues in our health care system may affect average hospital stay in the future?

33. Using the terms *increasing, decreasing,* and *concave,* describe the shape of the graphs of functions of the forms $y = ax^2 + bx + c$ and $y = ax^3 + bx^2 + cx + d$.

34. Discuss how to use end-behavior analysis in determining the differences among linear and polynomial functions and exponential, log, and logistic functions.

2.5   Choosing a Function to Fit Data

In the preceding sections, we discussed several types of models that we use in this book to describe two-variable data: linear, quadratic, cubic, exponential, logarithmic, and logistic. Although most of our work so far has been concerned with applying a particular type of equation to a given data set, it is important to understand that in many real-life situations, it is not always clear which model to apply. Sometimes, none of the models is appropriate. However, there are some general, commonsense guidelines that we should keep in mind.

Shifting Data

Before we consider how to choose among the six different types of functions, recall that shifting the data may turn an ill-fitting function into one that better fits the data. This is true when a data set strongly suggests one of the models that we have studied, but the technology-generated function is a poor fit to the data.

We have discussed aligning data input values (horizontal shifts) to reduce the magnitude of coefficients in model equations and to prevent numerical computation error. This alignment is vital for exponential and logistic models and sometimes desirable for log models. Aligning the output values (vertical shifts) may be desirable for exponential and logistic models.

Some of the functions that we have considered have both vertical and horizontal shifts built into them; that is, if they fit a certain set of data well, then they will fit those data equally well when the data are shifted vertically or horizontally. This is true of all the polynomial functions—linear, quadratic, and cubic. We list the functions we have studied, along with the types of shifts for which they account.

Function		*Shift*
Polynomial functions:		
Linear	$f(x) = ax + b$	Horizontal and vertical
Quadratic	$f(x) = ax^2 + bx + c$	Horizontal and vertical
Cubic	$f(x) = ax^3 + bx^2 + cx + d$	Horizontal and vertical
Log function	$f(x) = a + b \ln x$	Vertical
Exponential function	$f(x) = ab^x$	Horizontal
Logistic function	$f(x) = \dfrac{L}{1 + Ae^{-Bx}}$	Horizontal

The addition of a constant term in a model indicates that the function accounts for a vertical shift. Note that the polynomial and log functions all have built-in vertical shifts. The exponential and logistic functions account for horizontal shifts. As you proceed through the following discussion, remember that you may need to shift those functions that do not account for both horizontal and vertical shifts.

Examining Scatter Plots

The first step in determining which model to use for a given set of data is to *examine a scatter plot carefully* with an eye toward its general underlying shape relative to the shapes of the functions that you know. Try to imagine a smooth curve fitted to the scatter plot:

1. Does this curve appear to be a straight line? If so, try a linear model.

2. If the curve does not appear to be a line, then as you look from left to right, does the curve appear to be always concave up or concave down? If so, then a quadratic, exponential, or log model may be appropriate.

3. What if the curve does not appear to be a line and is not always concave in one direction (up or down)? If there is a single change in concavity, then a cubic or logistic model may be suggested because the graphs of these functions have a concavity change (inflection point).

4. In the case of exponential and logistic models, vertically shifting the data (by aligning the output values) may produce a better-fitting equation. Horizontally shifting data (by aligning the input values) may produce a better-fitting log equation.

5. If the data demonstrate a dramatic change in behavior, then it may be appropriate to combine two or more functions to form a piecewise continuous model.

After taking into account the five considerations above, suppose that you have narrowed your choices to two models. Which one should you choose? If one equation fits the data significantly better than the other one, it should be chosen. If the two equations appear to fit the data equally well, then the key is to examine the end behavior of the scatter plot and to see whether limiting values are suggested in the context. Remember, exponential models have one limiting value, logistic models have two limiting values, and linear, log, cubic, and quadratic models have none.

Look at the scatter plots in Figure 2.45. None of them appear to follow linear, quadratic, log, or exponential patterns. Can you determine which are cubic and which are logistic?

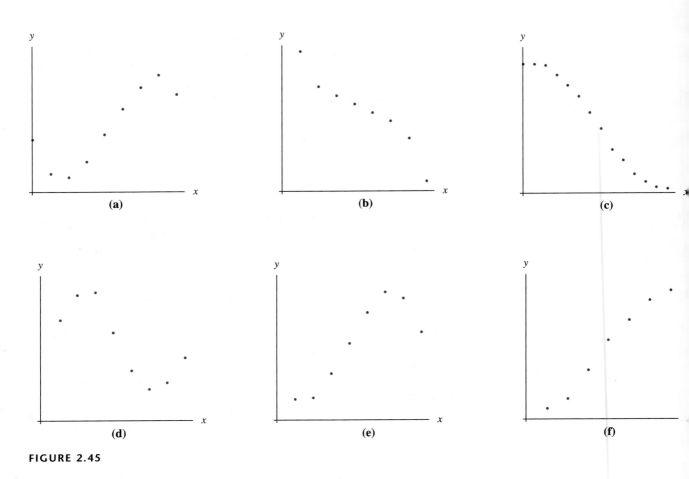

FIGURE 2.45

Plots *a*, *b*, *d*, and *e* suggest cubic functions because they do not indicate limiting values. Plots *c* and *f* could be either logistic or cubic. Here you must ask yourself two questions:

1. Are you more concerned with end behavior or with how well the equation fits the data?

2. If you are concerned with end behavior, what would the end behavior be?

If you are mainly interested in interpolating, then it would be appropriate to choose the equation that better fits the data points. However, if you wish to extrapolate or examine end behavior, you should choose the equation that you think would better follow the end behavior pattern, provided that the fit is satisfactory. In general, if the context or scatter plot suggests that the data have two limiting values, then use a logistic model. If the data are likely to rise or fall again, then a cubic model would be appropriate.

Consider the scatter plots in Figure 2.46, and see whether you can determine which types of functions are appropriate.

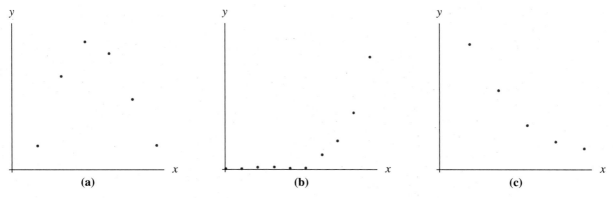

FIGURE 2.46

Plot *a* seems to follow a quadratic pattern, whereas plot *b* appears to be exponential. Plot *c* could be exponential, quadratic, or logarithmic. Again ask yourself those two questions: (1) Are you more concerned with end behavior or with how well the equation fits? (2) If you are concerned with end behavior, what would the end behavior be? Recall that a quadratic curve quickly becomes infinitely large (either negatively or positively) at each end, whereas both exponential and logarithmic curves tend quickly to infinity (or negative infinity) at one end and exhibit slow growth or decline at the other end. The exponential curve has a limiting value, whereas the log curve has no limiting value even though it exhibits increasingly slow growth or decline.

Finally, consider the two scatter plots shown in Figure 2.47.

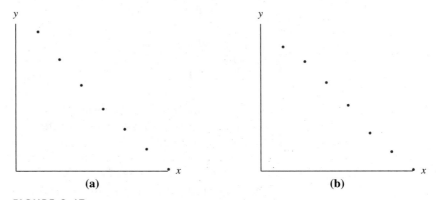

FIGURE 2.47

They both appear to be linear, so it makes sense to try linear functions. Examine lines graphed on the scatter plots, as shown in Figure 2.48, and notice that slight curvature can be seen in both cases. You may wish to try one of the other functions, but do so with caution. You may sacrifice an easy-to-use linear model without gaining much in the way of increased precision.

When you are considering higher-order polynomial models, our advice is *the simpler the better*. For instance, although scatter plots *a* and *c* in Figure 2.46 could be modeled by either a quadratic or a cubic, you should choose the quadratic. Do not consider a cubic model unless the data indicate an inflection point.

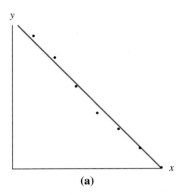

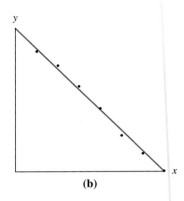

FIGURE 2.48

The Steps in Choosing a Model

The steps to employ when determining which function to use to model data are summarized below:

Steps in Choosing a Model

1. **Look at the curvature of a scatter plot of the data.**
   - If the points appear to lie in a straight line, try a linear model.
   - If the scatter plot is curved but has no inflection point, try a quadratic, an exponential, or a log model.
   - If the scatter plot appears to have an inflection point, try a cubic and/or a logistic model.
   - If the scatter plot indicates curvature not described by one of the functions we have studied, then consider combining two or more functions to form a piecewise continuous model.

   Note: If the input values are equally spaced, it might be helpful to look at first differences versus percentage differences to decide whether a linear or an exponential model would be more appropriate, or to compare second differences with percentage differences to decide whether a quadratic or an exponential model would be better.

2. **Look at the fit of the possible equations.** In Step 1, you should have narrowed the possible models to at most two choices. Compute these equations, and graph them on a scatter plot of the data. The one that comes closest to the most points (but does not necessarily go through the most points) is normally the better model to choose. Vertically shifting the data may produce a better-fitting exponential or logistic equation. Horizontally aligning the data may produce a better-fitting log equation.

3. **Look at the end behavior of the scatter plot.** If Step 2 does not reveal that one model is obviously better than another, consider the end behavior of the data, and choose the appropriate model.

4. **Consider that there may be two equally good models for a particular set or data.** If that is the case, then you may choose either.

The following general rules should be used when fitting a piecewise equation to data:

1. The data should exhibit a distinct change in behavior at an obvious "split point."

2. The data should be split so that no portion of data contains too few points.

3. The "split point" in the data set should be included in both portions of data when fitting the equations.

Let us look at some examples where we must apply these steps in choosing a model.

EXAMPLE 1   *Using Curvature and Fit to Choose a Model*

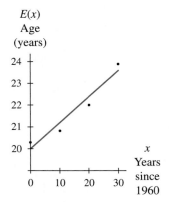

$E(x)$
Age
(years)

Exponential model
FIGURE 2.50

Age of Marriage   *The 1999 World Almanac* lists the median ages at first marriage of women in the United States as shown in Table 2.19.

TABLE 2.19

Year	1960	1970	1980	1990
Age	20.3	20.8	22.0	23.9

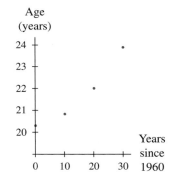

Age
(years)

FIGURE 2.49

a. Find the best model for the data.

b. Use the model in part *a* to estimate the median age in 1997. Compare this estimate with the reported age of 25.0 years.

c. Use the model in part *a* to predict the median age at first marriage of women in the year 2005.

d. Is the part *c* prediction valid?

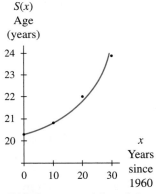

$S(x)$
Age
(years)

Shifted exponential model
FIGURE 2.51

Solution

a. *Step 1:* Consider a scatter plot of the data.

The scatter plot shown in Figure 2.49 is increasing and concave up, so we should try either an exponential model or a quadratic model.

Step 2: Compare the fit of the equations.

An exponential model for the data is

$$E(x) = 20.00024(1.00547^x) \text{ years of age}$$

where x is the number of years after 1960. However, when graphed on the scatter plot, it is obviously a poor fit to the data (see Figure 2.50). If we shift the data down by 20, we obtain the model

$$S(x) = 0.32145(1.08993^x) + 20 \text{ years of age}$$

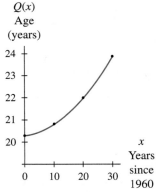

$Q(x)$
Age
(years)

Quadratic model
FIGURE 2.52

x years after 1960. This exponential equation is a much better fit to the data than the original one. See Figure 2.51. Other vertical shifts may result in even better-fitting equations.

A quadratic model for the data is

$$Q(x) = 0.0035x^2 + 0.015x + 20.3 \text{ years of age}$$

where x is the number of years since 1960. As Figure 2.52 shows, this model fits the data perfectly.

We could have avoided the process of comparing exponential and quadratic models if we had first calculated second differences and percentage differences.

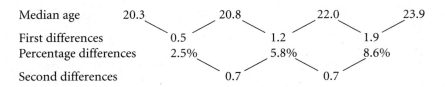

Median age	20.3		20.8		22.0		23.9
First differences		0.5		1.2		1.9	
Percentage differences		2.5%		5.8%		8.6%	
Second differences			0.7		0.7		

The constant second differences indicate that the data are perfectly quadratic. Thus there is no need to consider an exponential model.

Steps 3 and 4 are not necessary here, because we found the best model in Step 2. However, we must keep in mind that the quadratic model may not be valid for the years before 1960 or after 1990 because its end behavior does not describe that of the context.

b. The median age of marriage in 1997 is estimated to have been

$$Q(37) = 0.0035(37)^2 + 0.015(37) + 20.3 \approx 25.6 \text{ years}$$

Note that we report the answer to one decimal place because that is the accuracy to which the data were reported. The estimate exceeds the known value by 0.6 year. We can speculate that the model will overestimate in any extrapolations. Note that the shifted exponential model gives the estimate as $S(37) \approx 27.8$ years. This model estimates even higher ages than the quadratic model beyond the range of the data. This is further confirmation that the quadratic model is the better choice.

c. According to the model Q, the median age of marriage in the year 2005 will be

$$Q(45) = 0.0035(45)^2 + 0.015(45) + 20.3$$
$$\approx 28.1 \text{ years of age}$$

d. This prediction is probably not valid. Even though the model fits the data perfectly, there is no indication that it is a good predictor of future events and trends. There are many factors other than time that affect the median age of first marriage. ●

Sometimes it is not quite so easy to choose which model is best, as the next example illustrates.

EXAMPLE 2   Using Curvature, Fit, and End Behavior to Choose a Model

TABLE 2.20

Years	Number of countries	Years	Number of countries
1840	1	1865	30
1845	3	1870	34
1850	9	1875	36
1855	16	1880	37
1860	24		

(Source: "The Curve of Cultural Diffusion," *American Sociological Review*, August 1936, pp. 547–556.)

Stamps   Table 2.20 gives the numbers of European, North American, and South American countries that issued postage stamps from 1840 through 1880. Find an appropriate model for the data.

Solution

Step 1: Consider the curvature of a scatter plot.

The scatter plot in Figure 2.53 appears to be concave up between 1840 and 1850 or 1855 and then to be concave down between 1855 and 1880. Thus there is an apparent inflection point, so either a cubic or a logistic model could be appropriate.

Countries

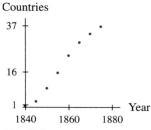

FIGURE 2.53

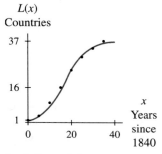

Logistic model
FIGURE 2.54

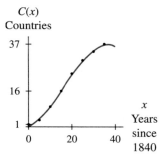

Cubic model
FIGURE 2.55

Step 2: Compare the fit of the two equations.

A logistic model is

$$L(x) = \frac{37.19488}{1 + 21.37367e^{-0.18297x}} \text{ countries}$$

where x is the number of years since 1840. It appears to fit the data reasonably well (see Figure 2.54).
   A cubic model is

$$C(x) = -0.00120x^3 + 0.06027x^2 + 0.40435x + 0.44444 \text{ countries}$$

where x is the number of years since 1840. The cubic model also appears to fit the data well (see Figure 2.55).
   If all you are concerned about is fit, you should choose the cubic model. However, if you would like to predict how many countries may have issued stamps in 1890, you should also consider end behavior.

Step 3: Consider end behavior.

Because the only countries under consideration are those in Europe, North America, and South America, there must be an upper limit on the number of countries issuing stamps. Thus it would be logical to think that a model for these data should approach some limiting value. This end behavior would be better described by the logistic function.
   In conclusion, if your only concern is the better-fitting equation, and you do not wish to use the model to extrapolate, then you should choose

$$C(x) = -0.00120x^3 + 0.06027x^2 + 0.40435x + 0.44444 \text{ countries}$$

where x is the number of years since 1840.
   If you are concerned about end behavior more than about a better fit, you should choose

$$L(x) = \frac{37.19488}{1 + 21.37367e^{-0.18297x}} \text{ countries}$$

where x is the number of years since 1840. ●

Be open to the possibility that none of the models we have studied is appropriate for a set of data. Indeed, many sets of data in the real world do not fit neatly into one of our categories. Modeling, even in the simplest sense, is as much an art as a science. Sometimes a creative mind can be as helpful as a mathematical one. In any case, the most beneficial way to learn how to determine which model to use for a data set is by practicing.
   Although the first two chapters have focused on the construction of elementary mathematical models, it is important that you understand that modeling is *not* our primary objective. In reality, mathematical modeling can be a complicated and highly sophisticated endeavor. We are using basic modeling with elementary curve fitting strictly as a way to obtain functional relationships between variables. Because most information in the real world is collected in discrete form (tables of data), it is important in the study of calculus to obtain continuous functional relationships between the variables in order to study their changing behavior. Our main objective is to apply concepts and methods of calculus to these functions.

2.5 Concept Inventory

○ *Choosing a model:*

    Consider curvature and numerical differences

    Compare the fit of possible equations

    Consider shifting or realigning the data

    Consider end behavior

    Consider that there may be two valid choices

○ *Sometimes no model we have studied is appropriate.*

2.5 Activities

For each of the scatter plots in Activities 1 through 6, state which function or functions are candidates to fit the data. Explain why those functions are appropriate, and also explain why the other types of functions are not appropriate.

1.   **2.**

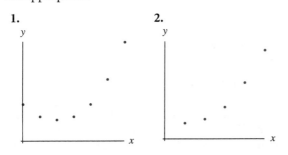

3.   **4.**

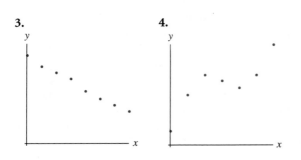

5.   **6.**

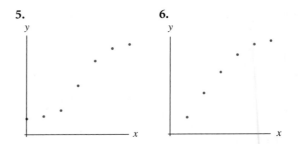

7. Given a data set in which the input values are evenly spaced, discuss how you could determine whether the data are

    a. linear.

    b. quadratic.

    c. exponential.

8. Rental Vacancies The following table gives the percentage of rental-housing units in the United States that were vacant during the third quarter of each of the selected years.

Year	Rental vacancy rate (percent)
1978	5.0
1980	5.7
1985	6.8
1989	7.3
1991	7.6
1996	8.0
1998	8.2
2001	8.4

(Source: *Housing Vacancy Survey,* August 22, 2002, U.S. Census Bureau.)

    a. Find both quadratic and log models for the percentage of vacant rental housing as a function of the number of years since 1900. Which function better fits the aligned data?

    b. Align the input as the number of years after 1974 and find both quadratic and log models for the aligned data. Does one function fit better than the other? Explain.

    c. Use both functions in part *b* to predict the rental-housing vacancy rate during the third quarter of 2002. In which estimate would you have more confidence? Explain your reasoning.

9. Swim Time   The following table gives the winning times for the 100-Meter Butterfly swimming competition at selected Summer Olympic Games between 1956 and 2000.

Year	Time (seconds)
1956	71.00
1960	69.50
1968	65.50
1972	63.34
1980	60.42
1988	59.00
1992	58.62
2000	56.61

(Source: *Statistical Abstract*, 2001.)

a. Basing your decision on a scatter plot, determine which functions could be used to model these data.

b. Find both quadratic and log models for the winning times as a function of the number of years since 1900. Which function better fits the aligned data?

c. Align the input as the number of years after 1946 and find a log model for the aligned data. How well does this equation fit the data?

d. If the end behavior were of no particular concern, would you choose the quadratic function in part b or the log function in part c to model the data? If the end behavior is important, would you choose the quadratic function in part b or the log function in part c to model the data?

10. Brain Weight   The accompanying table gives the average weight of the brain in human males as a percentage of body weight.

a. Describe in words what happens to the percentage of body weight accounted for by the brain as a male child grows.

b. Find quadratic, exponential, and log equations to fit the data. Note that you must shift the ages to the right in order to use a log equation. Shifting the percentages down will produce a better-fitting exponential equation. Write models using each equation.

Age (years)	Percentage of body weight accounted for by brain
0	11.0
2	8.0
4	7.0
6	6.0
8	5.0
10	4.5
12	4.0
14	3.5
16	3.25

c. Use each model in part b to determine the age (in months) at which brain weight is 10% of body weight.

d. Use each model in part b to predict the percentage brain weight for an 18-year-old man.

e. Discuss which equation best models the data and why.

11. CD Albums   The table gives the value of manufacturers' shipments of CD recordings in millions of dollars (based on list price) between 1990 and 2000.

Shipments of CD albums	
Year	Value (millions of dollars)
1990	3452
1992	5327
1994	8465
1996	9935
1998	11,416
2000	13,215

(Source: *Statistical Abstract*, 2001.)

a. Find a model for this set of data.

b. Use the model to estimate values of shipments of CD albums in 2001 and 2005.

12. Population   The table on the next page shows the population of South Carolina between 1790 and 2000.

a. Examine a scatter plot of the data, and discuss its curvature. What types of curves might be appropriate for such curvature?

Year	Population (thousands)
1790	249
1900	1340
1950	2117
1980	3122
1990	3486
2000	4012

(Source: *Statistical Abstract,* 2001.)

b. Find the two models that are appropriate according to your analysis in part *a*.

c. Discuss the fit of each equation from part *b* to the data.

d. Which of the models from part *b* would be more appropriate to model the South Carolina population? Discuss any issues (besides curvature) that lead you to this decision.

13. **Jeep Value**   The following table shows the 2002 private party resale value of a 2000 Jeep Grand Cherokee Laredo in excellent condition as a function of the mileage.

Mileage (thousands)	Resale value (dollars)
20	18,520
40	17,120
60	14,670
80	13,295
100	12,745
120	12,270

(Source: www.kbb.com. Accessed 9/30/01.)

a. Examine a scatter plot of the data, and discuss its curvature. What type of equation should you try to fit to these data? What is the model?

b. Use your model from part *a* to estimate the resale value for this Jeep at 52,000 miles.

14. **Marriage Age**   The median age at first marriage of males in the United States from 1950 to 2000 is given in the accompanying table.

a. Find an appropriate model for the data.

b. Explain why you chose the type of function you used.

Year	Median marriage age (years)	Year	Median marriage age (years)
1950	22.8	1990	26.1
1960	22.8	1994	26.7
1970	23.2	1997	26.8
1980	24.7	2000	26.8

(Source: www.infoplease.com. Accessed 10/9/02.)

15. **Expenditures**   The personal consumption expenditure for magazines, newspapers, and sheet music in the United States between 1990 and 2001 is found in the accompanying table.

Year	Expenditure (billions of dollars)
1990	21.6
1994	24.9
1995	26.2
1996	27.6
1997	29.1
1998	31.0
1999	32.5
2000	34.2
2001	35.2

(Sources: *Statistical Abstract,* 2001; and www.bea.doc.gov. Accessed 9/26/02).

a. Find an appropriate model for the data.

b. Explain why you chose the type of function you used.

16. **Fisheries**   The domestic catch of fish by U.S. fisheries for human food in selected years between 1970 and 1990 are given in the accompanying table.

a. Discuss the curvature of the data. Why is a cubic function not appropriate? Why is a single quadratic function not appropriate?

b. Divide the data into two subsets that share one data point. Find equations to fit each set of data, and combine them to form a piecewise continuous model.

c. Discuss factors that might have contributed to the behavior exhibited by the data.

d. Use your model to estimate domestic catch of fish in 1973, 1983, and 1993.

Year	Amount (millions of pounds)
1970	2537
1972	2435
1975	2465
1977	2952
1980	3654
1982	3285
1985	3294
1987	3946
1990	7041

(Source: *Statistical Abstract,* 1994.)

17. Population   Population data for Iowa in the 1980s and 1990s are given in the table.

Year	Population (thousands)
1980	2914
1985	2830
1987	2767
1989	2771
1991	2791
1993	2820
1995	2841
1997	2852

(Source: *Statistical Abstract,* 1998).

a. Examine a scatter plot of the data. Discuss the curvature.

b. Discuss the appropriateness or inappropriateness of using a cubic equation to fit this data set.

c. Find a cubic equation to fit the data, and graph it on the scatter plot. Does the graph support or contradict your conclusion in part *b*?

d. Divide the data into two subsets that share one point. Find a piecewise continuous model for the population of Iowa.

WEB/CD e. Update the data to the most recent year for which population data are available. Is the model you chose appropriate for the expanded data set?

Source

18. Suppose that you are required to model a set of data which has evenly spaced input values and for which second differences and percentage differences in the outputs are constant. How might you determine which model is best?

19. For which models is reducing the magnitude of the input values important and why?

20. Discuss when it is necessary to shift data vertically.

21. In this section, most of the examples and activities that involve vertical shifts use the process of shifting the data down to obtain a better fit for exponential and logistic functions. Data can also be shifted up to find a better fit. Consider the data given in Table 2.20 and the logistic function *L* shown in the solution of Example 2. Discuss how you would find a better-fitting logistic model for these data than the one that is given by the function *L*.

22. Incarceration Rate   The *incarceration rate* is the number of inmates in prison per 100,000 U.S. residents. The table for this activity (located on the *Calculus Concepts* CD-ROM and web site) lists the national and regional incarceration rates for prisoners with sentences of more than 1 year who were imprisoned under state and federal jurisdiction at year end from 1977 through 2001.

a. Construct scatter plots of the incarceration rates for the Northeast, Midwest, South, and West.

b. Does one region of the United States have a higher incarceration rate than the other regions between 1977 and 2001? Does this necessarily mean that it is more dangerous to live in that region of the United States?

c. Find a cubic model for each of the incarceration rates for the Northeast, Midwest, South, and West. In each case, determine whether the cubic function fits the data. If the fit is not good, discuss what type of function could be used to model the data.

23. Subscribers   The table for this activity (located on the *Calculus Concepts* CD-ROM and web site) lists the total revenue and revenue from basic cable for cable TV subscribers between 1975 and 2000.

a. Plotting the year on the horizontal axis, observe a scatter plot of the data in each of these columns. Discuss the curvature of each plot.

b. Find a quadratic model for the data in the total revenue column. Discuss the fit.

c. Find a cubic model for the data in the basic cable revenue column. Discuss the fit.

d. Using the results of parts b and c, find a model for cable TV revenue that is not from basic cable. Use this model to predict the cable TV revenue that is not from basic cable in 2003.

24. Incarceration Rate The table for this activity (located on the *Calculus Concepts* CD-ROM and web site) lists the national and regional incarceration rates for prisoners with sentences of more than 1 year who were imprisoned under state and federal jurisdiction at year end from 1977 through 2001.

WEB/CD a. Find U.S. resident population and state and federal prison population data for two of the years

Source

in the table. Explain how the values in the U.S. incarceration rate column are calculated for these 2 years. Give possible reasons for any discrepancies in the values you calculated and the values in the table.

b. Consider the year and U.S. incarceration rate columns in the table. Divide the data into two subsets that have one point in common. Find an equation to fit each set of data, and combine the two to form a piecewise continuous model. Graph your function and discuss how well it fits the data.

c. Use the function in part b to estimate the U.S. incarceration rate in 2005. How reliable is this prediction?

SUMMARY

Exponential Functions and Models

Second in importance to linear functions and models are exponential functions and models. Based on the familiar idea of repeated multiplication by a fixed positive multiplier b (the base), the basic exponential function is of the form

$$f(x) = ab^x$$

The parameter a appearing in the equation is the output when the input is zero.

Exponential functions model constant percentage change. In terms of the function $f(x) = ab^x$, exponential growth occurs when b is greater than 1, and exponential decline (decay) takes place when b is between 0 and 1. The constant percentage growth or decline is given by $(b - 1)100\%$.

Logarithmic Functions and Models

The basic form of the log function that we use is

$$f(x) = a + b \ln x$$

The input of this function must be a value greater than zero. The log function is useful for situations in which

the output grows or declines at an increasingly slow rate. When fitting a log equation to data, you must sometimes align the input data to ensure that the input values are greater than zero or to obtain a better fit. Aligning input data has the effect of shifting the data horizontally.

Logistic Functions and Models

Initial exponential growth followed by a leveling-off approach toward a limiting value L is characteristic of logistic growth, modeled by the logistic equation

$$f(x) = \frac{L}{1 + Ae^{-Bx}}$$

If the parameter B is positive, the model indicates growth. If the parameter B is negative, the model indicates decline in output toward the horizontal axis as the input values increase.

When fitting exponential and logistic equations to data, it is sometimes helpful to shift the output data. This vertical shift is particularly useful when the data appear to approach a value other than zero. The goal in shifting is to move the data closer to the horizontal axis.

Polynomial Functions and Models

Polynomial functions and models have a well-established role in calculus. In this text, we consider linear functions, quadratic functions, and cubic functions.

Quadratic equations have graphs known as parabolas. The parabola with equation $f(x) = ax^2 + bx + c$ opens upward (is concave up) if a is a positive and opens downward (is concave down) if a is negative.

Cubic equations have graphs that resemble one of the four types shown in Figure 2.39 on page 125. Like logistic models, cubic models show a change of concavity at an inflection point. We must be especially careful in using cubic models when extrapolating beyond the range of data values from which the models are constructed.

Choosing a Model

Although it is not always clear which (if any) of the functions we have discussed apply to a particular real-life situation, it helps to keep in mind a few general, commonsense guidelines. (1) Given a set of discrete data, begin with a scatter plot. The plot will often reveal general characteristics that point the way to an appropriate model. (2) If the scatter plot does not appear to be linear, consider the suggested concavity. One-way concavity (up or down) suggests a quadratic, exponential, or log model. (3) When a single change in concavity seems apparent, think in terms of cubic or logistic models. But remember, the graphs of logistic models tend to become flat on each end, whereas cubics do not. Never consider a cubic or a logistic model if you cannot identify an inflection point.

CONCEPT CHECK

Can you

○ Find an exponential model and determine its percentage change?

○ Solve exponential growth and decay problems?

○ Find and use a log model?

○ Horizontally shift data?

○ Find and use a logistic model?

○ Vertically shift data?

○ Find and use a quadratic model?

○ Find and use a cubic model?

○ Determine end behavior?

○ Find and use a piecewise continuous model?

○ Identify concavity and inflection points?

○ Observe a scatter plot and choose a model?

To practice, try

Section 2.1	Activity 19
Section 2.1	Activities 25, 27
Section 2.2	Activity 9
Section 2.2	Activity 13
Section 2.3	Activity 17
Section 2.3	Activity 25
Section 2.4	Activity 15
Section 2.4	Activity 25
Section 2.5	Activity 9
Section 2.5	Activity 16
Section 2.5	Activity 5
Section 2.5	Activities 12, 15

REVIEW TEST

1. **Rearing Children** According to the U.S. Census Bureau, the number of children living with their grandparents was 2.2 million in 1970 and 3.9 million in 1997. Assume that the number grew exponentially between 1970 and 1997.

   a. Find an exponential model for the number of children living with grandparents as a function of the number of years since 1970.

   b. What is the percentage change indicated by the model?

   c. According to the model, when will the number of children living with their grandparents reach 5 million?

   d. Do you believe an exponential model accurately reflects future change in the number of children living with grandparents? Explain. If not, what do you think would be a more appropriate model?

2. **Stocks** The high stock prices for Microsoft Corporation in June for selected years between 1989 and 2000 are shown in the accompanying table.

Year	Price (dollars)	Year	Price (dollars)
1989	0.74	1996	15.73
1990	2.11	1998	54.28
1992	4.38	2000	119.94
1994	6.83		

   (Source: Hoover's Online Guide, October 2002.)

   a. Examine a scatter plot of the data. Discuss the curvature of the data and the possible models that could be used for this data set.

   b. Find the model that you believe is best for this data set.

   c. What issues in our current economy do you believe will affect the price of Microsoft stock in the future? Do you think the model you chose is a good description of the future price?

3. **Temperature** On August 28, 1993, the *Philadelphia Inquirer* reported temperatures from the previous day. The following table lists these temperatures (in degrees Fahrenheit) from 5 A.M. to 5 P.M.

Time	Temperature (°F)	Time	Temperature (°F)
5 A.M.	76	noon	90
6 A.M.	75	1 P.M.	91
7 A.M.	75	2 P.M.	93
8 A.M.	77	3 P.M.	94
9 A.M.	79	4 P.M.	95
10 A.M.	83	5 P.M.	93
11 A.M.	87		

   a. Examine a scatter plot of the data. Explain why the curvature indicates that a cubic model is appropriate.

   b. Find a cubic model for the data.

   c. According to the model, what was the temperature at 5:30 P.M.?

   d. According to the model, at approximately what times (on August 27) was the temperature 90°F?

4. **Population** A 1998 United Nations study projected that the world population would stabilize at 10.73 billion just after the year 2200. The study's projections include those shown in the following table.

Year	Population (billions)
1999	6
2011	7
2025	8
2041	9
2071	10

a. Is the following statement true or false? Explain. "Because the first differences of the data are constant, a linear model is appropriate."

b. Examine a scatter plot of the data. Discuss the curvature suggested by the scatter plot and the functions that could be used to model the data.

c. Align the input as the number of years since 1900, and find quadratic, logistic, and log models for the projected population. Discuss how well each equation fits the data.

d. Use limit notation to express the end behavior of the three equations in part *c.* Which equation best exhibits the end behavior suggested by the United Nations study?

Project 2.1 Compulsory School Laws

Setting

In 1852, Massachusetts became the first state to enact a compulsory school attendance law. Sixty-six years later, in 1918, Mississippi became the last state to enact a compulsory attendance law. The table below lists the first 48 states to enact such laws and the year each state enacted its first compulsory school law.

State	Year	State	Year	State	Year
MA	1852	SD	1883	IA	1902
NY	1853	RI	1883	MD	1902
VT	1867	ND	1883	MO	1905
MI	1871	MT	1883	TN	1905
WA	1871	IL	1883	DE	1907
NH	1871	MN	1885	NC	1907
CT	1872	ID	1887	OK	1907
NM	1872	NE	1887	VA	1908
NV	1873	OR	1889	AR	1909
KS	1874	CO	1889	TX	1915
CA	1874	UT	1890	FL	1915
ME	1875	KY	1893	AL	1915
NJ	1875	PA	1895	SC	1915
WY	1876	IN	1897	LA	1916
OH	1877	WV	1897	GA	1916
WI	1879	AZ	1899	MS	1918

(Source: Richardson, "Variation in Date of Enactment of Compulsory School Attendance Laws," *Sociology of Education* vol. 53, July 1980, pp. 153–163.)

Tasks

1. Tabulate the cumulative number of states with compulsory school laws for the following 5-year periods:

1852–1856	1887–1891
1857–1861	1892–1896
1862–1866	1897–1901
1867–1871	1902–1906
1872–1876	1907–1911
1877–1881	1912–1916
1882–1886	1917–1921

2. Examine a scatter plot of the data in Task 1. Do you believe a logistic model is appropriate? Explain.

3. Find a logistic model for the data in Task 1.

4. What do most states in the third column of the original data have in common? Why would these states be the last to enact compulsory education laws?

5. The 17 states considered to be southern states (below the Mason-Dixon Line) are AL, AR, DE, FL, GA, KY, LA, MD, MO, MS, NC, OK, SC, TN, TX, VA, and WV. Tabulate cumulative totals for the southern states and the northern/western states for these dates:

Northern/Western States Years	Southern States Years
1852–1856	1891–1895
1857–1861	1896–1900
1862–1866	1901–1905
1867–1871	1906–1910
1872–1876	1911–1915
1877–1881	1916–1920
1882–1886	
1887–1891	
1892–1896	
1897–1901	
1902–1906	

Compulsory School Laws, *Continued*

6. Examine scatter plots for the two data sets in Task 5. Do you believe that logistic models are appropriate for these data sets? Explain.

7. Find logistic models for each set of data in Task 5. Comment on how well each equation fits the data.

8. It appears that the northern and western states were slow to follow the lead established by Massachusetts and New York. What historical event may have been responsible for the time lag?

9. One way to reduce the impact of unusual behavior in a data set (such as that discussed in Task 8) is to group the data in a different way. Tabulate the cumulative northern and western state totals for the following 10-year periods:

   1852–1861
   1862–1871
   1872–1881
   1882–1891
   1892–1901
   1902–1911

10. Find a logistic model for the data in Task 9. Comment on how well the function fits the data. Compare models for the data grouped in 10-year intervals and the data grouped in 5-year intervals (Task 6). Does grouping the data differently significantly affect how well the equation fits? Explain.

11. Find an equation that fits the data for the southern states better than the logistic equation. Write the model using the better-fitting equation. Explain your reasoning.

Reporting

1. Prepare a written report of your work. Include scatter plots, models, and graphs. Include discussions of each of the tasks in this project.

2. (Optional) Prepare a brief (15-minute) presentation on your work.

Project 2.2 Fund-Raising Campaign

Setting

In order to raise funds, the mathematics department in your college or university is planning to sell T-shirts before next year's football game against the school's biggest rival. Your team has volunteered to conduct the fundraiser. Because several other student groups have also volunteered to head this project, your team is to present its proposal for the fund drive, as well as predictions about its outcome, to a panel of mathematics faculty.

Tasks

1. Develop a slogan and a design for the T-shirt. Keep in mind that good taste is a concern. Decide on a target market, and determine a strategy to survey (at random) at least 100 students who represent a cross section of the target market to determine the demand for T-shirts (as a function of price) within that market. It is important that your sample survey group properly represent your target market. If, for example, you polled only near campus dining facilities at lunch time, your sample would be biased toward students who eat lunch at such facilities.

   The question you should ask is "How much would you be willing to spend on a T-shirt promoting the big football rivalry, $20, $18, $16, $14, $12, $10, $8, or not interested?" Keep an accurate tally of the number of students who answer in each category. In your report on the results of your poll, you should include information such as your target market; where, when, and how you polled within that market; and why you believe that your polled sample is likely to be a representative cross section of the market.

2. a. From the data you have gathered, determine how many students from your sample survey group would buy a T-shirt at $8, $10, $12, and so on.

   b. Devise a marketing strategy, and determine how many students within your target market you can reasonably expect to reach. Assuming that your poll is an accurate indicator of your target population, determine the number of students from your target market who will buy a T-shirt at each of the given prices.

Note: This project is also used as a portion of Project 5.2 on page 365.

Fund-Raising Campaign, *Continued*

   c. Taking into account the results of your poll and your projected target market, develop a model for demand as a function of price. Keep in mind that your model must make sense for all possible input values.

3. Use the partial price listing table (located on the *Calculus Concepts* CD-ROM and web site) to model a function for the cost to you when ordering T-shirts. Use the demand function from Task 2 to create equations for revenue, total cost, and profit as functions of price. (Revenue, total cost, and profit may not be one of the basic models that were discussed in class. They are sums and/or products of the demand function with other functions.)

Reporting

1. Prepare a written report summarizing your survey and modeling. The report should include your slogan and design, your target market, your marketing strategy, the results from your poll (as well as the specifics of how you conducted your poll), a discussion of how and why you chose the model of the demand function, a discussion of the accuracy of your demand model, and graphs and equations for all of your models. Attach your questionnaire and data to the report as an appendix.

2. Prepare a 15-minute oral presentation of your survey, modeling, and marketing strategy to be delivered before a panel of mathematics faculty. You will be expected to have overhead transparencies of all graphs and equations as well as any other information that you consider appropriate as a visual aid. Remember that you are trying to sell the mathematics department on your campaign idea.

Describing Change: Rates

Change is everywhere around us and affects our lives on a daily basis. In this chapter we consider several ways to describe change. Starting from the actual change in a quantity over an interval, it is a simple step to describe the change as a percentage change or as an average rate of change over that interval. The notion of average rate of change, when examined carefully in light of the underlying geometry of graphs, leads to the more subtle and challenging concept of instantaneous change. Indeed, the precise description of instantaneous change in terms of mathematics is one of the principal goals of calculus.

We therefore turn our attention to determining instantaneous rates of change. Because instantaneous rates of change are slopes of tangent lines, we first consider the numerical estimation of these slopes. Next we generalize the numerical method to an algebraic method that gives us an analytic description—a formula of sorts—for the derivative of an arbitrary function.

Concept Objectives

This chapter will help you understand the concepts of

- ○ Change, percentage change, and average rate of change
- ○ Instantaneous rate of change and derivative
- ○ Secant line and tangent line and their relationship
- ○ Percentage rate of change

and you will learn to

- ○ Find and interpret descriptions of change using data, graphs, and equations
- ○ Sketch and find slopes of secant and tangent lines
- ○ Relate the location of a tangent line to the concavity of a graph
- ○ Recognize where a derivative does not exist
- ○ Use limits to find rate-of-change formulas algebraically
- ○ Estimate rates of change using symmetric difference quotients
- ○ Interpret derivatives
- ○ Numerically estimate rates of change

Concept Application

The Centers for Disease Control (CDC) in Atlanta, Georgia, is concerned with following the spread of diseases in the United States. CDC information is valuable to doctors, hospitals, pharmaceutical companies, and medical researchers. The CDC has closely followed the AIDS epidemic. Mathematics can help the CDC answer such questions as

- On average, how quickly did the number of AIDS patients increase between 1985 and 2000?
- How quickly was the number of AIDS patients growing in 1990?
- By what percentage was the number of AIDS patients growing in 1996?

This chapter will provide you with some of the tools that make it possible to answer questions like these. You will have the opportunity to estimate the answers to questions similar to those above in Activity 22 of Section 3.3.

3.1   Change, Percentage Change, and Average Rates of Change

One of the primary goals of calculus is the description of change. In preparation for learning how calculus describes change, we introduce three numerical ways of reporting change. We begin with a simple business example.

The yearly revenue for a large department store declined from $1.4 billion in 2000 to $1.1 billion in 2003. There are three common ways to express this change in revenue. We could use a negative sign to indicate the decrease in revenue and say that the **change** in revenue was -$0.3 billion over 3 years. We could also express this change by saying that the revenue declined by $0.3 billion (or $300 million) over 3 years.

It is often helpful to express the change as a percent of the 2000 revenue. This **percentage change** (or percent change) is found by dividing the change by the revenue in 2000 and multiplying by 100:

$$\text{Percentage change} = \frac{-\$0.3 \text{ billion}}{\$1.4 \text{ billion}} \cdot 100\%$$
$$\approx -21.4\%$$

The company saw a 21.4% decline in revenue between 2000 and 2003. You should already be familiar with percentage change from the discussion of exponential models. This method of calculating percentage change is the same as the way we calculated percentage differences using data in Section 2.1. However, you should note that even though constant percentage difference over equally spaced intervals are unique to exponential models, the percentage change over some given interval may be calculated and interpreted in almost any context.

The third way to express change involves spreading the change over the time interval to obtain the **average rate of change.** This is done by dividing the change by the length of the interval. In the case of the department store revenue, the length of the interval is 3 years:

$$\text{Average rate of change} = \frac{-\$0.3 \text{ billion}}{3 \text{ years}}$$
$$= -\$0.1 \text{ billion per year}$$

On average, the revenue declined at a rate of $0.1 billion per year (or $100 million per year) between 2000 and 2003.

We summarize these three ways to describe change in the box on the following page.

Change, Percentage Change, and Average Rate of Change

If a quantity changes from a value of m to a value of n over a certain interval from a to b, then

○   The change in the quantity is found by subtracting the first value from the second.

$$\text{Change} = n - m$$

○   The percentage change is the change divided by the first value and then multiplied by 100.

$$\text{Percentage change} = \frac{\text{change}}{\text{first value}} \cdot 100\% = \frac{n - m}{m} \cdot 100\%$$

○   The average rate of change is the change divided by the length of the interval.

$$\text{Average rate of change} = \frac{\text{change}}{\text{length of interval}} = \frac{n - m}{b - a}$$

Interpreting Descriptions of Change

Correctly calculating these descriptions of change is important, but being able to state your result in a meaningful sentence in the context of the situation is equally important. When interpreting a description of change, you should answer the questions *when, what, how,* and *by how much.*

Interpreting Descriptions of Change

When describing change over an interval, be sure to answer the following questions:

○   When? Specify the interval.

○   What? Specify the quantity that is changing.

○   How? Indicate whether the change is an increase or a decrease.

○   By how much? Give the numerical answer labeled with proper units:

Description	*Units*
change	output units
percentage change	percent
average rate of change	output units per input unit

When stating an average rate of change, use the word *average* or the phrase *on average.*

EXAMPLE 1 *Describing Change Using a Table*

Temperature Consider Table 3.1, which shows temperature values on a typical May day in a certain midwestern city. Find the following descriptions of change, and write a sentence giving the real-life meaning of (that is, interpret) each result.

TABLE 3.1

Time	Temperature (°F)	Time	Temperature (°F)
7 A.M.	49	1 P.M.	80
8 A.M.	58	2 P.M.	80
9 A.M.	66	3 P.M.	78
10 A.M.	72	4 P.M.	74
11 A.M.	76	5 P.M.	69
noon	79	6 P.M.	62

a. The change in temperature from 7 A.M. to 1 P.M.

b. The percentage change in temperature between 3 P.M. and 6 P.M.

c. The average rate of change in temperature between 8 A.M. and 5 P.M.

Solution

a. To find the change in temperature, subtract the temperature at 7 A.M. from the temperature at 1 P.M.

$$\text{Change} = 80°F - 49°F = 31°F$$

The interpretation statement answers the questions *when* (1 P.M. and 7 A.M.), *what* (the temperature), *how* (was greater than), and *by how much* (31°F): The temperature at 1 P.M. was 31°F greater than the temperature at 7 A.M.

b. To find percentage change, first find the change in temperature between 3 P.M. and 6 P.M.

$$\text{Change} = 62°F - 78°F = -16°F$$

Next, divide the change by the temperature at 3 P.M. (the beginning of the time interval under consideration), and multiply by 100.

$$\text{Percentage change} = \frac{-16°F}{78°F} \cdot 100\% \approx -20.5\%$$

The temperature declined by about 20.5% between 3 P.M. and 6 P.M. In this statement, "between 3 P.M. and 6 P.M." answers the question *when*, "the temperature" again identifies *what*, and "declined by about 20.5%" tells *how* and *by how much*.

c. Begin the calculation of the average rate of change by finding the change in the temperature from 8 A.M. to 5 P.M.

$$\text{Change} = 69°F - 58°F = 11°F$$

Next, divide the change by the length of the time interval (9 hours).

$$\text{Average rate of change} = \frac{11°F}{9 \text{ hours}} \approx 1.2°F \text{ per hour}$$

In interpreting this average rate of change, we must use the word *average* in our sentence in addition to answering the four questions. We state the interpretation as follows: Between 8 A.M. and 5 P.M., the temperature rose at an average rate of 1.2°F per hour.   ●

Although these descriptions of change are useful, they have limitations. It appears from the answer to part *c* of Example 1 that the temperature rose slowly throughout the day. However, the average rate of change does not describe the 22°F rise in temperature followed by the 11°F drop in temperature that occurred between 8 A.M. and 5 P.M.

Finding Percentage Change and Average Rate of Change Using Graphs

You may have noticed that calculating the average rate of change is the same as calculating slope. This observation allows for the easy calculation of average rates of change if you are given a graph. For instance, when plotted, the May daytime temperatures in Example 1 fall in the shape of a parabola (see Figure 3.1).

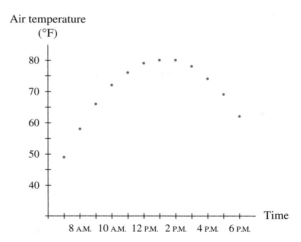

FIGURE 3.1

To find the average rate of change between 9 A.M. and 4 P.M., first use a straightedge to draw carefully a line connecting the points at 9 A.M. and 4 P.M. (see Figure 3.2). We call this line connecting two points on a scatter plot or a graph a **secant line** (from the Latin *secare*, "to cut"). Next, approximate the slope of this secant line by estimating the rise and the run for a portion of the line (see Figure 3.3).

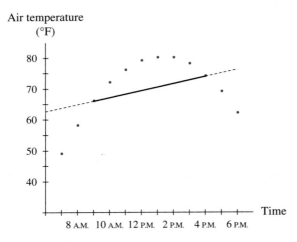

FIGURE 3.2

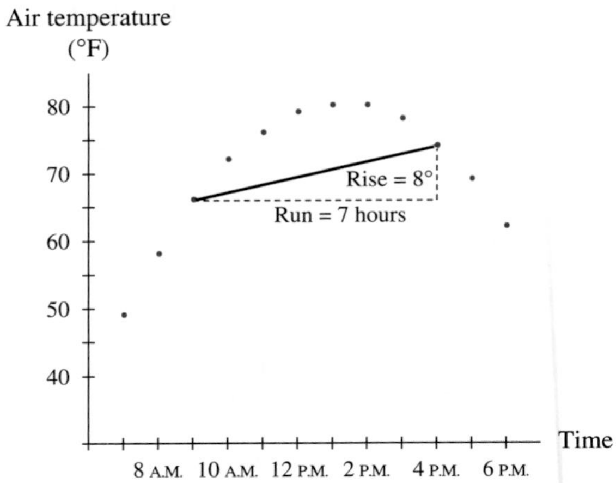

FIGURE 3.3

$$\text{Average rate of change} = \frac{\text{rise}}{\text{run}} = \frac{\text{change in temperature}}{\text{change in time}}$$

$$= \frac{8°\text{F}}{7 \text{ hours}} \approx 1.1°\text{F per hour}$$

Between 9 A.M. and 4 P.M., the temperature rose at an average rate of 1.1°F per hour. Note that the rise is the change in temperature. To calculate percentage change from the graph, divide the rise by the estimated output of the first point.

$$\text{Percentage change} = \frac{\text{change in temperature}}{\text{temperature at 9 A.M.}} \approx \frac{8°\text{F}}{66°\text{F}} \cdot 100\% \approx 12.1\%$$

It is important to note that this graphical method of calculating descriptions of change is imprecise if you are given only a scatter plot or a graph. It gives only approximations to change, percent change, and average rate of change. The method depends on drawing the secant line accurately and then correctly identifying two points on the line. Slight variations in sketching are likely to result in slightly different answers. This does not mean that the answers you obtain are incorrect. It simply means that descriptions of change obtained from graphs are approximations.

Example 1 and the subsequent discussion of air temperature use the term *between* two time values. There are other ways to describe intervals on the input axis, and we take a moment now to discuss one of them.

When we use data that someone else has collected, we often do not know when the data were reported or recorded. It seems logical to assume that yearly (or monthly or hourly and so forth) totals are reported at the *end* of the intervals representing those periods. For instance, a 2002 total covers the period of time from the end of 2001 through the end of 2002 (see Figure 3.4).

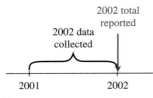

FIGURE 3.4

Therefore, we adopt the convention that "from *a* through *b*" on the input axis refers to the interval beginning at *a* and ending at *b*. If *a* and *b* are years and the output represents a quantity that can be considered to have been measured at the end of the year, then "from *a* to *b*" means the same thing as "from the end of year *a* through the end of year *b*." For our purposes the phrases "between *a* and *b*" and "from *a* to *b*" have the same meaning as "from *a* through *b*." We use this terminology in the remainder of the text.

EXAMPLE 2   *Describing Change Using a Graph*

Social Security   The Social Security assets of the federal government between 2002 and 2030 as estimated by the Social Security Advisory Board are shown in Figure 3.5. A smooth curve connecting the points is also shown.

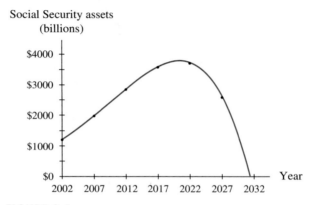

FIGURE 3.5

(Source: www.ssab.gov. Accessed on 10/1/02.)

a. Estimate the change, percentage change, and average rate of change in Social Security assets between 2002 and 2017. Write a sentence interpreting each answer in context.

b. Estimate the change, percentage change, and average rate of change in Social Security assets between 2022 and 2027. Write a sentence interpreting each answer in context.

Solution

a. Begin by estimating from the graph the Social Security assets in 2002 and 2017. One possible estimate is $1200 billion in 2002 and $3600 billion in 2017. To calculate change, subtract the 2002 value from the 2017 value.

$$Change = 3600 - 1200 = \$2400 \text{ billion}$$

Social Security assets are expected to rise about $2400 billion between 2002 and 2017.

Convert this change to percentage change by dividing by the estimated assets in 2002.

$$\text{Percentage change} = \frac{\$2400 \text{ billion}}{\$1200 \text{ billion}} \cdot 100\% \approx 200\%$$

Social Security assets are expected to increase about 200% between 2002 and 2017.

Convert the change to an average rate of change by dividing the change by the length of the interval between 2002 and 2017.

$$\text{Average rate of change} = \frac{\$2400 \text{ billion}}{15 \text{ years}}$$
$$= \$160 \text{ billion per year}$$

Social Security assets are expected to increase by an average of about $160 billion per year between 2002 and 2017. The average rate of change is the slope of the secant line through the point corresponding to 2002 and the point corresponding to 2017 shown in Figure 3.6.

b. Begin by estimating the output values in 2022 and 2027. We estimate $3700 billion in 2022 and $2600 billion in 2027. Following the same procedure as in part *a*, we have

$$\text{Change} \approx 2600 - 3700 = -\$1100 \text{ billion}$$

$$\text{Percentage change} \approx \frac{-\$1100 \text{ billion}}{\$3700 \text{ billion}} \cdot 100\% \approx -29.7\%$$

$$\text{Average rate of change} \approx \frac{-\$1100 \text{ billion}}{5 \text{ years}} = -\$220 \text{ billion per year}$$

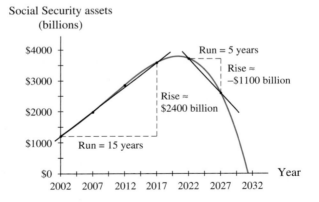

FIGURE 3.6

Between 2022 and 2027, Social Security assets are expected to decrease by about $1100 billion. This represents a percentage decrease of about 29.7%. On average, Social Security assets are expected to decrease by about $220 billion per year during this time. Again, the average rate of change is the slope of the secant line connecting the points on the graph that correspond to 2022 and 2027. (See Figure 3.6.) ●

Determining Percentage Change and Average Rate of Change Using an Equation

It is also possible to determine descriptions of change when we are given only an equation. A model for the temperature data on a typical May day in a certain midwestern city is

$$\text{Temperature} = -0.8t^2 + 2t + 79 \text{ °F}$$

where t is the number of hours after noon.

Calculate the percentage change and average rate of change between 11:30 A.M. and 6 P.M. as follows:

1. Note that at 11:30 A.M., $t = -0.5$ and that at 6 P.M., $t = 6$.

2. Substitute $t = -0.5$ and $t = 6$ into the equation to obtain the corresponding temperatures.

   At 11:30 A.M.: Temperature $= -0.8(-0.5)^2 + 2(-0.5) + 79 = 77.8°F$

   At 6 P.M.: Temperature $= -0.8(6)^2 + 2(6) + 79 = 62.2°F$

3. To calculate percentage change, divide the change in temperature by the temperature at 11:30 A.M. and multiply by 100.

$$\frac{62.2°F - 77.8°F}{77.8°F} \cdot 100\% \approx -20\%$$

4. Divide the change in temperature by the change in time, subtracting the first temperature from the second temperature.

$$\frac{62.2°F - 77.8°F}{6 - (-0.5)} = \frac{-15.6°F}{6.5 \text{ hours}} = -2.4°F \text{ per hour}$$

Thus, between 11:30 A.M. and 6 P.M., the temperature fell by 15.6°F. This represents a 20% decline and an average rate of decline of 2.4°F per hour. Note that the temperature fell 15.6°F in 6.5 hours; however, when finding an average rate of change, we state the answer (in this case) as the number of degrees per *one* hour.

EXAMPLE 3   *Describing Change Using an Equation*

3.1.1a, b

Population   In 2000 Nevada was the fastest growing state in the country. The population density of Nevada* from 1960 through 2000 can be approximated by the model

$$p(t) = 0.1536 \,(1.04892^t) \text{ people per square mile}$$

where t is the number of years since 1900.

a. Find the average rate of change of the population density from 1960 through 1980. Interpret your answer.

*Based on data from *Statistical Abstract*, 1992, 1995, and 2001.

b. Find the average rate of change of the population density between 1980 and 2000.

c. Use the fact that the land area of Nevada is 110,540 square miles to convert both average rates of change to people per year.

d. On the basis of your answers, what can you conclude about growth in Nevada?

e. Calculate the percentage change in the population density between 1960 and 2000. Interpret your answer.

Solution

a. The average rate of change from 1960 through 1980 is

As always, in calculating these answers, we used unrounded values from the function p. Of necessity, we show rounded intermediate values.

$$\frac{p(80) - p(60)}{80 - 60} \approx \frac{7.01 - 2.70 \text{ people/mi}^2}{20 \text{ years}}$$
$$\approx 0.22 \text{ person per square mile per year}$$

Between 1960 and 1980, the population density of Nevada increased at an average rate of 0.22 person per square mile per year.

b. The average rate of change between 1980 and 2000 is

$$\frac{p(100) - p(80)}{100 - 80} \approx \frac{18.22 - 7.01 \text{ people/mi}^2}{20 \text{ years}}$$
$$\approx 0.56 \text{ person per square mile per year}$$

c. From 1960 through 1980:

$$\frac{0.22 \text{ person/mi}^2}{\text{year}} \cdot 110,540 \text{ mi}^2 \approx 23,842 \text{ people per year}$$

From 1980 through 2000:

$$\frac{0.56 \text{ person/mi}^2}{\text{year}} \cdot 110,540 \text{ mi}^2 \approx 61,971 \text{ people per year}$$

d. In the 20-year period from 1980 through 2000, the population grew by over two-and-a-half times as much as it had grown in the 20-year period from 1960 through 1980.

e. The percentage change between 1960 and 2000 is

$$\frac{18.22 - 2.70}{2.70} \cdot 100\% = 575.6\%$$

The population density of Nevada increased by about 575.6% between 1960 and 2000. ●

All three descriptions of change (change, percentage change, and average rate of change) are valuable, but the concept of average rate of change will be the bridge between the algebraic descriptions of change examined in this section and the calculus description of change that we begin exploring in the next section. The graphical interpretation of the average rate of change as a slope of a secant line will be particularly useful.

3.1   Concept Inventory

○ *Change*
○ *Percentage change*
○ *Average rate of change*
○ *Secant line*
○ *Slope of a secant line = average rate of change*
○ *Interpreting descriptions of change*

3.1   Activities

Rewrite the sentences in Activities 1 through 5 to express how rapidly, on average, the quantity changed over the given interval.

1. In five trading days, the stock price rose $2.30.

2. The nurse counted 32 heart beats in 15 seconds.

3. The company lost $25,000 during the past 3 months.

4. In 6 weeks, she lost 17 pounds.

5. The unemployment rate has risen 4 percentage points in the past 3 years.

For Activities 6 through 9, calculate the change, percentage change, and average rate of change over the interval specified. Write a sentence interpreting each description of change.

6. **Insurance**   The average cost for a 20-year, level-premium, term life insurance policy for a 35-year-old male non-smoker was $448 in 1991 and $180 in 1999.
   (Source: *Insurance Information,* based on analysis of more than 600 companies.)

7. **ACT Scores**   The national ACT college test composite average for females was 20.3 in 1990 and 20.7 in 2002.
   (Source: ACT, Inc.)

8. **Population**   The American Indian, Eskimo, and Aleut population in the United States was 362 thousand in 1930 and 2434 thousand in 2000.
   (Source: U.S. Bureau of the Census.)

9. **Internet**   The number of Internet users in China grew from 0.9 million in 1997 to 12.0 million in 2000.
   (Source: BDA (China), The Strategis Group.)

10. **October Madness**

    a. On October 1, 1987, 193.2 million shares were traded on the stock market. On October 30, 1987, 303.4 million shares were traded. Find the

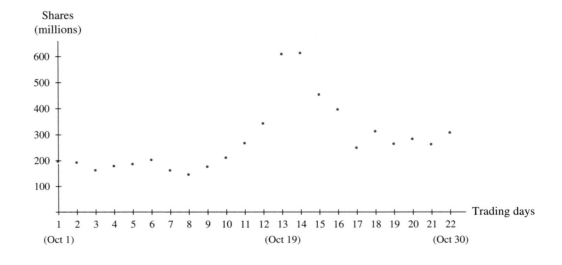

Figure for Activity 10

percent change and average rate of change in the number of shares traded per trading day between October 1 and October 30.

b. The scatter plot in the figure shows the number of shares traded each day during October of 1987. On the scatter plot, sketch a line whose slope is the average rate of change between October 1 and October 30.

c. The behavior of the graph on October 19th and 20th has been referred to as "October Madness." Write a sentence describing how the number of shares traded changed throughout the month. How well does the average rate of change you found in part *a* reflect what occurred throughout the month?

(Source: Phyllis S. Pierce, *The Dow Jones Averages 1885–1990*, ed. Homewood, IL: Business One Irwin, 1991.)

11. Lake Level   The graph in the accompanying figure shows the highest elevations above sea level attained by Lake Tahoe (located on the California-Nevada border) from 1982 through 1996.

(Source: Data from Federal Watermaster, U.S. Department of the Interior.)

a. Sketch a secant line connecting the beginning and ending points of the graph. Find the slope of this line.

b. Write a sentence interpreting the slope in the context of Lake Tahoe levels.

c. Write a sentence summarizing how the level of the lake changed from 1982 through 1996. How well does your answer to part *b* describe

the change in the lake level as shown in the graph?

12. Jobs   Refer once again to the roofing jobs example in Section 2.4. The data are shown in the following table.

Month	Number of jobs
January	117
February	140
March	224
April	368
May	575
June	842

a. Use the data to find the change and percent change in the number of roofing jobs from January through June.

b. Use the data to find the average rate of change in the number of roofing jobs from January through June.

c. Find a model for the data.

d. Use the equation to find the change and the average rate of change in the number of roofing jobs from January through June.

e. Are the answers obtained from the data more accurate than those obtained from the equation?

13. Bank Account   Imagine that 6 years ago you invested $1400 in an account with a fixed interest rate and with interest compounded continuously.

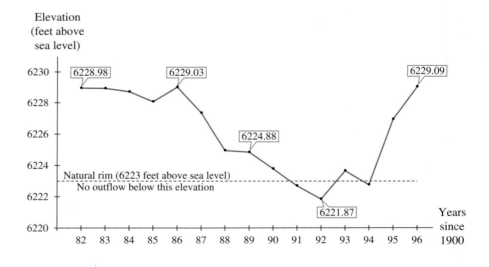

Figure for Activity 11

You do not remember the interest rate, but your end-of-the-year statements for the first 5 years yield the data shown in the table.

End of year	Amount at end of year
1	$1489.55
2	$1584.82
3	$1686.19
4	$1794.04
5	$1908.80

a. Use the data to find the change and percent change in the balance from the end of year 1 through the end of year 5.

b. Use the data to find the average rate of change of the balance from the end of year 1 through the end of year 5. Interpret your answer.

c. Using the data, is it possible to find the average rate of change in the balance from the middle of the fourth year through the end of the fourth year? Explain how it could be done or why it cannot be done.

d. Find a model for the data, and use the equation to find the average rate of change in the balance over the last half of the fourth year.

14. Profit   The table below gives the price in dollars of a round trip flight from Denver to Chicago on a certain airline and the corresponding monthly profit (in millions of dollars) for that airline on that route.

Ticket price (dollars)	Profit (millions of dollars)
200	3.08
250	3.52
300	3.76
350	3.82
400	3.70
450	3.38

a. Find a model for the data.

b. Estimate the average rate of change of profit when the ticket price rises from $200 to $325.

c. Estimate the average rate of change of profit when the ticket price rises from $325 to $450.

15. Life Expectancy   The life expectancy of black males in the United States at various ages for 1998 are as shown in the following table.

Age (years)	Life expectancy (years)	Age (years)	Life expectancy (years)
At birth	68.3	40	32.3
10	59.6	50	24.3
20	50.0	60	17.5
30	41.1	70	11.8

(Source: *National Vital Statistics Reports*, vol. 49, no. 12, Oct. 9, 2001.)

a. How rapidly (on average) does the life expectancy change between birth and the 70th year of life for black males in the United States?

b. Compare the average rates of change of life expectancy for the 10-year periods between ages 10 and 20 and ages 20 and 30.

16. Cruise Tickets   A travel agent vigorously promotes cruises to Alaska for several months. The number of cruise tickets sold during the first week and the total (cumulative) sales every 3 weeks thereafter are given in the table.

Week	Total tickets sold
1	71
4	197
7	524
10	1253
13	2443
16	3660
19	4432
22	4785
25	4923

a. Find the first differences in the numbers of tickets sold, and convert them to average rates of change.

b. When were ticket sales growing most rapidly? How rapidly (on average) were they growing at that time?

c. If the travel agent made a $25 commission on every ticket sold, how rapidly on average did the agent's commission revenue increase between weeks 7 and 10?

17. **Population**   The population of Mexico between 1921 and 2000 is given by the model

    Population $= 7.567(1.02639^t)$ million persons

    where t is the number of years since 1900.
    (Source: Based on data from www.inegi.gob.mx. Accessed 9/20/02.)

    a. How much did the population change from 1940 through 1955? Convert the change to percentage change.

    b. How rapidly was the population changing on average from 1983 through 1985?

18. **AIDS**   The number of persons living with AIDS from 1993 through 2001 can be modeled by

    Cases diagnosed
    $$= -151.281 + 214.725 \ln x \text{ thousand cases}$$

    where x is the number of years since 1990. Find the percent change and the average rate of change in cases diagnosed between 1995 and 1997.
    (Source: Based on data from *HIV/AIDS Surveillance Report*, vol. 13, no. 2.)

19. **Missile Path**   The graph in the accompanying figure shows the path of the misfired missile from Activity 7 of Section 2.4.

    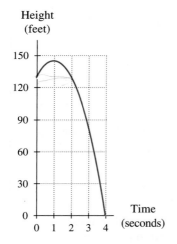

    a. Use a secant line to estimate the average rate of change in position between 0 seconds and 2 seconds.

    b. Use a secant line to estimate the average rate of change in position between 2 seconds and 3 seconds.

    c. Convert your answer in part *b* to miles per hour.

20. **Marriage Age**   The graph in the accompanying figure shows the median age at first marriage for men in the United States. Estimate by how much and how rapidly the median marriage age grew from 1980 through 1990. What is true about the average rate of change between any two points on a linear graph?

    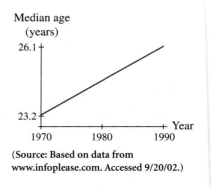

    (Source: Based on data from www.infoplease.com. Accessed 9/20/02.)

21. **Kelly Services**   A graph of the equation for a model for the sales of services between 1991 and 2001 by Kelly Services, Inc., a leading global provider of staffing services, is shown below.

    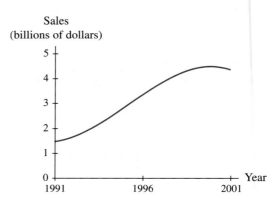

    (Source: Based on data from Kelly Services, Inc., *Annual Reports, 1996–2001.*)

    a. Use the graph to estimate the average rate of change in Kelly's sales of service between 1996 and 2001. Interpret your answer.

    b. Estimate the percentage change in the service sales between 1996 and 2001.

22. **P.T.A.**   The accompanying graph models the number of states associated with the national P.T.A. organization from 1895 through 1931.

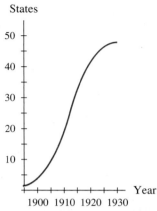

States

(Source: Based on data from Hamblin, Jacobson, and Miller, *A Mathematical Theory of Social Change.* New York: Wiley, 1973.)

a. Approximately how rapidly was the membership growing from 1905 through 1915?

b. Approximately how rapidly was the membership growing from 1920 through 1925?

23. **Surcharges**   The percentage of all banks between 1996 and 2001 that levied surcharges on the use of automated teller machines can be modeled by the equation

$$s(t) = \frac{95.98}{1 + 24{,}612.643e^{-1.415771t}} \text{ percent}$$

t years after 1990. Find the percentage change in the percent of banks assessing surcharges on ATMs between 1996 and 2001.
(Source: Based on data from *U.S. Public Interest Research Group National Survey.*)

24. **Refuse CPI**   The consumer price index values $(1982 - 1984 = 100)$ for refuse collection between 1990 and 2000 can be modeled by the equation

$$R(t) = 132.876 + 54.492 \ln t$$

t years after 1988. Find the average rate of change in the CPI values for refuse collection between 1992 and 2000.
(Source: Based on data from *Statistical Abstract,* 1998.)

25. Consider the linear function $y = 3x + 4$.

a. Find the average rate of change of *y* over each of the following intervals:

   i. From $x = 1$ to $x = 3$

   ii. From $x = 3$ to $x = 6$

   iii. From $x = 6$ to $x = 10$

b. Find the percentage change in *y* for each of the following intervals:

   i. From $x = 1$ to $x = 3$

   ii. From $x = 3$ to $x = 5$

   iii. From $x = 5$ to $x = 7$

c. On the basis of the results in part *a* and your knowledge of linear functions from Chapter 1, what generalizations can you make about percentage change and average rate of change for a linear function?

26. Consider the exponential function $y = 3(0.4^x)$.

a. Find the percentage change and average rate of change of *y* for each of the following intervals:

   i. From $x = 1$ to $x = 3$

   ii. From $x = 3$ to $x = 5$

   iii. From $x = 5$ to $x = 7$

b. On the basis of the results in part *a* and your knowledge of exponential functions from Chapter 2, what generalizations can you make about percentage change and average rate of change for an exponential function?

27. **Medicare**   The annual costs of President Clinton's 1999 proposed prescription drug plan for Medicare patients are given in the table.

Year	Cost (billions of dollars)
2000	0.57
2002	8.57
2004	12.86
2006	15.00
2008	16.43
2009	17.14

(Source: Congressional Budget Office.)

a. Draw a scatter plot of the data, and discuss the curvature.

b. Find a model for the data.

c. Find the change in the proposed annual costs between 2000 and 2005.

d. How quickly are the proposed costs rising on average between 2000 and 2005?

e. What is the percentage change in the proposed annual costs between 2000 and 2005?

28. Funding   Funds awarded to arthritis researchers by the Arthritis Foundation for selected years are given in the table below.

Year	Research funds (millions of dollars)
1990	9.8
1992	10.9
1994	14.6
1996	15.6
1997	18.4
2000	37.5*

*Proposed
(Source: Arthritis Foundation.)

a. Use the data to find the average rate of change of research funds awarded between 1994 and 1997. Interpret your answer.

b. Find a model for the data.

c. Use the equation to estimate the percentage change in research funds awarded between 1998 and 2000.

d. If the Arthritis Foundation's goal for research dollars is determined by the model in part *b*, in what year did the actual amount of funds awarded exceed the projected goal by the greatest amount? By how much did it exceed the goal in that year?

29. Births   The multiple-birth rate for births involving more than twins jumped 19.7% between 1995 and 1996 and 312.4% between 1980 and 1996.

a. If the birth rate for births involving three or more babies was 152.6 per 100,000 births in 1996, find the multiple-birth rates in 1995 and 1980.

b. Use the information presented in the table to find a model for the multiple-birth rate between 1971 and 2000.

Year	Multiple-birth rate (births per 100,000)
1971	29.1
1976	31.7
1981	39.4
1986	49.9
1991	80.2
1996	152.6
2000	180.5

(Sources: *The Greenville News*, July 1, 1998, p. A1; *National Vital Statistics Reports*, vol. 50, no. 5, February 12, 2002.)

c. Use the equation to estimate the multiple-birth rates in 1995 and 1980. How close are those values to the results of part *a*? Are these estimates found with interpolation or extrapolation?

d. Suggest reasons why the multiple-birth rate has been rising rapidly.

30. Incarceration Rate   The table for this activity (located on the *Calculus Concepts* CD-ROM and web site) lists the national and regional incarceration rates for prisoners with sentences of more than 1 year who were imprisoned under state and federal jurisdiction at year end from 1977 through 2001.

a. Determine the average rates of change of the incarceration rates for the four regions of the United States for each of the 5-year periods 1977–1981, 1982–1986, . . . , 1997–2001.

b. Discuss how the average rates of change in the incarceration rates within each region changed between 1977 and 2001.

c. Discuss any similarities and differences in the average rates of change in the incarceration rate between 1977 and 2001 across the four regions.

31. Explain how average rate of change, percentage change, and change relate and how they differ.

32. Give a graphical interpretation of change and average rate of change.

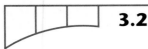

3.2 Instantaneous Rates of Change

TABLE 3.2

Time	Mile marker
1:00 P.M.	0
1:17 P.M.	19
1:39 P.M.	42
1:54 P.M.	56
2:03 P.M.	66
2:25 P.M.	80
2:45 P.M.	105

In Section 3.1 we considered the average rate of change over an interval. Now we consider the concept of the rate of change at a point. The most common example of an instantaneous rate of change is as close as the nearest steering wheel. Suppose that you begin driving north on highway I-81 at the Pennsylvania–New York border at 1:00 P.M. As you drive, you note the time at which you pass each of the indicated mile markers (see Table 3.2).

These data can be used to determine average rates of change. For example, between mile 0 and mile 19, the average rate of change of distance is 67.1 mph. In this context, the average rate of change is simply the average speed of the car. Average speed between any of the mile markers in the table can be determined in a similar manner. Average speed will not, however, answer the following question:

> If the speed limit is 65 mph and a highway patrol officer with a radar gun clocks your speed at mile post 17, were you exceeding the speed limit by more than 10 mph?

The only way to answer this question is to know your speed at the instant that the radar locked onto your car. This speed is the **instantaneous rate of change** of distance with respect to time, and your car's speedometer measures that speed in miles per hour.

Just as an average rate of change measures the slope between two points, an instantaneous rate of change measures the slope at a single point. Figure 3.7 shows a continuous graph of air temperature as a function of time.

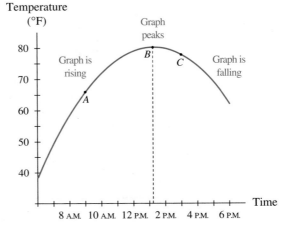

FIGURE 3.7

The graph reaches its peak value at approximately 1:15 P.M. Reading from left to right, the graph is rising until it reaches its peak at 1:15 P.M. and then is falling after 1:15 P.M. The slope of the graph is positive at each point on the left side of the peak and is negative at each point on the right side. The slope of the graph is zero at the top of the parabola. Note that the graph levels off as you move from 7 A.M. to 1:15 P.M. It is not as steep at 1 P.M. as it is at 7 A.M., so the slope at 1 P.M. is smaller than the slope at 7 A.M. In fact, at each point on the graph there is a different slope, and we need to be able to measure that slope in order to find the instantaneous rate of change at each point.

Instantaneous Rate of Change

The **instantaneous rate of change** at a point on a curve is the slope of the curve at that point.

In precalculus mathematics, the concept of slope is intrinsically linked with lines. In terms of lines, slope is a measure of the tilt of a line. Now we wish to measure how tilted any graph is at a point. We can consider the slope of a graph at a point as long as the graph is continuous and not sharp at that point. Intuitively, a continuous curve contains a *sharp point* when the pattern of the curve suddenly changes at that point. We call a continuous function **smooth** over an interval if it has no sharp points in the interval.

Local Linearity and Tangent Lines

For any smooth, continuous graph, we will eventually see a line as we look closer and closer. For example, consider the temperature graph in Figure 3.7. Using the number of hours after 6 A.M. as input, close-ups of the graph 1/100 unit away from each labeled point in both horizontal and vertical directions are shown in Figure 3.8.

3.2.1a, b

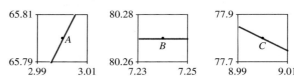

FIGURE 3.8

The graphs in Figure 3.8 illustrate the principle of **local linearity.** As we zoom in on a point *P* on a smooth (no sharp points), continuous curve, the curve will look more and more like a line. We call this line the **tangent line,** and its slope is the instantaneous rate of change of the curve at the point *P*, the **point of tangency.** A tangent line (from the Latin word *tangere,* "to touch") at a point on a graph touches that point and is tilted exactly the way that the graph is tilted at the point of tangency. The slope of the tangent line at a point is a measure of the slope of the graph at that point.

Local Linearity

If we look closely enough near any point on a smooth curve, the curve will look like a line, which is called the tangent line at that point.

The tangent lines at points *A*, *B*, and *C* are shown in Figure 3.9. Do you see that these are the same as the lines in Figure 3.8?

> The slope of a graph at a point is the slope of the tangent line at that point.

In Figure 3.10, tangent lines are drawn on the temperature graph at 7 A.M., noon, and 4 P.M. The tangent lines are tilted to match the tilt of the graph at each point. The tangent lines at points D and E are tilted up, so the slope at these points is positive. The slope at point F is negative, because the tangent line at point F is tilted down. Even though the tangent line at point F has the least slope of these three tangents, it is *steeper* than the tangent line at point E because the magnitude (absolute value) of its slope is larger than that of the line tangent to the curve at point E. That is, the temperature is falling faster at 4 P.M. than it is rising at noon.

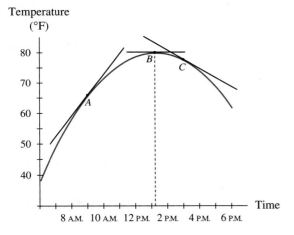

FIGURE 3.9

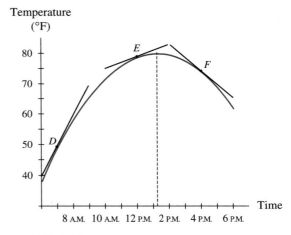

FIGURE 3.10

Examine Figure 3.10 carefully. The slope of the tangent line at point D is 10°F per hour. (A method for calculating this slope will be discussed later.) Therefore, the slope of the graph at 7 A.M. is also 10°F per hour. This is the same as saying that the instantaneous rate of change of the temperature at 7 A.M. is 10°F per hour. In other words, at 7 A.M., the temperature is rising 10°F per hour.

Similarly, the following statements can be made.

● The slope of the tangent line at point E is 2°F per hour.

● The slope of the graph at noon is 2°F per hour.

● The instantaneous rate of change of the temperature at noon is 2°F per hour.

● At noon, the temperature is rising 2°F per hour.

And,

● The slope of the tangent line at point F is -4.4°F per hour.

● The slope of the graph at 4 P.M. is -4.4°F per hour.

● The instantaneous rate of change of the temperature at 4 P.M. is -4.4°F per hour.

● At 4 P.M., the temperature is falling 4.4°F per hour.

We summarize the results of this discussion in the following way:

> Given a function f and a point P on the graph of f, the instantaneous rate of change at point P is the slope of the graph at P and is the slope of the line tangent to the graph at P (provided the slope exists).

EXAMPLE 1   *Comparing Slopes*

NRA Membership   Figure 3.11 shows the National Rifle Association membership between 1990 and 1995. Consider the following statements:

- The slope of the line tangent to the graph at A is -0.24 million members per year.
- The slope of the graph is zero at point B.
- The instantaneous rate of change of the NRA membership at point C is 340,000 members per year.
- NRA membership is increasing the fastest at point D. The rate of fastest increase is 0.42 million members per year.
- The slope of the line tangent to the graph at point E is 260,000 members per year.

NRA members (millions)

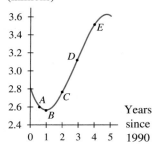

FIGURE 3.11

Using this information, answer the following questions.

a. At which of the indicated points is the slope of the graph (i) the greatest? (ii) the least?

b. At which of the indicated points is the steepness of the graph (i) the greatest? (ii) the least?

c. What is the instantaneous rate of change of the NRA membership at point A? at point E?

d. What is the slope of the tangent line at point C? at point D?

Solution

a. The numerical values of the slopes, in million members per year, at the indicated points are

$$A: -0.24 \qquad B: 0 \qquad C: 0.34 \qquad D: 0.42 \qquad E: 0.26$$

The greatest value occurs at point D (the inflection point), and the least where the only negative slope occurs, at point A.

b. The steepness of the graph is a measure of how much the graph is tilted at a particular point. The direction of tilt is considered in the slope but not when describing steepness. Thus the steepness at each of the indicated points is

$$A: 0.24 \qquad B: 0 \qquad C: 0.34 \qquad D: 0.42 \qquad E: 0.26$$

The graph is steepest at point D. The steepness is least at point B.

c. The instantaneous rate of change at A is -0.24 million, or -240,000, members per year. At E the rate of change is 0.26 million, or 260,000, members per year.

d. The slope of the tangent line at C is 0.34 million members per year. The slope at D is 0.42 million members per year. ●

Secant and Tangent Lines

In addition to understanding tangent lines in terms of local linearity, it is helpful to understand the relationship between secant lines and tangent lines.

Recall that a tangent line at a point on a graph touches that point and is tilted exactly the way the graph is tilted at that point. A secant line, on the other hand, is a line that passes through two points on a graph. We illustrate the relationship between secant lines and tangent lines with an example using points on a simple curve.

EXAMPLE 2   *Drawing Secant and Tangent Lines*

3.2.2

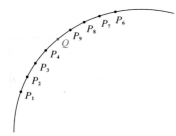

FIGURE 3.12

a. On the curve shown in Figure 3.12, draw secant lines through P_1 and Q, P_2 and Q, P_3 and Q, and P_4 and Q.

b. Which of the four secant lines drawn in part *a* appears to be tilted most like the curve at Q?

c. Again, on the curve shown in Figure 3.12, draw secant lines through P_6 and Q, P_7 and Q, P_8 and Q, and P_9 and Q. Which of these four secant lines appears to most closely match the tilt of the curve at Q?

d. Where could you place a point P_5 so that the secant line through P_5 and Q would be even closer than the secant lines in parts *a* and *c* to tilting the same way the curve does at Q?

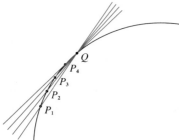

FIGURE 3.13

Solution

a. The four secant lines are shown in Figure 3.13.

b. The secant line through P_4 and Q appears almost to hide the curve near Q. That is, it appears to match the tilt of the curve at Q better than any of the other three secant lines.

c. The four secant lines are shown in Figure 3.14. Again, we see that the secant line through Q and the point closest to Q (P_9 in this case) appears to match the tilt of the curve at Q most closely.

d. The point P_5 should be placed even closer to Q than P_4 and P_9 in order for the tilt of the secant line to match the tilt of the curve at Q more closely. ●

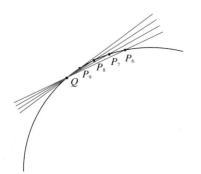

FIGURE 3.14

We generalize the results of Example 2 by saying that the tilt of the secant line through P and Q becomes closer and closer to the tilt of the curve at Q as P becomes closer and closer to Q. Indeed, if you draw a secant line through the point Q and a point P on the curve near Q, then the closer P is to Q, the more closely the secant line approximates the tangent line at Q. You can think of the tangent line at Q as the limiting form of the secant lines between P and Q as P gets closer and closer to Q.

Line Tangent to a Curve

The tangent line at a point Q on a smooth, continuous graph is the limiting position of the secant lines between point Q and a point P as P approaches Q along the graph (if the limiting position exists).

Sketching Tangent Lines

Sketching tangent lines may seem tedious, but this skill is required in later chapters.

Although thinking of a tangent line as a limiting position of secant lines is vital to your understanding of calculus (and is a subject to which we will return), it is important for you to have an intuitive feel for tangent lines and to be able to sketch them without first drawing secant lines. In drawing a line with the same tilt as a curve at a point, you will find that in general, the line lies very close to the curve near the point but does not cut through the curve.

General Rule for Tangent Lines

Lines tangent to a smooth non-linear curve do not "cut through" the graph of the curve at the point of tangency and lie completely on *one side* of the graph near the point of tangency except at an inflection point.

It is important to note that for the graph of a linear function, the only way to draw a line tangent to the graph at any point is to draw again the graph of the linear function.

Drawing a line tangent to a curve at an inflection point is dealt with after Example 3. For cases in which this exception does not apply, we can determine on which side of the curve the tangent line should lie by noting the concavity. If the curve is concave up at the point of tangency, then the tangent line will lie below the curve near the point of tangency. If the curve is concave down at the point of tangency, then the tangent line will lie above the curve near the point of tangency. See Figure 3.15.

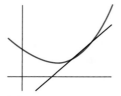   

(a) If the curve is *concave up* at the point of tangency, then the tangent line will lie *below* the curve near the point of tangency.
FIGURE 3.15

(b) If the curve is *concave down* at the point of tangency, then the tangent line will lie *above* the curve near the point of tangency.

EXAMPLE 3   *Using a Tangent Line to Estimate Slope*

Weight Loss   A woman joins a national weight-loss program and begins to chart her weight on a weekly basis. Figure 3.16 shows the graph of a continuous function of her weight from when she began the program through 7 weeks into the program.

FIGURE 3.16

a.   Carefully sketch a line tangent to the curve at 5 weeks.

b.   Estimate the slope of the tangent line at 5 weeks.

c.   How quickly was the woman's weight declining 5 weeks after the beginning of the program?

d.   What is the slope of the curve at 5 weeks?

Solution

a.
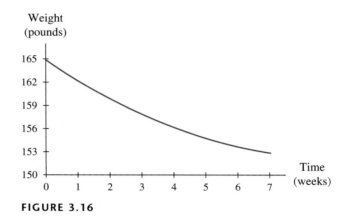

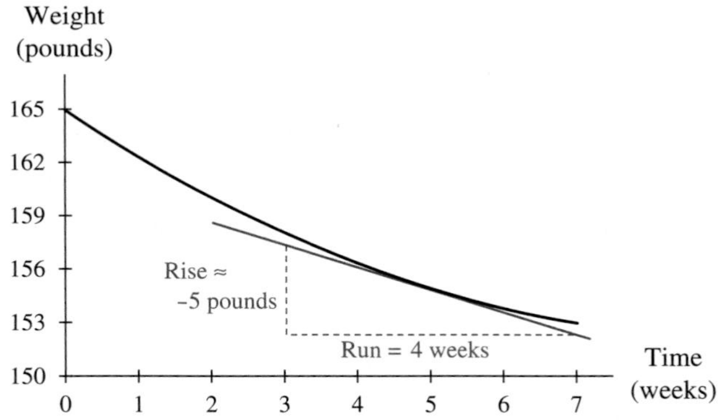

FIGURE 3.17

b.   Slope $= \frac{\text{rise}}{\text{run}} \approx \frac{-5 \text{ pounds}}{4 \text{ weeks}} = -1.25$ pounds per week (See Figure 3.17.)

c.   The woman's weight was declining by approximately 1.25 pounds per week after 5 weeks in the program.

d.   The slope of the curve at 5 weeks is approximately -1.25 pounds per week.   ●

As we mentioned in the general rule for tangent lines, there are exceptions to the principle that tangent lines do not cut through the graph and lie on only one side of the graph. At a point of inflection, the graph is concave up on one side and concave down on the other. As you might expect, the tangent line lies above the concave-down portion of the graph and below the concave-up part. To do this, the tangent line must cut through the graph. It does so at the point of inflection. When drawing tangent lines at inflection points, be careful to make sure that the tangent line is tilted to match the tilt of the graph at the point of tangency. See Figure 3.18.

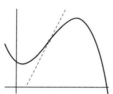

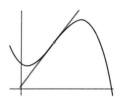

(a) Although this line lies above the concave-down portion and below the concave-up part of the graph, it is not a tangent line because it is not tilted in the same way that the graph is tilted at the point.

(b) This tangent line is correctly drawn at an inflection point.

FIGURE 3.18

Where Does the Instantaneous Rate of Change Exist?

Our discussion of tangent lines would not be complete without a mention of piecewise functions. Consider Table 3.3 showing the population of Indiana by official census from 1950 through 2000 and the graph of a model for the population data. (See Figure 3.19.)

TABLE 3.3

Year	Population (millions)
1950	3.934
1960	4.662
1970	5.195
1980	5.490
1990	5.544
2000	6.080

(Source: *World Almanac and Book of Facts,* ed. William A. McGeveran Jr. New York: World Almanac Education Group, Inc., 2003.)

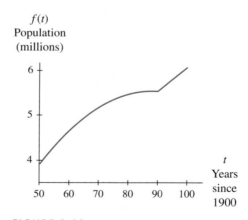

FIGURE 3.19

The equation of the graph in Figure 3.19 is

$$f(t) = \begin{cases} -0.001133t^2 + 0.199t - 3.193 \text{ million people} & \text{when } 50 \leq t < 90 \\ 0.0536t + 0.72 \text{ million people} & \text{when } 90 \leq t \leq 1000 \end{cases}$$

where t is the number of years after 1900.

Consider the tangent line at the point where the function f is not continuous (1990). If we use the idea of a limiting position of secant lines, then we would have to conclude that we cannot draw a line tangent to the graph at $t = 90$ because the graph jumps at this input.

If we use the principle of local linearity and zoom in close to the point at $t = 90$, then we see something similar to Figure 3.20. Recall that if we zoom close enough to a point on a smooth curve, we see a line that is, indeed, the tangent line. In this case, we do not have a smooth curve, and we see two lines. Again we conclude that there is no tangent line at $t = 90$. Does this mean that there is no instantaneous rate of change in the population of Indiana in 1990? No, it simply means that we cannot use our piecewise continuous model to calculate the rate of change in 1990.

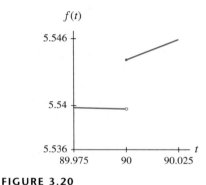

FIGURE 3.20

To help clarify the relationship between continuous and noncontinuous functions and rates of change, consider the graphs shown in Figure 3.21. The graph in Figure 3.21a is continuous everywhere and has a rate of change at every point. You may think that the same thing is true for the graph in Figure 3.21b; however, because a tangent line drawn at P is vertical, the run is zero. Therefore, the slope of the tangent line does not exist at that point. The graph in Figure 3.21c is also continuous; however, the graph is not smooth at P because of the sharp point there. We call a point P on the graph of a continuous function a **sharp point** when secant lines joining P to close points on either side of it have different limiting positions. That is, we cannot draw a tangent line at P, because secant lines drawn with points on the right and left do not approach the same slope.

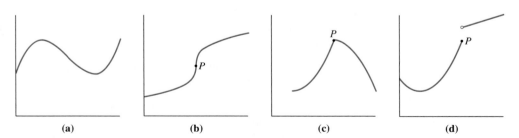

(a)   (b)   (c)   (d)

FIGURE 3.21

The graph in Figure 3.21d has a break at *P* and, therefore, is not continuous at *P*. This graph is similar to the Indiana population graph, and the slope does not exist at the break in the function. The slope does exist at all other points on the graph.

The graph in Figure 3.21d illustrates a general rule relating continuity and rates of change.

If a function is not continuous at a point, then the instantaneous rate of change does not exist at that point.

3.2.3

If you keep in mind the relationships among instantaneous rates of change, slopes of tangent lines, slopes of secant lines, and local linearity, then you should have little difficulty determining the times when the instantaneous rate of change does not exist.

Recall from our discussion in Chapter 1 that calculus is applied to continuous portions of functions. As we have just seen, instantaneous rates of change do not exist where a function is not continuous. Modeling is a tool by which we transform discrete data into a continuous function. Discrete data are useful for finding change, percentage change, and average rates of change, but a continuous or piecewise continuous function is necessary in order to find instantaneous rates of change.

Approximating Instantaneous Rates of Change

Concept Development: Symmetric Difference Quotient   Although a continuous or piecewise continuous model is necessary to calculate an instantaneous rate of change, it is possible to use a **symmetric difference quotient** to *estimate* a rate of change from a set of data without taking the time to construct a model. Quick approximations are sometimes as helpful as a more involved calculation.

TABLE 3.4

Year	1980	1985	1990	1992	1994	1995	1996	1997	1998	1999	2000
Population (thousands)	5490	5459	5544	5649	5746	5792	5835	5872	5908	5943	6080

(Source: *Statistical Abstract,* 2001.)

Consider the data in Table 3.4, which show the population of Indiana between 1980 and 2000. To estimate how quickly the population was changing in 1996 using only the data, simply find the slope of the line between the points on either side of 1996.

$$\frac{5872 - 5792}{1997 - 1995} = \frac{80 \text{ thousand people}}{2 \text{ years}} = 40{,}000 \text{ people per year}$$

In 1996, the population of Indiana was growing at a rate of approximately 40,000 people per year.

A scatter plot of the data and the secant line whose slope is our symmetric difference quotient are shown in Figure 3.22.

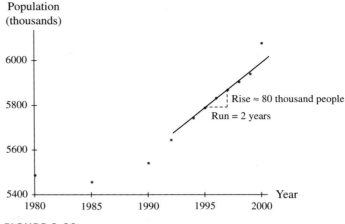

FIGURE 3.22

If you picture a smooth curve through these data points, then the slope of the secant line in Figure 3.22 should be close to the slope of the curve at 1996.

Although *symmetric difference quotient* is a cumbersome name, each word in the phrase is descriptive and should help you remember how to calculate a symmetric difference quotient. We use the term *symmetric* because we choose a close point on either side of our point of interest and the same distance away. (This technique should be used with care if the data are not equally spaced around the point at which we desire to estimate the rate of change). We use the term *difference* because we find the difference in the inputs and outputs of the two points and the term *quotient* because we divide to find the slope that is used to estimate the rate of change.

Again, consider the population data in Table 3.4 and the task of using a symmetric difference quotient to estimate the instantaneous rate of change in 1994. Note that the points on either side of 1994 are not equal distances away from 1994. In order to use a symmetric difference quotient, you would have to calculate the slope of the secant line using the years 1992 and 1996:

$$\text{Slope} = \frac{5835 - 5649}{1996 - 1992} = \frac{186 \text{ thousand people}}{4 \text{ years}} = 46.5 \text{ thousand people per year}$$

To summarize, symmetric difference quotients are useful if you want a fast estimate, if you cannot construct a model for a given data set, or if you need to estimate a rate of change that cannot be calculated from a given model.

Using Symmetric Difference Quotients

A symmetric difference quotient may be used to approximate an instantaneous rate of change if

○   You have a set of data but do not wish to take the time to find a model for the data.

○   You have a set of data that cannot be modeled by one of the functions we have studied or whose model does not have a tangent line at the point at which you need to know the instantaneous rate of change.

3.2 Concept Inventory

○ *Instantaneous rate of change*

○ *Local linearity*

○ *Tangent line*

○ *Slope of tangent line = instantaneous rate of change*

○ *Slope and steepness of a tangent line*

○ *The tangent line is the limiting position of secant lines*

○ *Tangent lines lie beneath a concave-up graph*

○ *Tangent lines lie above a concave-down graph*

○ *Situations in which the instantaneous rate of change does not exist*

○ *Symmetric difference quotient*

3.2 Activities

1. In your own words, describe the difference between

   a. discrete and continuous.

   b. average rate of change and instantaneous rate of change.

   c. secant lines and tangent lines.

2. What are some advantages of using a continuous model instead of discrete data? What are some disadvantages?

3. How are average rates of change and instantaneous rates of change measured graphically?

4. Explain in your own words how to tell visually whether a line is tangent to a smooth graph.

5. Using Table 3.2, the time/mileage table given in this section, verify that the average speed of the car from mile marker 0 to mile marker 19 is 67.1 mph.

6. **a.** Using Table 3.2, determine the average speed (in mph) from:

   i. milepost 66 to milepost 80.

   ii. milepost 80 to milepost 105.

   b. What might account for the difference in speed?

7. **a.** At each labeled point on the graph, determine whether the instantaneous rate of change is positive, negative, or zero.

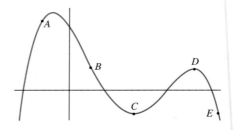

   b. Is the graph steeper at point *A* or at point *B*?

8. At each labeled point on the graph, determine whether the slope is positive, negative, or zero.

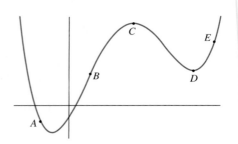

9. **a.** In the graph below, estimate where the output is falling most rapidly. Mark that point on the graph.

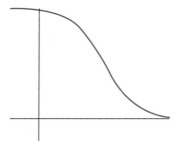

   b. In graph *c* of Activity 10, estimate where the slope is greatest. Mark that point on the graph.

10. Discuss the slopes of the following graphs.

a.

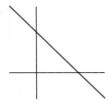

b.

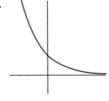

c.

11. Which of the lines drawn on the graph are *not* tangent lines?

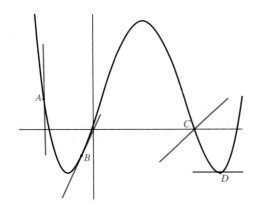

12. On the graph, draw secant lines through P_1 and Q, P_2 and Q, and P_3 and Q. Repeat for the points P_4 and Q, P_5 and Q, and P_6 and Q. Then draw the tangent line at Q.

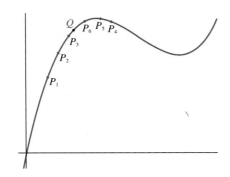

13. Draw secant lines through P_1 and Q, P_2 and Q, and P_3 and Q on the graph. Repeat for the points P_4 and Q, P_5 and Q, and P_6 and Q. Then draw the tangent line at Q.

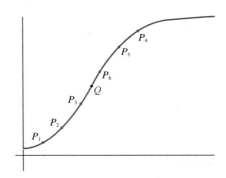

14. Explain using your own words the relationship between secant lines and tangent lines.

15. a. Is the graph shown concave up, concave down, or neither (an inflection point) at A, B, C, and D?

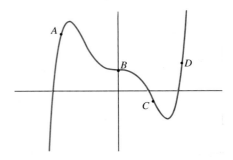

b. Should the tangent lines lie above or below the curve at each of the indicated points?

c. Carefully draw tangent lines at the labeled points on the figure.

d. At which of the labeled points is the slope of the tangent line positive? At which of the labeled points is the slope of the tangent line negative?

16. a. Is the accompanying graph concave up, concave down, or neither at A, B, C, and D?

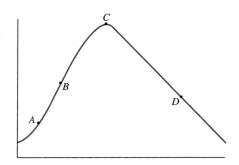

b. Should the tangent lines lie above or below the curve at each of the indicated points?

c. Carefully draw tangent lines at the labeled points.

d. At which of the labeled points is the slope of the curve positive? At which of the labeled points is the slope of the curve negative? Do any of the labeled points appear to be inflection points?

Use carefully drawn tangent lines to estimate the slopes at the labeled points in Activities 17 through 20.

17.

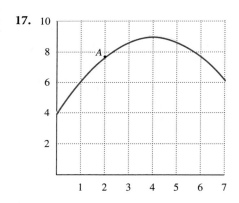

18.

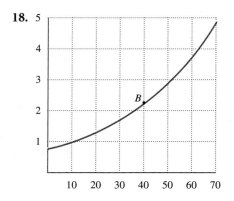

19.

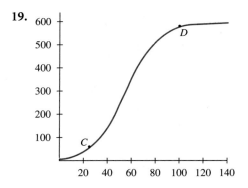

20.

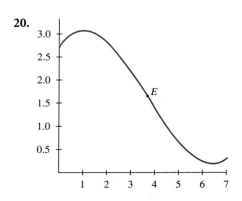

21. Subscribers The graph shows the total number of cellular phone subscribers from 1996 to 2001. The slope of the graph at point *A* is 23.1.

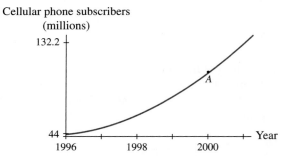

(Source: Based on data from The Cellular Telecommunications and Internet Association.)

a. What are the units on the slope at point *A*?

b. How rapidly was the number of subscribers growing in 2000?

c. What is the slope of the tangent line at *A*?

d. What was the instantaneous rate of change of the number of cell phone subscribers in 2000?

22. Phone Bill The graph shows the average monthly cellular phone bill since 1987. The slope of the curve at point A is -6.23.

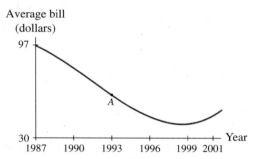

Average bill (dollars)

(Source: Based on data from The Cellular Telecommunications and Internet Association.)

a. What should be the units on the slope at point A?

b. How rapidly was the dollar amount growing in 1993?

c. What is the slope of the tangent line at point A?

d. What is the instantaneous rate of change of the amount at point A?

23. Growth Rate The growth of a pea seedling as a function of time can be modeled by two quadratic functions as shown. The slopes at the labeled points are (in ascending order) -4.2, 1.3, and 5.9.

Growth rate (mm/day)

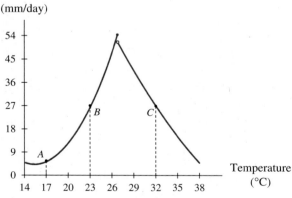

(Source: Based on data in *Elements of Ecology*, George L. Clarke, New York: Wiley, 1954.)

a. Match the slopes with the points, A, B, and C.

b. What are the units on the slopes for each of these points?

c. How quickly is the growth rate changing with respect to temperature at 23°C?

d. What is the slope of the tangent line at 32°C?

e. What is the instantaneous rate of change of the growth rate of pea seedlings at 17°C?

24. Beetles The graph shows the survival rate (percentage surviving) of three stages in the development of a flour beetle (egg, pupa, and larva) as a function of the relative humidity.

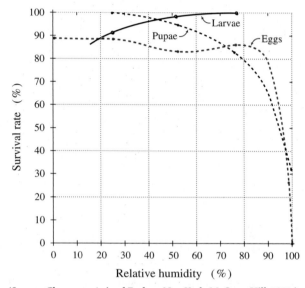

(Source: Chapman, *Animal Ecology*. New York: McGraw-Hill, 1931.)

In parts a through g, fill in each of the following blanks with the appropriate stage (eggs, pupae, or larvae).

a. At 60% relative humidity, the instantaneous rate of change of the survival rate of _____ is approximately zero.

b. An increase in relative humidity improves the survival rate of _____ and reduces the survival rate of _____.

c. At 97% relative humidity, the survival rate of _____ is declining faster than that of _____.

d. Any tangent lines drawn on the survival curve for _____ will have negative slope.

e. Any tangent lines drawn on the survival curve for _____ will have positive slope.

f. At 30% relative humidity, the survival rates for _____ and _____ are changing at approximately the same rate.

g. At 65% relative humidity, the survival curves for _____ and _____ have approximately the same slope.

25. Sun Declination   The figure shows a graph of the declination of the sun (the angle of the sun from the equator) throughout the year.

Declination of sun

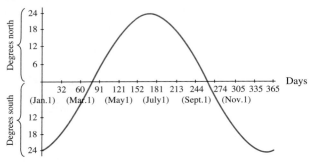

(Source: *The Mathematics Teacher*, March 1997, p. 238.)

a. A *solstice* is a time when the angle of the sun from the equator is greatest. Identify the summer and winter solstices on the graph. What is the slope of the graph at these points?

b. Identify the two steepest points on the graph. Sketch tangent lines at these two points, and estimate their slopes. What is the significance of a negative slope in this context?

c. An *equinox* is a time when the sun crosses the equator, resulting in a day and night of equal length. Identify the points on the graph that correspond to the spring and fall equinoxes.

26. Grasshoppers   The effects of temperature on the percentage of grasshoppers' eggs from West Australia that hatch is shown in the graph.

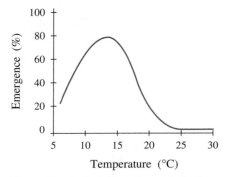

Temperature (°C)

(Source: Figure adapted from George L. Clarke, *Elements of Ecology.* New York: Wiley, 1954.)

a. What is the optimum hatching temperature?

b. What is the slope of the tangent line at the optimum temperature?

c. Sketch tangent lines at 10°C, 17°C, and 22°C, and estimate the slopes at these points.

d. Where does the inflection point appear to be on this graph?

27. Population   Predictions for the U.S. resident population from 1997 through 2050, as reported by *Statistical Abstract* for 1994, can be approximated by the model

$$p(t) = 2370.15t + 39{,}789.957 \text{ thousand people}$$

where t is the number of years since 1900. A graph of p is shown.

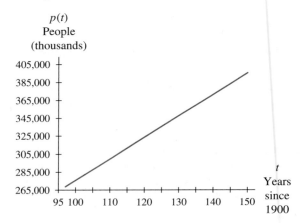

a. Sketch a tangent line at $t = 120$, and find its slope.

b. What is true about any line tangent to the graph of the function p?

c. What is the slope of any line tangent to this graph?

d. What is the slope at every point on the graph of this model?

e. According to the model, what is the instantaneous rate of change of the predicted population in any year from 1997 through 2050?

28. Population   Predictions for the U.S. resident population from 2001 through 2050, as reported by the *Statistical Abstract* for 2001, can be approximated by the model

$$P(x) = 2563.42x + 274{,}721.331 \text{ thousand people}$$

where x is the number of years since 2000.

a. Compare the model in this activity with the one in Activity 27.

b. With the additional information about population available in *Statistical Abstract* for 2001, were the population projections adjusted up or down?

c. Was the growth rate adjusted up or down?

d. Find the slope of the graph of *P* at $t = 20$.

e. Describe the tangent line at $t = 20$.

f. How rapidly does the model predict the population will be changing in 2020?

29. **Employees**   The number of Houghton Mifflin Company employees from 1993 through 2000 can be modeled by the graph.

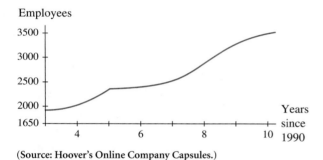

Employees

(Source: Hoover's Online Company Capsules.)

Draw tangent lines, if possible, to estimate how quickly the number of employees was changing in the indicated years. If it is impossible to do so, explain why.

a. 1994   **b.** 1995   **c.** 1998

30. **Employment**   The graph at the bottom of the page shows employment in Slovakia from 1948 through 1988.

a. Estimate how rapidly employment in agriculture and forestry was declining in 1958.

b. Estimate the instantaneous rate of change in industry employment in 1962.

c. Why is it not possible to sketch a tangent line to the industry graph at 1974?

31. **Subscribers**   The number of analog cellular phone subscribers between 1996 and 2001 are as shown in the table.

Year	1996	1997	1998	1999	2000	2001
Analog subscribers (millions)	39	45	47	40	31	17

(Source: The Cellular Telecommunications and Internet Association.)

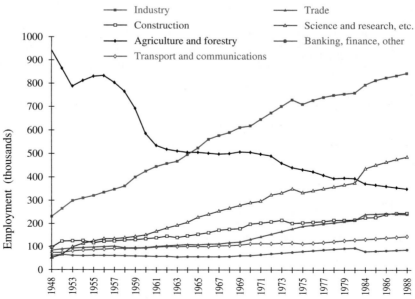

Figure for Activity 30
(Source: Figure from A. Smith, "From Convergence to Fragmentation," *Environment and Planning*, vol. 28, 1996. Pion Limited, London. Reprinted by permission. Data elaborated from *Statisticka röcenta* SUSR, various dates.)

a. Use a symmetric difference quotient to estimate the instantaneous rate of change of revenue in 1999.

b. The figure shows the graph of an equation for the subscriber data.

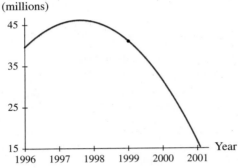

Analog subscribers (millions)

Use the graph and a tangent line to estimate the rate of change of subscribers in 1999. Compare your answer to the estimate in part *a*.

32. Investment   Values of cumulative capital investment in the cellular phone industry, beginning in 1987, are shown.

Year	Cumulative capital investment (billions of dollars)
1987	2.2
1989	4.5
1991	8.7
1993	14.0
1995	24.1
1997	46.1
1999	71.3
2001	105.0

(Source: The Cellular Telecommunications and Internet Association.)

a. Use a symmetric difference quotient to estimate the instantaneous rate of change of capital investment in the cellular phone industry in 1995.

b. A graph of an equation for the data is shown below. Use the graph and a tangent line to estimate the rate of change of capital investment in 1995. Compare your answer to the estimate in part *a*.

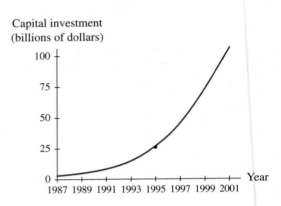

Capital investment (billions of dollars)

33. Most of the piecewise continuous models we have seen thus far have discontinuities at their break points. Consider piecewise continuous functions that are continuous at their break points. Is it possible to draw a tangent line at a break point for such a function? Discuss how and why this might or might not happen. You may find it helpful to use these two functions as examples:

$$f(x) = \begin{cases} -x^2 + 8 & \text{when } x \le 2 \\ x^3 - 9x + 14 & \text{when } x > 2 \end{cases}$$

$$g(x) = \begin{cases} x^3 + 9 & \text{when } x \le 3 \\ 5x^2 - 3x & \text{when } x > 3 \end{cases}$$

3.3   Derivatives

Derivative Terminology and Notation

By now, you should be comfortable with the concepts of average rate of change and instantaneous rate of change. Let's summarize the differences between these two rates of change.

Average Rates of Change

- measure how rapidly (on average) a quantity changes over an interval
- can be obtained by calculating the slope of the secant line between two points
- require discrete data points, a continuous curve, or a piecewise continuous curve

Instantaneous Rates of Change

- measure how rapidly a quantity is changing at a point
- can be obtained by calculating the slope of the tangent line at a single point
- require a continuous or piecewise continuous curve to calculate

---

Because instantaneous rates of change are so important in calculus, we commonly refer to them simply as **rates of change.** The calculus term for instantaneous rate of change is **derivative.** It is important to understand that the following phrases are equivalent.

Equivalent Terminology

All of the following phrases have the *same* meaning.
- instantaneous rate of change
- rate of change
- slope of the curve
- slope of the tangent line
- derivative

Even though we consider all these phrases synonymous, we must keep in mind that the last three phrases have specific mathematical definitions and so may not exist at a point on a function. However, the rate of change of the underlying situation does have an interpretation at that point in context. In such cases, we have to estimate the rate of change by using a symmetric difference quotient, as we saw in Section 3.2, or by using some other estimation technique.

There are also several symbolic notations that are commonly used to represent the rate of change of a continuous function f with input t. In this book, we use three different, but equivalent, symbolic notations:

Equivalent Notation

$\dfrac{df}{dt}$ This is read, *"dee f-dee t,"* *"the rate of change of f with respect to t,"* or *"the derivative of f with respect to t."*

(or)

$f'(t)$ This is read, *"f prime of t,"* or *"the rate of change of f with respect to t,"* or *"the derivative of f with respect to t."*

(or)

$\dfrac{d}{dt}[f(t)]$ This is read, *"dee-dee-t of f of t,"* *"the rate of change of f with respect to t,"* or *"the derivative of f with respect to t."*

Suppose that $G(t)$ is your grade out of 100 points on the next calculus test when you study t hours during the week before the test. The graph of G may look like that shown in Figure 3.23.

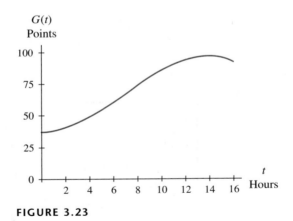

FIGURE 3.23

Note how the grade changes as your studying time increases. The grade slowly improves during the first 2 hours. The longer you study, the more rapidly the grade improves until you have studied approximately 7 hours. After 7 hours, the grade improves at a slower rate. Your grade peaks after 14 hours of studying and then actually declines. What might explain the decline?

Let us compare the rates at which your grade is increasing when $t = 1$ hour and when $t = 4$ hours of study. Tangent lines at $t = 1$ and $t = 4$ are shown in Figure 3.24. After 1 hour of studying, your grade is increasing at a rate of approximately 1.7 points per hour. (We will show you how to calculate, not estimate, this rate in a later section.) This value is the slope of the curve when $t = 1$ hour. After 4 hours of studying, your grade is increasing at a rate of approximately 5.2 points per hour. Can you see that the graph is steeper when $t = 4$ than when $t = 1$? The grade is improving more rapidly after 4 hours than it is after 1 hour. In other words, a small amount of additional study is more beneficial if you have already studied 4 hours than it is when you have studied only 1 hour.

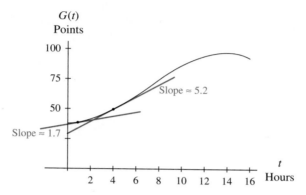

FIGURE 3.24   Tangent lines at $t = 1$ and $t = 4$

These rates can be summarized with the following notation:

$$\frac{dG}{dt} = 1.7 \text{ points per hour when } t = 1 \text{ hour, or}$$

$G'(1) = 1.7$ points per hour, and

$$\frac{dG}{dt} = 5.2 \text{ points per hour when } t = 4 \text{ hours, or}$$

$G'(4) = 5.2$ points per hour.

Interpreting Derivatives

As already mentioned, mathematical results are not very useful in real-world settings unless they are stated in a form that anyone can understand. For this reason, an **interpretation** of a result should be stated using a simple, nontechnical sentence. As in the case of interpreting descriptions of change discussed in Section 3.1, you should answer the questions *when, what, how,* and *by how much* when interpreting a rate of change.

Again, consider your score $G(t)$ out of 100 points on the next calculus exam as a function of the number of hours t that you have studied for the exam. Which of the following is a valid interpretation of $\frac{dG}{dt} = 5.2$ points per hour when $t = 4$ hours?

a.   The rate of change of my grade after 4 hours is 5.2 points per hour.

b.   The slope of the line tangent to the grade curve at $t = 4$ is 5.2.

c.   My grade increased 5.2 points after I studied 4 more hours.

d.   When I have studied for 4 hours, my grade is improving by 5.2 points per hour.

Choice *a* only restates the mathematical symbols in words. It does not give the meaning of the derivative in the real-life context. Choice *b* is a correct statement, but it uses technical words that a person who has not studied calculus probably would not understand. Also, the symbol *t* is used with no meaning attached to it, and units are not included with the value 5.2. The use of the word *increased* in choice *c* refers to an interval of time, not change at a point in time. It is an incorrect statement. Choice *d* is the only valid interpretation.

You should note that because a rate of change is measured at a point, it describes something that is in the process of changing. Therefore, we must use the progressive tense (verbs ending with *-ing*) to refer to rates of change. For example, we say that "after 1 hour of studying, your grade *is increasing* by approximately 1.7 points per hour." It is incorrect to say that your grade *increased* or *increases* at a specific point. These verbs refer to change over an interval rather than at a point.

EXAMPLE 1   *Interpreting Derivatives*

Study Time   Interpret the following four mathematical statements in the context of studying time and grades according to the function *G* whose graph is shown in Figure 3.23.

a. $\frac{dG}{dt} = 6.4$ points per hour when $t = 7$ hours.

b. $G'(12) = 3.0$ points per hour.

c. The derivative of *G* with respect to *t* is 0 points per hour when $t = 14$ hours.

d. The slope of the tangent line when $t = 15$ hours is approximately -2 points per hour.

Solution

a. The first statement says that when you have studied 7 hours, your grade is improving by 6.4 points per hour. As we later learn, this is the point of greatest slope—that is, the time when a small amount of additional study will benefit you the most.

b. The second statement says that after 12 hours of study, your grade is improving by 3.0 points per hour. Does this mean that at 12 hours of study, your grade is less than at 7 hours of study? No! It simply means that a small amount of additional study time beyond 12 hours may not result in as many extra points on your test as the same amount of time produces after you have studied only 7 hours.

c. The third statement says that after you have studied 14 hours, your grade will no longer be improving. A glance back at Figure 3.24 shows that you have reached your best possible score; more study will not improve your grade.

d. The fourth statement tells you that after 15 hours of study, your grade is actually declining by 2 points per hour. Additional study will only hurt your grade.

Make sure that you understand that these statements tell you nothing about what your grade is—they tell you only how quickly it is changing. ●

EXAMPLE 2   *Sketching Function Graphs Using Derivative Information*

Medicine $C(h)$ is the average concentration (in nanograms per milliliter, ng/mL) of a drug in the blood stream h hours after the administration of a dose of 360 mg. On the basis of the following information, sketch a graph of C.

$$C(0) = 125 \text{ ng/mL} \qquad C'(0) = 0 \text{ ng/mL per hour}$$
$$C(4) = 215 \text{ ng/mL} \qquad C'(4) = 37 \text{ ng/mL per hour}$$

The concentration of the drug is increasing most rapidly after 4 hours. The maximum concentration, 380 ng/mL, occurs after 10 hours. Between $h = 10$ and $h = 24$, the concentration declines at a constant rate of 16 ng/mL per hour. The concentration after 24 hours is 31 ng/mL higher than it was when the dose was administered.

Solution   The information about $C(h)$ at various values of h simply locates points on the graph of C. Plot the points $(0, 125)$, $(4, 215)$, $(10, 380)$, and $(24, 156)$.

Because $C'(0) = 0$, the curve has a horizontal tangent at $(0, 125)$. The point of most rapid increase, $(4, 215)$, is an inflection point. The graph is concave up to the left of that point and concave down to the right. The maximum concentration occurs after 10 hours, so the highest point on the graph of C is $(10, 380)$. Concentration declining at a constant rate between $h = 10$ and $h = 24$ means that that portion of C is a line with slope $= -16$.

One possible graph is shown in Figure 3.25. Compare each statement about $C(h)$ to the graph.

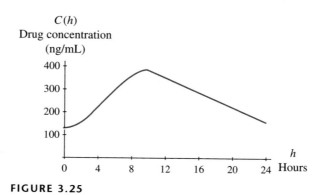

FIGURE 3.25

Note in Figure 3.25 that we cannot assign a value to $C'(10)$. However, the maximum concentration occurs at 10 hours, and on the basis of that, we can estimate that the rate of change of the concentration at that time is zero (even though a horizontal tangent line at $h = 10$ cannot be drawn on the graph in Figure 3.25).   ●

Approximating with Derivatives

Remember that derivatives are simply slopes of tangent lines. Return to Figure 3.23, and consider the point of the grade function graph between 8 and 15 hours studied and the tangent line at $t = 11$ hours (see Figure 3.26). If we magnify the boxed-in portion of this graph (between $t = 11$ and $t = 12$ hours), as seen in Figure 3.27, then we obtain the view of the grade function graph and tangent line at $t = 11$ shown in Figure 3.28.

It so happens that after 11 hours, the derivative (slope) is 4.2 points per hour, and the grade is 90.5 points. What is the grade after 12 hours? It is tempting to reason that if the grade after 11 hours is 90.5 and is increasing by 4.2 points per hour, then after 1 more hour of study, the grade would be $90.5 + 4.2 = 94.7$ points. However, this is not correct, because as Figure 3.28 shows, the grade after 12 hours is 94.2 points. It is the tangent line, not the grade graph, that reaches 94.7 points at 12 hours.

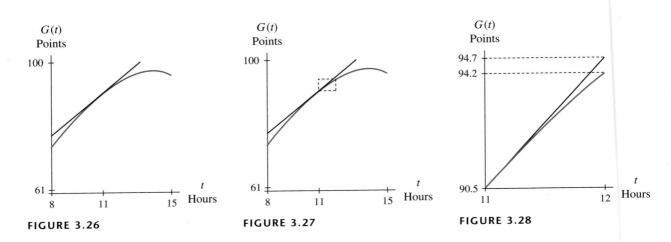

FIGURE 3.26   **FIGURE 3.27**   **FIGURE 3.28**

It is common practice to use tangent lines to estimate function outputs, but this must be done carefully. For example, it is certainly proper to say that, on the basis of a score of 90.5 points and a slope of 4.2 points per hour at 11 hours, the grade will be *approximately* $90.5 + 4.2 = 94.7$ points at 12 hours. You may also say that at 11.5 hours, the grade will be *approximately* $90.5 + 4.2\left(\frac{1}{2}\right) = 92.6$ points. But you must also be certain, in making such statements, to make it clear that the values are only *approximations* and are not exact values. As you can see from the graph in Figure 3.28, such approximations should be made only at points that are relatively close to the point of tangency. We have more to say about approximating with derivatives in a later chapter.

EXAMPLE 3   *Approximating Function Values Using Tangent Line Slopes*

Population   Figure 3.29 shows the graph of a model for the population of Iowa during the 1990s.

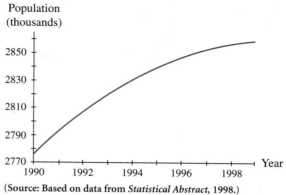

(Source: Based on data from *Statistical Abstract,* 1998.)

FIGURE 3.29

a.   Sketch the tangent line at the point corresponding to 1996, and estimate the slope.

b.   Estimate the population of Iowa in 1996.

c.   Using only your answers to parts *a* and *b*, estimate the population in 1997.

Solution

a. 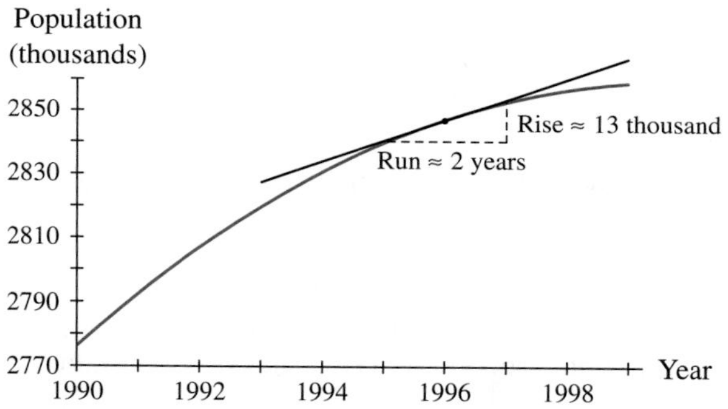   Population
     (thousands)

FIGURE 3.30

$$\text{Slope} = \frac{\text{rise}}{\text{run}} \approx \frac{13 \text{ thousand people}}{2 \text{ years}} = 6.5 \text{ thousand people per year}$$

b.   From the graph, it appears that the population was approximately 2848 thousand people in 1996.

c.   The slope at 1996 is approximately 6.5 thousand people per year, so between 1996 and 1997, the population rose by approximately 6500 people.

$$2848 + 6.5 = 2854.5 \text{ thousand people}$$

In 1997, the population was approximately 2,854,500 people.   ●

You should be aware that in part *c* of Example 3, we were using the tangent line at the point corresponding to 1996 to estimate the population in 1997. Figure 3.31 shows a close-up of the graph and tangent line near the point of tangency. Because the tangent line lies above the graph, the approximation that the tangent line yields overestimates the actual value. The population, according to the graph, was close to 2,852,900.

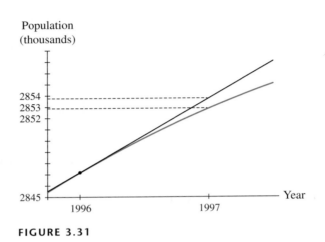

FIGURE 3.31

Does *Instantaneous* Refer to Time?

We saw that the instantaneous rate of change at a point *P* on the graph of a continuous function is the slope of the tangent line to the graph at *P*. If the function inputs are measured in units of time, then it is certainly natural to use the word *instantaneous* when describing rates of change, because each point *P* on the graph of the function corresponds to a particular instant in time. In fact, the use of the word *instantaneous* in this context arose precisely from the historical need to understand how rapidly distance traveled changes as a function of time. Today, we are accustomed to referring to the rate of change of distance traveled as a function of time as *speed*.

You should be aware, however, that the use of the word *instantaneous* in connection with rates of change does not necessarily mean that time units are involved. For example, suppose that a graph depicts profit (in dollars) resulting from the sale of a certain number of used cars. In this case, the slope of the tangent line at any particular point (the instantaneous rate of change) expresses how rapidly profit is changing per car. The unit of change is dollars per car; no time units are involved.

You should also remember that units for instantaneous rates of change, like average rates of change, are always expressed in output units per input unit. Without proper units, a number that purports to describe a rate of change is meaningless.

EXAMPLE 4   *Writing Derivative Notation and Slope Units*

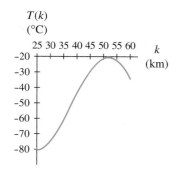

FIGURE 3.32

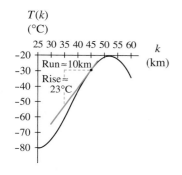

FIGURE 3.33

Temperature   The graph in Figure 3.32* shows the temperature, $T(k)$, of the polar night region (in °C) as a function of k, the number of kilometers above sea level.

a.   Sketch the tangent line at 45 km, and estimate its slope.

b.   What is the derivative notation for the slope of a line tangent to the graph of T?

c.   Write a sentence interpreting in context the meaning of the slope found in part *a*.

Solution

a.   From Figure 3.33 we calculate

$$\text{Slope} = \frac{\text{rise}}{\text{run}} \approx \frac{23°C}{10 \text{ km}} = 2.3°C \text{ per kilometer}$$

b.   Correct derivative notations include $\frac{dT}{dk}$, $\frac{d}{dk}[T(k)]$, and $T'(k)$.

c.   At 45 km above sea level, the temperature of the atmosphere is increasing 2.3°C per kilometer. In other words, the temperature rises by approximately 2.3°C between 45 and 46 km above sea level.   ●

Percentage Rate of Change

Recall from Section 3.1 that percentage change is found by dividing change over an interval by the output at the beginning of the interval and multiplying by 100. Similarly, **percentage rate of change** can be found by dividing the rate of change at a point by the function value at the same point and multiplying by 100. The units of a percentage rate of change are percent per input unit.

> ### Percentage Rate of Change
>
> $$\text{Percentage rate of change} = \frac{\text{rate of change at a point}}{\text{value of the function at that point}} \cdot 100\%$$

3.3.1

Percentage rates of change are useful in describing the relative magnitude of a rate of change. For example, suppose you are a city planner and estimate that the city's population is increasing at a rate of 50,000 people per year. Any growth in population will affect your planning activities, but just how significant is a growth rate of 50,000 people per year? If the current population is 200,000 people, then the percentage rate of change of the population is $\frac{50,000 \text{ people per year}}{200,000 \text{ people}} \cdot 100\% = 25\%$ per year. Growth of 25% per year in population is fast growth. However, if the current population is 2 million, then the percentage rate of change of the population is

---

*"Atmospheric Exchange Processes and the Ozone Problem," in *The Ozone Layer*, ed. Asit K. Biswas, Institute for Environmental Studies, Toronto. Published for the United Nations Environment Program by Pergamon Press, Oxford, 1979.

$\frac{50,000 \text{ people per year}}{2,000,000 \text{ people}} \cdot 100\% = 2.5\%$ per year. The steps that you as a city planner must take to accommodate growth if the city is growing by 25% per year are different from the steps you must take if the city is growing by 2.5% per year. Expressing a rate of change as a percentage puts the rate in the context of the current size and adds more meaning to the interpretation of the rate of change.

EXAMPLE 5   *Graphically Estimating Percentage Rate of Change*

Sales   The graph in Figure 3.34 shows sales (in thousands of dollars) for a small business from 1995 through 2003.

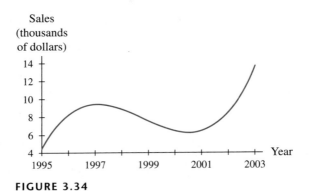

FIGURE 3.34

a. Estimate the rate of change of sales in 1999 and interpret the result.

b. Estimate and interpret the percentage rate of change of sales in 1999.

Solution

a. A tangent line is drawn at 1999 as shown in Figure 3.35.

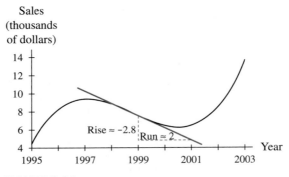

FIGURE 3.35

The slope is estimated to be

$$\frac{-\$2.8 \text{ thousand}}{2 \text{ years}} = -\$1.4 \text{ thousand per year} = -\$1400 \text{ per year}$$

In 1999, sales were falling at a rate of approximately $1400 per year.

b.   We can express this rate of change as a percentage rate of change if we divide it by the sales in 1999. It appears from the graph that the sales in 1999 were approximately $7.5 thousand dollars, or $7500. Therefore, the percentage rate of change in 1999 was approximately

$$\frac{-\$1400 \text{ per year}}{\$7500} \cdot 100\% \approx -0.187 \cdot 100\% \text{ per year} = -18.7\% \text{ per year}$$

In 1999, sales were falling approximately 18.7% per year. Expressing the rate of change of sales as a percentage of sales gives a much clearer picture of the impact of the decline in sales. The business was experiencing a reduction in sales of nearly 20% per year in 1999.   ●

3.3   Concept Inventory

○ *Derivative = rate of change = slope of tangent line*

○ *Derivative notation*

○ *Interpreting derivatives*

○ *Relating the graph of f to statements about f'*

○ *Approximating with derivatives*

○ *Percentage rate of change*

3.3   Activities

1. **Distance**   Suppose that $P(t)$ is the number of miles from an airport that a plane has flown after t hours.

   a.   What are the units on $\frac{d}{dt}[P(t)]$?

   b.   What common word do we use for $\frac{d}{dt}[P(t)]$?

2. **Mutual Fund**   Let $B(t)$ be the balance, in dollars, in a mutual fund t years after the initial investment. Assume that no deposits or withdrawals are made during the investment period.

   a.   What are the units on $\frac{dB}{dt}$?

   b.   What is the financial interpretation of $\frac{dB}{dt}$?

3. **Typing**   Let $W(t)$ be the number of words per minute that a student in a typing class can type after t weeks in the course.

   a.   Is it possible for $W(t)$ to be negative? Explain.

   b.   What are the units on $W'(t)$?

   c.   Is it possible for $W'(t)$ to be negative? Explain.

4. **Corn Crop**   Suppose that $C(f)$ is the number of bushels of corn produced on a tract of farm land when f pounds of fertilizer are used.

   a.   What are the units on $C'(f)$?

   b.   Is it possible for $C'(f)$ to be negative? Explain.

   c.   Is it possible for $C(f)$ to be negative? Explain.

5. **Profit**   Suppose that $F(p)$ is the weekly profit (in thousands of dollars) that an airline makes on its Boston to Washington D.C. flights when the ticket price is p dollars. Interpret the following:

   a. $F(65) = 15$ b. $F'(65) = 1.5$

   c. $\frac{dF}{dp} = -2$ when $p = 90$

6. **Sales** Let $T(p)$ be the number of tickets from Boston to Washington D.C. that a certain airline sells in 1 week when the price of each ticket is p dollars. Interpret the following:

   a. $T(115) = 1750$

   b. $T'(115) = -20$

   c. $\dfrac{dT}{dp} = -2$ when $p = 125$

7. On the basis of the following information, sketch a possible graph of t with input x.

   - $t(3) = 7$
   - $t(4.4) = t(8) = 0$
   - $\dfrac{dt}{dx} = 0$ at $x = 6.2$
   - The graph of t has no concavity changes.

8. Using the information that follows, sketch a possible graph of m with input t.

   - $m(0) = 3$
   - $\dfrac{d}{dt}[m(t)] = 0.34$

9. **Weight Loss** Suppose that $W(t)$ is your weight t weeks after you begin a diet. Interpret the following:

   a. $W(0) = 167$

   b. $W(12) = 142$

   c. $\dfrac{dW}{dt} = -2$ when $t = 1$

   d. $\dfrac{dW}{dt} = -1$ when $t = 9$

   e. $W'(12) = 0$

   f. $W'(15) = 0.25$

   g. On the basis of the information in parts a through f, sketch a possible graph of W.

10. **Fuel Efficiency** Suppose that $G(v)$ equals the fuel efficiency in miles per gallon of a car going v miles per hour. Give the practical meaning of the following statements.

    a. $G(55) = 32.5$

    b. $\dfrac{dG}{dv} = -0.25$ when $v = 55$

    c. $G'(45) = 0.15$

    d. $\dfrac{d}{dv}[G(51)] = 0$

11. **Births** $P(b)$ is the percentage of all births to single mothers in the United States in year b from 1940 through 2000. Using the following information, sketch a graph of P.

    (Source: Based on data from L. Usdansky, "Single Motherhood: Stereotypes vs. Statistics," *The New York Times*, February 11, 1996, Section 4, page E4; and on data from *Statistical Abstract*, 1998.)

    - $P(1940) \approx 4\%$
    - $P'(b)$ is never zero.
    - $P(b) \approx 12\%$ when $b = 1970$
    - $P(2000)$ is about 21 percentage points more than $P(1970)$.
    - The average rate of change of P between 1970 and 1980 is 0.6 percentage points per year.
    - Lines tangent to the graph of P lie below the graph at all points between 1940 and 1990 and above the graph between 1990 and 2000.

12. **Enrollment** $E(x)$ is the public secondary school enrollment (in millions of students) in the United States between 1940 and 2008 x years after 1940. Use the following information to sketch a graph of E.

    (Sources: Based on data appearing in *Datapedia of the United States*, Bernan Press, 1994; and in *Statistical Abstract*, 1998 and 2001.)

    - $E(40) = 13.2$
    - The graph of E is always concave down.
    - Between 1980 and 1990, enrollment declined at an average rate of 0.19 million students per year.
    - The projected enrollment for 2008 is 14,400,000 students.
    - It is not possible to draw a line tangent to the graph of E at $x = 50$.

13. **Profit** Let $P(x)$ be the profit in dollars that a fraternity makes selling x T-shirts.

    a. Is it possible for $P(x)$ to be negative? Explain.

    b. Is it possible for $P'(x)$ to be negative? Explain.

    c. If $P'(200) = -1.5$, is the fraternity losing money? Explain.

14. **Politics** Let $M(t)$ be the number of members in a political organization t years after its founding. What are the units on $\dfrac{d}{dt}[M(t)]$?

15. **Doubling Time**   Let $D(r)$ be the time in years that it takes for an investment to double if interest is continuously compounded at an annual rate of $r\%$. (Here r is expressed as a percentage, not a decimal.)

   a. What are the units on $\frac{dD}{dr}$?

   b. Why does it make sense that $\frac{dD}{dr}$ is always negative?

   c. Give the practical interpretation of the following:

   i. $D(9) = 7.7$

   ii. $\frac{dD}{dr} = -2.77$ when $r = 5$

   iii. $\frac{dD}{dr} = -0.48$ when $r = 12$

16. **Unemployment**   Let $U(t)$ be the number of people unemployed in a country t months after the election of a new president.

   a. Draw and label an input/output diagram for U.

   b. Is U a function? Why or why not?

   c. Interpret the following facts about $U(t)$ in statements describing the unemployment situation:

   i. $U(0) = 3,000,000$

   ii. $U(12) = 2,800,000$

   iii. $U'(24) = 0$

   iv. $\frac{dU}{dt} = 100,000$ when $t = 36$

   v. $\frac{dU}{dt} = -200,000$ when $t = 48$

   d. On the basis of the information in part c, sketch a possible graph of the number of people unemployed during the first 48 months of the president's term. Label numbers and units on the axes.

17. **Raindrop**   The accompanying graph shows the terminal speed (in meters per second) of a raindrop as a function of the size of the drop measured in terms of its diameter.

   a. Sketch a secant line connecting the points for diameters of 1 and 5 mm, and estimate its slope. What information does this secant line slope give?

   b. Sketch a line tangent to the curve at a diameter of 4 mm. What information does the slope of this line give?

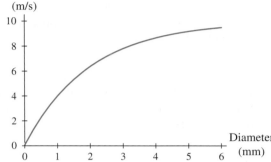

(Source: R. R. Rogers and M. K. Yau, *A Short Course in Cloud Physics*. White Plains, NY: Elsevier Science, 1989.)

   c. Estimate the derivative of the speed for a diameter of 4 mm. Interpret your answer.

   d. Estimate the rate at which the speed is rising for a raindrop with diameter 2 mm. Use this estimate to approximate the terminal speed of a raindrop with diameter 2.5 mm.

   e. Find and interpret the percentage rate of change of speed for a raindrop with diameter 2 mm.

18. **Customers**   The scatter plot and graph depict the number of customers that a certain fast-food restaurant serves each hour on a typical weekday.

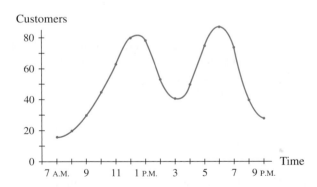

   a. Estimate the average rate of change of the number of customers between 7 A.M. and 11 A.M. Interpret your answer.

   b. Estimate the instantaneous rate of change and percentage rate of change of the number of customers at 4:00 P.M. Interpret your answer.

c. List the factors that might affect the accuracy of your answers to parts *a* and *b*.

d. Use your estimate in part *b* to approximate the number of customers at 5 P.M.

19. Study Time Refer once more to the function *G*, your grade out of 100 points on the next calculus test when you study *t* hours during the week before the test. The graph of *G* is shown below.

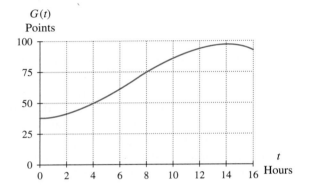

G(t)
Points

a. Carefully draw tangent lines at 4 hours and 11 hours. Estimate the slope of each tangent line.

b. Compare your answers with the slopes given on pages 193 and 196. How accurate are your estimates?

c. Estimate the average rate of change between 4 hours and 10 hours. Interpret your answer.

d. Estimate the percentage rate of change of the grade after 4 hours of study. Interpret your answer.

e. Use $G'(4)$ to estimate your grade after 4.6 hours of study.

20. Mortality Consider the accompanying graph of rates of death from cancer among U.S. males.

a. Estimate how rapidly the number of deaths due to lung cancer was increasing in 1970 and in 1980.

b. Estimate the percentage rate of change of deaths due to liver cancer in 1980.

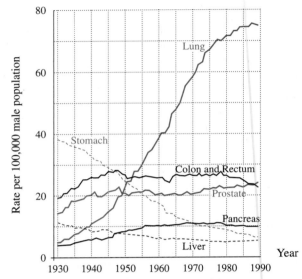

Cancer death rates by site, males, United States, 1930–89
(Rates are adjusted to the 1970 U.S. census population)
(Source: Figure courtesy of the American Cancer Society, Inc.)

c. Estimate the slope of the stomach cancer curve in 1960.

d. Describe in detail the behavior of the lung cancer curve from 1930 to 1990. Explain why the lung cancer curve differs so radically from the other curves shown.

e. List as many factors as you can that might affect a cancer death rate curve.

21. Mortality Consider the graph of rates of death from cancer among U.S. females.

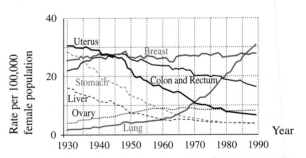

Cancer death rates by site, females, United States, 1930–89
(Rates are adjusted to the 1970 U.S. census population)
(Source: Figure courtesy of the American Cancer Society, Inc.)

a. Estimate how rapidly the number of deaths due to lung cancer was increasing in 1970 and 1980. Estimate the percentage rates of change in each year.

b. Compare the lung cancer death rate curve for females to the curve for males given in Activity 20. What do you think will happen to the two curves during the next 30 years?

22. AIDS The function C gives the cumulative number of AIDS cases since 1985 diagnosed at the end of year t as represented by the curve.

Cumulative AIDS
cases since 1985
(millions)

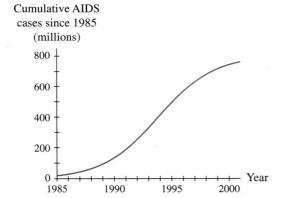

(Source: Based on data in *Statistical Abstract*, 2001.)

a. Estimate and interpret $\frac{dC}{dt}$ at the end of 1993.

b. Estimate and interpret $C'(1996)$.

c. Estimate and interpret $C(1996)$.

d. Estimate the percentage rate of change of the cumulative number of AIDS cases diagnosed by the end of 1996.

e. Estimate the number of AIDS cases diagnosed in 1996. (*Hint:* The graph shows *cumulative* cases since 1984.)

f. Estimate the derivative of C at the end of 1990.

g. Estimate the average rate of change in the cumulative number of AIDS cases between 1985 and 2000.

23. Mortality The accompanying table shows the cumulative totals since 1989 for deaths due to alcohol-related causes.

Year	Cumulative deaths (thousands)
1991	39.0
1993	78.2
1995	118.6
1997	158.0
1999	196.7

(Source: Centers for Disease Control.)

Use a symmetric difference quotient to estimate the rate of change in the cumulative number of alcohol-related deaths in 1995. Interpret your answer.

24. Mortality The table shows the death rates (deaths per 100,000 people) for alcohol-related deaths.

Year	Death rate (deaths per 100,000)	Year	Death rate (deaths per 100,000)
1981	8.1	1991	7.6
1983	7.4	1993	7.6
1985	7.5	1995	7.7
1987	7.4	1997	7.3
1989	8.0	1999	7.1

(Source: Centers for Disease Control.)

a. Use a symmetric difference quotient to estimate the rates of change in the death rate in 1983 and 1993. Interpret your answers.

b. Does a negative rate of change in the death rate indicate that the number of alcohol-related deaths was declining? Explain.

25. What is the meaning of the word *derivative* in the world's financial markets? To find out, see the article entitled "Derivatives? What Are Derivatives?" in *Newsweek,* March 13, 1995, page 50.

26. Explain how percentage change and percentage rate of change relate and how they differ.

27. Describe the process of using tangent lines to approximate function values. Include a discussion of when this technique is most accurate and when it is least accurate.

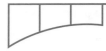

3.4 Numerically Finding Slopes

Finding Slopes by the Numerical Method

By now you should have a firm graphical understanding of rates of change. However, sketching tangent lines is an imprecise method of determining these rates. Although approximations are often sufficient, there are times when we need to find a more precise answer.

Consider the relatively simple problem of finding the slope of the graph of $f(x) = 2\sqrt{x}$ at $x = 1$. Part of the graph of $f(x) = 2\sqrt{x}$ is shown in Figure 3.36a. Take a few moments to sketch carefully a line tangent to the graph at $x = 1$ and estimate its slope. You should find that the tangent line at $x = 1$ has slope approximately 1. See Figure 3.36b.

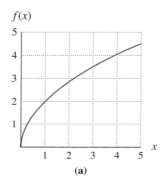
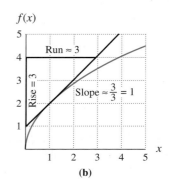

FIGURE 3.36

Another, more precise, method of estimating the slope of the graph of $f(x) = 2\sqrt{x}$ at $x = 1$ uses a technique introduced in Section 3.2. Recall that the tangent line at a point is the limiting position of secant lines through the point of tangency and other increasingly close points. In other words, *the slope of the tangent line is the limiting value of the slopes of secant lines* drawn through the point of tangency.

To illustrate, we begin by finding the slope of the secant line on the graph of $f(x) = 2\sqrt{x}$ through $x = 1$ and $x = 1.5$. (Note that $x = 1.5$ is an arbitrarily chosen value that is close to $x = 1$.) A graph of the secant line is shown in Figure 3.37. Its slope is calculated as follows:

Point at $x = 1$: $\left(1, 2\sqrt{1}\right) = (1, 2)$

Point at $x = 1.5$: $\left(1.5, 2\sqrt{1.5}\right) \approx (1.5, 2.449489743)$

Slope $\approx \dfrac{2.449489743 - 2}{1.5 - 1} = 0.8989794856$

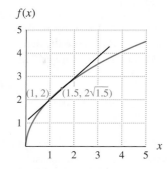

FIGURE 3.37 A secant line through $x = 1$ and $x = 1.5$

This value is an approximation to the slope of the tangent line at $x = 1$. To obtain a better approximation, we must choose a point closer to $x = 1$ than $x = 1.5$, say $x = 1.1$. (This is also an arbitrary choice.) See Figure 3.38.

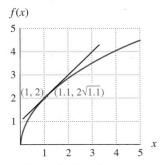

$f(x)$

FIGURE 3.38 A secant line between $x = 1$ and $x = 1.1$

Point at $x = 1$: $\left(1, 2\sqrt{1}\right) = (1, 2)$

Point at $x = 1.1$: $\left(1.1, 2\sqrt{1.1}\right) \approx (1.1, 2.097617696)$

$$\text{Slope} \approx \frac{2.097617696 - 2}{1.1 - 1} = 0.9761769634$$

This is a better approximation to the slope of the tangent line at $x = 1$ than the value from the previous calculation. To get an even better approximation, we need only choose a closer point, such as $x = 1.01$.

Point at $x = 1$: $\left(1, 2\sqrt{1}\right) = (1, 2)$

Point at $x = 1.01$: $\left(1.01, 2\sqrt{1.01}\right) \approx (1.01, 2.009975124)$

$$\text{Slope} \approx \frac{2.009975124 - 2}{1.01 - 1} = 0.9975124224$$

We also use $x = 1.001$.

Point at $x = 1$: $\left(1, 2\sqrt{1}\right) = (1, 2)$

Point at $x = 1.001$: $\left(1.001, 2\sqrt{1.001}\right) \approx (1.001, 2.00099975)$

$$\text{Slope} \approx \frac{2.00099975 - 2}{1.001 - 1} = 0.9997501248$$

As we choose points increasingly close to $x = 1$, what do you observe about the slopes of the secant lines shown in Figure 3.39?

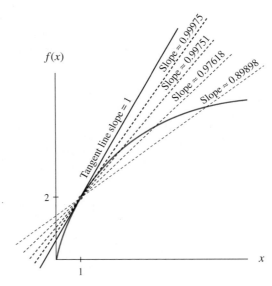

FIGURE 3.39

$x = 1.5$	Secant line slope ≈ 0.8989794856
$x = 1.1$	Secant line slope ≈ 0.9761769634
$x = 1.01$	Secant line slope ≈ 0.9975124224
$x = 1.001$	Secant line slope ≈ 0.9997501248

The pattern in the slope values continues as we choose closer points: for $x = 1.0001$, the slope to six decimal places is 0.999975, for $x = 1.00001$, the slope to seven decimal places is 0.9999975, and so on. Thus the slopes of the secant lines when we use points to the right of $x = 1$ appear to be approaching 1. You may have noticed by now that we are numerically estimating the limit of the slopes of secant lines. Recall from Section 1.3 that a limit exists only if the limit from the left and the limit from the right are equal. For this reason, in order to conclude that the slope of the tangent line at $x = 1$ is 1, we must also consider the limit of the slopes of secant lines using points to the left of $x = 1$. Choosing the x-values 0.5, 0.9, 0.99, and 0.999, we obtain the following slopes of secant lines between these x-values and $x = 1$:

$x = 0.5$	Secant line slope ≈ 1.171572875
$x = 0.9$	Secant line slope ≈ 1.026334039
$x = 0.99$	Secant line slope ≈ 1.002512579
$x = 0.999$	Secant line slope ≈ 1.000250125

Again, note the pattern in the slope values: for $x = 0.9999$, the slope to six decimal places is 1.000025. For $x = 0.99999$, the slope to seven decimal places is 1.0000025. Thus the slopes of secant lines using points to the left of $x = 1$ appear to be approaching 1. Because the limit of slopes using points to the left of $x = 1$ appears to be the same as the limit using points to the right of $x = 1$, we estimate that the slope of the line tangent to the graph of $f(x) = 2\sqrt{x}$ at $x = 1$ is 1. We use algebraic methods in the next section to verify that $f'(1) = 1$.

In this case, the graphical and numerical methods for estimating the slope of the tangent line yield similar results. However, Example 1 shows that calculating the slopes of nearby secant lines generally yields a much more precise result than sketching a tangent line and estimating its slope.

EXAMPLE 1   *Numerically Estimating a Rate of Change*

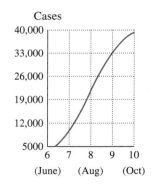

Cases

FIGURE 3.40

Epidemic   The number of polio cases in 1949 can be modeled* by the equation

$$\text{Cases} = \frac{42{,}183.911}{1 + 21{,}484.253e^{-1.248911t}}$$

where $t = 1$ on January 31, 1949, $t = 2$ on February 28, 1949, etc. How rapidly was the number of polio cases growing at the end of August 1949? Report the answer to the nearest whole number.

Solution   First, note that the question asks for the slope of the tangent line at $t = 8$. One method of approximating the slope of the tangent line is to sketch the tangent line and estimate its slope. On the graph in Figure 3.40, draw the tangent line at $t = 8$, and estimate the slope. If you accurately sketched the tangent line and were careful in reading two points off that line, you should have found that the slope of the tangent line is some value between 12,000 and 14,000 cases per month.

To obtain a more precise estimate of the slope of the tangent line, we calculate slopes of nearby secant lines. We choose increasingly close points to both the right

*Based on data from National Foundation for Infantile Paralysis, *Twelfth Annual Report,* 1949.

3.4.1a, b

and the left of the point where $t = 8$: (8, 21,262.9281). We calculate slopes until they remain constant to one decimal place beyond the desired accuracy for two or three calculations. You should verify each of the computations in Table 3.5.

TABLE 3.5

Points to the left	Points to the right
Point at $t = 7.9$: (7.9, 19,946.95986)	Point at $t = 8.1$: (8.1, 22,577.56655)
Slope $= \dfrac{19{,}946.95986 - 21{,}262.9281}{7.9 - 8}$	Slope $= \dfrac{22{,}577.56655 - 21{,}262.9281}{8.1 - 8}$
Slope $\approx 13{,}159.68$	Slope $\approx 13{,}146.38$
Point at $t = 7.99$: (7.99, 21,131.22196)	Point at $t = 8.01$: (8.01, 21,394.62098)
Slope $= \dfrac{21{,}131.22196 - 21{,}262.9281}{7.99 - 8}$	Slope $= \dfrac{21{,}394.62098 - 21{,}262.9281}{8.01 - 8}$
Slope $\approx 13{,}170.62$	Slope $\approx 13{,}169.28$
Point at $t = 7.999$: (7.999, 21,249.75795)	Point at $t = 8.001$: (8.001, 21,276.09819)
Slope $= \dfrac{21{,}249.75795 - 21{,}262.9281}{7.999 - 8}$	Slope $= \dfrac{21{,}276.09819 - 21{,}262.9281}{8.001 - 8}$
Slope $\approx \underline{13{,}170.19}$	Slope $\approx 13{,}170.05$
Point at $t = 7.9999$: (7.9999, 21,261.61112)	Point at $t = 8.0001$: (8.0001, 21,264.24514)
Slope $= \dfrac{21{,}261.61112 - 21{,}262.9281}{7.9999 - 8}$	Slope $= \dfrac{21{,}264.24514 - 21{,}262.9281}{8.0001 - 8}$
Slope $\approx \underline{13{,}170.13}$	Slope $\approx \underline{13{,}170.12}$
Point at $t = 7.99999$: (7.99999, 21,262.79643)	Point at $t = 8.00001$: (8.00001, 21,263.05983)
Slope $= \dfrac{21{,}262.79643 - 21{,}262.9281}{7.99999 - 8}$	Slope $= \dfrac{21{,}263.05983 - 21{,}262.9281}{8.00001 - 8}$
Slope $\approx \underline{13{,}170.12} \approx 13{,}170$	Slope $\approx \underline{13{,}170.12} \approx 13{,}170$

Whether points to the left or right of $t = 8$ are chosen, it seems clear that the slopes are approaching approximately 13,170. That is, the limit of slopes of secant lines using points to the left of $t = 8$ is approximately 13,170, and the limit of slopes of secant lines using points to the right of $t = 8$ is approximately 13,170. Thus we conclude that the slope of the line tangent to the graph at $t = 8$ is approximately 13,170 cases per month. (The actual slope correct to three decimal places is 13,170.122.)

Note that this numerical method of calculating slopes of nearby secant lines in order to estimate the slope of the tangent line at $t = 8$ gives a much more precise answer than graphically estimating the slope of the tangent line. (Be certain that you keep all decimal places in your calculations and enough decimal places in your recorded slope values to be able to see the limit.) We conclude that at the end of August 1949, the number of polio cases was growing at the rate of 13,170 cases per month. ●

Finding Slopes of Piecewise Functions

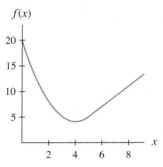

FIGURE 3.41

Remember that when we model data with a piecewise function, we usually end up with one or more points where the graph of the function is not continuous. Even if a piecewise function is continuous, it probably will not be smooth because the pieces of the function generally do not have the same slope where they join. The derivative of the function does not exist at a point where the function is not continuous or at a point where the function is continuous but is not smooth. In order for the derivative to exist at a point where the function is continuous, the slopes that we find by approximating with secant lines through points to the left and through points to the right must be equal. For example, the function represented in Figure 3.41 is defined by

$$f(x) = \begin{cases} x^2 - 8x + 20 & \text{when } 0 \le x \le 5 \\ 2x - 5 & \text{when } x > 5 \end{cases}$$

To determine whether the rate of change of f exists at $x = 5$, we must first determine whether the function is continuous at $x = 5$. We do this by evaluating the function at $x = 5$:

$$f(5) = (5)^2 - 8(5) + 20 = 5$$

Evaluating the linear portion at $x = 5$ yields

$$2(5) - 5 = 5$$

Because the pieces of the function join at $(5, 5)$, the function is continuous at that point.

Next we must determine if the function is smooth at $x = 5$. That is, we need to see if the point with input $x = 5$ is a sharp point on the graph of the function f. We must check the limiting value of the slopes of secant lines for points to the left of $x = 5$ and the limiting value of the slopes of secant lines for points to the right of $x = 5$ to see whether these limiting values are equal (see Table 3.6). The equation to the left of $x = 5$ is quadratic, and we use this in our approximations from the left. The limit from the left appears to be 2.

Recall from Section 3.2 that a continuous function is smooth over an interval if it has no sharp points in the interval.

TABLE 3.6

Close point to left	Slope
(4.9, 4.81)	1.9
(4.99, 4.9801)	1.99
(4.999, 4.998001)	1.999

Now we determine the slope from the right of $x = 5$ by using the equation $f(x) = 2x - 5$. This portion of the function is a line whose slope is 2.

The limit from the left appears to be 2, which is the same as the limit from the right, so our numerical investigation suggests that the derivative is 2 at $x = 5$. In this case, even though the function is defined as piecewise, it is continuous and smooth where the split occurs. The slope of a tangent line drawn at $x = 5$ is 2. In Example 2, we see a function that is not continuous where the two pieces of the function are split.

EXAMPLE 2   *Using Numerical Methods to Investigate Slope at a Break Point*

Employees   At the beginning of the twenty-first century, Comcast Corporation was the third largest U.S. cable communications and broadcast operator. The numbers of Comcast employees between 1994 and 2001 are given in Table 3.7.

a.   Align the input data in Table 3.7 as the number of years since 1990, and find a piecewise model for the data. Split the data in 1998.

TABLE 3.7

Year	1994	1995	1996	1997	1998	1999	2000	2001
Employees (thousands)	6.7	12.2	16.4	17.6	17.0	25.7	35.0	38.0

(Source: Hoover's Online Guide.)

    b. Sketch a graph of the piecewise model.

    c. Do the two pieces of the graph meet in 1998? Is the piecewise function continuous at $x = 8$?

    d. Does the derivative of the piecewise model exist at $x = 8$?

    e. Estimate the rate of change of the number of employees in 1998.

    f. Numerically estimate the rate of change in 2000.

Solution

    a. The number of Comcast employees can be modeled by the equation

$$E(x) = \begin{cases} -1.086x^2 + 15.629x - 38.534 \text{ thousand people} & \text{when } 4 \le x < 8 \\ -1.425x^2 + 34.305x - 166.585 \text{ thousand people} & \text{when } 8 \le x \le 11 \end{cases}$$

where x represents the number of years since 1990.

    b.

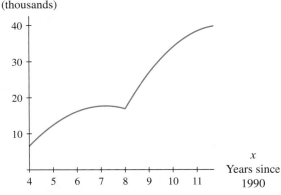

FIGURE 3.42

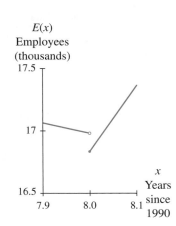

FIGURE 3.43

The unrounded equation is used to calculate the output at 8.

    c. If we look at a magnified view of the graph with input values from $x = 7.9$ to $x = 8.1$ and output values from 16.5 to 17.5, we see how the graphs of the two pieces of the equation behave near $x = 8$. From Figure 3.43, we see that the two pieces of the graph do not meet at $x = 8$. The graph suggests that the function is not continuous at $x = 8$.

        We algebraically verify that the function is not continuous at this input by substituting $x = 8$ into the left and right portions of E.

The left portion of the function evaluated at 8 yields

$$-1.086(8)^2 + 15.629(8) - 38.534 \approx 17.009$$

and the right portion of the function yields

$$E(8) = -1.425(8)^2 + 34.305(8) - 166.585 = 16.655$$

The function is not continuous at $x = 8$ because the left portion and the right portion do not join at a common point.

d. Because the function is not continuous at $x = 8$, the derivative of the function does not exist at that point.

e. Even though the derivative does not exist at $x = 8$ because our piecewise model is not continuous at that point, we can still estimate the rate of change in the number of employees in 1998 by using a symmetric difference quotient (introduced in Section 3.2). Even though a model has been found, we use the data for this estimate because we are now considering the original situation. We use the slope of the line through the data points (1993, 5.4) and (1999, 25.7) to estimate the rate of change in 1994:

$$\text{Slope} = \frac{25.7 - 17.6 \text{ thousand people}}{1999 - 1997} = 4.05 \text{ thousand people per year}$$

This result tells us that the number of employees was increasing by approximately 4.05 thousand people per year in 1998.

It is important that you understand that this value is not an estimate of the rate of change of the function, because that quantity does not exist. The symmetric difference quotient is an estimate of the rate of change of the underlying situation—in this case, the number of employees.

f. Because the model is continuous in 2000, we use the second portion of the function to estimate the rate of change numerically. We use the point (10, 33.965) and the points shown in Table 3.8 to calculate secant line slopes.

TABLE 3.8

We choose to report this slope to one decimal place. Consequently, we look for a pattern in the pertinent digits that are underlined in the table.

Points to the left	Secant line slope	Points to the right	Secant line slope
(9.99, 33.9068)	5.819	(10.01, 34.0229)	5.791
(9.999, 33.9592)	5.806	(10.001, 33.9708)	5.804
(9.9999, 33.9644)	5.805	(10.0001, 33.9656)	5.805

It appears that from the left and right, the slopes of secant lines are approaching 5.8. We therefore estimate that in 2000 the number of employees was growing at a rate of 5.8 thousand employees per year, or 5,800 employees per year. ●

3.4 Concept Inventory

○ *Slope of a tangent line = the limiting value of slopes of secant lines*

○ *Numerical method of estimating the slope of a tangent line*

○ *Slopes of piecewise functions*

○ *The derivative does not exist at a point if the function is not continuous at that point or if the function is continuous but not smooth, that is, if the limiting value of slopes of nearby secant lines from the left does not equal the limiting value of slopes of secant lines from the right.*

3.4 Activities

1. a. Sketch a line tangent to the graph of $y = 2^x$ at the point corresponding to $x = 2$, and estimate its slope.

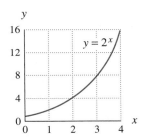

b. Use the equation $y = 2^x$ to estimate numerically the slope of the line tangent to the graph at $x = 2$.

2. a. Sketch a line tangent to the graph of $y = -x^2 + 4x$ at the point corresponding to $x = 3$, and estimate its slope.

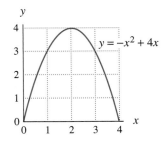

b. Use the equation $y = -x^2 + 4x$ to estimate numerically the slope of the line tangent to the graph at $x = 3$.

3. a. Numerically estimate the limit of slopes of secant lines on the graph of $g(x) = x^3 + 2x^2 + 1$ between the point corresponding to $x = 2$ and close points to the left of $x = 2$.

b. Numerically estimate the limit of slopes of secant lines on the graph of $h(x) = x^2 + 16x$ between the point corresponding to $x = 2$ and close points to the right of $x = 2$.

c. Combine the functions g and h to form the piecewise continuous function

$$f(x) = \begin{cases} x^3 + 2x^2 + 1 & \text{when } x \le 2 \\ x^2 + 16x & \text{when } x > 2 \end{cases}$$

Is the graph of f continuous at $x = 2$? Is the graph of f smooth at $x = 2$?

d. Is it possible to sketch a line tangent to the graph of f at $x = 2$? If so, what is its slope? If not, why not?

4. a. What is the slope of $f(t) = 4t + 15$ at $t = 3$?

b. Numerically estimate the limit of slopes of secant lines on the graph of $h(t) = 8(1.5^t)$ between the point corresponding to $t = 3$ and close points to the right of $t = 3$.

c. Combine the functions f and h to form the piecewise continuous function

$$g(t) = \begin{cases} 4t + 15 & \text{when } t \le 3 \\ 8(1.5^t) & \text{when } t > 3 \end{cases}$$

Is the graph of g continuous at $t = 3$? Is the graph of g smooth at $t = 3$?

d. Is it possible to sketch a line tangent to the graph of g at $t = 3$? If so, what is its slope? If not, why not?

5. AIDS The graph depicts the number of AIDS cases diagnosed between 1994 and 2000. The equation for the graph is Cases $= \dfrac{36.631}{1 + 0.051e^{1.123965x}} + 42$ thousand where x is the number of years since 1994.

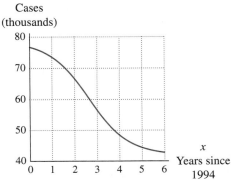

(Source: Based on data from *Statistical Abstract*, 2001.)

a. Sketch the tangent line at $x = 4$, and estimate its slope.

b. Numerically estimate the slope at $x = 4$ using at least three carefully chosen, increasingly close points on either side of $x = 4$.

c. Interpret your answer to part *b* as a rate of change.

6. Bank Account   The balance in a savings account is shown in the graph and is given by the equation Balance = $1500(1.0407^t)$ dollars, where t is the number of years since the principal was invested.

Account balance
(dollars)

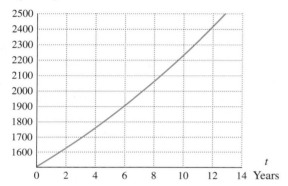

a. Using only the graph, estimate how rapidly the balance is growing after 10 years.

b. Use the equation to investigate numerically the rate of change of the balance when $t = 10$.

c. Which of the two methods (part *a* or part *b*) is more accurate? Support your answer by listing the different estimations that had to be made during each method.

7. VCR Homes   The percentage of households between 1980 and 2001 with VCRs can be modeled by the equation

$$P(t) = \frac{84.430}{1 + 33.605e^{-0.483921t}} \text{ percent}$$

where t is the number of years since 1980. A graph of the function is shown.

a. Use the graph to estimate $\frac{dP}{dt}$ when $t = 7$.

b. Use the equation to investigate $P'(7)$ numerically.

c. Interpret your answer to part *b*.

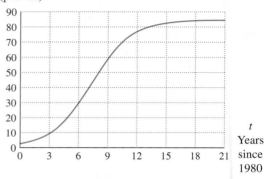

(Sources: Based on data from *Statistical Abstract*, 1998, and Television Bureau of Advertising.)

d. Discuss the advantages and disadvantages of using the two methods (in parts *a* and *b*) for finding derivatives.

8. Swim Time   The time it takes an average athlete to swim 100 meters freestyle at age x years can be modeled by the equation

$$T(x) = 0.181x^2 - 8.463x + 147.376 \text{ seconds.}$$

(Source: Based on data from *Swimming World*, August 1992.)

a. Use the numerical method to find the rate of change of the time for a 13-year-old swimmer to swim 100 m freestyle.

b. Is a 13-year-old swimmer's time improving or getting worse as the swimmer gets older?

9. Sales   Annual U.S. factory sales (in billions of dollars) of consumer electronics goods to dealers from 1990 through 2001 can be modeled by the equation

$$S(t) = 0.0388t^3 - 0.495t^2 + 5.698t + 43.6$$
billion dollars

where t is the number of years since 1990.

(Sources: Based on data from *Statistical Abstract*, 2001, and Consumer Electronics Association.)

a. Estimate the derivative of S when $t = 10$.

b. Interpret your answer to part *a*.

10. Cotton Gins   The accompanying table gives the number of cotton gins in operation in the United States during the early 1900s.

a. Use a symmetric difference quotient to estimate the rate of change in the number of cotton gins in 1915.

b. Find a quadratic model for the data.

Year	Cotton gins	Year	Cotton gins
1900	29,214	1910	26,234
1901	29,254	1911	26,349
1902	30,948	1912	26,279
1903	30,218	1913	24,749
1904	30,337	1914	24,547
1905	29,038	1915	23,162
1906	28,709	1916	21,624
1907	27,592	1917	20,351
1908	27,598	1918	19,259
1909	26,669	1919	18,815

(Source: *Cotton Ginnings in the United States Crop of 1970,* U.S. Department of Commerce, Economics and Statistics Administration, A30–90.)

c. Use the equation to estimate numerically the derivative in 1915.

d. Compare your answers to parts *a* and *c*, and discuss whether part *a* or part *c* gives a more accurate representation of the change that was occurring in 1915.

11. Crime Victims   The number of violent crimes per 1000 persons aged 12 and over where the victim is male is given for 1995 through 2001 in the table. Violent crimes included are homicide, rape, robbery, and both simple and aggravated assault.

Year	Male victim rate (crimes per 1000 persons)
1995	55.7
1996	49.9
1997	45.8
1998	43.1
1999	37.0
2000	32.9
2001	27.3

(Source: Bureau of Justice Statistics, FBI.)

a. Use a symmetric difference quotient to estimate the rate of change of violent crimes against males in 1999.

b. Find a linear model for this data set. What is the slope of the model? Compare this slope to your answer in part *a*, and discuss the information

each contributes to your understanding of the change in violent crimes against males in 1999.

12. Computer Use   The table below gives the number of students per computer in U.S. public schools from 1987 through 2001.

Year	Students per computer	Year	Students per computer
1987	37	1995	10.5
1989	25	1997	7.8
1991	20	1999	5.7
1993	16	2001	4.4

(Sources: *World Almanac and Book of Facts,* ed. Robert Famighetti, Mahwah, NJ: PRIMEDIA Reference, Inc., 1999; *Statistical Abstract,* 2001.)

a. Use a symmetric difference quotient to estimate how rapidly the number of students per computer was declining in 1998.

b. Find a log model as a function of the number of years after 1986 for the data.

c. Use the equation to estimate numerically how rapidly the number of computers in classrooms was growing in 1998.

d. Compare your answers to parts *a* and *c*. Which answer is more accurate? Support your choice.

13. Tuition CPI   The consumer price index (CPI) for college tuition between 1990 and 2000 is shown in the table (1982–1984 = 100).

a. Use a symmetric difference quotient to approximate the rate of change of the consumer price index for college tuition in 1998.

b. Find quadratic and log models for the data.

c. Use both models and at least three carefully chosen close points from each direction to estimate numerically the rate of change of the consumer price index for college tuition in 1998.

d. Compare and contrast the ease and accuracy of the two methods you used in parts *a* and *c*.

(Source: *Statistical Abstract,* 2001)

Year	CPI
1990	175.0
1991	192.8
1992	213.5
1993	233.5
1994	249.8
1995	264.8
1996	279.8
1997	294.1
1998	306.5
1999	318.7
2000	331.9

14. Discuss the advantages and disadvantages of finding rates of change graphically and numerically. Include in your discussion a brief description of when each method might be appropriate to use.

15. **Price** Cattle prices (for choice 450-pound steer calves) from April 1998 through December 1998 can be modeled by the equation

$$p(m) = \begin{cases} 0.3172m^3 - 2.0820m^2 \\ \quad -1.7895m + 98.6398 & \text{when} \\ \quad \text{dollars per 100 pounds} & 0 \le m \le 5 \\ 1.669m + 68.904 \text{ dollars} & \text{when} \\ \quad \text{per 100 pounds} & 0 < m \le 8 \end{cases}$$

where m is the number of months since April 1998.
(Source: Based on data from USDA: Cattle and Beef Industry Statistics, March 1999.)

a. What is the limiting value of slopes of secant lines on the graph of p from the left of $m = 5$?

b. What is the slope of the portion of the graph of p to the right of $m = 5$?

c. What do your answers to parts a and b tell you about the derivative of p at $m = 5$?

d. Estimate the rate of change of cattle prices in September of 1998.

16. **Jails** The capacity of jails in a southwestern state has been increasing since 1990. The average daily population of one of the jails can be modeled by

$$j(t) = \begin{cases} 8.10t^3 - 55.53t^2 + 128.8t + 626.8 \\ \quad \text{inmates} & \text{when} \le t \le 5 \\ 18.8t + 800.6 \text{ inmates} & \text{when } t > 5 \end{cases}$$

where t is the number of years since 1990.
(Source: Based on data from Washoe County Jail, Reno, Nevada.)

a. What is the limiting value of slopes of secant lines on the graph of j from the left of $t = 5$?

b. What is the slope of the portion of the graph of j to the right of $t = 5$?

c. What do your answers to parts a and b tell you about the derivative of j at $t = 5$?

d. Estimate the rate of change of jail population in 1995.

17. **Profit** Let $P(x) = 1.02^x$ Canadian dollars be the profit from the sale of x mountain bikes. On November 25, 2002, P Canadian dollars were worth $C(P) = \frac{P}{1.5786}$ American dollars. Assume that this conversion applies today.

a. Write a function for profit in American dollars from the sale of x mountain bikes.

b. What is the profit in Canadian and in American dollars from the sale of 400 mountain bikes?

c. How quickly is profit (in American dollars) changing when 400 mountain bikes are sold?

18. **Profit** Refer to the functions P and C in Activity 17.

a. Write a function giving average profit per mountain bike for the sale of x mountain bikes in Canadian dollars.

b. Write a function for average profit in American dollars.

c. How quickly is average profit (in American dollars) changing when 400 mountain bikes are sold?

19. **Social Security** The table for this activity (located on the *Calculus Concepts* CD-ROM and web site) gives the number of retired workers and survivors receiving OASI benefits. Consider the year and widow/widower beneficiaries columns in this table.

a. Divide the data into two subsets that have the 1996 data point in common. Find an equation to fit each set of data, and combine the two to form a piecewise continuous model W with input x years after 1975.

b. Use the model in part a and at least five carefully chosen close points from each direction to estimate numerically the rate of change of W at $x = 23$. Interpret this result.

c. What is the limiting value of the slopes of secant lines of the graph of W from the left of $x = 21$? What is the slope of the portion of the graph of W to the right of $x = 21$?

d. What do your answers to part c tell you about the derivative of W at $x = 21$?

e. Estimate the rate of change of the number of widow/widower beneficiaries in 1996.

20. **Population** The table for this activity (located on the *Calculus Concepts* CD-ROM and web site) gives the U.S. population for ages 0 through 85 by gender between 1990 and 1999 with projections from 2000 through 2020.

a. Construct a column of data showing by how much the female population exceeds the male population between 1990 and 2020.

b. Divide the data in the constructed column into two subsets that have one data point in common. Find an equation to fit each set of data, and combine the two to form a piecewise continuous model P with input t years after 1990.

c. What is the limiting value of the slopes of secant lines of the graph of P from the left of the input at the break point? What is the slope of the portion of the graph of P to the right of the input at the break point?

d. What do your answers to part c tell you about the derivative of P at the break point?

e. Use a symmetric difference quotient to estimate the rate of change of P at the break point. Interpret this result.

21. Explain why there may be differences between the numerical estimate of a rate of change of a modeled function at a point and the actual rate of change that occurred in the underlying real-world situation.

3.5   Algebraically Finding Slopes

When we have enough carefully chosen close points to observe the trend in the secant line slopes, the numerical method is a fairly good way to find a slope to a specified accuracy. For this reason, the process of numerically estimating slope is valuable, but keep in mind that it gives only an estimation of the actual slope. However, we can generalize the numerical method to develop an algebraic method that will give the exact slope of a tangent line at a point.

Finding Slopes Using the Algebraic Method

Consider finding the slope of $f(x) = \frac{1}{7}x^2 + 3x$ at $x = 2$. We begin with the point $\left(2, 6\frac{4}{7}\right)$. Instead of choosing close x-values such as 1.9, 2.1, 2.01, and so on, we simply call the close value $x = 2 + h$. (Note that if $h = 0.1$, then $x = 2.1$; if $h = 0.01$, $x = 2.01$; if $h = -0.01$, $x = 1.99$, and so on.) The output value that corresponds to $x = 2 + h$ is

$$f(2 + h) = \frac{1}{7}(2 + h)^2 + 3(2 + h)$$

$$= \frac{1}{7}(4 + 4h + h^2) + 6 + 3h = 6\frac{4}{7} + 3\frac{4}{7}h + \frac{1}{7}h^2$$

Next we find the slope of the secant line between the point of tangency $\left(2, 6\frac{4}{7}\right)$ and the close point $\left(2 + h, 6\frac{4}{7} + 3\frac{4}{7}h + \frac{1}{7}h^2\right)$.

$$\text{Slope of secant line} = \frac{\left(6\frac{4}{7} + 3\frac{4}{7}h + \frac{1}{7}h^2\right) - 6\frac{4}{7}}{(2 + h) - 2} = \frac{3\frac{4}{7}h + \frac{1}{7}h^2}{h}$$

We now have a formula for the slope of the secant line through the points at $x = 2$ and $x = 2 + h$. We can apply the formula to obtain slopes at points increasingly close to $x = 2$ (see Table 3.9).

TABLE 3.9

Close point to left	h	Slope $= \dfrac{3\frac{4}{7}h + \frac{1}{7}h^2}{h}$	Close point to right	h	Slope $= \dfrac{3\frac{4}{7}h + \frac{1}{7}h^2}{h}$
1.9	−0.1	3.55714	2.1	0.1	3.58571
1.99	−0.01	3.57	2.01	0.01	3.57286
1.999	−0.001	3.57129	2.001	0.001	3.57157
1.9999	−0.0001	3.57141	2.0001	0.0001	3.57144
1.99999	−0.00001	3.57143	2.00001	0.00001	3.57143

We arbitrarily chose three-decimal place accuracy for the answer in this illustration. We underline the digits that give one decimal position beyond our desired accuracy once we see that these digits do not change. Once we have two or three underlined values (in Table 3.9, two values) in each column, we stop calculating the left and right slopes and report an answer.

The slopes become increasingly close to approximately 3.571.

Numerically evaluating the secant line slope formula for smaller and smaller values of h gives us a good picture of where the slopes are headed, but it does not give us the exact answer. By noting that h approaches zero as the close point approaches 2, we can find the limit of the secant line slope formula as h approaches zero. That is, we need to find

$$\lim_{h \to 0} \frac{3\frac{4}{7}h + \frac{1}{7}h^2}{h}$$

The secant line slope formula is not continuous at $h = 0$ because it is not defined there. However, a rule called the Cancellation Rule for limits states:

> If the numerator and denominator of a rational function share a common factor, then the new function obtained by algebraically canceling the common factor has all limits identical to those of the original function.

If we factor h out of the numerator and cancel this common factor, then we can find the limit by evaluating the new expression at $h = 0$ because the new expression $\left(3\frac{4}{7} + \frac{1}{7}h\right)$ is continuous at $h = 0$.

$$\lim_{h \to 0} \frac{h\left(3\frac{4}{7} + \frac{1}{7}h\right)}{h} = \lim_{h \to 0}\left(3\frac{4}{7} + \frac{1}{7}h\right) = 3\frac{4}{7} + \frac{1}{7}(0) = 3\frac{4}{7}$$

On the basis of this limit calculation, we state that the slope of the line tangent to $f(x) = \frac{1}{7}x^2 + 3x$ at $x = 2$ is exactly $3\frac{4}{7}$.

This method of finding a formula for the slope of a secant line in terms of h and then determining the limiting value of the formula as h approaches zero is called the **algebraic method.** It is important because it always yields the *exact* slope of the tangent line.

EXAMPLE 1   *Using the Algebraic Method to Find the Slope at a Point*

Foreign-Born U.S. Residents   The percentage of people residing in the United States who were born abroad can be modeled* by the equation

$$f(x) = 0.000044x^3 - 0.0027x^2 - 0.16x + 14.93 \text{ percent}$$

where x is the number of years since 1910. (*Note:* The model is based on data from 1910 through 2000.)

a.   Find $\frac{df}{dx}$ at $x = 80$ by writing an expression for the slope of a secant line in terms of h and then evaluating the limit as h approaches 0.

b.   Interpret $\frac{df}{dx}$ at $x = 80$ in the context given.

Solution

a.   First, we find the output of f when $x = 80$.

$$f(80) \approx 7.4$$

Second, we write an expression for $f(x)$ when $x = 80 + h$.

$$f(80 + h) = 0.000044(80 + h)^3 - 0.0027(80 + h)^2 - 0.16(80 + h) + 14.93$$

We simplify this expression as much as possible, using the facts that

$$(80 + h)^2 = 6400 + 160h + h^2$$

and

$$(80 + h)^3 = (80 + h)^2 (80 + h) = 512{,}000 + 19{,}200h + 240h^2 + h^3$$

Thus

$$f(80 + h) = 7.378 + 0.2528h + 0.00786h^2 + 0.000044h^3$$

Third, we find the slope of the secant line through the two close points $(80, f(80))$ and $(80 + h, f(80 + h))$ as

$$\text{Slope of secant} = \frac{f(80 + h) - f(80)}{(80 + h) - 80}$$

$$= \frac{(7.378 + 0.2528h + 0.00786h^2 + 0.000044h^3) - 7.378}{(80 + h) - 80}$$

---

*Based on data from *World Almanac and Book of Facts,* ed. Robert Famighetti, Mahwah, NJ: PRIMEDIA Reference Inc., 1999; and U.S. Census Bureau.

Again, simplify this as much as possible.

$$\text{Slope of secant line (continued)} = \frac{0.2528h + 0.00786h^2 + 0.000044h^3}{h}$$

$$= \frac{h(0.2528 + 0.00786h + 0.000044h^2)}{h}$$

Finally, we find the slope of the tangent line (the derivative) at $x = 80$ by evaluating the limit of the slope of the secant line as h approaches 0.

$$\frac{df}{dx} = \text{slope of tangent line} = \lim_{h \to 0} \frac{h(0.2528 + 0.00786h + 0.000044h^2)}{h}$$

$$= \lim_{h \to 0} (0.2528 + 0.00786h + 0.000044h^2)$$

$$= 0.2528$$

b. In 1990, the percentage of the U.S. population who were foreign-born was increasing at a rate of 0.2528 percentage point per year. ●

A General Formula for Derivatives

The real value of the algebraic method is not in finding a slope at a particular point but in finding general formulas for derivatives. These rate-of-change (or slope) formulas can be used to find rates of change for many input values.

To illustrate, consider $y = x^2$. Because we desire a general equation for any x-value, we use (x, x^2) as the point of tangency. This is the same idea as the algebraic method in Example 1 and the preceding discussion, but there we worked with a numerical value for x. Next we choose a close point. We use $x + h$ as the x-value of the close point and find the y-value by substituting $x + h$ into the function: $y = (x + h)^2 = x^2 + 2xh + h^2$. Thus the original point is (x, x^2), and a close point is $(x + h, x^2 + 2xh + h^2)$. Now we find the slope of the secant line between these two points.

$$\text{Slope of the secant line} = \frac{(x^2 + 2xh + h^2) - x^2}{(x + h) - x} = \frac{2xh + h^2}{h}$$

Finally, we determine the limiting value of the secant line slope as h approaches 0.

$$\lim_{h \to 0} \frac{2xh + h^2}{h} = \lim_{h \to 0} (2x + h) = 2x + 0 = 2x$$

Therefore, the slope formula for $y = x^2$ is $\frac{dy}{dx} = 2x$. Using this formula, we find that the slope of the graph of $y = x^2$ is 6 at $x = 3$, -12 at $x = -6$, 0 at $x = 0$, 1 at $x = 0.5$, and so on.

This method can be generalized to obtain a formula for the derivative of an arbitrary function.

Four-Step Method to Find $f'(x)$

Given a function f, the equation for the derivative with respect to x can be found as follows:

1. Begin with a typical point $(x, f(x))$.

2. Choose a close point $(x + h, f(x + h))$.

3. Write a formula for the slope of the secant line between the two points.

$$\text{Slope} = \frac{f(x + h) - f(x)}{(x + h) - x} = \frac{f(x + h) - f(x)}{h}$$

It is important at this step to simplify the slope formula.

4. Evaluate the limit of the slope as h approaches 0.

$$\lim_{h \to 0} \frac{f(x + h) - f(x)}{h}$$

This limiting value is the derivative formula at each input where the limit exists.

Thus we have the following derivative formula (slope formula, rate-of-change formula) for an arbitrary function:

Derivative Formula

If $y = f(x)$, then the derivative $\frac{dy}{dx}$ is given by the formula

$$\frac{dy}{dx} = \lim_{h \to 0} \frac{f(x + h) - f(x)}{h}$$

provided the limit exists.

Example 2 illustrates the Four-Step Method for finding a derivative formula.

EXAMPLE 2   *Using the Four-Step Method to Find a Slope Formula*

Labor   The number of women in the U.S. civilian labor force* from 1930 through 1990 can be modeled as

$$W(t) = 0.011t^2 + 0.101t + 10.817 \text{ million women}$$

where t is the number of years since 1930.

Information Please Almanac, Atlas, and Yearbook, 1996 (Boston: Houghton Mifflin, 1996).

a. Use the limit definition of the derivative (the Four-Step Method) to develop a formula for the rate of change of the number of women in the labor force.

b. How quickly was the number of women in the U.S. civilian labor force growing in 1975?

Solution

a. ***Step 1.*** A typical point is $(t, 0.011t^2 + 0.101t + 10.817)$.

   Step 2. A close point is $(t + h, 0.011 (t + h)^2 + 0.101(t + h) + 10.817)$. We rewrite the output of the close point before we proceed:

$$0.011(t + h)^2 + 0.101(t + h) + 10.817$$
$$= 0.011(t^2 + 2th + h^2) + 0.101(t + h) + 10.817$$
$$= 0.011t^2 + 0.022th + 0.011h^2 + 0.101t + 0.101h + 10.817$$

   Step 3. Write the slope of the secant line between the two points from Steps 1 and 2.

$$\text{Slope} = [(0.011t^2 + 0.022th + 0.011h^2 + 0.101t + 0.101h + 10.817) - (0.011t^2 + 0.101t + 10.817)] \div [(t + h) - t]$$
$$= \frac{0.022th + 0.011h^2 + 0.101h}{h}$$
$$= \frac{h(0.022t + 0.011h + 0.101)}{h}$$

   Step 4. Evaluate the limit of the secant line slope as h approaches 0.

$$W'(t) = \lim_{h \to 0} \frac{h(0.022t + 0.011h + 0.101)}{h} = \lim_{h \to 0}(0.022t + 0.011h + 0.101)$$
$$= 0.022t + 0.101$$

   Thus the number of women in the civilian labor force was increasing by

$$W'(t) = 0.022t + 0.101 \text{ million women per year}$$

   t years after 1930.

b. We find that $W'(45) \approx 1.1$ million women per year. Thus the number of women in the U.S. civilian labor force was growing by approximately 1.1 million women per year in 1975. ●

Using the Four-Step Method to find a derivative formula for a non-polynomial function may require more thoughtful algebra.

EXAMPLE 3   *Algebraically Finding a Slope Formula*

3.5.1

Consider the function $f(x) = 2\sqrt{x}$.

a. Use the Four-Step Method to find a formula for the slope graph of f.

b.   Use the formula from part *b* to find the slope of the graph of *f* at $x = 1$. Compare this answer with the one found by numerical estimation in Section 3.4.

Solution

a.   To find a slope formula, begin with a general point $\left(x, 2\sqrt{x}\right)$ and a close point $\left(x + h, 2\sqrt{x + h}\,\right)$. Next find the slope between these two points:

$$\text{Secant line slope} = \frac{2\sqrt{x + h} - 2\sqrt{x}}{x + h - x} = \frac{2\sqrt{x + h} - 2\sqrt{x}}{h}$$

Now find the limit of this formula as *h* approaches zero:

$$\lim_{h \to 0} \frac{2\sqrt{x + h} - 2\sqrt{x}}{h}$$

Unlike the case for polynomial functions that always contain *h* as a common factor, we cannot cancel the *h* here without rewriting the numerator. The key to this cancellation is to rewrite the numerator by multiplying the numerator and denominator by the term $2\sqrt{x + h} + 2\sqrt{x}$. Observe how this multiplication enables us to cancel the *h* term:

$$\lim_{h \to 0} \frac{2\sqrt{x + h} - 2\sqrt{x}}{h} \cdot \frac{2\sqrt{x + h} + 2\sqrt{x}}{2\sqrt{x + h} + 2\sqrt{x}}$$

$$= \lim_{h \to 0} \frac{4(x + h) - 4x}{h(2\sqrt{x + h} + 2\sqrt{x})}$$

$$= \lim_{h \to 0} \frac{4h}{h(2\sqrt{x + h} + 2\sqrt{x})}$$

$$= \lim_{h \to 0} \frac{4}{2\sqrt{x + h} + 2\sqrt{x}}$$

$$= \frac{4}{2\sqrt{x + 0} + 2\sqrt{x}} = \frac{4}{4\sqrt{x}} = \frac{1}{\sqrt{x}}$$

Thus the slope formula is $f'(x) = \dfrac{1}{\sqrt{x}}$.

b.   Using this slope formula, we find that $f'(1) = \dfrac{1}{\sqrt{1}} = 1$. That is, the slope of the line tangent to the graph of $f(x) = 2\sqrt{x}$ at $x = 1$ is 1. This calculation confirms that the numerical estimate in Section 3.4 is correct.   ●

The definition of the derivative of a function gives us a formula for the slope graph of the function, which enables us to calculate exact rates of change quickly. Unfortunately, this method is primarily for polynomial functions and not for exponential, logarithmic, or logistic functions. However, in Chapter 4, we will use the algebraic method as a powerful tool to help us develop some general rules for derivative formulas.

3.5   Concept Inventory

○ *Using the algebraic method to find the slope of a graph at a given point*

○ *Using the Four-Step Method to find a rate-of-change formula*

○ *Limit definition of a derivative*

3.5   Activities

1. **Swim Time**   The time it takes an average athlete to swim 100 meters freestyle at age x years can be modeled by the equation $T(x) = 0.181x^2 - 8.463x + 147.376$ seconds.

   a. Find the swim time when $x = 13$.

   b. Write a formula for the average swim time when $x = 13 + h$.

   c. Write a simplified formula for the slope of the secant line connecting the points at $x = 13$ and $x = 13 + h$.

   d. What is the limiting value of the slope formula in part c as h approaches 0? Interpret your answer.

2. **Sales**   Annual U.S. factory sales of consumer electronics goods to dealers between 1990 and 2001 can be modeled by the equation

   $$S(x) = 0.0388x^3 - 0.495x^2 + 5.698x + 43.6$$
   billion dollars

   where x is the number of years since 1990.
   (Sources: Based on data from *Statistical Abstract*, 2001, and Consumer Electronics Association.)

   a. Find the sales cases when $x = 10$.

   b. Write an expression for the sales when $x = 10 + h$.

   c. Write a simplified formula for the slope of the secant line connecting the points at $x = 10$ and $x = 10 + h$.

   d. What is the limiting value of the slope formula in part c as h approaches 0? Interpret your answer.

   e. How do your answers to parts a and b of Activity 9 in Section 3.4 and part d of this activity compare? Which method is the most accurate and why?

3. **Tuition CPI**   The CPI for college tuition between 1990 and 2000 can be modeled by the equation

   $$c(t) = -0.498t^2 + 20.603t + 174.458$$

   where t is the number of years since 1990.
   (Source: Based on data from *Statistical Abstract*, 2001.)

   a. Find the consumer price index for college tuition in 1998.

   b. Write a formula in terms of h for the consumer price index of college tuition a little after 1998.

   c. Write a simplified formula for the slope of the secant line connecting the points at 1998 and a little after 1998.

   d. What is the limiting value for the slope formula as h approaches 0?

   e. Interpret your answer to part d.

   f. How do your answers to part c of Activity 13 in Section 3.4 and part d of this activity compare? Which method should you use in each of the following situations?

      i.   You are concerned most with accuracy.

     ii.   You want a quick, rough estimate.

    iii.   You want a fairly good estimate without taking much time.

4. Discuss the advantages and disadvantages of finding rates of change graphically, numerically, and algebraically. Include in your discussion a brief description of when each method might be appropriate to use.

5. Consider the function $g(t) = -6t^2 + 7$. Use the algebraic method to find $\frac{dg}{dt}$ at $t = 4$ by evaluating the limit of an expression for the slope of a secant line.

6. Consider the function $m(p) = 4p + p^2$. Use the algebraic method to find $\frac{dm}{dp}$ at $p = -2$ by evaluating the limit of an expression for the slope of a secant line.

7. **Falling Object** An object is dropped off a building. Ignoring air resistance, we know from physics that its height above the ground t seconds after being dropped is given by

$$\text{Height} = -16t^2 + 100 \text{ feet}$$

   a. Use the Four-Step Method to find a rate-of-change equation for the height.

   b. Use your answer to part a to determine how rapidly the object is falling after 1 second.

8. **Distance** Clinton County, Michigan, is mostly flat farmland partitioned by straight roads (often gravel) that run either north/south or east/west. A tractor driven north on Lowell Road from the Schafer's farm is

$$d(m) = 0.28m + 0.6 \text{ miles}$$

   north of Howe Road m minutes after leaving the farm's drive.

   a. How far is the Schafer's drive from Howe Road?

   b. Use the Four-Step Method to show that the tractor is moving at a constant speed.

   c. How quickly (in miles per hour) is the tractor moving?

9. **Drivers** The number of licensed drivers between the ages of 16 and 21 in 1997 is given below.

Age (years)	Number of drivers (millions)
16	0.85
17	1.24
18	1.41
19	1.47
20	1.54
21	1.51

(Source: U.S. Department of Labor and Transportation.)

   a. Find a quadratic model for the number of licensed drivers as a function of age. Round the

coefficients of the equation to three decimal places.

   b. Use the limit definition of the derivative to develop a formula for the derivative of the rounded model.

   c. Use the derivative formula in part b to find the rate of change of the function in part a for an age of 20 years. Interpret your answer.

   d. Find the percentage rate of change in the number of licensed drivers 20 years old. Interpret this result.

10. **Drivers** The data give the percentage of females of a certain age who were licensed drivers in 1997.

Age (years)	Licensed drivers (percent)
15	0.4
16	43.4
17	61.9
18	72.7
19	73.8

(Source: U.S. Department of Labor and Transportation.)

   a. Find a quadratic model for these data. Round the coefficients in the equation to three decimal places.

   b. Use the limit definition of the derivative to develop the derivative formula for the rounded equation.

   c. Use the derivative formula in part b to find the rate of change of the equation in part a when the input is 16 years of age. Interpret your answer.

   d. Find the percentage rate of change in the number of female licensed drivers 20 years old. Interpret this result.

In Activities 11 through 15, use the Four-Step Method outlined in this section to show that each statement is true.

11. The derivative of $y = 3x - 2$ is $\frac{dy}{dx} = 3$.

12. The derivative of $y = 15x + 32$ is $\frac{dy}{dx} = 15$.

13. The derivative of $y = 3x^2$ is $\frac{dy}{dx} = 6x$.

14. The derivative of $y = -3x^2 - 5x$ is $\frac{dy}{dx} = -6x - 5$.

15. The derivative of $y = x^3$ is $\frac{dy}{dx} = 3x^2$.

    (*Hint*: $(x + h)^3 = x^3 + 3x^2h + 3xh^2 + h^3$)

16. **Labor Force**   Recall from Example 2 the model for the number of women in the U.S. civilian labor force between 1930 and 1990 and the derivative formula for the model:

    $W(t) = 0.011t^2 + 0.101t + 10.817$ million women

    $W'(t) = 0.022t + 0.101$ million women per year

    t years after 1930. Using more current data, the number of women in the U.S. civilian labor force between 1970 and 2000 can be modeled by the equation

    $C(x) = -0.0121x^2 + 1.4925x + 31.5727$ million women

    x years after 1970.
    (**Source: Based on data from *Statistical Abstract*, 2001.**)

    a. Use the function W and its derivative to estimate the number of women in the civilian labor force in the years 1945, 1995, and 2003 and to estimate the rates of change in those years.

    b. Use the Four-Step Method to find the derivative formula for the function C.

    c. Use the function C and its derivative to estimate the number of women in the U.S. civilian labor force in the years 1945, 1995, and 2003 and to estimate the rates of change in those years.

    d. Compare the answers to parts *a* and *c*. Which answers do you believe are more accurate? Why?

    WEB/CD **e.** Find the most recent data available for the number of women in the U.S. civilian labor force. Use all the available data to construct a model. Round the coefficients in the equation to three decimal places.

    Source

    f. Use the Four-Step Method to find the derivative formula for the function in part *e*.

    g. Use the model in part *e* and its derivative to estimate the number of women in the U.S. civilian labor force in 1945, 1995, and 2003 and the rates of change in those years. Compare your answers to those you obtained in parts *a* and *c*. Do you believe the answers based on the model in part *e* are more reliable than the ones in parts *a* and *c*? Explain.

17. Explain from a graphical viewpoint how algebraically finding a slope formula relates to numerically estimating a rate of change.

SUMMARY

This chapter is devoted to describing change: the underlying concepts, the language, and proper interpretations.

Change, Average Rate of Change, and Percentage Change

The change in a quantity over an interval is a difference of output values. Apart from describing the actual change in a quantity that occurs over an interval, change can be described as the average rate of change over an interval or as a percentage change. The numerical description of an average rate of change has an associated geometric interpretation—namely, the slope of the secant line joining two points on a graph.

Instantaneous Rates of Change

Whereas average rates of change indicate how rapidly a quantity changes (on average) over an interval, instantaneous rates of change indicate how rapidly a quantity is changing at a point. The instantaneous rate of change at a point on a graph is simply the slope of the line tangent to the graph at that point. It describes how quickly the output is increasing or decreasing at that point.

Tangent Lines

The principle known as local linearity guarantees that the graph of any continuous function looks like a line if you are close enough. A line tangent to a graph at a point is the line you see when you zoom in on the graph closer and closer to that point.

The line tangent to a graph at a point P can also be thought of as the limiting position of nearby secant lines—that is, secant lines through P and nearby points on the graph. It reflects the tilt, or slope, of the graph at the point of tangency. We can estimate the instantaneous rate of change at a point on a curve by sketching a tangent line at that point and approximating the tangent line's slope.

Derivatives and Percentage Rate of Change

Derivative is the calculus term for (instantaneous) rate of change. Accordingly, all of the following terms are synonymous: derivative, instantaneous rate of change, rate of change, slope of the curve, and slope of the line tangent to the curve.

Three common ways of symbolically referring to the derivative of a function G with respect to x are $\frac{dG}{dx}$,

$\frac{d}{dx}[G(x)]$, and $G'(x)$. The proper units on derivatives are output units per input unit. Rates of change also can be expressed as percentages. A percentage rate of change describes the relative magnitude of the rate.

We can quickly estimate a rate of change at a point by finding the slope between points an equal distance away from, and on either side of, our point of interest. This slope, which is called a symmetric difference quotient, usually gives a good approximation to the rate of change.

Numerically and Algebraically Finding Slopes

When we have an equation $y = f(x)$ to associate with the curve, we can improve our graphical approximations of the slope of the tangent line with numerical approximations of the limit of slopes of secant lines. The method of numerically estimating slopes can be generalized to provide a valuable algebraic method for finding exact slopes at points, as well as formulas for slopes at any input value. We call this method the Four-Step Method of finding derivatives. This method yields the formal definition of a derivative: If $y = f(x)$, then

$$\frac{dy}{dx} = f'(x) = \lim_{h \to 0} \frac{f(x+h) - f(x)}{h}$$

provided that the limit exists.

CONCEPT CHECK

Can you

To practice, try

○ Find and interpret change, percentage change, and
average rates of change

using data?	Section 3.1	Activity 13
using graphs?	Section 3.1	Activity 21
using equations?	Section 3.1	Activity 17

○ Find the slope of a secant line? — Section 3.1 — Activity 19

○ Understand the relationship between secant lines
and tangent lines? — Section 3.2 — Activities 1, 3

○ Accurately sketch tangent lines?
○ Use tangent lines to estimate rates of change?
○ Use symmetric difference quotients to estimate rates of change?
○ Understand derivative notation?
○ Correctly interpret derivatives?
○ Find and interpret percentage rate of change?
○ Determine if a function is continuous and/or smooth?
○ Graphically and numerically estimate rates of change?
○ Use the algebraic method to find a rate of change at a point?
○ Use the Four-Step Method to find a rate-of-change formula?

REVIEW TEST

1. **Speed**   Answer the following questions about the graph:

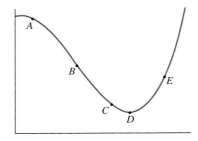

   a. List the labeled points at which the slope appears to be (i) negative, (ii) positive, and (iii) zero.

   b. If *B* is the inflection point, what is the relationship between the steepness at *B* and the steepness at the points *A*, *C*, and *D* on the graph?

   c. For each of the labeled points, will a tangent line at the point lie above or below the graph?

   d. Sketch tangent lines at points *A*, *B*, and *E*.

   e. Suppose the graph represents the speed of a roller coaster (in feet per second) as a function of the number of seconds after the roller coaster reached the bottom of the first hill.

      i. What are the units on the slopes of tangent lines? What common word is used to describe the quantity measured by the slope in this context?

      ii. When, according to the graph, was the roller coaster slowing down?

      iii. When, according to the graph, was the roller coaster speeding up?

      iv. When was the roller coaster's speed the slowest?

      v. When was the roller coaster slowing down most rapidly?

2. **Visa Cards**   The number of Visa cards worldwide for 1996 through 2002 are shown in the table.

Year	1996	1998	2000	2002
Visa cards (millions)	510	656	1000	1200

(Source: *Visa Press Release*, www.visa.com. (Access dates: 11/23/99 and 10/7/02.)

   a. Find the average rate of change in the number of Visa cards between 1996 and 2002. Interpret your answer.

   b. Find the percentage change in the number of Visa cards between 2000 and 2002. Interpret your answer.

3. Employees   The graph of a model for the number of Dell Computer Corporation employees between 1992 and 2002 is shown below.

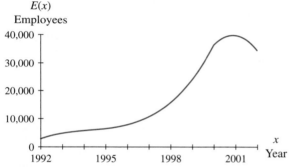

(Source: Based on data in Hoover's Online Guide Company Capsules.)

   a. Sketch a secant line between the points with inputs 1993 and 1997. Describe the information that the slope of this line provides.

   b. Sketch a tangent line at the point corresponding to 1998. Describe the information that the slope of this line provides.

   c. Estimate the rate of change and percentage rate of change in the number of employees in 1998. Interpret your answers.

   d. Estimate the average rate of change in the number of employees between 1993 and 1997. Interpret your answer.

4. Swim Time $T(x)$ is the number of seconds that it takes an average athlete to swim 100 meters free style at age x years.

   a. Write sentences interpreting $T(22) = 49$ and $T'(22) = -0.5$.

   b. What does a negative derivative indicate about a swimmer's time?

5. Stock Value   The yearly high stock price for Microsoft Corporation for selected years between 1989 and 2000 is shown.

Year	Price per share (dollars)	Year	Price per share (dollars)
1989	0.74	1996	15.73
1990	2.11	1997	33.73
1992	4.38	1998	54.28
1994	6.83	2000	119.94

(Source: Hoover's Online Guide.)

   a. Find an exponential model for the data.

   b. Numerically investigate the rate of change of the stock price in 1999. Choose at least three increasingly close points. In table form, record the close points, the slopes with four decimal places, and the limiting value with two decimal places.

   c. Interpret the limiting value in part *b.*

   d. Give the formula for the derivative of the equation in part *a.* Evaluate the derivative in 1999.

6. In your own words, outline the Four-Step Method for calculating derivatives. Illustrate the method for the function $f(x) = 7x + 3$.

Project 3.1 Fee-Refund Schedules

Setting

Some students at many colleges and universities enroll in courses and then later withdraw from them. Such students may have part-time status upon withdrawing. Part-time students have begun questioning the fee-refund policy, and a public debate is taking place. Recently, the student senate passed a resolution condemning the current fee-refund schedule. Then, the associate vice president issued a statement claiming that further erosion of the university's ability to retain student fees would reduce course offerings. The Higher Education Commission has scheduled hearings on the issue. The Board of Trustees has hired your firm as consultants to help them prepare their presentation.

Tasks

1. Examine the current fee-refund schedule for your college or university. Present a graph and formula for the current refund schedule. Critique the refund schedule.

   Create alternative fee-refund schedules that include at least two quadratic plans (one concave up and one concave down), an exponential plan, a logistic plan, a no-refund plan, and a complete-refund plan. (*Hint:* Linear models have constant first differences. What is true about quadratic and exponential models?) For each plan, present the refund schedule in a table, in a graph, and with an equation. Critique each plan from the students' viewpoint and from that of the administration.

   Select the nonlinear plan that you believe to be the best choice from both the students' and the administration's perspectives. Outline the reasons for your choice.

2. Estimate the rate of change of your selected equation for withdrawals after 1 week, 3 weeks, and 5 weeks. Include any other times that are indicated by your school's schedule. Interpret the rates of change in this context. How might the rate of change influence the administration's view of the model you chose? Would the administration consider a different model more advantageous? If so, why? Why did you not propose it as your model of choice?

Reporting

1. Prepare a written report of your results for the Board of Trustees. Include scatter plots, models, and graphs. Include in an appendix the reasoning that you used to develop each of your models.

2. Prepare a press release for the college or university to use when it announces the adoption of your plan. The press release should be succinct and should answer the questions *who, what, when, where,* and *why.* Include the press release in your report to the Board.

3. (Optional) Prepare a brief (15-minute) presentation on your work. You will be presenting it to members of the Board of Trustees of your college or university.

Project 3.2 Doubling Time

Setting

Doubling time is defined as the time it takes for an investment to double. Doubling time is calculated by using the compound interest formula $A = P\left(1 + \frac{r}{n}\right)^{nt}$ or the continuously compounded interest formula $A = Pe^{rt}$. An approximation to doubling time can be found by dividing 72 by 100r. This approximating technique is known as the **Rule of 72.**

Dr. C. G. Bilkins, a nationally known financial guru, has been criticized for giving false information about doubling time and the Rule of 72 in seminars. Your team has been hired to provide mathematically correct information for Dr. Bilkins to use.

Tasks

1. Construct a table of doubling times for interest rates from 2% to 20% (in increments of 0.25%) when interest is compounded annually, semiannually, quarterly, monthly, weekly, and daily. Construct a table of doubling-time approximations for interest rates of 2% through 20% when using the Rule of 72. Devise similar rules for 71, 70, and 69. Then construct tables for these rules. Examine the tables and determine the best approximating rule for interest compounded semiannually, quarterly, monthly, weekly, and daily. Justify your choices.

    For each interest compounding listed above, compare percent errors when using the Rule of 72 and when using the rule you choose. Percent error is $\frac{\text{estimate} - \text{true value}}{\text{true value}}$ 100%. Comment on when the rules overestimate, when they underestimate, and which is preferable.

2. Dr. Bilkins is interested in knowing how sensitive doubling time is to changes in interest rates. Estimate rates of change of doubling times at 2%, 8%, 14%, and 20% when interest is compounded quarterly. Interpret your answers in a way that would be meaningful to Dr. Bilkins.

Reporting

1. Prepare a written report for Dr. Bilkins in which you discuss your results in Tasks 1 and 2. Be sure to discuss whether Dr. Bilkins should continue to present the Rule of 72 or present other rules that depend on the number of times interest is compounded.

2. Prepare a summary document for Dr. Bilkins. It should include (a) a brief summary of how to estimate doubling time using an approximation rule and (b) a statement about the error involved in using the approximation. Also include a brief statement summarizing the sensitivity of doubling time to fluctuations in interest rates. Include the document in your written report.

3. (Optional) Prepare a brief (15-minute) presentation of your study. You will be presenting it to Dr. Bilkins. Your presentation should be only a summary, but you need to be prepared to answer any technical questions that may arise.

Determining Change: Derivatives

We have described change in terms of rates: average rates, instantaneous rates, and percentage rates. Of these three, instantaneous rates are the most important in our study of calculus.

In Chapter 3 we presented an algebraic method, using the definition of the derivative, that allows us to find derivative models for certain functions.

In this chapter, we consider some rules for derivatives: the Simple Power Rule, the Constant Multiplier Rule, the Sum and Difference Rules, the Chain Rule, the Product Rule, and the Quotient Rule. These rules provide the foundation needed to work with more complicated functions that we often encounter in the course of real-life investigations of change.

Concept Objectives

This chapter will help you understand the concepts of

O   Slope graph
O   Points of undefined slope
O   Rules for finding rate-of-change formulas

and you will learn to

O   Draw rate-of-change graphs
O   Apply derivative rules to many types of functions to find rate-of-change formulas quickly
O   Write derivative formulas for piecewise continuous functions
O   Find the exact value of the rate of change at a point using derivative rules

Concept Application

The aging of the American population may be the demographic change that has the greatest impact on our society over the next several decades. Given a model for the projected number of senior Americans (65 years of age or older), the function and its derivative can be used to answer questions such as

○ What is the projected number of senior Americans in 2030?

○ How rapidly will that number be changing in 2030?

○ What is the estimated percentage rate of change in the number of senior Americans in 2030?

You will be able to answer these questions using the model given in Activity 21 of Section 4.2 and the derivative rules presented in this chapter.

4.1   Drawing Rate-of-Change Graphs

In Chapter 3 we considered the rate of change of a function at a given point. We learned how to estimate the instantaneous rate of change graphically, numerically, and algebraically. At the end of Section 3.5 we saw that the algebraic method for determining the rate of change at a specific point can be generalized to a formula that when evaluated gives the rate of change at any valid input. This algebraic formula involving limits is a very powerful tool that we will use in Section 4.2 to develop general rules about slope formulas. However, it is important that we have a good intuitive understanding of the relationship between functions and their slope formulas before we begin the symbolic manipulation of functions. To assist with this understanding, we consider the relationship between the graph of a function and its slope graph.

Extracting Rate-of-Change Information from a Function Graph

Every smooth, continuous curve with no vertical tangent lines has a slope associated with each point on the curve. When these slopes are plotted, they also form a smooth, continuous curve. We call the resulting curve a **slope graph, rate-of-change graph,** or **derivative graph.**

What do we know about the slopes of the graph shown in Figure 4.1? Sketch lines tangent to the curve at the points where $x = A$ and $x = C$ and at several other points on the curve, as shown in Figure 4.2.

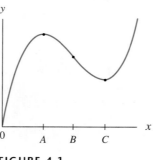

FIGURE 4.1

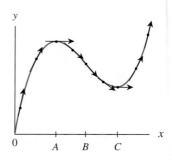

FIGURE 4.2

We deduce the following facts:
- The tangent lines at the points with inputs A and C are horizontal, so the slope is zero at those points.
- Between 0 and A, the graph is increasing, so the slopes are positive. The tangent lines become less steep moving from 0 to A, so the slopes start off large and become smaller as we approach A from the left.
- Between A and C, the graph is decreasing, so the slopes are negative.
- At B the graph has an inflection point. This is the point at which the graph is decreasing most rapidly—that is, the point at which the slope is most negative.

Because the vertical axis units of the slope graph are different from those of the function graph, we do not draw the slope graph on the same set of axes as the original graph. Doing so may be convenient, but it is sloppy mathematics.

- To the right of *C*, the graph is again increasing, so the slopes are positive. The tangent lines become steeper as we move to the right of *C*, so the slopes become larger as the input increases beyond *C*.

We record this information as indicated in Figure 4.3. On the basis of this information about the slopes, we sketch the shape of the slope graph, which appears in Figure 4.4.

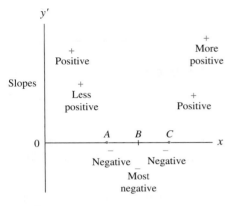

FIGURE 4.3

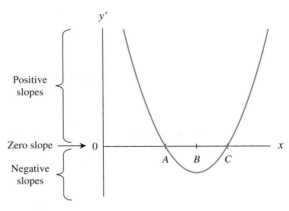

FIGURE 4.4

We do not know the specifics of the slope graph—how far below the horizontal axis it dips, where it crosses the vertical axis, how steeply it rises to the right of *C*, and so on. However, we do know its basic shape.

EXAMPLE 1   *Sketching the Slope Graph of a Logistic Curve*

The graph in Figure 4.5 is a logistic curve. Sketch its slope graph.

Solution   First we note that the logistic curve in Figure 4.5 is always increasing. Thus its slope graph is always positive (above the horizontal axis). We also note that even though there is no relative maximum or relative minimum, the logistic curve does level off at both ends. Thus its slope graph will be near zero at both ends. (See Figure 4.6.) Finally, we note that the logistic curve has its steepest slope at *A* because this is the location of the inflection point. Therefore, the slope graph is greatest (has a maximum) at this point. (See Figure 4.7.) We sketch the slope graph as shown in Figure 4.8.

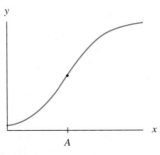

FIGURE 4.5

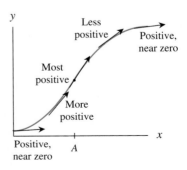

FIGURE 4.6

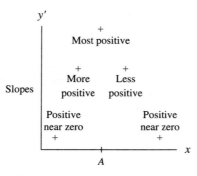

FIGURE 4.7

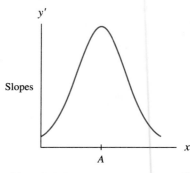

FIGURE 4.8   ●

EXAMPLE 2   *Relating Function and Rate-of-Change Graphs*

a. Sketch a graph that is always decreasing but has slopes that are always increasing. Also sketch its slope graph.

b. Sketch a graph that is always decreasing with slopes that are always decreasing. Also sketch its slope graph.

Solution

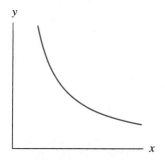

FIGURE 4.9

a. If a graph is decreasing, it is falling from left to right and has negative slopes. If the negative slopes are increasing, they are becoming less and less negative (moving toward zero) as the input increases. That means that the graph is becoming less steep. Such a graph must be concave up, as shown in Figure 4.9.

   The slopes of this graph are always negative, so a slope graph must lie completely below the input axis. As Figure 4.10 shows, the slopes are increasing. This means that they are rising closer to zero, but because they will never be positive, the graph must curve downward. A slope graph with these characteristics is shown in Figure 4.11.

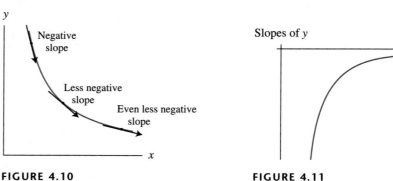

FIGURE 4.10   **FIGURE 4.11**

b. Like the graph in part *a*, this graph is falling from left to right and has negative slopes. However, the slopes are decreasing. When negative numbers decrease, they become more and more negative. This means the graph becomes more steep as input increases. A declining graph that becomes increasingly steep looks like the one shown in Figure 4.12.

Again, the slopes are always negative, so the slope graph will lie completely below the input axis. Instead of rising toward the input axis, this slope graph will fall away from the input axis as the slopes of the original graph become more and more negative. See Figure 4.13. The slope graph is shown in Figure 4.14.

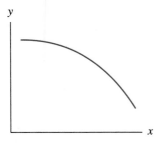

FIGURE 4.12

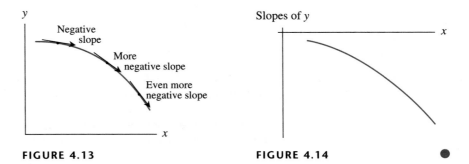

FIGURE 4.13

FIGURE 4.14

In the preceding discussion and example, the graphs we used had no labeled tick marks on the horizontal or vertical axes. In such cases, it is not possible to estimate the value of the slope of the graph at any given point. Instead, we sketch the general shape of the slope graph by observing the important points and general behavior of the original graph, such as

1. Points at which a tangent line is horizontal
2. The regions over which the graph is increasing or decreasing
3. Points of inflection
4. Places at which the graph appears to be horizontal or leveling off

As the previous examples indicate, sketching lines tangent to a curve helps us determine the relative magnitude of the slopes. As this process becomes more familiar, you should be able to visualize the tangent lines and consider the steepness of the curve itself. This technique is illustrated in Example 3.

EXAMPLE 3   *Using Relative Magnitudes to Sketch Slope Graphs*

Growth Rate   The height of a plant often follows the general trend shown in Figure 4.15. Draw a graph depicting the growth rate of the plant.

Solution   The slopes at *A*, *B*, and *C* are all positive. Is the slope at *A* smaller or larger than that at *B*? It is larger, so the slope graph at *A* should be higher than it is at *B*. The graph at *C* is not as steep as it is at either *A* or *B*, so the slope graph should be lower at *C* than at *B*. (See Figure 4.16.)

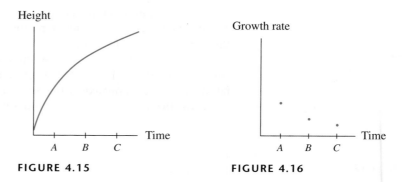

FIGURE 4.15   **FIGURE 4.16**

Where on the graph is the slope steepest? The answer is at the left endpoint. Where is the graph least steep? You can see that it is at the right endpoint. Add these observations to your plot. (See Figure 4.17.)

Now sketch the slope graph according to your plot, and be sure to include the appropriate labels on the horizontal and vertical axes. (See Figure 4.18.)

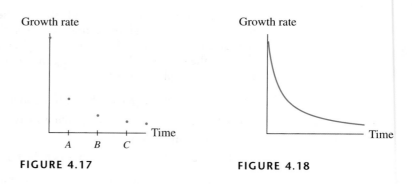

FIGURE 4.17   **FIGURE 4.18**

A Detailed Look at the Slope Graph

When a graph has labeled tick marks on both the horizontal and the vertical axes or an equation for the graph is known, it is possible to estimate the values of slopes at certain points on the graph. However, it would be tedious to calculate the slope graphically or numerically for every point on the graph. In fact, because there are infinitely many points on a continuous curve, it is not possible. Instead, we calculate the slope at a few special points, such as inflection points, in order to obtain a more accurate slope graph.

Consider again a graph with a maximum, an inflection point, and a minimum similar to the one we saw at the beginning of this section in Figure 4.1. Figure 4.19 shows such a graph, but this time the graph has labeled tick marks on both the horizontal and the vertical axes.

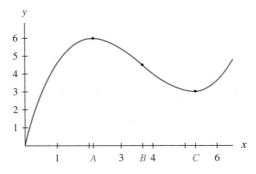

FIGURE 4.19

We know that the slope graph crosses the horizontal axis at *A* and *C* and that a minimum occurs on the slope graph below the horizontal axis at *B*. Before, we did not know how far below the axis to draw this minimum. Now that there is a numerical scale on the axes, we can graphically estimate the slope at the inflection point and use that estimate to help us sketch the slope graph.

By drawing the tangent line at *B* and estimating its slope, we find that the minimum of the slope graph is approximately 1.4 units below the horizontal axis. (See Figure 4.20.)

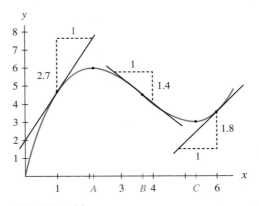

FIGURE 4.20

If we estimate the slopes at two additional points, say at $x = 1$ and $x = 6$, then we can produce a fairly accurate sketch of the slope graph. Table 4.1 shows a list of estimated slope values.

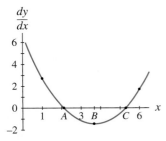

FIGURE 4.21

TABLE 4.1

x	1	A	B	C	6
Slope	2.7	0	-1.4	0	1.8

Plotting these points and sketching the slope graph give us the graph in Figure 4.21.

If we do not have a continuous curve but have a scatter plot, then we can sketch a rate-of-change graph by first sketching a smooth curve that fits the scatter plot. Then we can draw the slope graph of that smooth curve as shown in Example 4.

EXAMPLE 4 *Using a Curve Through Data to Sketch a Slope Graph*

Population In 1797, the Lorenzo Carter family built a cabin on Lake Erie where today the city of Cleveland, Ohio, is located. Table 4.2 gives population data* for Cleveland from 1810 through 1990.

a. Sketch a smooth curve representing population. Your curve should have no more inflection points than the number suggested by the scatter plot.

b. Sketch a graph representing the rate of change of population.

Solution

a. Draw a scatter plot of the population data, and sketch the smooth curve. (See Figure 4.22.)

b. The population graph is fairly level in the early 1800s, so the slope graph will begin near zero. The smooth sketched curve increases during the 1800s and early 1900s until it peaks in the 1940s. Thus the slope graph will be positive until the mid-1940s, at which time it will cross the horizontal axis and become negative. Population decreased from the mid-1940s onward.

There appear to be two inflection points. The point of most rapid growth appears around 1910, and the point of most rapid decline appears near 1975. These are the years in which the slope graph will be at its maximum and at its minimum, respectively.

TABLE 4.2

Year	Population
1810	57
1820	606
1830	1076
1840	6071
1850	17,034
1860	43,417
1870	92,829
1880	160,146
1890	261,353
1900	381,768
1910	560,663
1920	796,841
1930	900,429
1940	878,336
1950	914,808
1960	876,050
1970	750,879
1980	573,822
1990	505,616

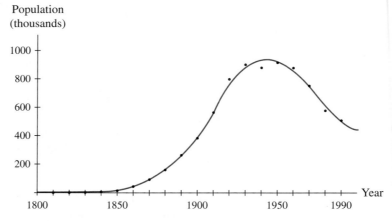

FIGURE 4.22

By drawing tangent lines at 1910 and at 1975 and estimating their slopes, we find that population was increasing by approximately 22,500 people per year in 1910 and was decreasing by about 12,500 people per year in 1975. See Figure 4.23.

*U.S. Department of Commerce, Bureau of the Census.

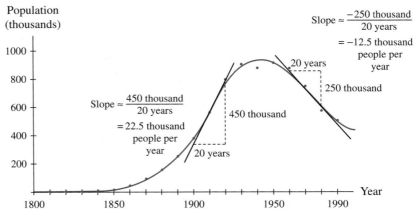

FIGURE 4.23

Now we use all the information from this analysis to sketch the slope graph shown in Figure 4.24.

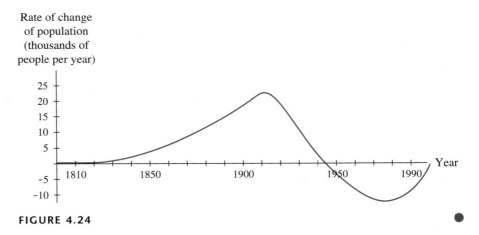

FIGURE 4.24 ●

Of course, if we had a formula for the graph, we could estimate the slope numerically at a few points instead of graphically estimating it with tangent lines. Even so, we still need to understand curvature and horizontal-axis intercepts to adequately sketch the rate-of-change graph.

Points of Undefined Slope

It is possible for the graph of a function to have a point at which the slope does not exist. Remember that the slope of a tangent line is the limit of slopes of approximating secant lines and that we should be able to use secant lines through points either to the left or to the right of the point at which we are estimating the slope to find this limit.* Recall that if the limits from the left and from the right are not the same, then

---

*It is worth noting that in the case in which a function is defined only to the right or only to the left of a point, then the derivative at that point can be found using secant lines through points only to the right or left.

the derivative does not exist at that point. We depict the nonexistence of the derivative at such points on the slope graph by drawing an open dot on each piece of the slope graph. This is illustrated in Figure 4.25a.

It is possible for there to be a point on a graph at which the derivative does not exist although the slope from the left and the slope from the right are the same. This occurs when the function is not continuous at that point (see Figure 4.25b). If a function is not continuous at a point, then its slope is undefined at that point.

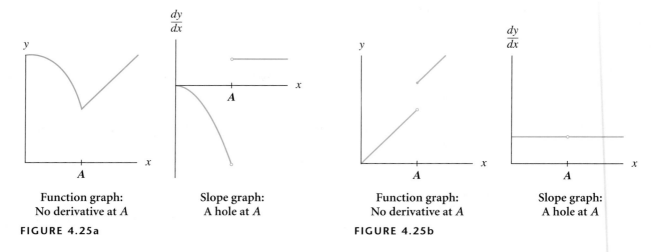

Function graph: Slope graph: Function graph: Slope graph:
No derivative at *A*   A hole at *A*   No derivative at *A*   A hole at *A*

FIGURE 4.25a   **FIGURE 4.25b**

Also, points at which the tangent line is vertical (that is, the slope calculation results in a zero in the denominator) are considered to have an undefined slope. The graph of one such function is shown in Figure 4.26.

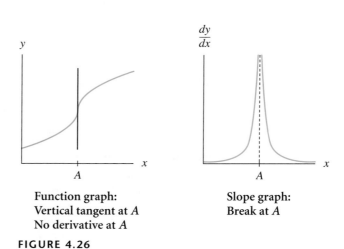

Function graph: Slope graph:
Vertical tangent at *A*   Break at *A*
No derivative at *A*

FIGURE 4.26

Most of the time when there is a break in the slope graph, it is because the original function is piecewise as shown in Figure 4.25. You should be careful when drawing slope graphs of piecewise continuous functions. Figure 4.26 shows a smooth, continuous function with a point at which the derivative is undefined. We do not often encounter such phenomena in real-life applications, but they could happen.

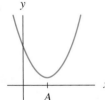

4.1 Concept Inventory

○ *Slope graph = rate-of-change graph = derivative graph*

○ *Increasing function ⇒ positive slopes*

○ *Decreasing function ⇒ negative slopes*

○ *Maximum or minimum of function ⇒ zero slope*

○ *Inflection point ⇒ maximum or minimum point on slope graph or point of undefined slope*

○ *Points of undefined slope*

4.1 Activities

In Activities 1 through 10, list as many facts as you can about the slopes of the graphs. Then, on the basis of those facts, sketch the slope graph of each function.

1.

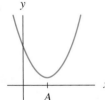

2.

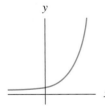

3.

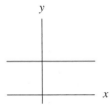

4.

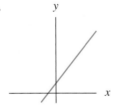

5.

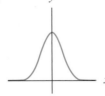

6.

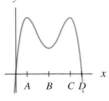

7.

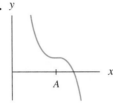

8.

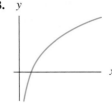

9.

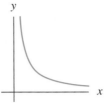

10.

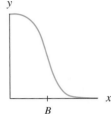

11. a. Sketch a graph that is increasing with increasing slopes. Also sketch its slope graph.

 b. Sketch a graph that is increasing with decreasing slopes. Also sketch its slope graph.

12. Investment The graph shows the cumulative amount of capital invested in the cellular phone industry since 1987.

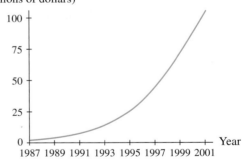

(Source: Based on data from The Cellular Telecommunication and Internet Association.)

a. Sketch tangent lines for the input values shown below, and estimate and record their slopes.

Year	1987	1991	1995	1999
Slope of tangent line				

b. Use the information in part *a* to sketch an accurate rate-of-change graph for the cumulative amount of capital invested in the cellular phone industry. Label the axes with units as well as values.

13. Phone Bill   The graph shows the average monthly cell phone bill in the U.S. between 1987 and 2001.

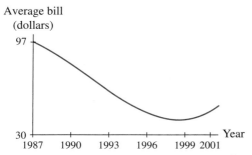

Average bill (dollars)

(Source: Based on data from The Cellular Telecommunication and Internet Association.)

a. Sketch tangent lines for the input values shown below, and estimate and record their slopes.

Year	1991	1993	1997	1999	2001
Slope of tangent line					

b. Use the information in part *a* to sketch an accurate rate-of-change graph for the average monthly cell phone bill. Label the axes with units as well as values.

14. Population   The graph shows the population of Iowa between 1990 and 1998.

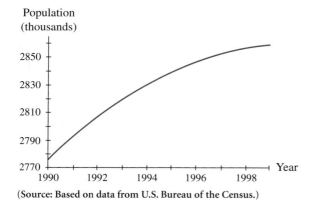

Population (thousands)

(Source: Based on data from U.S. Bureau of the Census.)

a. Sketch tangent lines for the input values shown below, and estimate and record their slopes.

Year	1990	1992	1994	1996	1998
Slope of tangent line					

b. Use the information in part *a* to sketch an accurate rate-of-change graph for the population of Iowa. Label the axes with units as well as values.

15. AIDS   The graph shows the cumulative number of AIDS cases between 1985 and 2000 diagnosed in the United States since 1984.

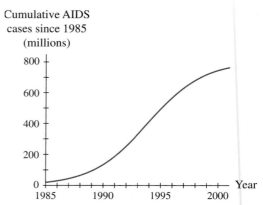

Cumulative AIDS cases since 1985 (millions)

(Source: Based on data from Centers for Disease Control.)

a. Sketch tangent lines for the input values shown below, and estimate and record their slopes.

Year	1985	1990	1995	1997	2000
Slope of tangent line					

b. Use the information in part *a* to sketch an accurate rate-of-change graph for the cumulative number of AIDS cases diagnosed in the United States. Label the axes with units as well as values.

16. Fuel   The graph shows the average annual fuel consumption of vehicles in the United States between 1970 and 1995.

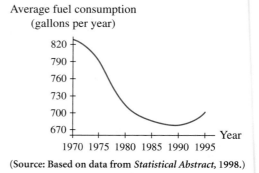

Average fuel consumption (gallons per year)

(Source: Based on data from *Statistical Abstract*, 1998.)

a. Sketch tangent lines for the input values shown in the table, and estimate their slopes. Record the slopes.

Year	1970	1975	1980	1985	1990	1995
Slope of tangent line						

b. Use the information in part *a* to sketch an accurate rate-of-change graph for the average annual fuel consumption. Label the axes with units as well as values.

17. **Membership** The graph gives the membership in a campus organization during its first year.

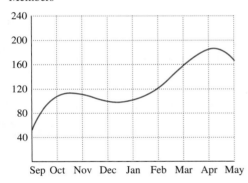

a. Estimate the average rate of change in the membership during the academic year.

b. Estimate the instantaneous rates of change in September, November, February, and April.

c. On the basis of your answers to part *b*, sketch a rate-of-change graph. Label the units on the axes.

d. The membership of the organization was growing most rapidly in September. Not including that month, when was the membership growing most rapidly? What is this point on the membership graph called?

e. Why was the result of the calculation in part *a* of no use in part *c*?

18. **Police Calls** The scatter plot depicts the number of calls placed each hour since 2 A.M. to a sheriff's department.

a. Sketch a smooth curve through the scatter plot with no more inflection points than the number suggested by the scatter plot.

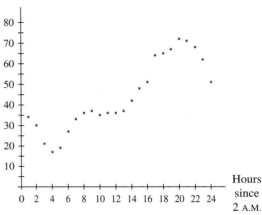

(Source: Sheriff's Office of Greenville County, South Carolina.)

b. At what time(s) is the number of calls a minimum? a maximum?

c. Are there any other times when the graph appears to have a zero slope? If so, when?

d. Estimate the slope of your smooth curve at any inflection points.

e. Use the information in parts *a* through *d* to sketch a graph depicting the rate of change of calls placed each hour. Label the units on both axes of the rate-of-change graph.

19. **Jails** The capacity of jails in a southwestern state has been increasing since 1990. The average daily population of one jail between 1990 and 2000 can be modeled by

$$j(t) = \begin{cases} 8.101t^3 - 55.53t^2 + \\ \quad 128.8t + 626.8 \text{ inmates} & \text{when } 0 \le t \le 5 \\ 18.8t + 800.6 \text{ inmates} & \text{when } t > 5 \end{cases}$$

where *t* is the number of years since 1990.

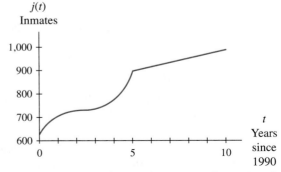

(Source: Based on data from Washoe County Jail, Reno, Nevada.)

Sketch the slope graph of j. (*Hint:* Numerically estimate $j'(t)$ at $t = 0$ and $t = 4.5$ in order to sketch the slope graph accurately.) Label both the horizontal and the vertical axes.

20. **Price**   Cattle prices (for choice 450-pound steer calves) from October 1994 through May 1995 can be modeled by

$$p(m) = \begin{cases} -0.0025m^2 + 0.0305m + \\ \quad 0.8405 \text{ dollars per pound} \quad 0 \leq m < 3 \\ -0.028m + 0.996 \qquad \text{when} \\ \quad \text{dollars per pound} \qquad 3 \leq m \leq 7 \end{cases}$$

where m is the number of months since October 1994.

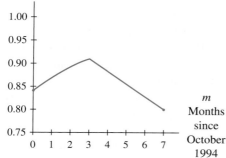

$p(m)$
Cattle price
(dollars per pound)

(Source: Based on data from the National Cattleman's Association.)

Sketch a slope graph of p. (*Hint:* Numerically estimate $p'(m)$ at $m = 0$ and $m = 2.5$ in order to sketch the slope graph accurately.) Label the horizontal and vertical axes.

21. **Profit**   A graph depicting the average monthly profit for Slim's Used Car Sales for the previous year is shown at the top of the next column.

a. Estimate the average rate of change in Slim's average monthly profit if the number of cars he sells increases from 40 to 70 cars.

b. Estimate the instantaneous rates of change at 20, 40, 60, 80, and 100 cars.

c. On the basis of your answers to part *b*, sketch a rate-of-change graph. Label the units on the axes.

d. For what number of cars sold between 20 and 100 is average monthly profit increasing most rapidly? For what number of cars sold is average

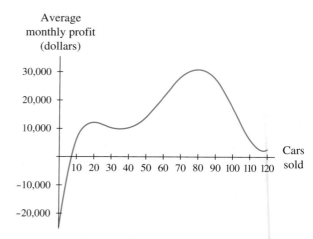

Average
monthly profit
(dollars)

monthly profit decreasing most rapidly? What is the mathematical term for these points?

e. Why was the result of the calculation in part *a* of no use in part *c*?

22. **Mortality**   Use the graph in Activity 20 on page 204 to carefully estimate the rate of change in deaths of males due to lung cancer in 1940, 1960, and 1980. Use this information to sketch an accurate rate-of-change graph for deaths due to lung cancer. Label the units on both axes of the derivative graph.

In Activities 23 through 26, indicate the input values for which the graph has no derivative. Explain why the derivative does not exist at those points. Sketch a derivative graph for each of the function graphs.

23.

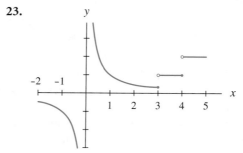

24.

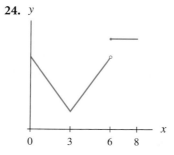

25.

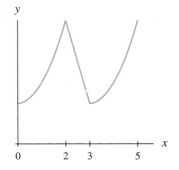

26.

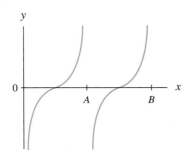

27. Sketch the slope graph of a function f with input t that meets these criteria: $f(-2) = 5$, the slope is positive for $t < 2$, the slope is negative for $t > 2$, and $f'(2)$ does not exist.

28. Sketch the slope graph of a function g with input x that meets these criteria: $g(3)$ does not exist, $g'(0) = -4$, $g'(x) < 0$ for $x < 3$, g is concave down for $x < 3$, $g'(x) > 0$ for $x > 3$, g is concave up for $x > 3$, $\lim\limits_{x \to 3^+} g(x) \to \infty$, and $\lim\limits_{x \to 3^-} g(x) \to -\infty$.

29. Construct the graphs of a function h and its slope h', with input x, such that $h'(1)$ is significantly different from the percentage rate of change of h at $x = 1$.

30. The figure shows a graph of the function $q(t) = \frac{1}{t}$.

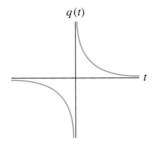

a. Sketch a slope graph of q.

b. Find a formula for the slope graph. [*Hint:* Multiply the numerator and denominator of the secant line slope formula by $t(t + h)$.] Compare a graph of this formula with the graph you sketched in part *a*.

31. The figure below shows a graph of the function $p(m) = m + \sqrt{m}$.

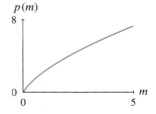

a. Sketch a slope graph of p.

b. Find a formula for the slope graph. Compare a graph of this formula with the graph you sketched in part *a*. (*Hint:* Multiply the numerator and denominator of the secant line slope formula by $\sqrt{m + h} + \sqrt{m}$.)

32. The figure below shows a graph of the function $k(x) = x + \frac{1}{x}$.

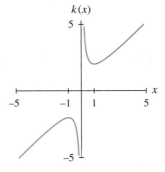

a. Sketch a slope graph of k.

b. Find a formula for the slope graph. Use the same hint as in Activity 30. Compare a graph of this formula with the graph you sketched in part *a*.

33. Why is it important to understand curvature and horizontal-axis intercepts in order to adequately sketch a rate-of-change graph?

34. What elements of a function graph are of specific importance when sketching a rate-of-change graph for that function? Explain why these elements are important.

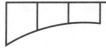

4.2   Simple Rate-of-Change Formulas

By now you should have a thorough understanding of the concept of instantaneous rate of change as the slope of a line tangent to a curve at a point, where that slope is defined as a limiting value of slopes of secant lines. We now rely on this conceptual understanding in order to present formulas for rapid calculation of rates of change as we use them in future sections, but we urge you not to forget the graphical and numerical roots of the algebraic work we are about to begin.

You already know two rate-of-change formulas from our study of linear functions in Chapter 1. A horizontal line has slope zero, and a nonhorizontal line of the form $y = ax + b$ has slope a. We know that the rate of change, or derivative, of a line is its slope, so we can state the following derivative formulas:

Derivative of a Linear Function

If $y = ax + b$, then $\dfrac{dy}{dx} = a$.

Constant Rule

If $y = b$, then $\dfrac{dy}{dx} = 0$.

EXAMPLE 1   *Finding the Derivative of a Linear Function*

Cricket's Chirping   The frequency of a cricket's chirp is affected by air temperature and can be modeled by

$$C(t) = 0.212t - 0.309 \text{ chirps per second}$$

when the temperature is $t°$F. Write a formula for the rate of change of a cricket's chirping speed with respect to a change in temperature.

Solution   The frequency of a cricket's chirp is changing by

$$C'(t) = 0.212 \text{ chirps per second per degree Fahrenheit}$$

no matter what the temperature is.   ●

The Simple Power Rule

Next, consider quadratic and cubic functions. In Section 3.5, we determined that the rate-of-change formula for $y = x^2$ is $\dfrac{dy}{dx} = 2x$. In Activity 15 of that same section, you were asked to show that the rate-of-change formula for $y = x^3$ is $\dfrac{dy}{dx} = 3x^2$. Note that in each instance, the power on x in the derivative formula is one less than the power on x in the original function. Also, the x-term in the derivative formula is multiplied by the power on x in the original function. This is no coincidence. These are two special cases of one of the most important rules that we use for quickly finding derivative formulas, the **Simple Power Rule.**

Simple Power Rule for Derivatives

If $y = x^n$, then $\dfrac{dy}{dx} = nx^{n-1}$ where n is any nonzero real number.

The Constant Multiplier and Sum and Difference Rules

In order to find derivative formulas for any polynomial, we need two rules in addition to the Simple Power Rule. The first derivative rule we illustrate is the **Constant Multiplier Rule.** Each of the figures in Figure 4.27 shows the graph of a function and a graph of a constant multiple of that function, and the slope graphs of the function and the constant multiple of that function.

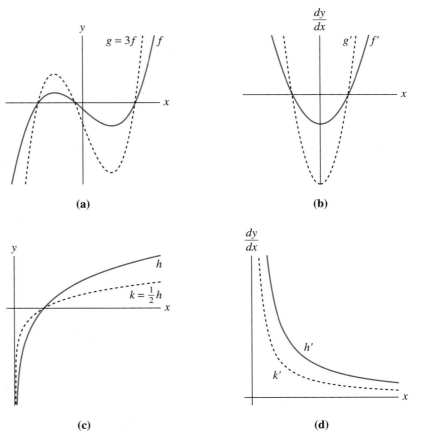

FIGURE 4.27

Notice that the effect of a constant multiplier is to amplify (or if the multiplier has magnitude between 0 and 1, to diminish) the behavior of a function. Thus, we expect that the behavior of the rate of change of the function will be amplified or diminished by the same factor.

If the constant multiplier is negative, it has the additional effect of reflecting the function over the horizontal axis. The rate-of-change function is likewise reflected, as illustrated in Figures 4.28a and b.

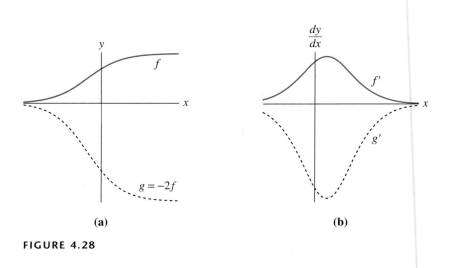

FIGURE 4.28

The graphs in Figures 4.27 and 4.28 are basic illustrations of the following general rule:

Constant Multiplier Rule for Derivatives

If $y = kf(x)$, then $\dfrac{dy}{dx} = kf'(x)$.

This rule allows us to calculate quickly the rate-of-change formula for a function such as $y = 5x^4$: Leave the 5 alone, and apply the Simple Power Rule to x^4. This process gives $\dfrac{dy}{dx} = 5(4x^3) = 20x^3$.

The final rules needed for rapid calculation of the rate-of-change formula for any polynomial are the **Sum and Difference Rules.**

Concept Development: Sum and Difference Rules   Even though the Sum and Difference Rules can be proven algebraically using the limit definition of the derivative, we choose to present two graphical illustrations of these rules. Figure 4.29 illustrates the graphs of $f(x) = x$, $g(x) = x^2$ and the graph of $(f + g)(x) = x + x^2$, the function that is the sum of f and g.

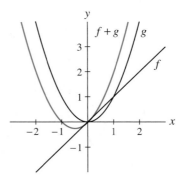

FIGURE 4.29

The minimum of the sum function $f + g$ appears to occur at $x = -\frac{1}{2}$, indicating that the slope graph of the sum function crosses the horizontal axis at $x = -\frac{1}{2}$. Moving from left to right along the horizontal axis, the outputs of $f + g$ are decreasing before $x = -\frac{1}{2}$ and increasing again after $x = -\frac{1}{2}$. Thus, the sum function slope graph is negative to the left of $x = -\frac{1}{2}$ and positive to the right of $x = -\frac{1}{2}$. Further, note that the graph of the sum function has the same basic parabolic shape as the function f except that it is shifted and appears to be stretched a bit taller.

We now investigate the basic shape and magnitude of the slope graph by looking at a few arbitrarily chosen inputs ($x = -2$, $x = 3$, and $x = 7$). By first using the Linear Function Rule and the Simple Power Rule to obtain the slopes of f and g, we complete the second and third rows of Table 4.3. Numerically investigating the slope of $f + g$ at the inputs given in the first row of Table 4.3 yields the estimates in the fourth row.

Figure 4.30 shows the slope graphs of f, g, and the sum function $f + g$ (in teal). The graph of the sum function was obtained by plotting the outputs in the fourth row of the table and connecting them with a smooth curve.

TABLE 4.3

x	-2	$-\frac{1}{2}$	0	3	7
$\frac{df}{dx} = 1$	1	1	1	1	1
$\frac{dg}{dx} = 2x$	-4	-1	0	6	14
$\frac{d(f + g)}{dx} \approx$	-3	0	1	7	15

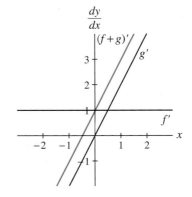

FIGURE 4.30

Note in both Table 4.3 and Figure 4.30, the slopes of the sum function $f + g$ can be obtained by summing the slopes of the individual functions f and g. A similar result is obtained if we investigate the difference function $f - g$. We would see that the slope of the difference function $f - g$ at a certain input could be obtained by subtracting the slope of g from the slope of f at that same input. See Figures 4.31a and b.

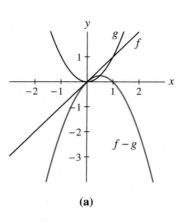

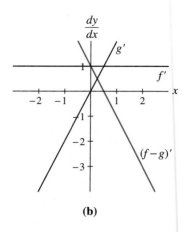

(a)   **(b)**

FIGURE 4.31

The second graphical illustration of the Sum and Difference Rules specifically illustrates the differences of two simple power functions, $k(x) = x^3$ and $h(x) = x^2$. Figure 4.32 shows graphs of these two functions and their difference function $(k - h)(x) = x^3 - x^2$.

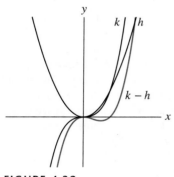

FIGURE 4.32

The difference function $k - h$ appears to have a minimum at $x = \frac{2}{3}$ and a maximum at $x = 0$. Therefore, the slope graph crosses the horizontal axis at these input values. Moving from left to right along the horizontal axis, the difference function is increasing before $x = 0$ and after $x = \frac{2}{3}$ and is decreasing between these two inputs. Thus the slope graph is positive to the left of $x = 0$ and to the right of $x = \frac{2}{3}$ and is negative between these values. Numerically evaluating the limiting value of the slope of $k - h$ at $x = -1$, $x = \frac{1}{3}$, and $x = 2$ yields the estimates in the second row of Table 4.4. Using these numerical estimates as additional points on the graph, we sketch the slope graph of $k - h$. See Figure 4.33.

TABLE 4.4

x	-1	$\frac{1}{3}$	2
$\dfrac{d(k - h)}{dx}$	5	-0.33333	8

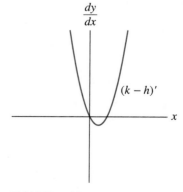

FIGURE 4.33

Table 4.5 and Figure 4.34 illustrate the relationship of the slopes of the functions k and h and the difference function $k - h$. Similar to the result we found with the sum function, notice that the slopes of the difference function are the differences of the slopes of the two individual functions.

TABLE 4.5

x	-1	0	$\frac{1}{3}$	$\frac{2}{3}$	2
$\frac{dk}{dx} = 3x^2$	3	0	$\frac{1}{3}$	$\frac{4}{3}$	12
$\frac{dh}{dx} = 2x$	-2	0	$\frac{2}{3}$	$\frac{4}{3}$	4
$\frac{d(k-h)}{dx} \approx$	5	0	$\frac{-1}{3}$	0	8

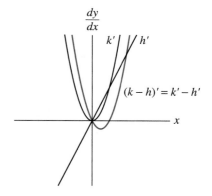

FIGURE 4.34

The illustrations of the slopes of $x^2 + x$ and $x^3 - x^2$ are two examples of the following general rule:

The Sum and Difference Rules for Derivatives

If $j(x) = f(x) + g(x)$, then $\dfrac{dj}{dx} = \dfrac{df}{dx} + \dfrac{dg}{dx}$.

If $j(x) = f(x) - g(x)$, then $\dfrac{dj}{dx} = \dfrac{df}{dx} - \dfrac{dg}{dx}$.

The Sum and Difference Rules also apply to sums and/or differences of more than two functions. With the rules we now have, we can find the rate-of-change formula for any polynomial function. For example, if $p(x) = 3.22x^3 - 0.15x^2 + 9.98x - 30$, we use the Constant Multiplier and the Simple Power Rules to find the derivative formula for each term, and then we combine the terms using the Sum and Difference Rules. The rate-of-change formula for p is

$$p'(x) = 3.22(3x^2) - 0.15(2x) + 9.98 - 0 = 9.66x^2 - 0.3x + 9.98$$

EXAMPLE 2   *Applying Derivative Rules*

Maintenance Costs   Table 4.6 gives the average yearly maintenance costs per vehicle for 15,000 miles of operation in the United States from 1993 through 2000.

TABLE 4.6

Year	1993	1994	1995	1996	1997	1998	1999	2000
Maintenance costs (cents per mile per vehicle)	2.4	2.5	2.6	2.8	2.8	3.1	3.6	3.9

(Source: Bureau of Transportation Statistics.)

Find a quadratic model for the data, and use it to approximate how rapidly maintenance costs were increasing in 1998.

Solution   A quadratic model for the data is

$$g(t) = 0.0304t^2 - 0.00417t + 2.446 \text{ cents per mile per vehicle}$$

where t is the number of years after 1993. Applying the Sum and Difference, Power, Constant Multiplier, and Constant Rules, we find that the derivative of g with respect to t is

$$\frac{dg}{dt} = 2(0.0304)t - 0.00417 + 0$$

so

$$\frac{dg}{dt} = 0.0608t - 0.00417 \text{ cents per mile per vehicle per year}$$

t years after 1993.

Evaluating the derivative at $t = 5$ gives 0.2998 cent per mile per vehicle per year as the rate of change of maintenance costs in 1998. Thus we estimate that in 1998 the average maintenance cost per vehicle was increasing at a rate of approximately 0.30 cent per mile per year.   ●

Derivatives of Piecewise Functions

Our final example illustrates how to use the rules presented in this section to find a derivative formula for a piecewise continuous function. If a piecewise continuous function is discontinuous at a point, then the derivative does not exist at the input value corresponding to the discontinuity. When you are writing the derivative formula, it is important to indicate that the derivative is not defined for this input value. If you need to estimate a rate of change of the underlying output quantity at that input value, you could use a symmetric difference quotient or some other estimation method.

EXAMPLE 3   *Writing the Derivative Formula for a Piecewise Continuous Function*

Population   On the basis of data from the U.S. Census Bureau for years between 1980 and 1997, the population of Iowa can be modeled by the equation

Piece wise function

$$I(x) = \begin{cases} -2.1x^2 - 6.3x + 2914 \text{ thousand people} & \text{when } 0 \leq x \leq 7 \\ -0.1979x^3 + 7.254x^2 - 74.98x + 3002.28 \text{ thousand people} & \text{when } 7 < x \leq 17 \end{cases}$$

where x is the number of years since 1980.

a. Write the derivative equation for the population model.

b. How quickly was the population of Iowa growing or declining in 1985? in 1995?

c. Estimate how quickly the population of Iowa was growing or declining in 1987.

Solution

a. The function is not continuous at $x = 7$ because the quadratic part of the function has a value of 2767 when $x = 7$, and the cubic part of the function has a value of 2764.9863 when $x = 7$. Therefore, the derivative formula must be undefined when $x = 7$. To find the formula, simply apply the derivative rules to each function individually.

$$I'(x) = \begin{cases} -4.2x - 6.3 \text{ thousand people per year} & \text{when } 0 \le x < 7 \\ -0.5937x^2 + 14.508x - 74.98 \text{ thousand people per year} & \text{when } 7 < x \le 17 \end{cases}$$

where x is the number of years since 1980. Note that the inequality $0 \le x \le 7$ in the top function becomes $0 \le x < 7$ in the function's derivative. Because there is no equals sign indicating $x = 7$ in either portion of the $I'(x)$ formula, the derivative is not defined at $x = 7$. Graphs of I and I' are shown in Figures 4.35a and b, respectively.

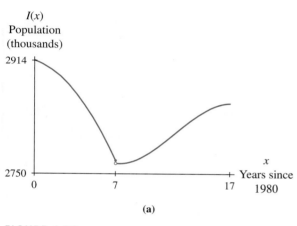

(a)

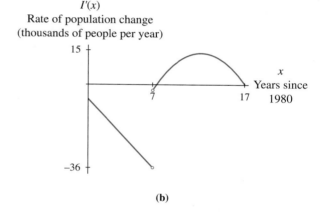

(b)

FIGURE 4.35

b. The rate of change of the population of Iowa in 1985 was $I'(5) = -4.2(5) - 6.3 = -27.3$ thousand people per year. In other words, the population of Iowa was declining by approximately 27,300 people per year in 1985.

   In 1995, the rate of change of the population was $I'(15) = -0.5937(15^2) + 14.508(15) - 74.98 \approx 9.06$ thousand people per year. The population in 1995 was increasing by approximately 9000 people per year.

c. The function I does not have a derivative at $x = 7$ because it jumps at the break point. However, we can still estimate the rate of change of the population of Iowa in 1987. Use the model that was given to estimate the population of Iowa in two different years that are equidistant from 1987, say 1986 and 1988. Then use those two population estimates to calculate the symmetric difference quotient (average rate of change). The model estimates the 1986 population as 2800.6 thousand people and the 1988 population as 2765.4 thousand people. The average rate of change between these two years is

$$\frac{2765.4 - 2800.6}{1988 - 1986} \approx \frac{-35.2 \text{ thousand people}}{2 \text{ years}} \approx -17.6 \text{ thousand people per year}$$

Be sure you understand that these estimates are not estimates of the derivative of the function I for $x = 7$. These are estimates of the rate of change of the underlying situation—in this case, the population of Iowa. ●

In summary, here is a list of the rate-of-change formulas you should know. Commit these formulas to memory through extensive practice.

Simple Derivative Rules

Rule Name	Function	Derivative
Constant Rule	$y = b$	$\dfrac{dy}{dx} = 0$
Linear Function Rule	$y = ax + b$	$\dfrac{dy}{dx} = a$
Power Rule	$y = x^n$	$\dfrac{dy}{dx} = nx^{n-1}$
Constant Multiplier Rule	$y = kf(x)$	$\dfrac{dy}{dx} = kf'(x)$
Sum Rule	$y = f(x) + g(x)$	$\dfrac{dy}{dx} = f'(x) + g'(x)$
Difference Rule	$y = f(x) - g(x)$	$\dfrac{dy}{dx} = f'(x) - g'(x)$

4.2 Concept Inventory

○ Derivative formulas
  For constants a, b, and n,

  If $y = b$, then $y' = 0$.
  If $y = ax + b$, then $y' = a$.
  If $y = x^n$, then $y' = nx^{n-1}$.
  If $y = kf(x)$, then $y' = kf'(x)$.
  If $y = f(x) \pm g(x)$, then $y' = f'(x) \pm gt(x)$.

○ *Derivatives of piecewise continuous functions*

4.2 Activities

For each of the functions whose graphs are given in Activities 1–6, first sketch the slope graph and then give the slope equation.

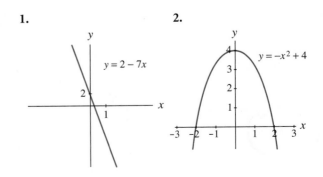

1. $y = 2 - 7x$

2. $y = -x^2 + 4$

3.

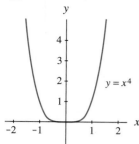

4.

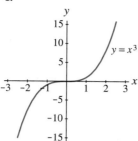

5.

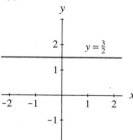

6.

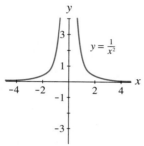

Give the derivative formulas for the functions in Activities 7 through 14.

Skills

7. $y = 7x^2 - 12x + 13$

8. $f(x) = 0.0127x^3 + 9.4861x^2 - 0.2649x + 128.98$

9. $y = 5x^3 + 3x^2 - 2x - 5$

10. $f(x) = \sqrt{x}$ (*Hint*: Rewrite as $x^{1/2}$.)

11. $y = \dfrac{-9}{x^2}$

12. $f(x) = 3x^{-2}$

13. $h(x) = 17.463 - 0.049\sqrt{x}$

14. $j(x) = \dfrac{3.694x^2 + 19.42x + 1.06}{x}$ (*Hint*: Rewrite as three separate terms.)

15. ATM Fee The average ATM transaction fee charged by U.S. banks between 1996 and 1999 can be modeled by the equation $A(t) = 0.1333t + 0.17$ dollars t years after 1990.
(Source: Based on data from the U.S. Public Interest Research Group.)

a. Write the derivative formula for A.

b. Estimate the transaction fee in 2000.

c. How quickly was the fee changing in 1999?

16. Metabolic Rate The table shows the metabolic rate of a typical 18- to 30-year-old male according to his weight.

Weight (pounds)	Metabolic rate (kilocalories per day)
88	1291
110	1444
125	1551
140	1658
155	1750
170	1857
185	1964
200	2071

(Source: L. Smolin and M. Grosvenor, *Nutrition: Science and Applications*. Philadelphia, PA: Saunders College Publishing, 1994.)

a. Find a formula for a typical man's metabolic rate.

b. Write the derivative of the formula in part *a*.

c. What does the derivative in part *b* tell you about a man's metabolic rate if that man weighs 110 pounds? if he weighs 185 pounds?

17. Population The population of Hawaii between 1970 and 1990 can be modeled by

$$P(t) = 15.48t + 485.4 \text{ thousand people}$$

t years after 1950.
(Source: Based on data from George T. Kurian, *Datapedia of the United States, 1790–2000*. Latham, MD: Bernan Press, 1994.)

a. Write the formula for P'.

b. How many people lived in Hawaii in 1970?

c. How quickly was Hawaii's population changing in 1990?

18. Veterans The number of military veterans serving in the U.S. House of Representatives between 1975 and 1997 can be modeled by the equation

$$V(t) = -7.736t + 291.786 \text{ veterans}$$

where t is the number of years after 1975.
(Source: Based on data from Dan Hoover, "Military Service No Longer Politically Necessary," *The Greenville News*, June 29, 1998, p. 1A).

a. Was the number of veterans increasing or decreasing during the time period modeled by the equation?

b. How rapidly was the number of veterans in the House of Representatives changing in 1975? in 1997?

c. What was the average rate of change of the number of veterans in the House between 1975 and 1995?

19. **Temperature**   The graph shows the temperature values (in °F) on a typical May day in a certain mid-western city.

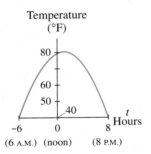

Temperature
(°F)

The equation of the graph is

$$\text{Temperature} = -0.8t^2 + 2t + 79°F$$

where t is the number of hours since noon. Use the derivative formula to verify each of the following statements.

a. The graph is not as steep at 1:30 P.M. as it is at 7 A.M.

b. The slope of the tangent line at 7 A.M. is 10°F per hour.

c. The instantaneous rate of change of the temperature at noon is 2°F per hour.

d. At 4 P.M. the temperature is falling by 4.4°F per hour.

20. **Study Time**   A graph representing a test grade (out of 100 points) as a function of hours studied is given in Figure 3.23 on page 192. The equation of the graph is

$$G(t) = -0.044t^3 + 0.918t^2 + 38.001 \text{ points}$$

after t hours of study. Use the equation to verify each of the following statements.

a. $\dfrac{dG}{dt} = 1.704$ points per hour when $t = 1$ hour.

b. $G'(4) = 5.232$ points per hour.

c. The slope of the tangent line when $t = 15$ hours is approximately -2 points per hour.

21. **Population**   The projected number of Americans age 65 or older for the years 1995 through 2030 can be modeled by the equation

$$N(x) = 0.03x^2 + 0.315x + 34.23 \text{ million people}$$

where x is the number of years after 2000.
(Source: Based on data from John Greenwald, "Elder Care: Making the Right Choice," *Time*, August 30, 1999, p. 52.)

a. What is the projected number of Americans 65 years of age and older in 1995? in 2030?

b. What is the rate of change of the projected number in 1995? in 2030?

c. Find the percentage rate of change in the projected number in 2030.

d. The Census Bureau predicts that in 2030, 20.1% of the U.S. population will be 65 years of age or older. Use this prediction and one of the unrounded answers to part *a* to estimate the total U.S. population in 2030.

22. **Sales**   A publishing company estimates that when a new book by a best-selling American author first hits the market, its sales can be predicted by the equation $n(x) = 68,952.921\sqrt{x}$, where $n(x)$ represents the total number of copies of the book sold in the United States by the end of the xth week. The number of copies of the book sold abroad by the end of the xth week can be modeled by

$$a(x) = -0.039x^2 + 125.783x \text{ copies of the book}$$

a. Write the formula for the total number of copies of the book sold in the United States and abroad by the end of the xth week.

b. Write the formula for the rate of change of the total number of books sold.

c. How many books will be sold by the end of the first year (that is, after 52 weeks)?

d. How rapidly are books selling at the end of the first year? Interpret your answer.

23. Births The number of live births to U.S. women 45 years and older between 1950 and 2000 can be modeled by the equation

$$B(x) = 0.2685x^3 - 15.6x^2 + 94.684x + 5378.03 \text{ births}$$

where x is the number of years since 1950.
(Source: Based on data from www.infoplease.com. Accessed 9/24/02.)

a. Was the number of live births rising or falling in 1970? in 1995?

b. How rapidly was the number of live births rising or falling in 1970? in 1995?

24. Profit An artisan makes hand-crafted painted benches to sell at a craft mall. Her weekly revenue and costs (not including labor) are given in the table.

Number of benches sold each week	Weekly revenue (dollars)	Weekly cost (dollars)
1	300	57
3	875	85
5	1375	107
7	1750	121
9	1975	143
11	1950	185
13	1700	213

a. Find models for revenue and cost.

b. Construct a model for profit from the revenue and cost models in part *a*.

c. Write the derivative formula for profit.

d. Find and interpret the rate of change of profit when the artisan sells 6 benches.

e. Repeat part *d* for 9 benches and 10 benches.

f. What does the information in part *e* tell you about the number of benches the artisan should produce each week?

25. Sales The accompanying table gives revenue from new car sales and associated advertising expenditures for franchised new car dealerships in the United States for selected years between 1980 and 2000.

Year	Advertising expenses (billions of dollars)	Revenue (billions of dollars)
1980	1.2	130.5
1985	2.8	251.6
1990	3.7	316.0
1995	4.6	456.2
1998	5.3	546.3
2000	6.4	646.8

(Source: *Statistical Abstract*, 2001.)

a. Find a model that describes revenue as a function of advertising expenditures.

b. Write the formula for the derivative of the equation in part *a*.

c. Use your model to find how rapidly the revenue was changing when $5 billion was spent on advertising.

d. Use the model to estimate the revenue from new car sales when advertising expenditures were $5 billion.

e. What was the percentage rate of change in the revenue when $4.6 billion was spent on advertising?

26. Costs Production costs (in dollars per hour) for a certain company to produce between 10 and 90 units per hour are given in the table.

Units	Cost (dollars per hour)
10	150
20	200
30	250
40	400
50	750
60	1400
70	2400
80	3850
90	5850

a. Graph a scatter plot of the data, and discuss its curvature.

b. Consider the cost for producing 0 units to be $0. Include (0, 0) in the data, and find the first differences of the cost data. Convert each of these differences to average rates of change. Use the average rates of change to argue that this data set has an inflection point.

c. Include the point $(0, 0)$, and find a cubic model for production costs.

d. Convert the model in part c to one for the average cost per unit produced.

e. Find the slope formula for average cost.

f. How rapidly is the average cost changing when 15 units are being produced? 35 units? 85 units? Interpret your answers.

27. **Profit**   The managers of Windolux, Inc., have modeled some cost data and found that if they produce x storm windows each hour, the cost (in dollars) to produce one window is given by the function

$$C(x) = 0.0146x^2 - 0.7823x + 46.9125 + \frac{49.6032}{x}$$

Windolux sells its storm windows for $175 each. (You may assume that every window made will be sold.)

a. Write the formula for the profit made from the sale of one storm window when Windolux is producing x windows each hour.

b. Write the formula for the rate of change of profit.

c. What is the profit made from the sale of a window when Windolux is producing 80 windows each hour?

d. How rapidly is profit from the sale of a window changing when 80 windows are produced each hour? Interpret your answer.

28. **Population**   The joint population of the United States and Canada for the period from 1980 through 2002 can be modeled by the equation

$$n(x) = -0.00256x^3 + 0.122x^2 + 1.586x + 3.267$$
million people

x years after 1980. The population of Mexico for the same period can be modeled by the equation

$$m(x) = 69.635(1.0188^x) \text{ million people}$$

x years after 1980.
(Source: Based on data from U.S. Census Bureau, International Data Base.)

a. Find the formula for the combined population of the United States, Canada, and Mexico from 1980 through 2002.

b. Find the formula for the rate of change of the combined population of the United States, Canada, and Mexico.

c. According to the models, what was the combined population of the United States, Canada, and Mexico in 2002?

d. How rapidly was the combined population of the United States, Canada, and Mexico changing in 2002?

29. **Fisheries**   The amount of fish produced for food by fisheries in the United States between 1970 and 1990 can be modeled 21 by the equation

$$f(x) = \begin{cases} 22.204x^2 - 108.431x - 2538.603 \text{ million lb} \\ \qquad\qquad \text{when } 0 \le x \le 10 \\ 85.622x^2 - 2253.95x + 17{,}772.39 \text{ million lb} \\ \qquad\qquad \text{when } 10 < x \le 20 \end{cases}$$

x years after 1970.
(Source: Based on data from *Statistical Abstract*, 1994.)

a. Write a formula for $\frac{df}{dx}$.

b. Find the rates of change of the amount of fish produced in 1970, 1980, and 1990.

c. What was the percentage rate of change in 1990?

30. **Emissions**   The atmospheric release of chlorofluorocarbons (CFCs) between 1958 and 1978 can be modeled by the equation

$$C(t) = \begin{cases} 0.708t^2 - 71.401t + 1832.4275 \text{ million kg} \\ \qquad\qquad \text{when } 58 \le x \le 10 \\ -2.6125t^2 + 377.775t - 13{,}230.7 \text{ million kg} \\ \qquad\qquad \text{when } 10 < x \le 20 \end{cases}$$

where t is the number of years since 1900.
(Source: Based on data from Ronald Bailey, ed., *The True State of the Planet*. New York: The Free Press for the Competitive Enterprise Institute, 1995.)

a. Write the derivative formula for C.

b. What was the atmospheric release of CFCs in 1968, and how quickly was the amount changing at that time?

c. Estimate the rate of change in the atmospheric release of CFCs in 1974.

31. Use your knowledge of the shape and end behavior of the graph of a cubic function to explain why the slope graph of a cubic function is the graph of a quadratic function. Use this argument to explain why the rate-of-change formula for a cubic function is the formula for a quadratic function.

32. Use the simple derivative rules presented in this section to explain why the rate-of-change formula for a cubic function of the form $y = ax^3 + bx^2 + cx + d$ is the formula for a quadratic function. Write your explanation in paragraph form.

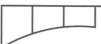

4.3   More Simple Rate-of-Change Formulas

In Section 4.2 we developed some general rules for rate-of-change formulas. In this section we will continue using those rules and develop others.

Exponential Rules

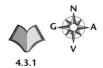

4.3.1

Our next formulas involve derivatives of exponential functions. Because the proof of these rules is beyond the scope of this book, we explore the functions graphically and numerically to develop an understanding of the behavior of the derivative of the function before stating the general derivative formula. We begin our exploration with the function $y = e^x$. First, we consider the concavity and end behavior of this function (see Figure 4.36):

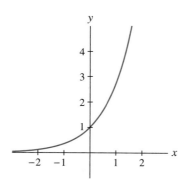

FIGURE 4.36

- $y = e^x$ approaches 0 as x decreases without bound ($x \to -\infty$); so, as $x \to -\infty$ the graph of $y = e^x$ seems to become horizontal (i.e., $y' \to 0$).

- $y = e^x$ increases without bound as x increases without bound ($x \to \infty$) and the graph of the function is concave up and has no vertical asymptotes. As $x \to \infty$, the graph of $y = e^x$ seems to become more and more vertical; that is, the slopes are increasing without bound as $x \to \infty$.

Next, to obtain an idea of magnitude, we numerically investigate the slope at a few points. Table 4.7 shows function values and slope values (rounded to three decimal places) for several inputs.

TABLE 4.7

x	-2	0	1	3
$y = e^x$	0.135	1.000	2.718	20.086
$y' = \dfrac{dy}{dx}$	0.135	1.000	2.718	20.086

This function is surprising in that the rate-of-change values are precisely the same as the function values. *This function is its own derivative!* In other words, if $y = e^x$, then $\frac{dy}{dx} = y' = e^x$. The slope graph of $y = e^x$ coincides with the graph of the original function.

Derivative of e^x

If $y = e^x$, then $\dfrac{dy}{dx} = e^x$.

Does this rule apply to exponential functions that have bases different from e? In other words, if $y = b^x$, is $\frac{dy}{dx} = b^x$? Consider the functions $y = 2^x$ and $y = 3^x$. Graphically, the descriptions of end behavior and curvature lead to the same conclusions about the shape of the derivative graphs for these exponential functions as did the analysis of the shape and end behavior of $y = e^x$. (See Figures 4.37a and b.) Any differences that occur should appear in a numerical investigation of the magnitude of the slopes.

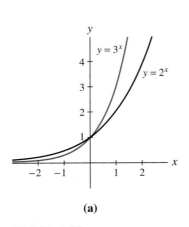

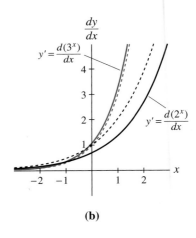

(a)   (b)

FIGURE 4.37

TABLE 4.8

x	$y = 2^x$	$y' = \dfrac{dy}{dx}$
-2	0.25	0.17329
0	1	0.69315
1	2	1.38629
3	8	5.54518

We begin by exploring the derivative of $y = 2^x$. See Table 4.8 (values are rounded to five decimal places for convenience). It is obvious from the table that the derivative of $y = 2^x$ is not $y' = 2^x$. If we use the definition of derivative for the function $f(x) = 2^x$, we obtain

$$f'(x) = \lim_{h \to 0} \frac{2^{x+h} - 2^x}{h}$$

You may recall that $2^{x+h} = 2^x 2^h$. Using this fact, we rewrite the derivative formula as

$$f'(x) = \lim_{h \to 0} \frac{2^x 2^h - 2^x}{h} = \lim_{h \to 0} \left[2^x \left(\frac{2^h - 1}{h} \right) \right]$$

Because the term 2^x is not affected by h approaching zero, we treat it as a constant:

$$f'(x) = 2^x \left[\lim_{h \to 0} \frac{2^h - 1}{h} \right]$$

This formula indicates that the derivative of $y = 2^x$ is 2^x times a constant. In Table 4.9 we numerically estimate the limiting value of the multiplier $\lim_{h \to 0} \frac{2^h - 1}{h}$.

TABLE 4.9

$h \to 0^+$	$\dfrac{2^h - 1}{h}$	$h \to 0^-$	$\dfrac{2^h - 1}{h}$
0.1	0.717735	-0.1	0.669670
0.01	0.695555	-0.01	0.690750
0.001	0.693387	-0.001	0.692907
0.0001	0.693171	-0.0001	0.693123
0.00001	0.693150	-0.00001	0.693145
0.000001	0.693147	-0.000001	0.693147
Limit ≈ 0.693147		Limit ≈ 0.693147	

It may seem that the multiplier 0.693147 is arbitrary because it is not a familiar number, but that is not the case. You should verify that $0.693147 \approx \ln 2$. In fact, it can be proven that the limit term in the derivative formula for $f(x) = 2^x$ is $\ln 2$. We state the formula for the derivative of $y = 2^x$ as $y' = (\ln 2)2^x$. A similar exploration suggests that the formula for the derivative of $y = 3^x$ is $y' = (\ln 3)3^x$.

The two derivative formulas $\frac{d}{dx}(2^x) = (\ln 2)2^x$ and $\frac{d}{dx}(3^x) = (\ln 3)3^x$ are special cases of the general derivative formula for exponential functions. The derivative of $y = b^x$ is $y' = (\ln b)b^x$ if $b > 0$. In fact, the rule $\frac{d}{dx}(e^x) = e^x$ is also a special case of this formula. You will be asked to verify this fact in the Activities.

Derivative of b^x

If $y = b^x$, where the real number $b > 0$, then $\dfrac{dy}{dx} = (\ln b)b^x$.

EXAMPLE 1   *Using Exponential Derivative Formulas*

Credit Cards   If credit card purchases are not paid off by the due date on the credit card statement, finance charges are applied to the remaining balance. In July 2001, one major credit card company had a daily finance charge of 0.05425% on unpaid balances. Assume that the unpaid balance is $1 and that no new purchases are made.

a. Find an exponential function for the credit balance d days after the due date.

b. How much is owed after 30 days?

c. Write the derivative formula for the function from part *a*.

d. How quickly is the balance changing after 30 days?

e. Repeat parts *a* through *d*, assuming that the unpaid balance is $2000.

Solution

a. Recall that the constant b in an exponential function $f(x) = ab^x$ is $(1 + \text{percent-age growth})$ and that the constant a is the value of $f(0)$. Thus we use the function

$$f(d) = 1(1.0005425^d) = 1.0005425^d \text{ dollars}$$

to represent the balance due d days after the due date.

b. Thirty days after the due date, the balance is $f(30) \approx \$1.02$.

c. According to the derivative rules we saw in this chapter, the derivative of $y = b^x$ is $y' = (\ln b)b^x$. Thus the derivative formula for our function f is

$$f'(d) = (\ln 1.0005425)1.0005425^d \text{ dollars per day}$$

after d days.

d. We use the derivative formula evaluated at $d = 30$ to find the rate of change of the balance. After 30 days, the balance is increasing at a rate of 0.0006 dollar per day.

e. If the unpaid balance is $2000, the balance-due function is

$$f(d) = 2000(1.0005425^d) \text{ dollars}$$

after d days. The amount due after 30 days is $f(30) = \$2032.81$.

According to the Constant Multiplier and the Exponential Rules, the derivative of the balance function is

$$f'(d) = 2000(\ln 1.0005425)1.0005425^d \text{ dollars per day}$$

After 30 days, the credit balance is increasing at a rate of $f'(30) = \$1.10$ per day. ●

Natural Logarithm Rule

As with exponential functions, we motivate the derivative rule for the natural log function graphically and numerically.

The natural logarithm function is not defined for negative input values or for an input of zero. But as x approaches 0 from the right ($x \to 0^+$), the outputs of the natural log function decrease without bound. The tilt of the function appears to become vertical. That is, the slope is increasing without bound as x approaches 0 from the right. Also, as x increases without bound, $\ln x$ increases without bound, but more and more slowly. The slope never becomes zero. Thus, as $x \to \infty$, the derivative function approaches 0. (See Figures 4.38a and b.)

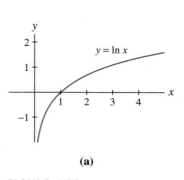

(a)   (b)

FIGURE 4.38

TABLE 4.10

x	Derivative of $y = \ln x$
$\frac{1}{2}$	2.000
1	1.000
2	0.500
4	0.250
10	0.100

Numerically investigating the slope (to three decimal places) for a few input values (to three decimal places) once again helps to establish magnitude. (See Table 4.10.) Unlike the case for the exponential function, whose derivative at a certain input value is dependent on the corresponding output value, these derivatives are dependent on the input values. Note that each derivative value is the reciprocal of the input value— that is, it is 1 divided by the input value. The derivative of $y = \ln x$ at $x = 2$ is $\frac{1}{2}$, the derivative at $x = 4$ is $\frac{1}{4}$, the derivative at $x = 10$ is $\frac{1}{10}$, and so on.

Derivative of ln x

If $y = \ln x$, then $\dfrac{dy}{dx} = \dfrac{1}{x}$ for positive x values.

Our next example illustrates the use of all of the derivative rules presented in this section and the previous section.

EXAMPLE 2   *Using Simple Derivative Rules*

4.3.2a, b, c

Find the derivatives of the following functions:

a. $f(x) = 12.36 + 6.2 \ln x$ b. $g(t) = 4e^t + 19$

c. $m(r) = \dfrac{8}{r} - 12\sqrt{r}$ d. $j(y) = 17{,}000\left(1 + \dfrac{0.025}{12}\right)^{12y}$

Solution

a. Apply the Constant Rule to the first term and the Constant Multiplier and the Natural Log Rules to the second term. Use the Sum Rule to add the two derivatives together.

$$\dfrac{df}{dx} = f'(x) = 0 + 6.2\left(\dfrac{1}{x}\right) = \dfrac{6.2}{x}$$

b. Apply the Constant Multiplier and the e^x Rules to the first term and the Constant Rule to the second term. Again, use the Sum Rule to add the two derivatives.

$$\dfrac{dg}{dt} = g'(t) = 4e^t + 0 = 4e^t$$

c. The key to finding the derivative formula for this function is rewriting the two terms using algebra rules for exponents. Recall that a negative exponent is used to indicate that a term is in the denominator and that an exponent of $\frac{1}{2}$ indicates a square root. Using these facts, rewrite m as

$$m(r) = 8r^{-1} - 12r^{1/2}$$

Now apply the Constant Multiplier, the Power, and the Sum Rules to obtain

$$\dfrac{dm}{dr} = m'(r) = -1(8)r^{-1-1} - \dfrac{1}{2}(12)r^{1/2-1} = -8r^{-2} - 6r^{-1/2} = \dfrac{-8}{r^2} - \dfrac{6}{\sqrt{r}}$$

d. Begin by calculating the number $\left(1 + \dfrac{0.025}{12}\right)^{12}$ in order to rewrite the formula in the form ab^x. We have rounded this number to six decimal places, but in using the derivative to calculate rates of change, you should keep all decimal places stored in your calculator.

$$j(y) = 17{,}000\left(1 + \dfrac{0.025}{12}\right)^{12y} \approx 17{,}000(1.025288^y)$$

Now apply the Constant Multiplier and the Exponential Rules.

$$\dfrac{dj}{dy} = j'(y) \approx 17{,}000(\ln 1.025288)(1.025288^y) \approx 424.5579(1.025288^y) \quad ●$$

In summary, we present a list of the rate-of-change formulas you should know. (The list includes the formulas from Section 4.2.) The formulas are best learned through practice. We urge you to work as many of the Activities in this section as possible. Although you may need to refer to this table for some of the beginning activities, you should attempt to work most of them without looking at this list.

Simple Derivative Rules

Rule Name	Function	Derivative
Constant Rule	$y = b$	$\dfrac{dy}{dx} = 0$
Linear Function Rule	$y = ax + b$	$\dfrac{dy}{dx} = a$
Power Rule	$y = x^n$	$\dfrac{dy}{dx} = nx^{n-1}$
Exponential Rule	$y = b^x, b > 0$	$\dfrac{dy}{dx} = (\ln b)b^x$
e^x Rule	$y = e^x$	$\dfrac{dy}{dx} = e^x$
Natural Log Rule	$y = \ln x, x > 0$	$\dfrac{dy}{dx} = \dfrac{1}{x}$
Constant Multiplier Rule	$y = kf(x)$	$\dfrac{dy}{dx} = kf'(x)$
Sum Rule	$y = f(x) + g(x)$	$\dfrac{dy}{dx} = f'(x) + g'(x)$
Difference Rule	$y = f(x) - g(x)$	$\dfrac{dy}{dx} = f'(x) - g'(x)$

4.3   Concept Inventory

○ *Derivative formulas*

For constants *a*, *b*, and *n*,

If $y = b$, then $y' = 0$.

If $y = ax + b$, then $y' = a$.

If $y = x^n$, then $y' = nx^{n-1}$.

If $y = b^x$, then $y' = (\ln b)b^x$.

If $y = e^x$, then $y' = e^x$.

If $y = \ln x$, then $y' = \dfrac{1}{x}$.

If $y = kf(x)$, then $y' = kf'(x)$.

If $y = f(x) \pm g(x)$, then $y' = f'(x) \pm g'(x)$.

4.3   Activities

For each of the functions whose graphs are given in Activities 1–6, first sketch the slope graph and then give the slope equation.

1.

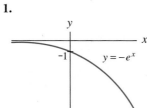

$y = -e^x$

2.

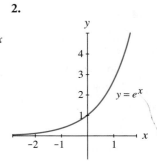

$y = e^x$

3.

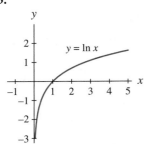

$y = \ln x$

4.

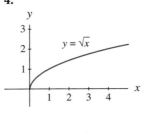

$y = \sqrt{x}$

5.

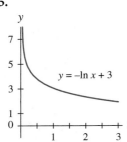

$y = -\ln x + 3$

6.

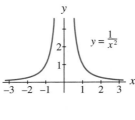

$y = \frac{1}{x^2}$

Give the derivative formulas for the functions in Activities 7 through 14.

WEB/CD

Skills

7. $h(x) = 7x^2 + 13 \ln x$

8. $y = 1 + 4.9876e^x$

9. $g(x) = 17(4.962^x)$

10. $y = \dfrac{-9}{x}$

11. $f(x) = 100,000\left(1 + \frac{0.05}{12}\right)^{12x}$ (*Hint:* Rewrite as ab^x.)

12. $j(x) = 4.2(0.8^x) + \dfrac{1.6}{x}$

13. $y = 9 - 4.2 \ln x + 3.3(2.9^x)$

14. $k(x) = 3\sqrt{x} + 9e^x$

15. Rising Dough   For the first couple of hours after yeast dough has been kneaded, it approximately doubles in volume every 42 minutes. If we prepare 1 quart of yeast dough and let it rise in a warm room, then its growth can be modeled by the function

$$V = e^h \text{ quarts}$$

where h is the number of hours the dough has been allowed to rise.

a. How many minutes will it take the dough to attain a volume of 2.5 quarts?

b. Write a formula for the rate of growth of the yeast dough.

c. How quickly is the dough expanding after 24 minutes, after 42 minutes, and after 55 minutes? Report your answers in quarts per minute.

16. Investment   The value of a $1000 investment in an account with 4.3% interest compounded continuously can be modeled as

$$A = e^{0.043t} \text{ thousand dollars after } t \text{ years}$$

a. Write the rate-of-change formula for the value of the investment. (*Hint:* Let $b = e^{0.043}$, and use the rule for $y = b^t$.)

b. How much is the investment worth after 5 years?

c. How quickly is the investment growing after 5 years?

d. What is the percentage rate of growth after 5 years?

17. We have seen that the derivative of $y = b^x$ is $\frac{dy}{dx} = (\ln b)b^x$ as long as $b > 0$. We have also seen that the derivative of $y = e^x$ is $\frac{dy}{dx} = e^x$.

a. Show that the derivative formula for $y = e^x$ is a special case of the derivative formula for $y = b^x$ by applying the formula for $y = b^x$ to $y = e^x$ and then reconciling the result with the known derivative formula for $y = e^x$.

b. Use the derivative formula for $y = b^x$ to find a formula for the derivative of an exponential function of the form $y = e^{kx}$, where k is some known constant.

18. Costs   Dairy company managers have found that it costs them approximately $c(u) = 3250 + 75 \ln u$ dollars to produce u units of dairy products each week. They also know that it costs them approximately $s(u) = 50u + 1500$ dollars to ship u units. Assume that the company ships its products once each week.

a. Write the formula for the total weekly cost of producing and shipping u units.

b. Write the formula for the rate of change of the total weekly cost of producing and shipping u units.

c. How much does it cost the company to produce and ship 5000 units in 1 week?

d. What is the rate of change of total production and shipping costs at 5000 units? Interpret your answer.

19. **Weight**   The weight of a laboratory mouse between 3 and 11 weeks of age can be modeled by the equation

$$w(t) = 11.3 + 7.37 \ln t \text{ grams}$$

where the age of the mouse is $(t + 2)$ weeks after birth (thus for a 3-week-old mouse, $t = 1$).

a. What is the weight of a 9-week-old mouse, and how rapidly is its weight changing?

b. How rapidly on average does the mouse grow between ages 6 and 11 weeks?

c. What happens to the rate at which the mouse is growing as it gets older? Explain.

20. **Web TV**   The projected number of homes with access to the Internet via cable television between 1998 and 2005 can be modeled by the equation

$$I(x) = -138.27 + 76.29 \ln x \text{ million homes}$$

x years after 1990.
(Source: Based on data from Paul Kagen Associates, Inc., Cable Television Technology.)

a. Give the rate-of-change formula for the projected number of such homes.

b. How many homes are projected to have Internet access via a cable TV system in 2005, and how rapidly is that number projected to be growing?

21. **Tuition CPI**   The consumer price index for college tuition between 1990 and 2000 is shown.

Year	CPI	Year	CPI	Year	CPI
1990	175.0	1994	249.8	1998	306.5
1991	192.8	1995	264.8	1999	318.7
1992	213.5	1996	279.8	2000	331.9
1993	233.5	1997	294.1		

(Source: *Statistical Abstract*, 2001.)

a. Use only the data to estimate the rate of change of the CPI in 1998.

b. Align the data as the number of years since 1980, and find a log model for the CPI.

c. Use the model to find the rate of change of the CPI in 1998.

22. **Investment**   An individual has $45,000 to invest: $32,000 will be put into a low-risk mutual fund averaging 6.2% interest compounded monthly, and the remainder will be invested in a bond fund averaging 9.7% interest compounded continuously.

a. Find an equation for the total amount in the two investments.

b. Give the rate-of-change equation for the combined amount.

c. How rapidly is the combined amount of the investments growing after 6 months? after 15 months?

23. **Income**   The Bureau of the Census reports the median family income since 1947 as shown in the table. (Median income means that half of American families make more than this value and half make less.)

Year	Median family income (constant 1997 dollars)
1947	20,102
1957	26,133
1967	35,076
1977	40,656
1987	43,756
1997	44,568

a. Find a model for the data.

b. Find a formula for the rate of change of the median family income.

c. Find the rates of change and percentage rates of change of the median family income in 1972, 1980, 1984, 1992, and 1996.

d. Do you think the above rates of change and percentage rates of change affected the reelection campaigns of Presidents Nixon (1972), Carter (1980), Reagan (1984), Bush (1992), and Clinton (1996)?

24. **Investment**   The accompanying table gives the value of a $2000 investment after 2 years in an account whose annual percentage yield is $100r\%$.

r	Value (dollars)	r	Value (dollars)
0.02	2080.80	0.08	2332.80
0.03	2121.80	0.09	2376.20
0.04	2163.20	0.10	2420.00
0.05	2205.00	0.11	2464.20
0.06	2247.20	0.12	2508.80
0.07	2289.80		

a. Find a model for the value of a $2000 investment after two years as a function of the interest rate (in decimals).

b. Find the rate of change of the value of the investment when the annual percentage yield is 4%. Interpret your answer.

c. Repeat part *a*; however, instead of using the interest rate in decimals, enter the rate in whole numbers so that $r = 2$ at 2%, $r = 3$ at 3%, etc. Compare this model with the model in part *a*.

d. Use the model from part *c* to find the rate of change of the value of the investment when the annual percentage yield is 4%. Compare this answer with that of part *b*.

25. **Investment**   The value of a $1000 investment after 10 years in an account whose interest rate is $100r$% compounded continuously is

$$A(r) = 1000e^{10r} \text{ dollars}$$

a. Write the rate-of-change function for the value of the investment.

b. Determine the rate of change of the value of the investment at 7% interest. Discuss why the rate of change is so large.

c. If the interest rate is input as a percentage instead of a decimal, the function for the value of a $1000 investment after 10 years in an account with interest compounded continuously is

$$A(r) = 1000e^{0.1r} \text{ dollars}$$

where $r = 1.00$ when the interest rate is 1%, $r = 1.25$ when the interest rate is 1.25%, etc. Write the rate-of-change function for the value of the investment. Compare this rate of change function with that in part *a*.

d. Using the function in part *c*, determine the rate of change of the value of the investment at 7% interest. Compare this answer to that of part *b*.

26. **VCR Homes**   The percentage of households with TVs that also have VCRs from 1990 through 2001 is shown.

Year	Households (percent)	Year	Households (percent)
1990	68.6	1996	82.2
1992	75.0	1998	84.6
1994	79.0	2001	86.2

(Sources: *Statistical Abstract*, 1998; Television Bureau of Advertising.)

a. Align the input data as the number of years since 1987, and find a log model for the data.

b. Write the rate-of-change formula for the model in part *a*.

c. According to the model, what was the percentage of households with TVs that also had VCRs in 2000? How rapidly was the percentage growing in that year? Interpret your answers.

27. **Population**   The population of Aurora, a Nevada ghost town, can be modeled as

$$p(t) = \begin{cases} -7.91t^3 + 120.96t^2 + 193.92t - 123.21 \\ \qquad \text{people} \qquad\qquad \text{when } 0.7 \le t \le 13 \\ 45{,}544(0.8474^t) \text{ people when } 13 < t \le 55 \end{cases}$$

where t is the number of years since the beginning of 1860.

(*Source:* Based on data from Don Ashbaugh, *Nevada's Turbulent Yesterday: A Study in Ghost Towns.* Los Angeles: Westernlore Press, 1963.)

a. Write the derivative equation for the population model.

b. How quickly was the population growing or declining in the beginning of the years 1870, 1873, and 1900?

28. **Population**   An alternative model for the population of Iowa in the 1980s and early 1990s is

$$p(t) = \begin{cases} -2.1t^2 - 63t + 2914 \text{ thousand people} \\ \qquad\qquad\qquad\qquad \text{when } 0 \le t \le 7 \\ 0.435(1.45479^t) \text{ thousand people} \\ \qquad\qquad\qquad\qquad \text{when } 7 < t \le 13 \end{cases}$$

where t is the number of years since 1980.
(Source: Based on data from *Statistical Abstract*, 1994.)

a. Write the derivative equation for the population model.

b. How quickly was the population growing or declining in 1985? in 1987? in 1989?

c. Can the model be used to answer reasonably the question "How quickly was the population of Iowa growing or declining in 1995?"

29. Use your knowledge of the shape and end behavior of the graph of a shifted exponential function of the form $y = ab^x + c$ as well as your knowledge of the simple derivative rules in Sections 4.2 and 4.3 to describe the shape and end behavior of the graph and mathematical form of the rate-of-change function of a shifted exponential model.

30. Use your knowledge of the shape and end behavior of the graph of a log function of the form $y = a + b \ln x$ as well as your knowledge of the simple derivative rules in Sections 4.2 and 4.3 to describe the shape and end behavior of the graph and mathematical form of the rate-of-change function of a log model.

4.4   The Chain Rule

We now present derivative formulas for more complicated functions than those considered in Sections 4.2 and 4.3. In particular, we introduce the method used for finding derivative formulas for composite functions.

The First Form of the Chain Rule

It is well known that high levels of carbon dioxide (CO_2) in the atmosphere are linked to increasing populations in highly industrialized societies. This is because large urban environments consume enormous amounts of energy and CO_2 is a natural by-product of the (often incomplete) consumption of that energy.

Imagine that in a certain large city, the level of CO_2 in the air is linked to the size of the population by the equation $C(p) = \sqrt{p}$, where the units of $C(p)$ are parts per million (ppm) and p is the population. Also suppose that the population is projected to grow quadratically between 2000 and 2015 according to the equation $p(t) = 400t^2 + 2500$ people, where t is the number of years since 2000. Note that C is a function of p, and p is a function of t. Thus indirectly, C is also a function of t. Suppose we want to know the rate of change of the CO_2 concentration *with respect to time* in 2013; that is, how rapidly the CO_2 concentration is rising or falling in 2013. The mathematical notation for this rate of change is $\frac{dC}{dt}$, and the units are ppm per year.

The derivative of C is $\frac{dC}{dp} = \frac{1}{2\sqrt{p}}$ ppm per person. But this is not the rate of change that we want because $\frac{dC}{dp}$ is the rate of change *with respect to population*, not time. The question now becomes "How do we transform ppm per person to ppm per year?" If we knew the rate of change of population with respect to time (people per year), then we could multiply as indicated to get the desired units:

$$\left(\frac{\text{ppm}}{\text{person}} \right)\left(\frac{\text{people}}{\text{year}} \right) = \frac{\text{ppm}}{\text{year}}$$

Population is given as a function of time, so its derivative is the rate of change that we need: $\frac{dp}{dt} = 800t$ people per year. This motivates

$$\left(\frac{dC}{dp}\right)\left(\frac{dp}{dt}\right) = \frac{dC}{dt} \quad \text{or}$$

$$\left(\frac{1}{2\sqrt{p}}\frac{\text{ppm}}{\text{person}}\right)\left(800t\frac{\text{people}}{\text{year}}\right) = \frac{dC}{dt}\frac{\text{ppm}}{\text{year}}$$

Because $\frac{dC}{dt}$ is a rate of change *with respect to time*, it is standard procedure to write the derivative formula in terms of t. Recall that $p(t) = 400t^2 + 2500$ people where t is the number of years since 1990. Substituting $400t^2 + 2500$ for p in the equation for $\frac{dC}{dt}$, we have

$$\frac{dC}{dt} = \frac{1}{2\sqrt{400t^2 + 2500}}(800t) \text{ ppm/year}$$

Now we substitute $t = 13$ (for 2013) to obtain our desired result:

$$\frac{dC}{dt} = \frac{1}{2\sqrt{400(13)^2 + 2500}}[800(13)] \approx 19.64 \text{ ppm/year}$$

In 2013, the CO_2 concentration will be increasing by approximately 19.64 ppm per year.

The method used to find $\frac{dC}{dt}$ in the situation above is called the **Chain Rule** because it links together the derivatives of two functions to obtain the derivative of their composite function.

The Chain Rule (Form 1)

If C is a function of p and p is a function of t, then

$$\frac{dC}{dt} = \left(\frac{dC}{dp}\right)\left(\frac{dp}{dt}\right)$$

EXAMPLE 1   *Using the First Form of the Chain Rule*

Violin Production   Let $A(v)$ denote the average cost to produce a student violin when v violins are produced, and let $v(t)$ represent the number (in thousands) of student violins produced t years after 2000. Suppose that 10 thousand student violins are produced in 2008 and that the average cost to produce a violin at that time is $142.10. Also, suppose that in 2008 the production of violins is increasing by 100 violins per year and the average cost of production is decreasing by 15 cents per violin.

a. Describe the meaning and give the value of each of the following in 2008:

    i. $v(t)$ ii. $v'(t)$ iii. $A(v)$ iv. $A'(v)$

b. Calculate the rate of change with respect to time of the average cost for student violins in 2008.

Solution

a.   i. There are $v(8) = 10{,}000$ violins produced in 2008.

   ii. The rate of change of violin production in 2008 is $v'(8) = 0.1$ thousand violins per year. That is, $\frac{dv}{dt} = 100$ violins per year.

   iii. The average cost to produce a violin is $A(10) = \$142.10$ when 10,000 violins are produced.

   iv. When 10,000 violins are produced, the average cost is changing at a rate of $A'(10) = -\$0.15$ per violin. That is, $\frac{dA}{dv} = -\$0.15$ per violin.

b.   The rate of change with respect to time of the average cost to produce a student violin in 2008 is

$$\frac{dA}{dt} = \frac{dA}{dv} \cdot \frac{dv}{dt}$$

$$= (-\$0.15 \text{ per violin})(100 \text{ violins per year}) = -\$15 \text{ per year}$$

In 2008 the average cost to produce a violin is declining by \$15 per year.   ●

The Second Form of the Chain Rule

Recall the discussion at the beginning of this section concerning CO_2 pollution and population. We were given two functions—p with input t and C, whose input corresponds to the output of p—and then asked to find the derivative $\frac{dC}{dt}$. You may wonder why we did not substitute the expression for population into the CO_2 equation before finding the derivative:

$$C(p) = \sqrt{p} \quad \text{with} \quad p(t) = 400t^2 + 2500 \quad \text{so} \quad C(p(t)) = \sqrt{400t^2 + 2500}$$

This process, called function composition (see Section 1.3), allows us to express C directly as a function of t. If we now take the derivative, we get $\frac{dC}{dt}$, which is exactly what we needed all along! The reason we did not do this before is that we did not know a formula for finding the derivative of a composite function. However, we can now use the Chain Rule to obtain a formula. First, we review some terminology from Section 1.2.

Because $p(t)$ was substituted into the formula for C to create the composite function $C \circ p$, we call p the inside function and C the outside function. Next, recall the Chain Rule process:

$$\frac{dC}{dt} = \left(\frac{dC}{dp}\right)\left(\frac{dp}{dt}\right)$$

$$= \left(\frac{1}{2\sqrt{p}}\right)(800t)$$

$$= \left(\frac{1}{2\sqrt{400t^2 + 2500}}\right)800t$$

The first term, $\frac{1}{2\sqrt{400t^2 + 2500}}$, is simply the derivative of $\sqrt{p}$ with p replaced by $400t^2 + 2500$. This is the derivative of the outside function with the inside function substituted for p. The second term, $800t$, is the derivative with respect to t of p, the inside function. This leads us to a second form of the Chain Rule. If a function is expressed as a result of function composition (that is, it is a combination of an inside function and an outside function), then its slope formula can be found as follows:

$$\text{Slope formula of composite function} = \begin{pmatrix} \text{derivative of the} \\ \text{outside function} \\ \text{with the inside} \\ \text{function untouched} \end{pmatrix} \begin{pmatrix} \text{derivative} \\ \text{of the inside} \\ \text{function} \end{pmatrix}$$

Mathematically, we state this form of the Chain Rule as follows:

The Chain Rule (Form 2)

If a function f can be expressed as the composition of two functions h and g; that is, if

$$f(x) = (h \circ g)(x) = h(g(x))$$

then its slope formula is

$$\frac{df}{dx} = f'(x) = h'(g(x)) \cdot g'(x)$$

In Example 2, we consider three somewhat different forms of composite functions, identify the inside function and the outside function for each, and use the Chain Rule to find formulas for the derivatives.

EXAMPLE 2   *Using the Second Form of the Chain Rule*

Write the derivatives (with respect to x) for the following three functions.

a. $y = e^{x^2}$ b. $y = (x^3 + 2x^2 + 4)^{1/2}$ c. $y = \dfrac{3}{4 - 2x^2}$

Solution

a. We can consider $y = e^{x^2}$ as composed of an outside function $y = e^p$ and an inside function $p = x^2$. The derivative of the outside function is e^p. (This exponential function is its own derivative.) Form 2 of the Chain Rule instructs us to leave the inside function untouched (that is, in its original form) so instead of e^p appearing in the derivative, the first expression in the slope formula is e^{x^2}. The second expression in the slope formula is the derivative, $2x$, of the inside function. The final answer is the product of these two derivatives.

$$\frac{dy}{dx} = (e^{x^2})(2x) = 2xe^{x^2}$$

b. The inside function of $y = (x^3 + 2x^2 + 4)^{1/2}$ is $p = x^3 + 2x^2 + 4$, and the outside function is $y = p^{1/2}$. The derivative of the outside function is $\frac{1}{2}p^{-1/2}$; with p

untouched, this becomes $\frac{1}{2}(x^3 + 2x^2 + 4)^{-1/2}$. The derivative of the inside function is $3x^2 + 4x$. Thus the Chain Rule gives

$$\frac{dy}{dx} = \frac{1}{2}(x^3 + 2x^2 + 4)^{-1/2}(3x^2 + 4x)$$

c. The function $y = \dfrac{3}{4 - 2x^2}$ can be thought of as the composition of the outside function $y = \dfrac{3}{p}$ and the inside function $p = 4 - 2x^2$. The derivative of the outside function is $\dfrac{-3}{p^2}$, or $\dfrac{-3}{(4 - 2x^2)^2}$. The derivative of the inside function is $-4x$. The derivative of the composite function is then

$$\frac{dy}{dx} = \left(\frac{-3}{(4 - 2x)^2}\right)(-4x) = \frac{12x}{(4 - 2x^2)^2} \quad\bullet$$

As illustrated in Example 3, one common use of the Chain Rule is to find the derivative of a logistic function.

EXAMPLE 3   *Using the Chain Rule to Find a Logistic Function Derivative*

4.4.1

VCR Homes   The percentage of households between 1980 and 2001 with VCRs can be modeled* by

$$P(t) = \frac{84.4}{1 + 33.6e^{-0.484t}} \text{ percent}$$

where t is the number of years since 1900. Find the rate-of-change formula for P with respect to t.

Solution   The function P can be rewritten as

$$P(t) = 84.4(1 + 33.6e^{-0.484t})^{-1}$$

In this form, it is easy to see that $p = 84.4u^{-1}$ is the outside function and that the inside function is $u = 1 + 33.6e^{-0.484t}$. Further, we can split u into an outside and inside function with $u = 1 + 33.6e^v$ as the outside function and $v = -0.484t$ as the inside function.

Now, the derivative of P with respect to t is

$$
\begin{aligned}
P'(t) &= (\text{derivative of } 84.4u^{-1})(\text{derivative of } u) \\
&= (\text{derivative of } 84.4u^{-1})[(\text{derivative of } 1 + 33.6e^v)(\text{derivative of } v)] \\
&= (-84.4u^{-2})[(33.6e^v)(-0.484)] \\
&= \frac{-84.4(33.6e^v)(-0.484)}{u^2} \\
&\approx \frac{1372.546e^v}{u^2}
\end{aligned}
$$

---

*Based on data from *Statistical Abstract*, 1998.

Next, substitute $u = 1 + 33.6e^{-0.484t}$ and $v = -0.484t$ back into the expression to obtain the derivative in terms of t.

$$\frac{dP}{dt} \approx \frac{1372.546e^{-0.484t}}{(1 + 33.6e^{-0.484t})^2} \text{ percentage points per year}$$

where t is the number of years since 1980. ●

4.4 Concept Inventory

○ *Function composition*

○ *Inside and outside portions of a composite function*

○ *Chain Rule:*

$$\frac{dC}{dt} = \frac{dC}{dp} \cdot \frac{dp}{dt} \qquad \text{(Form 1)}$$

$$C'(t) = C'(p(t)) \cdot p'(t) \quad \text{(Form 2)}$$

4.4 Activities

1. Let x be a function of t, and let f be a function whose input corresponds to the output of x. If $x(2) = 6, f(6) = 140, x'(2) = 1.3,$ and $f'(6) = -27,$ give the values of

   a. $f(x(2))$

   b. $\dfrac{df}{dx}$ when $x = 6$

   c. $\dfrac{dx}{dt}$ when $t = 2$

   d. $\dfrac{df}{dt}$ when $t = 2$

2. Let v be a function of x, and let g be a function whose input corresponds to the output of v. If $v(88) = 17, v'(88) = 1.6, g(17) = 0.04,$ and $g'(17) = 0.005,$ give the values of

   a. $g(v(88))$

   b. $\dfrac{dv}{dx}$ when $x = 88$

   c. $\dfrac{dg}{dv}$ when $x = 88$

   d. $\dfrac{dg}{dx}$ when $x = 88$

3. **Investment** An investor has been buying gold at a constant rate of 0.2 troy ounce per day. The investor currently owns 400 troy ounces of gold. If gold is currently worth \$323.10 per troy ounce, how quickly is the value of the investor's gold increasing per day?

4. **Leaking Tank** A gas station owner is unaware that one of the underground gasoline tanks is leaking. The leaking tank currently contains 600 gallons of gas and is losing 3.5 gallons per day. If the value of the gasoline is \$1.51 per gallon, how much potential revenue is the station losing per day?

5. **Revenue** Let $R(x)$ be the revenue in Canadian dollars from the sale of x units of a commodity, and let $C(r)$ be the U.S. dollar value of r Canadian dollars. On November 25, 2002, \$10,000 Canadian were worth \$633.47 U.S., and the rate of change of the U.S. dollar value was \$0.6335 U.S. per Canadian dollar. On the same day, sales were 476 units, producing a revenue of \$10,000 Canadian, and revenue was increasing by \$2.6 Canadian per unit. Identify the following values on November 25, 2002, and write a sentence interpreting each value.

   a. $R(476)$

   b. $C(10,000)$

   c. $\dfrac{dR}{dx}$

   d. $\dfrac{dC}{dr}$

   e. $\dfrac{dC}{dx}$

6. **Mail** Suppose that $V(t)$ is the volume of mail (in thousands of pieces) processed at a post office on the tth day of the current year and that $E(v)$ is the number of employee-hours needed to process v thousand pieces of mail. On January 1 of this year, 150 thousand pieces of mail were processed, and that number was decreasing by 200 pieces per day. The rate of change of the number of employee-hours is a constant 12 hours per thousand pieces of mail. Identify the following quantities on January 1 of this year, and write a sentence interpreting each value.

a. $V(1)$

b. $\dfrac{dV}{dt}$

c. $\dfrac{dE}{dv}$

d. $\dfrac{dE}{dt}$

7. Refuse The population of a city in the northeast is given by $p(t) = \dfrac{130}{1 + 12e^{-0.02t}}$ thousand people, where t is the number of years since 2000. The number of garbage trucks needed by the city can be modeled by the equation $g(p) = 2p - 0.001p^3$, where p is the population in thousands. Find the value of the following in 2010.

a. $p(t)$　　　**b.** $g(p)$　　　**c.** $\dfrac{dp}{dt}$

d. $\dfrac{dg}{dp}$　　　**e.** $\dfrac{dg}{dt}$

f. Interpret the answers to parts *a–e*.

8. Profit Let $p(x) = 1.019^x$ Canadian dollars be the profit from the sale of x mountain bikes. On November 25, 2002, p Canadian dollars were worth $C(p) = \dfrac{p}{1.5786}$ U.S. dollars. On the same day, sales were 476 mountain bikes. Identify the following quantities on November 25, 2002:

a. $p(x)$　　　**b.** $C(p)$　　　**c.** $\dfrac{dp}{dx}$

d. $\dfrac{dC}{dp}$　　　**e.** $\dfrac{dC}{dx}$

f. Interpret the answers to parts *a–e*.

Rewrite each pair of functions in Activities 9 through 16 as a single composite function, and then find the derivative of the composite function.

9. $c(x) = 3x^2 - 2$　　　$x(t) = 4 - 6t$

10. $f(t) = 3e^t$　　　$t(p) = 4p^2$

11. $h(p) = \dfrac{4}{p}$　　　$p(t) = 1 + 3e^{-0.5t}$

12. $g(x) = \sqrt{7 + 5x}$　　　$x(w) = 4e^w$

13. $k(t) = 4.3t^3 - 2t^2 + 4t - 12$　　　$t(x) = \ln x$

14. $f(x) = \ln x$　　　$x(t) = 5t + 11$

15. $p(t) = 7.9(1.046^t)$　　　$t(k) = 14k^3 - 12k^2$

16. $r(m) = \dfrac{9.1}{m^2} + 3m$　　　$m(f) = f^4 - f^2$

For each of the composite functions in Activities 17 through 37, identify an outside function and an inside function, and find the derivative with respect to x of the composite function.

WEB/CD

Skills

17. $f(x) = (3.2x + 5.7)^5$

18. $f(x) = (5x^2 + 3x + 7)^{-1}$

19. $f(x) = \sqrt{x^2 - 3x}$　　　**20.** $f(x) = \sqrt[3]{x^2 + 5x}$

21. $f(x) = \ln(35x)$　　　**22.** $f(x) = (\ln 6x)^2$

23. $f(x) = \ln(16x^2 + 37x)$　　　**24.** $f(x) = e^{3.785x}$

25. $f(x) = 72.378e^{0.695x}$　　　**26.** $f(x) = e^{4x^2}$

27. $f(x) = 1 + 58.32e^{0.0856x}$　　　**28.** $f(x) = \dfrac{8}{(x - 1)^3}$

29. $f(x) = \dfrac{350}{4x + 7}$

30. $f(x) = \dfrac{112}{1 + 18.370e^{0.695x}} + 7.39$

31. $f(x) = \dfrac{3706.5}{1 + 8.976e^{-1.243x}} + 89{,}070$

32. $f(x) = \left(\sqrt{x} - 3x\right)^3$

33. $f(x) = 3^{\sqrt{x} - 3x}$

34. $f(x) = 2^{\ln x}$

35. $f(x) = \ln(2^x)$

36. $f(x) = Ae^{-Bx}$

37. $f(x) = \dfrac{L}{1 + Ae^{-Bx}}$

38. Rearing Children The percentage of children living with their grandparents between 1970 and 2000 can be modeled by the equation

$$p(t) = 3 + 0.216e^{0.09263t} \text{ percent}$$

t years after 1970.

(Source: Based on data from the U.S. Census Bureau.)

a. Write the rate-of-change formula for p.

b. How rapidly was the percentage of children living with their grandparents growing in 1995?

c. How rapidly on average did the percentage of children living with their grandparents grow between 1970 and 1990?

d. Geometrically illustrate the answers to parts *b* and *c*.

39. Bank Account   Imagine that you invest $1500 in a savings account at 4% annual interest compounded continuously.

a. Write an equation for the balance in the account after *t* years.

b. Write an equation for the rate of change of the balance.

c. What is the rate of change of the balance at the end of 1 year? 2 years?

d. Do the rates of change in part *c* tell you how much interest your account will earn over the next year? Explain.

40. Tuition   The tuition at a private 4-year college from 2000 through 2010 is projected to grow as shown.

Year	Tuition (dollars)	Year	Tuition (dollars)
2000	14,057	2006	16,918
2001	14,434	2007	17,561
2002	14,847	2008	18,264
2003	15,298	2009	19,033
2004	15,790	2010	19,873
2005	16,329		

a. Find an exponential equation of the form $f(x) = ab^x$ to fit the data.

b. Convert the equation you found in part *a* to one of the form $f(x) = ae^{kx}$.

c. Find rate-of-change formulas for both equations.

d. Use both equations to find the rate of change in 2010. How do your answers compare?

41. Revenue   In October of 1999, iGo Corp. offered 5 million shares of public stock at $9 per share. Revenue for the two years preceding the stock offering can be modeled by the equation

$$R(q) = 2.9 + 0.0314e^{0.62285q} \text{ million dollars}$$

q quarters after the beginning of 1998.

(Source: Based on data from the Securities and Exchange Commission.)

a. Write the rate-of-change formula of *R*.

b. Complete the table below.

Quarter ending	June 1998	June 1999	June 2000
Revenue			
Rate of change of revenue			
Percentage rate of change of revenue			

42. Cable TV   The percentage of households with TVs that subscribe to cable from 1970 through 2002 can be modeled by the logistic equation

$$P(t) = 6 + \frac{62.7}{1 + 38.7e^{-0.258t}} \text{ percent}$$

where *t* is the number of years since 1970.

(Source: Based on data from Television Bureau of Advertising.)

a. Write the rate-of-change formula for the percentage of households with TVs who subscribe to cable.

b. How rapidly was the percentage growing in 2000?

c. According to the model, what will happen to the percentage of cable subscribers in the long run? Do you believe that the model is a correct predictor of the long-term behavior? Explain.

43. Police Calls   Dispatchers at a sheriff's office record the total number of calls received since 5 A.M. in 3-hour intervals. Total calls for a typical day are given in the table.

Time	Total calls since 5 A.M.
8 A.M.	81
11 A.M.	167
2 P.M.	301
5 P.M.	495
8 P.M.	738
11 P.M.	1020
2 A.M.	1180
5 A.M.	1225

(Source: Greenville County, South Carolina, Sheriff's Office.)

a. Is a cubic or a logistic model more appropriate for this data set? Explain.

b. Find the more appropriate model for the data.

c. Find the rate-of-change formula for the model.

d. Evaluate the rate of change at noon, 10 P.M., midnight, and 4 A.M. Interpret the rates of change.

e. Discuss how rates of change can help a sheriff's office schedule dispatchers for work each day.

44. **P.T.A.** A model for the number of states associated with the national P.T.A. organization is

$$m(x) = \frac{49}{1 + 36.0660e^{-0.206743x}} \text{ states}$$

x years after 1895.
(Source: Based on data from Hamblin, Jacobsen, and Miller, *A Mathematical Theory of Social Change*. New York: Wiley, 1973.)

a. Write the derivative of m.

b. How many states had national P.T.A. membership in 1902?

c. How rapidly was the number of states joining the P.T.A. growing in 1890? in 1915? in 1927?

45. **Flu** Civilian deaths due to the influenza epidemic in 1918 can be modeled as

$$c(t) = \frac{93,700}{1 + 5095.9634e^{-1.097175t}} \text{ deaths}$$

t weeks after August 31, 1918.
(Source: Based on data from A. W. Crosby Jr., *Epidemic and Peace 1918*. Westport, CT: Greenwood Press, 1976.)

a. How rapidly was the number of deaths growing on September 28, 1918?

b. What percentage increase does the answer to part *a* represent?

c. Repeat parts *a* and *b* for October 26, 1918.

d. Why is the percentage change for parts *b* and *c* decreasing even though the rate of change is increasing?

46. **Cost** A manufacturing company has found that it can stock no more than 1 week's worth of perishable raw material for its manufacturing process. When purchasing this material, however, the company receives a discount based on the size of the order. Company managers have modeled the cost data as $C(t) = 196.25 + 44.45 \ln t$ dollars to produce t units per week. Each quarter, improvements are made to

the automated machinery to help enhance production. The company has kept a record of the average units per week that were produced in each quarter since January 2000. These data are given below.

Quarter	Units per week
Jan–Mar 2000	2000
Apr–June 2000	2070
July–Sept 2000	2160
Oct–Dec 2000	2260
Jan–Mar 2001	2380
Apr–June 2001	2510
July–Sept 2001	2660
Oct–Dec 2001	2820
Jan–Mar 2002	3000
Apr–June 2002	3200
July–Sept 2002	3410
Oct–Dec 2002	3620
Jan–Mar 2003	3880
Apr–June 2003	4130
July–Sept 2003	4410
Oct–Dec 2003	4690

a. Find an appropriate model for production per week x quarters after January 2000.

b. Use the company cost model along with your production model to write an expression modeling cost per unit as a function of the number of quarters since January 2000.

c. Use your model to predict the company's cost per week for each quarter of 2004.

d. Carefully study a graph of the function in part *b* from January 2000 to January 2005. According to this graph, will cost ever decrease?

e. Find an expression for the rate of change of the cost function in part *b*. Look at the graph of the rate-of-change function. According to this graph, will cost ever decrease?

47. **Cost** A dairy company's records reveal that it costs about $C(u) = 3250.23 + 74.95 \ln u$ dollars per week for the company to produce u units each week. Consumer demand has been increasing, so the company has been increasing production to keep up with demand. The accompanying table

indicates the production of the company, in units per week, from 1990 through 2003.

Year	Production (units per week)
1990	5915
1991	5750
1992	5940
1993	6485
1994	7385
1995	8635
1996	10,245
1997	12,210
1998	14,530
1999	17,200
2000	20,230
2001	23,610
2002	27,345
2003	31,440

a. Describe the curvature of the scatter plot of the data in the table. What types of equations could be used to fit these data?

b. Find an appropriate model for production.

c. Use the company's cost model along with your production model to write an expression modeling cost per week as a function of the number of years since 1990.

d. Write the rate-of-change function of the cost function you found in part c.

e. Use your model to estimate the company's cost per week in 2002, 2003, 2004, and 2005. Also, estimate the rates of change for those same years.

f. Carefully study the cost graph from 1990 to 2010. According to this graph, will cost ever decrease? Why or why not?

g. Look at the slope graph of the cost function from 1990 to 2010. According to this graph, will cost ever decrease? Why or why not?

48. **Advertising** The marketing division of a large firm has found that it can model the effectiveness of an advertising campaign by $S(u) = 0.75\sqrt{u} + 1.8$, where $S(u)$ represents sales in millions of dollars when the firm invests u thousand dollars in advertising. The firm plans to invest $u(x) = -2.3x^2 + 53.2x + 249.8$ thousand dollars each year x years from now.

a. Write the formula for predicted sales x years from now.

b. Write the formula for the rate of change of predicted sales x years from now.

c. What will be the rate of change of sales in 2007?

49. When you are composing functions, why is it important to make sure that the output of the inside function agrees with the input of the outside function?

50. Use your knowledge of the shape and end behavior of the graph of a shifted exponential function of the form $y = ae^{bx} + c$ as well as your knowledge of the simple derivative rules and the Chain Rule from Sections 4.2 through 4.4 to describe the shape and end behavior of the graph and mathematical form of the rate-of-change function of a shifted exponential model.

4.5 The Product Rule

It is fairly common to construct a new function by multiplying two functions. For example, revenue is the number of units sold multiplied by price. If both units sold and price are given by functions, then revenue is given by the product of the two functions. How to find rates of change for product functions is the topic of this section.

Applying the Product Rule without Equations

Suppose that the enrollment in a university is given by a function E and the percentage (expressed as a decimal) of students who are from out of state is given by a function P. In both functions, the input is t, the year corresponding to the beginning of the school year (that is, for the 2003–04 school year, t is 2003) because school enrollment figures are stated for the beginning of the fall term. Note that the product function $N(t) = E(t) \cdot P(t)$ gives the number of out-of-state students in year t. For example, if in the current year, enrollment began at 17,000 students with 30% of those from out of state, then the number of out-of-state students is calculated as $(17,000)(0.30) = 5100$.

Suppose that, in addition to enrollment being 17,000 with 30% from out of state, the enrollment is decreasing at a rate of 1420 students per year $\left(\frac{dE}{dt} = -1420\right)$, and the percentage of out-of-state students is increasing at a rate of 1.5 percentage points per year $\left(\frac{dP}{dt} = 0.015\right)$. How rapidly is the number of out-of-state students changing?

Because $N(t)$, the number of out-of-state students, is the product of $E(t)$ and $P(t)$, there are two rates to consider. First, of the 1420 students per year by which enrollment is declining, 30% of the students are from out of state. The product $(-1420$ students per year$)(0.30) = -426$ gives the amount by which the number of out-of-state students is decreasing. Thus, as a consequence of the decline in enrollment, the number of out-of-state students is declining by 426 students per year.

On the other hand, of the 17,000 students enrolled, the percentage of out-of-state students is growing by 1.5% per year. Thus the product $(17,000$ students$) \cdot (0.015$ per year$) = 255$ gives the amount by which the number of out-of-state students is increasing. The increasing percentage of out-of-state students results in a rate of increase of 255 out-of-state students per year.

To find the overall rate of change in the number of out-of-state students, we add the rate of change due to decline in enrollment and the rate of change due to increase in percentage.

$$-426 \text{ students} + 255 \text{ students} = -171 \text{ students}$$
$$\text{per year} \quad\quad \text{per year} \quad\quad \text{per year}$$

We interpret this result as follows: As a result of the declining enrollment and the increasing percentage of students from out of state, the number of out-of-state students is declining by 171 per year.

The steps to obtain this rate of change can be summarized in the equation

Rate of change of out-of-state students
$$= (-1420 \text{ students per year})(0.30)$$
$$+ (17,000 \text{ students})(0.015 \text{ per year})$$
$$= -171 \text{ out-of-state students per year}$$

Expressed in terms of the functions N, E, and P, this equation can be written as

$$\frac{dN}{dt} = \left(\frac{dE}{dt}\right)P(t) + E(t)\left(\frac{dP}{dt}\right)$$

This result is known as the **Product Rule** and can be stated as follows: If a function is the product of two functions, that is,

$$\text{Product function} = \left(\begin{array}{c}\text{first}\\\text{function}\end{array}\right)\left(\begin{array}{c}\text{second}\\\text{function}\end{array}\right)$$

then

$$\begin{array}{c}\text{Derivative}\\\text{of product}\\\text{function}\end{array} = \left(\begin{array}{c}\text{derivative}\\\text{of first}\\\text{function}\end{array}\right)\left(\begin{array}{c}\text{second}\\\text{function}\end{array}\right) + \left(\begin{array}{c}\text{first}\\\text{function}\end{array}\right)\left(\begin{array}{c}\text{derivative}\\\text{of second}\\\text{function}\end{array}\right)$$

The Product Rule

If $f(x) = g(x) \cdot h(x)$, then $\dfrac{df}{dx} = \left(\dfrac{dg}{dx}\right)h(x) + g(x)\left(\dfrac{dh}{dx}\right)$.

EXAMPLE 1   *Using the Product Rule Without an Equation*

Egg Production   The industrialization of chicken (and egg) farming brought improvements to the production rate of eggs. Consider a chicken farm that has 1000 laying hens, each of which lays an average of 24 eggs each month. By selling or buying hens, the farmer can decrease or increase production. Also, by selective breeding and genetic research, it is possible that over a period of time the farmer can increase the average number of eggs that each hen lays.

a.   How many eggs does the farm produce in a month?

b.   Suppose the farmer increases the number of hens by 12 hens per month and increases the average number of eggs laid by each hen by 1 egg per month. By how much will the farmer's production be increasing?

Solution

a.   The monthly egg production is the product of h, the number of hens, and l, the number of eggs each hen lays in one month. Currently, $h = 1000$ hens and $l = 24$ eggs per month. The farmer's current monthly production is

$$h \cdot l = (1000 \text{ hens})(24 \text{ eggs per hen}) = 24{,}000 \text{ eggs}$$

b.   Let t be the number of months from now. We are told that $\dfrac{dh}{dt} = 12$ hens per month and that $\dfrac{dl}{dt} = 1$ egg per hen per month. Applying the Product Rule yields

$$\frac{d(hl)}{dt} = \frac{dh}{dt}l + h\frac{dl}{dt}$$

$$= \left(12\frac{\text{hens}}{\text{month}}\right)\left(24\frac{\text{eggs}}{\text{hen}}\right) + (1000 \text{ hens})\left(1\frac{\text{egg/hen}}{\text{month}}\right)$$

$$= 1288\frac{\text{eggs}}{\text{month}}$$

The farmer's egg production will be increasing by 1288 eggs per month.   ●

Applying the Product Rule with Equations

The next example illustrates using the product rule for quantities that are represented by mathematical functions.

EXAMPLE 2   *Using the Product Rule in a Business Setting*

Sales   A music store has determined from a customer survey that when the price of each CD is x dollars, the number of CDs sold in a four-week period can be modeled by the function

$$N(x) = 6250(0.92985^x) \text{ CDs}$$

Find and interpret the rates of change of revenue when CDs are priced at $10, $12, $13.75, and $15.

Solution   Revenue is the number of units sold times the selling price. In this case, the monthly revenue $R(x)$ is given by

$$R(x) = N(x) \cdot x = 6250(0.92985^x) \cdot x \text{ dollars}$$

where x dollars is the selling price. Using the Product Rule, we find that the rate of change equation is

$$\frac{dR}{dx} = \left[\frac{d}{dx} N(x) \right] x + N(x) \left[\frac{d}{dx}(x) \right]$$
$$= 6250(\ln 0.92985)(0.92985^x) \cdot x + 6250(0.92985^x)(1) \text{ dollars per dollar}$$

where x dollars is the selling price. The output units on the derivative, dollars per dollar, indicates dollars of revenue per dollar of price.

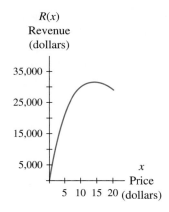

FIGURE 4.39

$R(x)$
Revenue
(dollars)

TABLE 4.11

Price	Rate of change of revenue (to nearest dollar)
$10.00	823
$12.00	332
$13.75	0
$15.00	-191

Evaluating $\frac{dR}{dx}$ at the indicated values of x yields Table 4.11. At $10, revenue is increasing by $823 per $1 of CD price. In other words, increasing the price results in an increase in revenue. Similarly, at $12, revenue is increasing by $332 per $1 of CD price. At $13.75, revenue is neither increasing nor decreasing. This is the price at which revenue has reached its peak. Finally, at $15, revenue is declining by $191 per $1 of CD price.

The graph of the revenue function is shown in Figure 4.39. Review the statements above about how the revenue is changing as they are related to the graph.   ●

Often product functions are formed by multiplying a quantity function by a function that indicates the proportion of that quantity for which a certain statement is true. This is illustrated in Example 3.

EXAMPLE 3   *Using the Product Rule When the Product Involves a Proportion*

Tourists   The number of overseas international tourists who traveled to the United States between 1995 and 2000 can be modeled* by the equation

$$f(t) = 0.148t^3 - 3.435t^2 + 26.673t - 45.44 \text{ million tourists,}$$

where t is the number of years since 1990. Suppose that during the same time period, the proportion (percentage expressed as a decimal) of foreign travelers to the United States who were from Europe is given by

$$p(t) = -0.00179t^3 + 0.0395t^2 - 0.275t + 1.039$$

where t is the number of years since 1990.

a.   Find a formula for the number of European tourists to the United States.

b.   Find the derivative of the formula in part a.

c.   Find the number of European tourists to the United States in 2000, and determine how rapidly that number was changing in that year.

Solution

a.   The number $N(t)$ of European tourists to the United States is given by the product function $N(t) = f(t) \cdot p(t)$.

$$N(t) = (0.148t^3 - 3.435t^2 + 26.673t - 45.44)(-0.00179t^3 + 0.0395t^2 - 0.275t + 1.039) \text{ million European tourists}$$

t years after 1990.

b.   To use the Product Rule, we need the derivatives f' and p':

$$f'(t) = 0.444t^2 - 6.87t + 26.673 \text{ million tourists per year, and}$$
$$p'(t) = -0.00537t^2 + 0.079t - 0.275$$

(Note that we did not label $p'(t)$ with units. If $p(t)$ had been expressed as a percentage, then the units would be percentage points per year. Expressed as a decimal, $p(t)$ is actually a proportion which is a unitless number. While it is possible to label $p'(t)$ as hundredths of a percentage point per year, we choose to state the derivative of p without a label.)

Thus, by the Product Rule,

$$N'(t) = (0.444t^2 - 6.87t + 26.673)(-0.00179t^3 + 0.0395t^2 - 0.275t + 1.039) + (0.148t^3 - 3.435t^2 + 26.673t - 45.44)(-0.00537t^2 + 0.079t - 0.275)$$

European tourists per year t years after 1990.

c.   The number of European tourists in 2000 is $N(10)$.

$$N(10) = (25.79)(0.449) \approx 11.6 \text{ million tourists}$$

*Based on data from U.S. Department of Commerce, Office of Travel and Tourism Industries.

The rate of change of the number of European tourists in 1995 is $N'(5)$.

$$N'(10) \approx (2.373)(0.449) + (25.79)(-0.022)$$

$$\approx 1.065 - 0.567$$

$$\approx 0.50 \text{ million European tourists per year}$$

In 2000, there were approximately 11.6 million tourists from Europe, and that number was growing by about 0.50 million tourists per year.   ●

The Quotient Rule

Occasionally functions are constructed as the quotients of other functions, that is, functions of the form $f(x) = \frac{g(x)}{h(x)}$. You may wish to rewrite $f(x)$ as $f(x) = g(x)[h(x)]^{-1}$ and use the Chain and Product Rules to write the derivative formula for such functions. Another approach is to use the **Quotient Rule:**

$$f'(x) = \frac{g'(x) \cdot h(x) - h'(x) \cdot g(x)}{[h(x)]^2}$$

We present this rule without illustration because it is an algebraic consequence of the Product and the Chain Rules.

4.5   Concept Inventory

○ *Product Rule*

If $f(x) = g(x) \cdot h(x)$, then $\frac{df}{dx} = \frac{dg}{dx} \cdot h(x) + g(x) \cdot \frac{dh}{dx}$.

○ *Quotient Rule*

4.5   Activities

1. Find $h'(2)$ if $h(x) = f(x) \cdot g(x)$, $f(2) = 6$, $f'(2) = -1.5$, $g(2) = 4$, and $g'(2) = 3$.

2. Find $r'(100)$ if $r(t) = p(t) \cdot q(t)$, $p(100) = 4.65$, $p'(100) = 0.5$, $q(100) = 160$, and $q'(100) = 12$.

3. **Computer Homes**   Let $h(t)$ be the number of households in a city, and let $c(t)$ be the proportion (expressed as a decimal) of households in that city that own a computer. In both functions, t is the number of years since 2005.

    a. Write sentences interpreting the following mathematical statements:

        i. $h(2) = 75,000$ iii. $c(2) = 0.52$

        ii. $h'(2) = -1200$ iv. $c'(2) = 0.05$

    b. If $N(t) = h(t) \cdot c(t)$, what are the input and output of N?

    c. Find the values of $N(2)$ and $N'(2)$. Interpret your answers.

4. **Demand**   Let $D(x)$ be the demand (in units) for a new product when the price is x dollars.

    a. Write sentences interpreting the following mathematical statements:

        i. $D(6.25) = 1000$

        ii. $D'(6.25) = -50$

    b. Give a formula for the revenue $R(x)$ generated from the sale of the product when the price is x dollars. (Assume demand = number sold.)

c. Find $R'(x)$ when $x = 6.25$. Interpret your answer.

5. Stock Value The value of one share of a company's stock is given by $S(x) = 15 + \frac{2.6}{x+1}$ dollars x weeks after it is first offered. An investor buys some of the stock each week and owns $N(x) = 100 + 0.25x^2$ shares after x weeks. The value of the investor's stock after x weeks is given by $V(x) = S(x) \cdot N(x)$.

a. Find and interpret the following:

  i. $S(10)$ and $S'(10)$

  ii. $N(10)$ and $N'(10)$

  iii. $V(10)$ and $V'(10)$

b. Give a formula for $V'(x)$.

6. Education Cost The number of students in an elementary school t years after 2002 is given by $S(t) = 100 \ln (t + 5)$ students. The yearly cost to educate one student can be modeled by $C(t) = 1500(1.05^t)$ dollars per student.

a. What are the input and output of the function $F(t) = S(t) \cdot C(t)$?

b. Find and interpret the following:

  i. $S(3)$ **iv.** $C'(3)$

  ii. $S'(3)$ **v.** $F(3)$

  iii. $C(3)$ **vi.** $F'(3)$

c. Find a formula for $F'(t)$.

7. Farming A wheat farmer is converting to corn because he believes that corn is a more lucrative crop. It is not feasible for him to convert all his acreage to corn at once. In the current year, he is farming 500 acres of corn and is increasing that number by 50 acres per year. As he becomes more experienced in growing corn, his output increases. He currently harvests 130 bushels of corn per acre, but the yield is increasing by 5 bushels per acre per year. When both the increasing acreage and the increasing yield are considered, how rapidly is the total number of bushels of corn increasing per year?

8. Basketball A point guard for an NBA team averages 15 free-throw opportunities per game. He currently hits 72% of his free throws. As he improves, the number of free-throw opportunities decreases by 1 free throw per game, while his percentage of hits increases by 0.5 percentage point per game. When his decreasing free throws and increasing percentage are taken into account, what is the rate of change in the number of free-throw points that this point guard makes per game?

9. Politics Two candidates are running for mayor in a small town. The campaign committee for candidate A has been taking weekly telephone polls to assess the progress of the campaign. Currently there are 17,000 registered voters, 48% of whom are planning to vote. Of those planning to vote, 57% will vote for candidate A. Candidate B has begun some serious mud slinging, which has resulted in increasing public interest in the election and decreasing support for candidate A. Polls show that the percentage of people who plan to vote is increasing by 7 percentage points per week, while the percentage who will vote for candidate A is declining by 3 percentage points per week.

a. If the election were held today, how many people would vote?

b. How many of those would vote for candidate A?

c. How rapidly is the number of votes that candidate A will receive changing?

Find derivative formulas for the functions in Activities 10 through 26.

WEB/CD

Skills

10. $f(x) = (x + 5)e^x$

11. $f(x) = (3x^2 + 15x + 7)(32x^3 + 49)$

12. $f(x) = 2.5(0.9^x)(\ln x)$

13. $f(x) = (12.8893x^2 + 3.7885x + 1.2548)[29.685(1.7584^x)]$

14. $f(x) = (5x + 29)^5 (15x + 8)$

15. $f(x) = (5.7x^2 + 3.5x + 2.9)^3(3.8x^2 + 5.2x + 7)^{-2}$

16. $f(x) = \dfrac{2.97x^3 + 3.05}{2.71x + 15.29}$

17. $f(x) = \dfrac{12.624(14.831^x)}{x^2}$

18. $f(x) = (8x^2 + 13)\left(\dfrac{39}{1 + 15.29e^{-0.0954x}}\right)$

19. $f(x) = (79.32x)\left(\dfrac{1984.32}{1 + 7.68e^{-0.859347x}} + 1568\right)$

20. $f(x) = [\ln(15.7x^3)](e^{15.7x^3})$

21. $f(x) = \dfrac{430(0.62^x)}{6.42 + 3.3(1.46^x)}$

22. $f(x) = (19 + 12\ln 2x)(17 - 3\ln 4x)$

23. $f(x) = 4x\sqrt{3x + 2} + 93$

24. $f(x) = \dfrac{4(3^x)}{\sqrt{x}}$

25. $f(x) = \dfrac{14{,}000x}{1 + 12.6e^{-0.73x}}$

26. $f(x) = \dfrac{1}{(x - 2)^2}(3x^2 - 17x + 4)$

27. Sales During the first 8 months of last year, a grocery store raised the price of a certain brand of tissue paper from $1.19 to $1.54. Consequently, sales declined. The price and number sold each month are shown in the table.

Month	Price	Number sold
Jan	$1.19	279
Feb	$1.25	277
Mar	$1.29	272
Apr	$1.34	266
May	$1.38	257
June	$1.45	247
July	$1.48	236
Aug	$1.54	221

a. Find models for price and number sold as functions of the month.

b. From the models in part *a*, construct an equation for revenue.

c. Use the equation to find the revenue in August and the projected revenue in September.

d. Would you expect the rate of change of revenue to be positive or negative in August? Why?

e. Give the rate-of-change formula for revenue.

f. How rapidly was revenue changing in February, August, and September?

28. Sales A music store has determined that the number of CDs sold monthly is approximately

$$\text{Number} = 6250(0.9286^x) \text{ CDs}$$

where *x* is the price in dollars.

a. Give an equation for revenue as a function of price.

b. If each CD costs the store $7.50, find an equation for profit as a function of price.

c. Find formulas for the rates of change of revenue and profit.

d. Complete the table below.

Price	Rate of change of revenue	Rate of change of profit
$13		
$14		
$20		
$21		
$22		

e. What does the table tell the store manager about the price corresponding to the highest revenue?

f. What is the price corresponding to the highest profit?

29. Population The population (in millions) of the United States as a function of the year is given in the table.

Year	Population (in millions)
1970	205.1
1980	227.7
1985	238.5
1990	248.8
1995	262.8
2000	281.4

(Source: *Statistical Abstract*, 2001.)

a. Determine the best model for the data.

b. The percentage of people in the United States who live in the midwest can be modeled by the equation $m(t) = 6.53(0.941163^t) + 22$ percent, where t is the number of years since 1970. Write an expression for the number of people who live in the midwest t years after 1970.

c. Find an expression for the rate of change of the population of the midwest.

d. According to the model, how rapidly was the population of the midwest changing in 1990, 1995, and 2000?

30. **Costs**   Costs for a company to produce between 10 and 90 units per hour are given in the table.

Units	Cost (dollars)	Units	Cost (dollars)
10	150	60	1400
20	200	70	2400
30	250	80	3850
40	400	90	5850
50	750		

a. Find an exponential model for production costs.

b. Find the slope formula for production costs.

c. Convert the model in part a to one for the average cost per unit produced.

d. Find the slope formula for average cost.

e. How rapidly is the average cost changing when 15 units are being produced? 35 units? 85 units?

f. Examine the slope graph for average cost. Is there a range of production levels for which average cost is decreasing?

g. Determine the point at which average cost begins to increase. (That is, find the point at which the rate of change of average cost changes from negative to positive.) Explain how you found this point.

31. **Poverty**   The accompanying table gives the number of men 65 or older in the United States and the percentage of men age 65 or older living below the poverty level.

Year	Men 65 years or older (millions)	Percentage below poverty level
1970	8.3	20.2
1980	10.3	11.1
1985	11.0	8.7
1990	12.6	7.8
1997	14.0	7.0
2000	14.4	7.5

(Sources: *Statistical Abstract*, 2001; *Current Population Survey*, March 2001.)

a. Using time as the input determine the best model for each set of data.

b. Write an expression for the number of men 65 or older who are living below the poverty level.

c. How rapidly was the number of male senior citizens living below the poverty level changing in 1990 and in 2000?

32. **VCR Homes**   The number of households with TVs in the United States is given for selected years between 1970 and 2002.

Year	Households (millions)	Year	Households (millions)
1970	59	1990	92
1975	69	1995	95
1980	76	2000	101
1985	85	2002	106

(Sources: *Statistical Abstract*, 2001; Television Bureau of Advertising.)

a. Find a model for the data given in the table above.

b. The table on page 289 gives (for selected years between 1978 through 2002) the percentages of U.S. households with TVs that also have VCRs.

Year	Percentage	Year	Percentage
1978	0.3	1995	81.0
1980	1.1	2000	85.1
1985	20.9	2002	91.2
1990	68.6		

(Sources: *Statistical Abstract*, 2001; Television Bureau of Advertising.)

Align the data so that the input values correspond with those in the model for part *a* (that is, if 1980 is $x = 10$ in part *a*, then you want 1980 to be $x = 10$ here also). Find a logistic model for the data.

c. Find a model for the number of U.S. households with VCRs.

d. Find an equation for the rate of change of the number of U.S. households with VCRs.

e. How rapidly was the number of households with VCRs in the U.S. growing in 1990? in 1995? in 2000?

33. **Childbirth** On the basis of data from a study conducted by the University of Colorado School of Medicine at Denver, the percentage of women receiving regional analgesia (epidural pain relief) during childbirth at small hospitals between 1981 and 1997 can be modeled by the equation

$$p(x) = 0.73(1.2912^x) + 8 \text{ percent}$$

x years after 1980.
(Source: Based on data from "Healthfile," *Reno Gazette Journal*, Oct. 19, 1999, p. 4.)

Suppose that a small hospital in southern Arizona has seen the yearly number of women giving birth decline as described by the equation

$$b(x) = -0.026x^2 - 3.842x + 538.868 \text{ women}$$
giving birth *x* years after 1980.

a. Give the equation and its derivative for the number of women receiving regional analgesia while giving birth at the Arizona hospital.

b. Was the percentage of women who received regional analgesia while giving birth increasing or decreasing in 1997?

c. Was the number of women who gave birth at the Arizona hospital increasing or decreasing in 1997?

d. Was the number of women who received regional analgesia during childbirth at the Arizona hospital increasing or decreasing in 1997?

e. If the Arizona hospital made a profit of $57 per woman for the use of regional analgesia, what was the profit for the hospital from this method of pain relief during childbirth in 1997?

34. **Funding** The amount of federal funds spent for agricultural research and services from 1990 through 2002 in the United States is given.

Year	Amount (billions of dollars)
1990	2.197
1992	2.539
1994	2.695
1996	2.682
1998	2.909
2000	3.189
2002	4.252

(Source: U.S. Office of Management and Budget.)

The purchasing power of the dollar, as measured by producer prices from 1988 through 2000 is given. (In 1982, one dollar was worth $1.00.)

Year	Purchasing power of $1
1988	0.93
1990	0.84
1992	0.81
1994	0.80
1997	0.76
2000	0.73

(Source: *Statistical Abstract*, 1998, 2001.)

a. Find models for both sets of data. (Remember to align such that both models have the same input values.)

b. Use these models to determine a new model for the amount, measured in constant 1982 dollars, spent on agricultural research and services.

c. Use your new model to find the rates of change and the percentage rates of change of the amount spent on agricultural research and services in 1992 and 2000.

d. Why might it be of interest to consider an expenditure problem in constant dollars?

35. **Dropouts**   The table shows the number of students enrolled in the ninth through twelfth grades and the number of dropouts from those same grades in South Carolina for each school year from 1980–1981 through 1989–1990.

School year	Enrollment	Dropouts
1980–81	194,072	11,651
1981–82	190,372	10,599
1982–83	185,248	9314
1983–84	182,661	9659
1984–85	181,949	8605
1985–86	182,787	8048
1986–87	185,131	7466
1987–88	183,930	7740
1988–89	178,094	7466
1989–90	172,372	5768

(Source: Compiled from *Rankings of the Counties and School Districts of South Carolina*.)

a. Find a model for enrollment and a cubic model for the number of dropouts.

b. Use the two models that you found in part *a* to construct an equation for the percentage of high school students who dropped out each year.

c. Find the rate-of-change formula of the percentage of high school students who dropped out each year.

d. Look at the rate of change for each school year from 1980–1981 through 1989–1990. In which school year was the rate of change smallest? When was it greatest?

e. Are the rates of change positive or negative? What does this say about high school attrition in South Carolina during the 1980s?

36. **Jobs**   A house painter has found that the number of jobs that he has per year is decreasing in inverse proportion to the number of years he has been in business. That is, the number of jobs he has each year can be modeled by $j(x) = \dfrac{104.25}{x}$, where x is the number of years since 1997. He has also kept a ledger of how much, on average, he was paid for each job. His income per job is presented below.

Year	Income per job (dollars)
1997	430
1998	559
1999	727
2000	945
2001	1228
2002	1597
2003	2075

a. Find an exponential model for his income per job.

b. Write the formula for the painter's total income per year.

c. Write the formula for the rate of change of the painter's income each year.

d. What was the painter's total income in 2003?

e. How rapidly was the painter's income changing in 2003?

37. When you are working with products of models, why is it important to make sure that the input values of the two models correspond? (That is, why must you align both models in the same way?)

38. We have discussed three ways to find rates of change: graphically, numerically, and algebraically. Discuss the advantages and disadvantages of each method. Explain when it would be appropriate to use each method.

SUMMARY

Drawing Slope Graphs

The smooth, continuous graphs that we use to model real-life data have slopes (derivatives) at every point on the graph except at points that have vertical tangent lines. When these slopes (derivatives) are plotted, they usually form a smooth, continuous graph—the slope graph (rate-of-change graph, or derivative graph) of the original graph. We can also obtain slope graphs for piecewise continuous models, although the slope graph is usually not defined at the points where the model is divided. Slope graphs tell us a great deal about the change that is occurring on the original graph.

Slope Formulas

The Four-Step Method of finding derivatives that was discussed in Chapter 3 is invaluable, but it is also cumbersome. For this reason, we desire formulas for derivatives of the most common functions we encounter. Here is a list of slope (derivative) formulas that you should know.

Function	Derivative
$y = b$	$\dfrac{dy}{dx} = 0$
$y = ax + b$	$\dfrac{dy}{dx} = a$
$y = x^n$	$\dfrac{dy}{dx} = nx^{n-1}$
$y = e^x$	$\dfrac{dy}{dx} = e^x$
$y = b^x$	$\dfrac{dy}{dx} = (\ln b)b^x$
$y = \ln x$	$\dfrac{dy}{dx} = \dfrac{1}{x}$
$y = kf(x)$	$\dfrac{dy}{dx} = kf'(x)$
$y = f(x) \pm g(x)$	$\dfrac{dy}{dx} = f'(x) \pm g'(x)$

The Chain Rule

The Chain Rule tells us how to calculate rates of change for a composite function. We present it in two forms. If C is a function of p and p is a function of t, then C can be regarded as a function of t, and

Form 1 $\quad \dfrac{dC}{dt} = \left(\dfrac{dC}{dp}\right)\left(\dfrac{dp}{dt}\right)$

Form 2 $\quad \dfrac{dC}{dt} = C'(p(t))\, p'(t)$

The Product Rule

The Product Rule tells us how to calculate rates of change for a product function. If $f(x) = g(x) \cdot h(x)$, then

$$\frac{df}{dx} = \left(\frac{dg}{dx}\right)h(x) + g(x)\left(\frac{dh}{dx}\right)$$

If you need to calculate a rate of change for a general quotient function, say $h(x) = \dfrac{f(x)}{g(x)}$, use the Quotient Rule or simply view the quotient as a product

$$h(x) = f(x)\left(\frac{1}{g(x)}\right) = f(x)[g(x)]^{-1}$$

and apply the Product and Chain Rules.

There are other formulas for derivatives that we have not given. However, we are providing the formulas that are most useful for the functions encountered in everyday situations associated with business, economics, finance, management, and the social and life sciences. You can look up other formulas (if you should ever need them) in a calculus book that emphasizes applications in science or engineering.

CONCEPT CHECK

Can you	To practice, try	
○ Sketch a general rate-of-change graph?	Section 4.1	Activities 3, 9
○ Use tangent lines to sketch an accurate slope graph?	Section 4.1	Activity 15
○ Sketch the slope graph of a piecewise continuous function?	Section 4.1	Activity 25
○ Apply simple derivative rules?	Sections 4.2 and 4.3	Any of Activities 7–15
○ Write the derivative of a piecewise continuous function?	Section 4.3	Activity 29
○ Use the Chain Rule?	Section 4.4	Activities 7 and any of 17–37
○ Use the Product Rule?	Section 4.5	Activities 5 and any of 10–26

REVIEW TEST

1. Consider the figure.

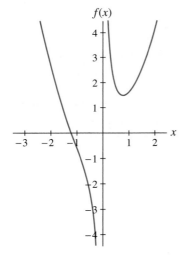

a. Note any input values for which the slope of the graph is zero.

b. For what input values is the slope of the graph positive? negative?

c. Are there any input values for which the slope of the graph does not exist?

d. Graphically estimate the slope of f at $x = -2$ and $x = 1$.

e. Sketch a slope graph for the function f.

2. Turkey   The average annual per capita consumption of turkey in the United States between 1980 and 2002 can be modeled by the equation

$$D(t) = \frac{8.101}{1 + 214.8e^{-0.797t}} + 10 \text{ pounds per person}$$

t years after 1980.

(Source: Based on data from Economic Research Service/USDA.)

a. Find a formula for $\frac{dD}{dt}$.

b. Find the value of $D'(10)$. Interpret your answer.

c. How quickly was the consumption of turkey growing in 2001?

Year	1980	1985	1990	1992	1994	1996	1998	2000
Total outstanding debt (billions of dollars)	1465	2378	3808	4073	4380	4865	5698	6890

Table for Activity 3   (Source: *Statistical Abstract*, 2001)

3. **Mortgage**   The total outstanding mortgage debt in the United States for selected years between 1980 and 2000 are shown in the table at the top of the page. Below are shown a model for the data and a graph of the model.

$$A(t) = 0.173t^4 - 6.24t^3 + 71.06t^2$$
$$- 32.2t + 1460.59 \text{ billion dollars}$$

   t years after 1980.

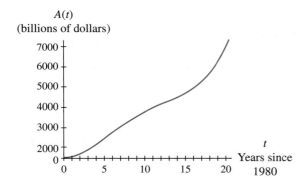

a. Use only the data to estimate the rate of change in the mortgage debt amounts in 1998.

b. Use the graph to estimate how quickly the mortgage debt amount was changing in 1998.

c. Find the derivative of the equation in 1998. Interpret your answer.

4. **Mortgage**   If $N(t)$ is the total amount (in billions of dollars) of new mortgages t years after 1980, we find the percentage of outstanding mortgage debt represented by new mortgages each year by dividing $N(t)$ by the equation in Question 3 and multiplying by 100.

$$P(t) = \frac{N(t)}{A(t)} \cdot 100 \text{ percent } t \text{ years after 1980}$$

a. Discuss how you find a formula for the rate of change of the percentage of total mortgage debt represented by new mortgages each year.

b. What are the input and output units of P'?

Project 4.1 Superhighway

Setting

The European Communities have decided to build a new superhighway that will run from Berlin through Paris and Madrid and end in Lisbon. This superhighway, like some others in Europe, will have no posted maximum speed so that motorists may drive as fast as they wish. There will be three toll stations installed, one at each border. Because these stations will be so far apart and motorists may not anticipate their need to stop, there has been widespread concern about the possibility of high-speed collisions at these stations. In response to the concern for safety, the Committee on Transportation has determined that flashing warning lights should be installed at an appropriate distance before each toll station. Your firm has been contracted to study known stopping distances and to develop a model for predicting where the warning lights should be installed and to suggest what other precautions could be taken to avoid accidents at the toll stations. (Bear in mind that you are a consultant to Europeans who wish to see all results in the metric system. However, because you work for an American-based company, you must also include the English equivalent.)

Tasks

1. Find data that give stopping distances as a function of speed and cite the source for the data. Present the data in a table and as a graph. Find a model to fit the data. Justify your choice. Before using your model to extrapolate, consult someone who could be considered an authority to determine whether the model holds outside of the data range. Consult a reliable source to determine probable speeds driven on such a highway. On the basis of your model, make a recommendation about where the warning lights should be placed. Justify your recommendation.

2. Find rates of change of your model for at least three speeds, one of which should be the speed that you believe to be most likely. Interpret these rates of change in this context. Would underestimating the most likely speed have a serious adverse affect? Support your answer.

Reporting

1. Prepare a written report of your study for the Committee on Transportation.

2. Prepare a press release for the Committee on Transportation to use when it announces the implementation of your safety precautions. The press release should be succinct and should answer the questions who, what, when, where, and why?

WEB/CD

Hint: Drivers' handbooks and Department of Transportation documents are possible sources of data on stopping distance.

Source

Project 4.2 Fertility Rates

Setting

The *Statistical Abstract of the United States* (2001 edition) reports fertility rates in the United States. Data for three fertility rates are located on the *Calculus Concepts* CD-ROM and web site.

Tasks

1. Find a table of fertility rate data in a current edition of the *Statistical Abstract*, and summarize in your own words the meaning of "fertility rate." Assign units to the given fertility rate data.

2. Find a piecewise model consisting of at most three pieces that fits the given white fertility rate data. Write the derivative function for this model. Construct a table of fertility rates and the rate of change of fertility rates for all years from 1970 through 2000. Discuss any points at which the derivative of your model does not exist (and why it does not exist), and explain how you estimated the rates of change at these points.

3. Using the data in a current edition of the *Statistical Abstract*, complete the same analysis as in Task 2 for the fertility rates for blacks in the United States. (Note: The *Statistical Abstract* "Total Black" data for 1984 and earlier includes races other than Black.)

4. Add any other recent data for the fertility rate of whites. (*Note:* Recent editions of the *Statistical Abstract* occasionally update older data points. Therefore, you should check the given data and change any updated values so that your data agree with the most recent *Statistical Abstract*.) Find a piecewise model for the updated data. Use this new model to calculate the rates of change that occurred in the years since 2000.

Reporting

Write a report discussing your findings and their demographic impact. Include your mathematical computations as an appendix.

CHAPTER

5

Analyzing Change: Applications of Derivatives

This chapter is devoted to exploring the ways in which rate-of-change information can be used to help analyze change. We begin by examining how a rate of change can be used to approximate change in a function's output.

Next, we consider the importance of those places at which the rate of change is zero. In many cases, these points correspond to local maximum or minimum function values. Absolute extrema may occur at these points. We then extend these optimization techniques to problems in which no equation or graph is given.

Finally, we consider situations in which we find a rate of change of a function with respect to a variable when the variable does not appear in the equation defining the function. The resulting equation, which is called a related-rates equation, is useful in describing how rates of change are interrelated.

Concept Objectives

This chapter will help you understand the concepts of

○ Using tangent lines to approximate change
○ Marginal analysis
○ Relative and absolute extreme points
○ Inflection points
○ Second derivative
○ Dependent and independent variables
○ Related-rates equations

and you will learn to

○ Approximate change using derivatives
○ Find relative extreme points
○ Find absolute extreme points
○ Find second derivatives
○ Use second derivatives to determine concavity
○ Locate and interpret inflection points
○ Set up and solve optimization problems
○ Find and solve related-rates equations

296

James Leynse/CORBIS SABA

Concept Application

Businesses often experience growth or decline as a result of changes in the economy, turnover in management, introduction of new products, or even whims of consumers. A company measures its performance by measuring quantities such as revenue, profit, productivity, and stock prices. In analyzing its performance, a company may examine the past behavior of these quantities and seek the answers to such questions as

○ When did the quantity exhibit highs and lows?

○ Was there a time when the trend of the quantity changed?

○ Is it possible to identify when the rate of change of the quantity was greatest?

Such analysis may be helpful as the company seeks to improve its performance. You will find the tools in this chapter useful in answering questions such as these. One such exploration for the Polo Ralph Lauren Corporation's revenue is found in Activity 24 of Section 5.3.

5.1   Approximating Change

Recall from our discussion of linear models that the slope of a line $y = ax + b$ can be thought of as how much y changes when x changes by 1 unit. For example, the amount spent on pollution control in the United States during the 1980s can be described by the equation*

Amount = $3788.65t - 252,216.11$ million dollars

where t is the number of years since 1900. The slope is $3788.65 million per year, so we can say that each year during the 1980s an *additional* $3788.65 million was spent on pollution control. It follows that every 2 years spending increased by (2)($3788.65) million, every 3 years spending increased by (3)($3788.65) million, and so on.

Using Rates of Change to Approximate Change

Because rates of change are slopes, we can apply a similar type of reasoning to functions other than lines. However, we must be careful in doing so for although the slope of a line is constant, the slopes of other functions can change at every point. Recall from the local linearity discussion in Section 3.2 that in a small enough interval around a point on a smooth, continuous function, the function and the line tangent to the function at that point appear to be the same. We call upon this similarity when we use the rate of change (slope of the tangent) to approximate the actual change in the function.

Concept Development: Approximating Change   For example, consider the average retail price (cost to the consumer) during the 1990s of a pound of salted, grade AA butter, which can be modeled by the equation†

$$p(t) = 0.0517t^2 - 0.2872t + 1.9487 \text{ dollars}$$

where t is the number of years since the end of 1990. We calculate how rapidly the average price was increasing at the end of 1998 as follows:

$$\frac{dp}{dt} = 2(0.0517)t - 0.2872 \text{ dollars per year}$$

When $t = 8$,

$$\frac{dp}{dt} = 2(0.0517)(8) - 0.2872 = \$0.54 \text{ per year}$$

Thus, at the end of 1998, the price of butter was rising by approximately $0.54 per year. On the basis of this rate of change and the fact that the average price at the end

---

*Based on data from *Statistical Abstract*, 1992.
†Based on data from *Statistical Abstract*, 1998.

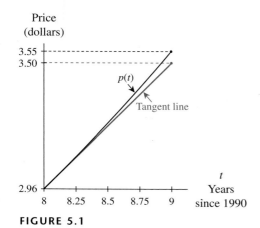

Price
(dollars)

3.55
3.50

$p(t)$

Tangent line

2.96

8   8.25   8.5   8.75   9

t
Years
since 1990

FIGURE 5.1

of 1998 was approximately $p(8) = \$2.96$, we estimate that during the following year (1999), the price increased by approximately $0.54 to a price of $3.50; and that during the first 6 months of 1999, the price increased by approximately $\frac{1}{2}(\$0.54) = \$0.27$ to a price of $3.23. Figure 5.1 illustrates these approximations. Notice in Figure 5.1 that the approximation of price using the tangent line is relatively close to the actual price. Of course, because the price function p is not linear, the farther we are from the point of tangency, the less accurate is our tangent-line approximation.

Compare the 6 months and 1 year approximations, as well as approximations for time periods of 3 months, 9 months, and 2 years with the actual values given by the model. These approximations and function values are listed in Table 5.1.

TABLE 5.1

Time from end of 1998	Approximated price	Price from equation	Difference between equation value and approximation
3 months	$\$2.96 + 0.25(\$0.54) = \$3.10$	$p(8.25) = \$3.10$	$0.00
6 months	$\$2.96 + 0.5(\$0.54) = \$3.23$	$p(8.5) = \$3.24$	$0.01
9 months	$\$2.96 + 0.75(\$0.54) = \$3.37$	$p(8.75) = \$3.39$	$0.02
1 year	$\$2.96 + 1(\$0.54) = \$3.50$	$p(9) = \$3.55$	$0.05
2 years	$\$2.96 + 2(\$0.54) = \$4.04$	$p(10) = \$4.25$	$0.21

Note that the shorter the time period, the closer the approximated price is to the price given by the model. This is no coincidence, because over a short time period, the rate of change is more likely to be nearly constant than over a longer period of time. Rates of change can often be used to approximate changes in a function, and they generally give good approximations over small intervals.

To summarize, consider the following statements:

- The change in a function f from x to $x + h$ can be approximated by the change in the line tangent to the graph of the function at x from x to $x + h$ when h is a small number.

- The change in the tangent line from x to $x + h$ is

  (Slope of the tangent line at x) $\cdot h = f'(x) \cdot h$

The mathematical notation for the statement "the change in a function from x to $x + h$ is approximately the change of the tangent line at x from x to $x + h$" is

$$f(x + h) - f(x) \approx f'(x) \cdot h$$

for small values of h. We illustrate this statement geometrically in Figure 5.2.

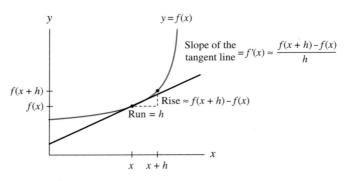

FIGURE 5.2

Approximating Change

The approximate change in *f* is the rate of change of *f* times a small change in *x*. That is,

$$f(x + h) - f(x) \approx f'(x) \cdot h$$

where *h* represents the small change in *x*.

It follows from this formula for approximating change that we can approximate the function value $f(x + h)$ by adding the approximate change to $f(x)$.

Approximating the Result of Change

When *x* changes by a small amount to $x + h$, the output of *f* at $x + h$ is approximately the value of *f* at *x* plus the approximate change in *f*.

$$f(x + h) \approx f(x) + f'(x) \cdot h$$

It is the formula $f(x + h) \approx f(x) + f'(x) \cdot h$ that we used to obtain the approximated price column in Table 5.1 in the butter price example.

EXAMPLE 1   *Using a Tangent Line Approximation to Estimate Outputs*

Temperature   The temperature for a 2-hour period during and after a thunderstorm can be modeled by

$$T(h) = 2.37h^4 - 5.163h^3 + 8.69h^2 - 9.87h + 78° \text{ Fahrenheit}$$

where *h* is the number of hours since the storm began.

a.   Use the rate of change of $T(h)$ at $h = 0.25$ to estimate by how much the temperature changed between 15 and 20 minutes after the storm began.

b. Find the temperature and rate of change of temperature at $h = 1.5$ hours.

c. Using only the answers to part b, estimate the temperature 1 hour and 40 minutes after the storm began.

d. Sketch the graph of T from $h = 0$ to $h = 1.75$ with lines tangent to the graph at $h = 0.25$ and $h = 1.5$. On the basis of the graph, determine whether the answers to parts a and c are overestimates or underestimates of the temperature given by the model.

Solution

a. $T'(0.25) = 9.48(0.25)^3 - 15.489(0.25)^2 + 17.38(0.25) - 9.87$

$\approx$ -6.34°F per hour

The change in the temperature between 15 and 20 minutes after the storm began is approximately $(-6.34°\text{F/hr})\left(\frac{1}{12}\text{ hr}\right) = -0.53°\text{F}$. The temperature fell approximately half a degree.

b. $T(1.5) \approx 77.3°\text{F}$ and $T'(1.5) \approx 13.3°\text{F}$ per hour

c. Note that 40 minutes $= \frac{2}{3}$ hour.

$$\begin{pmatrix} \text{Temperature at 1 hour} \\ \text{and 40 minutes} \end{pmatrix} \approx \begin{pmatrix} \text{temperature at} \\ 1\frac{1}{2}\text{ hours} \end{pmatrix} + \begin{pmatrix} \text{approximate change} \\ \text{in temperature} \end{pmatrix}$$

$$T\left(1 + \frac{2}{3}\right) \approx T(1.5) + T'(1.5)\left(\frac{1}{6}\right)$$

$$T\left(1 + \frac{2}{3}\right) \approx 77.3°\text{F} + (13.3°\text{F per hour})\left(\frac{1}{6}\text{ hour}\right)$$

$$\approx 77.3°\text{F} + 2.2°\text{F}$$

$$= 79.5°\text{F}$$

d. Temperature
(°F)

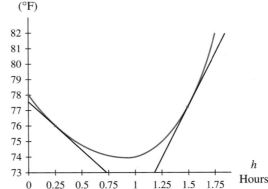

FIGURE 5.3

Because the tangent line at $h = 0.25$ hour is steeper than the graph between $h = 0.25$ (15 minutes) and $h \approx 0.33$ (20 minutes), the approximate change in temperature overestimates the actual change (see Figure 5.3). Thus, using the approximate change to estimate the temperature at $h \approx 0.33$ gives a temperature that underestimates the temperature given by the model at $h \approx 0.33$. The tangent line at $h = 1.5$ is not as steep as the graph to the right of $h = 1.5$, so our temperature approximation for $h = 1.67$ is an underestimate of the temperature given by the model.   ●

Sometimes it is necessary to use a tangent line to estimate a function output value because we do not have enough information available to develop a model. Other times, the tangent line approximation is used for short-term extrapolation. This is illustrated in Example 2.

EXAMPLE 2   *Using a Tangent Line Approximation to Extrapolate*

Population   The population of California from 1990 through 2000 is shown in Table 5.2. Population figures are for July 1 of each year. The data can be modeled by

$$P(t) = 6.713t^3 - 89.015t^2 + 624.007t + 29,854.538 \text{ thousand people}$$

t years after July 1, 1990. Figure 5.4 shows the data and model in graphical form.

TABLE 5.2

Year	Population (thousands)
1990	29,811
1991	30,414
1992	30,876
1993	31,147
1994	31,317
1995	31,494
1996	31,781
1997	32,218
1998	32,683
1999	33,145
2000	33,872

(Source: *Statistical Abstract*, 2001.)

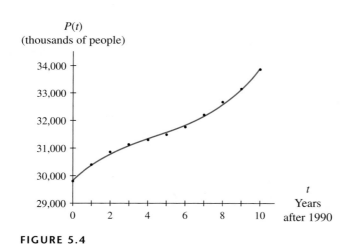

FIGURE 5.4

a. Use the last two data points given to estimate the population on July 1, 2001.

b. Use the model to estimate the population on July 1, 2001.

c. Find the output of the model and the derivative of the model on July 1, 2000. Use these values to estimate the population on July 1, 2001.

d. Discuss the assumptions one makes when using each estimation technique in parts *a*, *b*, and *c*.

Solution

a. The average rate of change between 1999 and 2000 is

$$\frac{33{,}872 - 33{,}145 \text{ thousand people}}{1 \text{ year}} = 727 \text{ thousand people per year}$$

Adding this value to the population on July 1, 2000, we obtain an estimate of the July 1, 2001, population:

$$33{,}872 + 727 = 34{,}599 \text{ thousand people}$$

b. The model estimates the population on July 1, 2001, as

$$P(11) \approx 34{,}884 \text{ thousand people}$$

c. The derivative of *P* is

$$P'(t) = 20.140t^2 - 178.030t + 624.007 \text{ thousand people per year}$$

t years after July 1, 1990. On July 1, 2000, the rate of change of the population was approximately

$$P'(10) \approx 857.8 \text{ thousand people per year}$$

and the population according to the model was 33,906,636. Thus

$$P(11) \approx P(10) + P'(10)$$
$$\approx 33{,}906.6 + 857.8$$
$$\approx 34{,}764 \text{ thousand people}$$

d. To estimate using only the last two data points (part *a*) is to assume that the average rate of change in the population from July 1, 2000, through July 1, 2001, will be approximately the same as it was from July 1, 1999, through July 1, 2000. To estimate using only the model (part *b*) is to assume that future growth will continue in the manner of the cubic model. To estimate using the derivative of the model (part *c*) is to assume that the rate of change on July 1, 2000 (as estimated by the model) is a good predictor of the change in the population during the following year.

   All three of these estimates are valid. If it were July 1, 2000, and someone needed an estimate for the population a year later, that person might use any one of these or many other techniques to make the prediction. ●

Marginal Analysis

In economics, it is customary to refer to the rates of change of cost, revenue, and profit with respect to the number of units produced or sold as **marginal cost, marginal revenue,** and **marginal profit.** These rates are used to approximate actual change in cost, revenue, or profit when the number of units produced or sold is increased by one. The term **marginal analysis** is often applied to this type of approximation.

EXAMPLE 3   *Understanding Marginal Cost*

Cost   Suppose a manufacturer of toaster ovens currently produces 250 ovens per day with a total production cost of $12,000 and a marginal cost of $24 per oven.

a.   What information does the marginal cost value give the manufacturer?

b.   If $C(x)$ is the cost to produce x toaster ovens, what is the notation for marginal cost?

Solution

a.   The marginal cost is the rate of change of cost with respect to the number of units produced. It is the approximate increase in cost, $24, that will result if production is increased from 250 ovens per day to 251 ovens per day.

b.   The marginal cost is $C'(x)$, or $\frac{dC}{dx}$. In this example, $C'(250) = \$24$ per oven.   ●

We know that profit = revenue − cost. If $p(x)$, $r(x)$, and $c(x)$ are respectively the profit, revenue, and cost associated with x units, then we have the relationship $p(x) = r(x) - c(x)$. If we take the derivative of this expression, we have $p'(x) = r'(x) - c'(x)$, or

Marginal profit = marginal revenue − marginal cost

From this equation, we see that if marginal profit is to be positive, so that increased sales will increase profit, then marginal revenue must be greater than marginal cost. Example 4 explores these relationships.

EXAMPLE 4   *Using a Model for Marginal Analysis*

Profit   A seafood restaurant has been keeping track of the price of its Monday night all-you-can-eat buffet and the corresponding number of nightly customers. These data are given in Table 5.3.

a.   Find a model for the data, and convert it to a model for revenue.

b.   If the cost to the restaurant is $4.50 per person regardless of the number of customers, find models for cost and profit.

c.   Find the marginal revenue, cost, and profit values for 50, 92, and 100 customers. Interpret the values in context.

Solution

a.   A linear model for these buffet price data is

$$B(x) = -0.078702x + 14.477879 \text{ dollars}$$

where x is the number of customers.
Revenue is equal to (price)(number of customers) so is given by the equation

$$R(x) = -0.078702x^2 + 14.477879x \text{ dollars}$$

where x is the number of customers.

TABLE 5.3

Number of customers	Buffet price (dollars)
86	7.70
83	7.90
80	8.20
78	8.30
76	8.60
73	8.80
70	8.90
68	9.10

b. The cost model is $c(x) = 4.5x$ dollars for x customers. The profit model is

$$p(x) = R(x) - c(x) = -0.078702x^2 + 14.477879x - 4.5x$$
$$= -0.078702x^2 + 9.977879x \text{ dollars}$$

where x is the number of customers.

c. The derivatives of revenue, cost, and profit are, respectively,

$$R'(x) = -0.157404x + 14.477879 \text{ dollars per customer}$$
$$c'(x) = \$4.50 \text{ per customer}$$
$$p'(x) = -0.157404x + 9.977879 \text{ dollars per customer}$$

where x is the number of customers. Evaluating the derivatives at 50, 92 and 100 customers gives the marginal values shown in Table 5.4.

TABLE 5.4

Demand (number of customers)	Marginal revenue (dollars per customer)	Marginal cost (dollars per customer)	Marginal profit (dollars per customer)
50	6.61	4.50	2.11
92	0.00	4.50	-4.50
100	-1.26	4.50	-5.76

What do these marginals tell us? If the buffet price is set on the basis of 50 customers, then revenue is increasing by $6.61 per customer. Because this value is greater than marginal cost, we see a positive marginal profit. In other words, increasing the number of customers to 51 (by lowering the price) will increase nightly revenue by approximately $6.61 and nightly profit by $2.11. It would benefit the restaurant to increase the number of customers by lowering price.

Similarly, we estimate that if the number of customers is increased from 92 to 93, then revenue will not change significantly and profit will, therefore, decline. With 92 customers, stimulating sales by lowering the price will not benefit the restaurant.

Finally, note that when price is set so that the restaurant expects 100 customers, the marginal revenue and profit are negative. Increasing the number of customers (by decreasing price) to 101 will result in an approximate decrease in nightly revenue of $1.26 and a decrease of nightly profit of $5.76. That is clearly undesirable. ●

We have seen that the change in a quantity over a small interval can be approximated using the rate of change of that quantity over the corresponding interval. This tangent line approximation technique can be especially useful when there is insufficient data to calculate change directly or when it is desirable to make short-term extrapolations. The derivative is a useful tool in marginal analysis for business and marketing applications.

5.1   Concept Inventory

○ *Change in function ≈ change in tangent line close to the point of tangency*

$$f(x + h) - f(x) \approx f'(x) \cdot h$$

○ $f(x + h) \approx f(x) + f'(x) \cdot h$ *for small values of h*

○ *Marginal cost, marginal profit, marginal revenue*

5.1   Activities

1. If the humidity is currently 32% and is falling at a rate of 4 percentage points per hour, estimate the humidity 20 minutes from now.

2. If an airplane is flying 300 mph and is accelerating at a rate of 200 mph per hour, estimate the airplane's speed in 5 minutes.

3. If $f(3) = 17$ and $f'(3) = 4.6$, estimate $f(3.5)$.

4. If $g(7) = 4$ and $g'(7) = -12.9$, estimate $g(7.25)$.

5. Interpret the following statements.

   a. At a production level of 500 units, marginal cost is $17 per unit.

   b. When weekly sales are 150 units, marginal profit is $4.75 per unit.

6. **Sales**   Interpret the marginal values given in the following statements.

   a. When weekly sales are 500 units, marginal revenue is $10 per unit and marginal cost is $13 per unit.

   b. When weekly sales are 10 units, marginal profit is -$3.46 per unit.

7. **Profit**   A fraternity currently realizes a profit of $400 selling T-shirts at the opening baseball game

of the season. If its marginal profit is -$4 per shirt, what action should the fraternity consider taking to improve its profit?

8. **Profit**   If the marginal profit is negative for the sale of a certain number of units of a product, is the company marketing the item losing money on the sale? Explain.

9. **Insurance**   A graph showing the annual premium for a one-million-dollar term life insurance policy as a function of the age of the insured person is given in the figure.

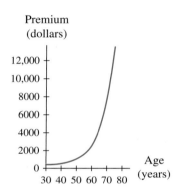

Sketch a tangent line at 70 years of age, and use it to predict the premium for a 72-year-old person.

10. **Life Expectancy**   A graph showing world life expectancy as a function of the number of decades since 1900 is given in the figure.

Life expectancy
(years)

(Source: Based on data in *The True State of the Planet*, ed. Ronald Bailey. New York: The Free Press for the Competitive Enterprise Institute, 1995.)

Sketch the tangent line at 1990, and use it to estimate the world life expectancy in 2000.

11. Sales A graph of revenue from new car sales and associated advertising expenditures for franchised new car dealerships in the United States between 1980 (when advertising expenditures were $1.2 billion) and 2000 (when advertising expenditures were $6.4 billion) is shown in the figure.

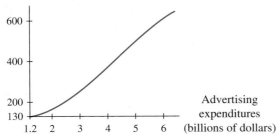

Revenue
(billions of dollars)

Advertising
expenditures
(billions of dollars)

(Source: Based on data from *Statistical Abstract*, 2001.)

a. Sketch a tangent line at the point where advertising expenditures were $6 billion, and use it to estimate the revenue from new car sales when $6.5 billion was spent on advertising.

b. The model graphed in the figure is

$$R(x) = -3.68x^3 + 47.958x^2 - 80.759x + 166.98 \text{ billion dollars of revenue}$$

when $x billion is spent on advertising. What does the model estimate as the revenue when $6.5 billion is spent on advertising?

c. Which estimate do you believe is the more valid one? Why?

12. Emissions In 1987, because of concern that CFCs have a detrimental effect on the stratospheric ozone layer, the Montreal Protocol calling for phasing out all chlorofluorocarbon (CFC) production was rati-

Release of
CFC-11
(millions of kilograms)

Years
since 1988

(Source: Based on data in *The True State of the Planet*, 1995.)

fied. The accompanying graph shows estimated releases of CFC-11 between 1988 and 1993.

a. Sketch a tangent line at 1992, and use it to estimate CFC-11 releases in 1993.

b. The model whose graph is shown in the figure is

$$C(x) = 6.107x^2 - 60.799x + 315.994$$
$$\text{million kilograms of CFC-11}$$

where x is the number of years since 1988. What estimate does the model give for CFC-11 releases in 1993?

c. The actual amount of CFCs released into the atmosphere in 1993 was 149 million kilograms. Which estimate is the more accurate one?

13. Population The population of South Carolina between 1790 and 2000 can be modeled by

$$\text{Population} = 268.79(1.013087^x) \text{ thousand people}$$
$$x \text{ years after } 1790.$$

(Source: Based on data from *Statistical Abstract*, 2001.)

a. Find the rate of change of the population of South Carolina in 2000.

b. On the basis of your answer to part *a*, approximate by how much the population changed between 2000 and 2003.

c. Write an explanation of the procedure you used to find the approximate change in the population between 2000 and 2003.

14. Revenue A pizza parlor has been lowering the price of a large one-topping pizza to promote sales. The average revenues from the sale of large one-topping pizzas on a Friday night (5 P.M. to midnight) are given.

Price (dollars)	9.25	10.50	11.75	13.00	14.25
Revenue (dollars)	1202.50	1228.50	1210.25	1131.00	1054.50

a. Find a model for the data.

b. Find and interpret the rate of change of revenue at a price of $9.25.

c. Approximate the change in revenue if the price is increased from $9.25 to $10.25.

d. Find and interpret the rate of change of revenue at a price of $11.50.

   e. Approximate the change in revenue if the price is increased from $11.50 to $12.50.

   f. Explain why the approximate change is an over-estimate of the change from $9.25 to $10.25 but an underestimate of the change from $11.50 to $12.50.

15. **Population**   The population of Mexico between 1921 and 2000 can be modeled by

   $$P(t) = 7.567e^{0.026048t} \text{ million people}$$

   where t is the number of years since 1900.
   (Source: Based on data from www.inegi.gob.mx. Accessed 9/20/02.)

   a. How rapidly was the population growing in 1998?

   b. On the basis of your answer to part a, determine by approximately how much the population of Mexico should have increased between 1998 and 1999.

16. **Study Time**   Suppose your test grade out of 100 points as a function of the time that you spend studying can be modeled by

   $$G(t) = -0.044t^3 + 0.918t^2 + 38.001 \text{ points}$$

   where t is the number of hours spent studying.

   a. Confirm the following assertions:
   i. After 11 hours of study, the slope is 4.2 points per hour, and the grade is 90.5 points.
   ii. The grade after 12 hours of study is 94.2 points.

   b. Use the information in the first part of a to estimate the grade after 12 hours. Is this an over-estimate or an underestimate of the grade given by the model? Explain.

17. **Mail**   A model for the number of pieces of first-class mail handled by the U.S. Postal Service between 1980 and 2000 is

   $$p(x) = 17.50 + 26.53 \ln x \text{ million pieces}$$

   where x is the number of years since 1975.
   (Source: Based on data from *Statistical Abstract,* 2000.)

   a. How rapidly was the amount of first-class mail growing in 1998?

   b. On the basis of the rate of change in 1998, what approximate increase would you expect between 1998 and 1999?

   c. On the basis of the equation, what was the actual increase between 1998 and 1999?

   d. According to the 2001 *Statistical Abstract,* the actual amounts of first-class mail handled by the U.S. Postal Service in 1998 and 1999 were 100.4 million pieces and 101.9 million pieces, respectively. What was the actual 1998 through 1999 increase?

   e. Compare the increases given by the derivative in part b, by the model in part c, and by the data in part d. Which of the three answers is most accurate? Explain.

18. **Costs**   Production costs for various hourly production levels of television sets are given in the table.

Hourly production	Cost (dollars)
5	740
10	1060
15	1210
20	1320
25	1420
30	1580
35	1900

   a. Find a model for the data.

   b. Find and interpret marginal cost at production levels of 5, 20, and 30 units.

   c. Find the cost to produce the 6th unit, the 21st unit, and the 31st unit.

   d. Why is the cost less than the marginal cost to produce the 6th unit but greater than the marginal cost to produce the 21st and 31st units?

   e. Find a model for average cost.

   f. Find and interpret the rate of change of average cost at production levels of 5, 20, and 30 units.

19. **Sales**   A concession stand owner finds that if he prices hot dogs so as to sell a certain number at each sporting event, then the corresponding revenues are those given in the table on the next page.

   a. Find a model for the data.

Number of hot dogs sold	Revenue (dollars)
100	195
400	620
700	875
1000	1000
1200	1020
1500	975

Table for Activity 19

Balls produced each hour (hundreds)	Cost (dollars)
2	248
5	356
8	432
11	499
14	532
17	567
20	625

Table for Activity 20

b. Each hot dog costs the owner of the concession stand $0.50. Use this fact and the model from part *a* to write models for cost and profit. Assume there are no fixed costs.

c. Find marginal revenue, cost, and profit for sales levels of 200, 800, 1100, and 1400 hot dogs. Interpret your answers.

d. Graph revenue, profit, and cost for sales values between 200 and 1400 hot dogs. How are the marginal values in part *c* related to the graphs?

20. Production A golf ball manufacturer knows that the cost associated with various hourly production levels are as shown in the accompanying table.

a. Find a model for the data.

b. If 1000 balls are currently being produced each hour, find and interpret the marginal cost.

c. Repeat part *b* for 300 golf balls and for 2100 golf balls.

d. Convert the model in part *a* to one for average cost.

e. Find and interpret the rate of change of average cost for production levels of 300 and 1700 golf balls.

21. CPI Rise in consumer prices is often used as a measure of inflation rate. The table at the bottom of the page shows the CPI during the 1980s for several different countries.

a. Find the best models for the data for the United States, Canada, Peru, and Brazil.

b. How rapidly were consumer prices rising in each of those four countries in 1987?

c. On the basis of your answers to part *b*, what would you expect the CPI to have been in the four countries in 1988?

Country					Year			
	1980	1981	1982	1983	1984	1985	1986	1987
United States	100	110.4	117.2	120.9	126.1	130.5	133.1	137.9
Canada	100	112.4	124.6	131.8	137.5	143.0	148.9	155.4
Mexico	100	127.9	203.3	410.2	679.0	1071.2	1994.9	4624.7
Japan	100	104.9	107.8	109.9	112.3	114.6	115.3	115.4
Israel	100	217	478	1174	5560	22,498	33,330	39,937
Peru	100	175.4	288.4	609	1280	3372	5999	11,150
Brazil	100	206	407	984	2924	9556	23,436	77,258
Argentina	100	204	541	2403	17,462	134,833	256,308	592,900

(Source: *International Marketing Data and Statistics, 1988–89.*)

Table for Activity 21

22. Investment   The amount in an investment after t years is given by

$$A(t) = 120{,}000(1.12682503^t) \text{ dollars}$$

a. Give the rate-of-change formula for the amount.

b. Find the rate of change of the amount after 10 years. Write a sentence interpreting the answer.

c. On the basis of your answer to part b, determine by approximately how much the investment will grow during the first half of the 11th year.

d. Find the percentage rate of change after 10 years. Given that A is exponential, what is the significance of your answer?

23. Advertising   A sporting goods company keeps track of how much it spends on advertising each month and of its monthly profit. From this information, the list of advertising expenditures and probable associated profit shown in the table below was compiled.

Advertising (thousands of dollars)	Profit (thousands of dollars)
5	150
7	200
9	250
11	325
13	400
15	450
17	500
19	525

a. Find a model for the data.

b. Find and interpret the rate of change of profit both as a rate of change and as an approximate change when $10,000 is spent on advertising.

c. Repeat part b for $18,000.

24. Newspapers   The circulation of daily English language newspapers in the United States as of September 30 of each year between 1986 and 2000 is as shown in the accompanying table.

a. Find a model for the data. Why did you choose this model?

Year	Circulation (millions)
1986	62.5
1988	62.7
1990	62.3
1992	60.1
1994	59.3
1996	57.0
1998	56.2
2000	55.8

(Source: *Statistical Abstract*, 1995, 2001.)

b. What should be the circulation of daily English language newspapers in 2007 according to your model? Is this reasonable?

c. Estimate how rapidly the newspaper circulation was changing in 1998.

d. Use the derivative to approximate the change in the newspaper circulation between 1990 and 1991.

25. Bank Account   Three hundred dollars is invested in an account that compounds 6.5% APR monthly.

a. Find an equation for the balance in the account after t years.

b. Rewrite the equation in part a to be of the form $A = Pb^t$.

c. How much is in the account after 2 years?

d. How rapidly is the value of the account growing after 2 years?

e. Use the answer in part d to approximate how much the value of the account changes during the first quarter of the third year.

26. Bank Account   Two thousand dollars is invested in an account that compounds 3.2% APR monthly.

a. Write an equation for the balance in the account after t years.

b. Rewrite the equation in part a to be of the form $A = Pb^t$.

c. How much is in the account after 5 years?

d. How rapidly is the value of the account growing after 5 years?

e. Use the answer in part *d* to approximate how much the value of the account changes during the first month of the fifth year. How close is this approximation to the actual change?

27. Write a brief essay that explains why, when rates of change are used to approximate change in a function, approximations over shorter time intervals generally give better answers than approximations over longer time intervals. Include graphical illustrations in your discussion.

28. Write a brief essay that explains why, when rates of change are used to approximate the change in a concave-up portion of a function, the approximation is an underestimate and, when rates of change

are used to approximate change in a concave-down portion of a function, the approximation is an overestimate. Include graphical illustrations in your discussion.

29. Recall that the derivative of a function *f* with input *x* is defined as

$$f'(x) = \lim_{h \to 0} \frac{f(x + h) - f(x)}{h}$$

provided that this limit exists. Starting with this equation, derive the formula for the approximation of change:

$$f(x + h) - f(x) \approx f'(x) \cdot h$$

5.2   Relative and Absolute Extreme Points

We use the terms *extreme points* and *optimal points* to refer to maxima and minima (either relative or absolute, depending on the context) but not to inflection points.

In this section, we turn our attention to finding high points (maxima) and low points (minima) on the graph of a function. Points at which maximum or minimum outputs occur are called **extreme points,** and the process of **optimization** involves techniques for finding them. Maxima and minima often can be found using derivatives, and they have important applications to the world in which we live.

Relative Extrema

We begin by examining a model for the population of Kentucky* from 1980 through 1993:

Population $= p(x) = 0.395x^3 - 6.674x^2 + 30.257x + 3661.147$ thousand people

where *x* is the number of years since the end of 1980. It is evident from the graph in Figure 5.5 that between 1980 and 1993, the population indicated by the model was smallest in 1980 (3661 thousand people) and greatest in 1993 (3794 thousand people). However, there are two other points of interest on the graph. Sometime near 1983, the population reached a peak. We call the peak a **relative** (or **local**) **maximum.** It does not represent the highest overall point, but it is a point to which the population

*Based on data from *Statistical Abstract,* 1994.

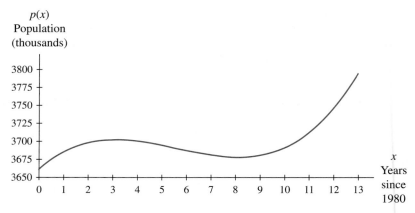

FIGURE 5.5

rises and after which it declines. Similarly, near 1988 the population reached a **relative** (or **local**) **minimum.** There are lower points on the graph, but in the region around this relative minimum, the population decreases and then increases as time increases.

It should be intuitively clear from the discussions in Chapter 3 that at a point where the graph of a smooth, continuous function reaches a relative maximum or minimum, the tangent line is horizontal and the slope is 0. We can consider this important link between such a function and its derivative in more detail by examining the relationship between the Kentucky population function and its slope graph. Horizontal tangent lines on the population function graph correspond to the points at which the slope graph crosses the horizontal axis. Figure 5.6 shows the graphs of the population function and its derivative.

Note that near $x = 3$, the slope graph crosses the x-axis. This is the x-value for which $p(x)$ is at a local maximum. Likewise, the slope graph crosses the x-axis at the x-value near 8 for which $p(x)$ is at a local minimum. The connection between the graphical and algebraic views of this situation is a key feature of optimization techniques.

> Finding the point at which the slope graph of a function crosses or touches the input axis is the same as finding *the point at which the derivative of the function is zero.* That is, the slope graph of a function f crosses or touches the input x-axis where $f'(x) = 0$.

The derivative of the population function is

$$\frac{dp}{dx} = 1.185x^2 - 13.348x + 30.257 \text{ thousand people per year}$$

Do not confuse extreme values with the inputs at which the extrema occur or with the extreme point. The extreme value is always an output value.

where x is the number of years since 1980. Setting this expression equal to zero and solving for x results in two solutions: $x \approx 3.14$ and $x \approx 8.12$. This information, together with the graph of p shown in Figure 5.6, tells us that, according to the model, the population peaked in early 1984 at approximately $p(3.14) = 3703$ thousand people. We also conclude that the population declined to a local minimum in early 1989. The population at that time was approximately $p(8.12) = 3678$ thousand people.

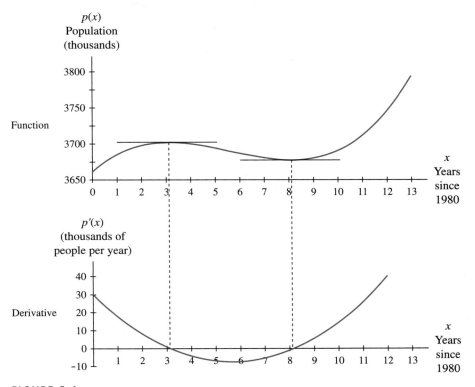

FIGURE 5.6

It is important to notice that relative extrema do not occur at the endpoints of an interval. A relative maximum is an output value that is larger than all other output values in some interval *around* the maximum. Similarly, a relative minimum is an output value that is smaller than all other output values in some interval *around* the minimum.

EXAMPLE 1   *Relating Zeros of a Derivative to Relative Extrema of the Function*

Baggage Complaints   The number of consumer complaints about baggage[*] on U.S. airlines between 1989 and 2000 can be modeled by the function

$$B(x) = 55.15x^2 - 524.09x + 1768.65 \text{ complaints}$$

where x is the number of years after 1989.

a. Consider the function out of its modeling context. Find any relative maxima and minima of B on the interval $0 \leq x \leq 11$.

b. Graph the function and its derivative. On each graph, clearly mark the input value that corresponds to the minimum found in part *a*.

[*]Based on data from *Statistical Abstract*, 2001.

c. Find the year between 1989 and 2000 in which the number of baggage complaints was at a relative minimum.

Solution

a. Because the graph of B is continuous and smooth over the interval $0 \leq x \leq 11$, we know that the relative extrema occur where the derivative is zero. Thus we set the derivative equal to zero and solve for x:

$$B'(x) = 110.3x - 524.09 = 0$$
$$x \approx 4.8$$

The value of B at $x \approx 4.8$ is approximately 523.5.

The equation for B is a quadratic with a positive leading coefficient; therefore, the graph of B is a concave-up parabola. Thus, we know that $(4.8, 523.5)$ is a relative minimum point. This is the only relative minimum.

b.

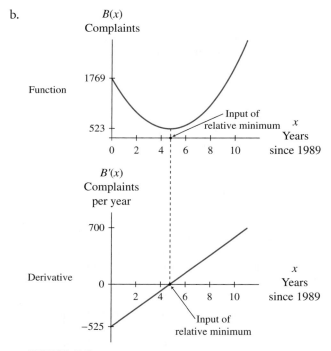

FIGURE 5.7

We see in Figure 5.7 that a relative minimum on the function graph corresponds to the point at which the derivative graph crosses the x-axis.

c. Because the model gives yearly complaint totals, it must be interpreted discretely. Thus the minimum number of complaints occurred in either 1993 ($x = 4$) or 1994 ($x = 5$). Checking the value of the function in each of these years, we find that the least number of complaints was approximately 527 in 1994. ●

Conditions When Relative Extrema Might Not Exist

We have just seen how derivatives can be used to locate relative maxima and minima. *You should use caution,* however, and not automatically assume that just because the derivative is zero at a point, there is a relative maximum or relative minimum at that point. When a function has a point at which the derivative of the function is zero (that is, when there is a horizontal tangent line on the graph of the function), one of the four situations depicted in Figure 5.8 occurs.

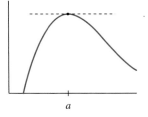

(a) Relative maximum:
To the left of input *a*, slopes
are positive, and to the right
of *a*, slopes are negative.

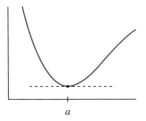

(b) Relative minimum:
To the left of input *a*, slopes
are negative, and to the right
of *a*, slopes are positive.

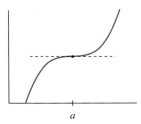

(c) Inflection point:
Slopes are positive to the
left and right of input *a*.

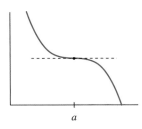

(d) Inflection point:
Slopes are negative to the
left and right of input *a*.

FIGURE 5.8

It is evident from Figures 5.8c and d that the graph of a function may have a horizontal tangent line at a point that is not an extreme point. It is therefore important that you graph the function when using derivatives to locate extreme points.

> Always begin the process of finding extreme points by graphing the function to see whether there are any relative maxima or minima before proceeding to work with derivatives.

Let us further investigate the graph shown in Figure 5.8d. Figure 5.9 shows the graph of the function that appears in Figure 5.8d and its slope (derivative) graph. If you carefully examine the slope graph, you see that it touches the *x*-axis but does not cross it. Thus *f* does not have a relative maximum or minimum at *a*. You may notice that the derivative graph reaches its maximum at *a* as it touches the *x*-axis. Do not confuse maxima and minima on the derivative graph with maxima and minima of the original function. In Section 5.3, we will see that maxima and minima of the derivative graph have other important interpretations in terms of the original function.

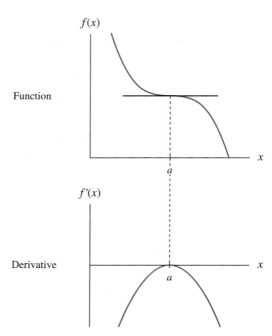

FIGURE 5.9 $f'(a) = 0$ but the graph of f has no relative maximum or minimum at *a*.

EXAMPLE 2   *Relating Derivative Intercepts to Relative Extrema*

5.2.1a–c

Revenue   Acme Cable Company actively promoted sales in a town that previously had no cable service. Once Acme saturated the market, it introduced a new 50-channel system and raised rates. As the company began to offer its expanded system, a different company, Bigtime Cable, began offering satellite service with more channels than Acme and at a lower price. A model for Acme's revenue for the 26 weeks after it began its sales campaign is

$$R(x) = -3x^4 + 160x^3 - 3000x^2 + 24{,}000x \text{ dollars}$$

where *x* is the number of weeks since Acme began sales. The graph of the model is shown in Figure 5.10. Some points on the graph of *R* are given in Table 5.5.

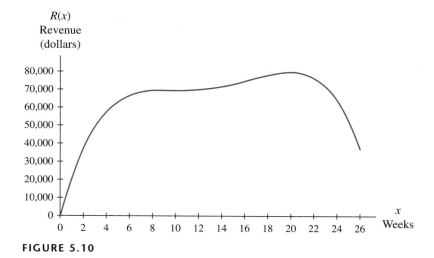

FIGURE 5.10

TABLE 5.5

Weeks	2	6	10	14	18	22	26
Revenue (dollars)	37,232	66,672	70,000	71,792	78,192	76,912	37,232

a.   Determine the point at which Acme's revenue peaked during this 26-week interval.

b.   At what other point is $R'(x) = 0$? Explain what happens to Acme's revenue at this point.

Solution

a.   From the graph, we know that revenue peaks at the relative maximum. A closer examination of the graphs of R and its derivative (see Figure 5.11) locates this point near 20 weeks. Solving the equation

$$R'(x) = -12x^3 + 480x^2 - 6000x + 24{,}000 = 0$$

gives two solutions, $x = 10$ weeks and $x = 20$ weeks. Revenue peaked at 20 weeks, with a value of $R(20) = \$80{,}000$. This appears to correspond to the time immediately prior to when Bigtime Cable's sales began negatively affecting the Acme company.

b.   The other point at which $R'(x) = 0$ is (10, 70,000). The fact that the rate-of-change equation is zero at two places indicates that there are two places on the graph with horizontal tangent lines. Indeed, at $x = 10$ weeks, the line tangent to the curve is horizontal because the curve has leveled off. This corresponds to the time when Acme had saturated the market. However, no local maximum occurs at this point, because the slope graph only touches—it does not cross—the input axis at $x = 10$. Note the relationships between the rate-of-change graph and the revenue graph shown in Figure 5.11 and how they connect to the slope descriptions given in Figure 5.8.

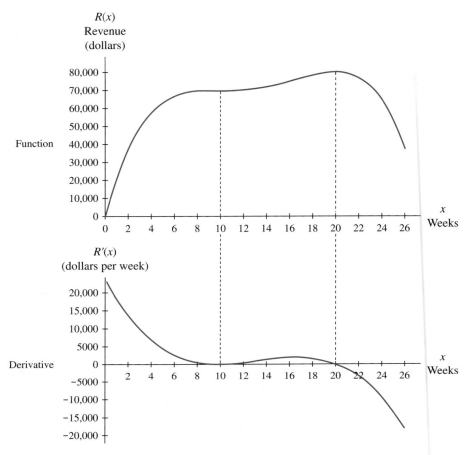

FIGURE 5.11

The maximum occurs where the derivative graph crosses the x-axis. The leveling-off point occurs where the derivative touches, but does not cross, the x-axis. ●

Relative Extrema on Functions That Are Not Smooth

Is it true for every continuous function that relative maxima and minima occur only where the derivative crosses the input axis? Although this seems to be the case for most of the functions we use, consider the piecewise continuous function describing the average concentration (in nanograms per milliliter) of a 360-mg dose of a blood pressure drug in a patient's blood during the 24 hours after the drug is given:

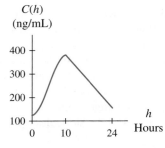

FIGURE 5.12

$$C(h) = \begin{cases} -0.51h^3 + 7.65h^2 + 125 \text{ ng/mL} & \text{when } 0 \leq h \leq 10 \\ -16.07143h + 540.71430 \text{ ng/mL} & \text{when } 10 < h \leq 24 \end{cases}$$

where h is the number of hours since the drug was given.

By calculating the left and right limits as h approaches 10 and comparing them with the function value at $h = 10$, we determine that C is continuous for all input values from 0 to 24. The two portions of the function join at the peak shown in Figure 5.12.

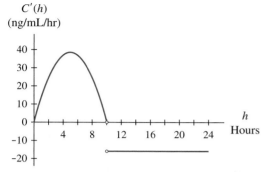

$C'(h)$
(ng/mL/hr)

FIGURE 5.13

It is evident from the graph that the highest concentration of the drug occurs 10 hours after the patient receives the initial dose. Thus $C(10) = 380$ ng/mL is the maximum concentration. However, is $C'(10) = 0$? Does the slope graph for C cross the horizontal axis at $h = 10$? The slope at $h = 10$ exists only if the graph is smooth—that is, only if the slopes of the two portions of the graph of C are the same at $h = 10$. The slope of the graph on the left is zero at $h = 10$, but the slope of the graph on the right is -16.07143 at all points. These differing slopes cause the sharp point on the graph of C, resulting in a graph that is continuous but not smooth at $h = 10$. Thus $C'(10)$ does not exist. There is no line tangent to the graph of C at $h = 10$. That $C'(10)$ does not exist is illustrated in the graph of C' shown in Figure 5.13. It is therefore possible for an extreme point to occur where the derivative of the function does not exist as long as the function has a value at that point.

Conditions Where Extreme Points Exist

For a function f with input x, a relative extremum can occur at $x = c$ only if $f(c)$ exists (is defined). Furthermore,

1. A relative extremum exists where $f'(c) = 0$ and the graph of $f'(x)$ crosses (not just touches) the input axis at $x = c$.

2. A relative extremum can exist where $f(x)$ exists but $f'(c)$ does not exist. (Further investigation is needed.)

Thus in order to find relative maxima and relative minima of a function f, first determine the input values for which the derivative of f is zero or undefined, then examine a graph of f to determine which of these input values correspond to relative maxima or relative minima.

Absolute Extrema

Recall the population of Kentucky example at the beginning of this section. The model for the population of Kentucky from 1980 through 1993 is

$$\text{Population} = p(x) = 0.395x^3 - 6.674x^2 + 30.257x + 3661.147 \text{ thousand people}$$

where x is the number of years since the end of 1980. Figure 5.14 shows a graph of p.

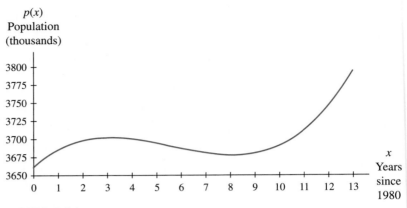

FIGURE 5.14

In the earlier discussion, it was noted that the population of Kentucky has a relative maximum of approximately 3703 thousand people in early 1984 ($x \approx 3.14$) and a relative minimum of approximately 3678 thousand people in early 1989 ($x \approx 8.12$). However, it is evident from the graph that between 1980 and 1993 there were years in which the population was greater than 3703 thousand (the relative maximum) and there was a time at which the population was less than 3678 thousand (the relative minimum).

When considering maxima and minima over an interval, it is important to consider not only relative extrema but also **absolute extrema.** A function can have several different local maxima (or minima) in a given interval. However there can be only one absolute maximum value and one absolute minimum value for that interval. An absolute extremum can occur at a relative extremum or it can occur at an endpoint of the interval. In the case of the Kentucky population example, over the period between 1980 and 1993, the population of Kentucky reached its **absolute maximum** (the greatest output) in 1993. The **maximum value** of the population function is approximately 3794 thousand people. On the other hand, Kentucky's population between 1980 and 1993 was at its **absolute minimum** (the least output) in 1980 when its **minimum value** was approximately 3661 thousand people. Notice that the absolute maximum and absolute minimum are stated in terms of the interval of years between 1980 and 1993. Obviously there was a time before 1980 that the population was even lower, and it has probably risen since 1993. We discuss the idea of absolute extrema over the entire set of real numbers following Example 3.

EXAMPLE 3   *Finding Absolute Extrema on a Given Interval*

Baggage Complaints   Recall from Example 1 that the number of consumer complaints about baggage* on U.S. airlines between 1989 and 2000 can be modeled by the function

$$B(x) = 55.15x^2 - 524.09x + 1768.65 \text{ complaints}$$

where x is the number of years after 1989. Consider the function out of its modeling context. Find the absolute maximum and the absolute minimum of B on the interval $0 \leq x \leq 11$.

*Based on data from *Statistical Abstract*, 2001.

Solution   As discussed in Example 1, the graph of B is a concave-up parabola with a minimum that occurs at $x \approx 4.8$ and is approximately 523.5.

To find the absolute maximum, we must determine the value of B at the endpoints of the given interval: $B(0) \approx 1768.7$ and $B(11) \approx 2676.8$. From our knowledge of the general shape of the graph of B and from these calculations, we conclude that the absolute maximum of B is approximately 2676.8 and occurs at the right endpoint when $x = 11$ and that the absolute minimum of B is approximately 523.5 and occurs at $x \approx 4.8$.   ●

What if we do not have a specified input interval and are asked to find the absolute extrema? Let us again consider the function in Example 3 apart from its context and determine the absolute extrema of the function $B(x) = 55.15x^2 - 524.09x + 1768.65$ over all real number inputs. Because B is a concave-up quadratic function, $\lim_{x \to \pm\infty} B(x) \to \infty$; that is, the graph continues to increase infinitely in both directions. Thus there is no absolute maximum over all real number inputs. We know that the absolute minimum is the relative minimum because of the end behavior of the function. Therefore, the least output value is approximately 523.5 at $x \approx 4.8$. This value is the absolute minimum of the function over all real number inputs.

In general, in order to determine if an absolute maximum or minimum exists for a function over all real number inputs, we must analyze the end behavior of the function as well as consider the outputs of the function at all of the input values for which the function is discontinuous or has relative extrema.

In order to help you be organized as you practice the concepts presented in this section, we conclude by outlining the steps for finding relative and absolute extrema.

Finding Extrema

To find the relative maxima and minima of a function f,

Step 1: Determine the input values for which $f' = 0$ or f' is undefined.

Step 2: Examine a graph of f to determine which input values found in Step 1 correspond to relative maxima or relative minima.

To find the absolute maximum and minimum of a function f on an interval from a to b,

Step 1: Find all relative extrema of f in the interval.

Step 2: Compare the relative extreme values in the interval with $f(a)$ and $f(b)$, the output values at the endpoints of the interval. The largest of these values is the absolute maximum, and the smallest of these values is the absolute minimum.

To find the absolute maximum and minimum of a continuous function f without a specified input interval,

Step 1: Find all relative extrema of f.

Step 2: Determine the end behavior of the function in both directions in order to consider a complete view of the function. The absolute extrema either do not exist or are among the relative extrema.

5.2 Concept Inventory

○ *Relative (local) maximum*

○ *Relative (local) minimum*

○ *An extreme point occurs at an input value*

○ *The extreme value is an output value*

○ *Conditions under which extreme points exist*

○ *Absolute maximum*

○ *Absolute minimum*

5.2 Activities

1. Which of the six basic models discussed thus far in this book could have relative maxima or minima?

2. Discuss in detail all of the options you have available for finding the relative maxima and relative minima of a function.

In Activities 3 through 8, mark the location of all relative maxima and minima with an X and all absolute maxima and minima with an O. For each extreme point that is not an endpoint, indicate whether the derivative at that point is zero or does not exist.

3. *y*

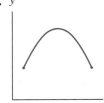

4. *y*

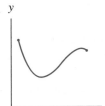

5. *y*

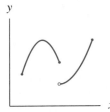

6. *y*

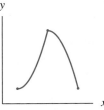

7. *y*

8. *y*

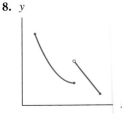

9. Sketch a graph of a function whose derivative is zero at a point but that does not have a relative maximum or minimum at that point.

10. Sketch the graph of a function with a relative minimum at a point at which the derivative does not exist.

11. Identify for which of the graphs *a* through *d* all of the following statements are true. For the other graphs, identify which statements are not true.

$$f'(x) > 0 \text{ for } x > 2$$

$$f'(x) > 0 \text{ for } x < 2$$

$$f'(x) = 0 \text{ for } x = 2$$

a. *f(x)*

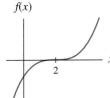

b. *f(x)*

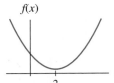

c. *f(x)*

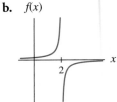

d. *f(x)*

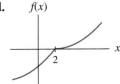

12. Identify for which of the graphs *a* through *d* all of the following statements are true. For the other graphs, identify which statements are not true.

$$f'(x) > 0 \text{ for } x < 2$$

$$f'(x) < 0 \text{ for } x > 2$$

$$f'(x) = 0 \text{ for } x = 2$$

a. $f(x)$

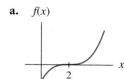

b. $f(x)$

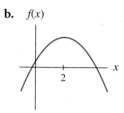

c. $f(x)$

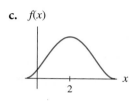

d. $f(x)$

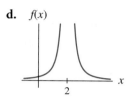

13. Sketch the graph of a function f such that

$f'(x) < 0$ for $x < -1$

$f'(x) > 0$ for $x > -1$

$f'(-1)$ does not exist.

14. Sketch the graph of a function f such that

$f'(x) < 0$ for $x < -1$

$f'(x) > 0$ for $x > -1$

$f'(-1) = 0$

15. Sketch the graph of a function f such that

f has a relative minimum at $x = 3$.

f has a relative maximum at $x = -1$.

$f'(x) > 0$ for $x < -1$ and $x > 3$

$f'(x) < 0$ for $-1 < x < 3$

$f'(-1) = 0$ and $f'(3) = 0$

16. Sketch the graph of a function f such that

$f'(x) > 0$ for $x < -1$ and $x > 3$

$f'(x) < 0$ for $-1 < x < 3$

$f'(-1) = 0$ and $f'(3) = 0$

17. Consider the function

$g(x) = 0.04x^3 - 0.88x^2 + 4.81x + 12.11$

a. Find the relative maximum and relative minimum of g.

b. Find the absolute maximum and the absolute minimum of g between $x = 0$ and $x = 14.5$.

c. Graph the function and its derivative. Indicate the relationship between the relative maximum and minimum of g and the corresponding points on the derivative graph.

18. Population   The U.S. Bureau of the Census' prediction for the percentage of the population 65 to 74 years old from 2000 through 2050 can be modeled by

$$p(x) = (1.619 \cdot 10^{-5})x^4 - (1.675 \cdot 10^{-3})x^3 + 0.050x^2 - 0.308x + 6.693 \text{ percent}$$

where x is the number of years after 2000. Find the absolute maximum and the absolute minimum percentages between 2000 and 2050. Give the years and the corresponding percentages.
(Source: Based on data from *Statistical Abstract*, 1998.)

19. Grasshoppers   The percentage of southern Australian grasshopper eggs that hatch as a function of temperature (for temperatures between 7°C and 25°C) can be modeled by

$$P(t) = -0.00645t^4 + 0.488t^3 - 12.991t^2 + 136.560t - 395.154 \text{ percent}$$

where t is the temperature in °C.
(Source: Based on information in *Elements of Ecology*, George L. Clarke, New York: Wiley, 1954, p. 170.)

a. Find the temperature between 7°C and 25°C that corresponds to the greatest percentage of eggs hatching.

b. Use the equation $°F = \frac{9}{5}(°C) + 32$ to convert your answer to °F.

20. Lake Level   Lake Tahoe lies on the California/Nevada border, and its level is regulated by a 17-gate concrete dam at the lake's outlet. By federal court decree, the lake level must never be higher than 6229.1 feet above sea level. The lake level is monitored every midnight. The level of the lake from October 1, 1995, through July 31, 1996, can be modeled by

$$L(d) = (-5.345 \cdot 10^{-7})d^3 + (2.543 \cdot 10^{-4})d^2 - 0.0192d + 6226.192 \text{ feet above sea level}$$

d days after September 30, 1995.
(Source: Based on data from the Federal Watermaster, U.S. Department of the Interior.)

According to the model, did the lake remain below the federally mandated level between October 1, 1995, and July 31, 1996?

21. **River Rate** The flow rate (in cubic feet per second, cfs) of a river in the 24 hours after the beginning of a severe thunderstorm can be modeled by

$$C(h) = \begin{cases} -0.865h^3 + 12.045h^2 - 8.952h \\ \quad + 123.02 \text{ cfs} & \text{when } 0 \leq h \leq 10 \\ -16.643h + 539.429 \text{ cfs} & \text{when } h > 10 \end{cases}$$

where h is the number of hours after the storm began.

a. What were the flow rates for $h = 0$ and $h = 24$?

b. Determine the absolute maximum and minimum flow rates between $h = 0$ and $h = 24$.

22. **Fisheries** The domestic catch of fish by fisheries in the United States for human food from 1970 through 2001 can be modeled by

$$F(t) = \begin{cases} 22.204t^2 - 108.431t + 2538.603 \\ \quad \text{million lb} & \text{when } 0 \leq t \leq 10 \\ 48.768t^2 - 1206.613t + 10{,}723.167 \\ \quad \text{million lb} & \text{when } 10 < t < 23 \\ -213.543t + 13{,}063.343 \\ \quad \text{million lb} & \text{when } 23 \leq t < 28 \\ 185.75t^2 - 10{,}908.95t + 166{,}990.65 \\ \quad \text{million lb} & \text{when } 28 \leq t \leq 31 \end{cases}$$

where t is the number of years since 1970.
(Sources: Based on data from *Statistical Abstract*, 2001, and *World Almanac and Book of Facts*, ed. William A. McGeveran Jr. New York, NY: World Almanac Education Group, Inc., 2003.)

a. Determine all relative and absolute extrema of F between $t = 0$ and $t = 31$. Consider the function out of its modeling context.

b. F must be interpreted discretely because it gives yearly totals. With this restriction in mind, determine when the domestic catch was greatest and when it was least from 1970 through 2001.

23. **Swim Time** *Swimming World* (August 1992) lists the time in seconds that an average athlete takes to swim 100 meters free style at age x years. The data are given in the accompanying table.

a. Find the best model for the data.

b. Using the model, find the age at which the minimum swim time occurs. Also find the minimum swim time.

Age x (years)	Time (seconds)	Age x (years)	Time (seconds)
8	92	22	50
10	84	24	49
12	70	26	51
14	60	28	53
16	58	30	57
18	54	32	60
20	51		

c. Compare the table values with the values in part b.

24. **Costs** A company analyzes the production costs for one of its products and determines the hourly operating costs when x units are produced each hour. The results are given in the accompanying table.

Production level, x (units per hour)	Hourly cost (dollars)
1	210
7	480
13	650
19	760
25	810
31	845
37	880
43	950
49	1070
55	1280
61	1590

a. Find a model for hourly cost in terms of production level.

b. Find and interpret the marginal cost when 40 units are produced.

c. On the basis of the model in part a, what is the equation for the average hourly cost per unit when x units are produced each hour?

d. Find the production level that minimizes average hourly cost. Give the average hourly cost and total cost at that level.

25. Sales *Consumer expenditure* and *revenue* are terms for the same thing from two perspectives. Consumer expenditure is the amount of money that consumers spend on a product, and revenue is the amount of money that businesses take in by selling the product. A street vendor constructs a table on the basis of sales data.

Price of a dozen roses (dollars)	Number of dozens sold per week
10	190
15	145
20	110
25	86
30	65
35	52

a. Find a model for quantity sold.

b. Convert the equation in part *a* to an equation for consumer expenditure.

c. What price should the street vendor charge to maximize consumer expenditure?

d. If each dozen roses costs $6, what price should the street vendor charge to maximize profit?

e. Why do the derivatives of the revenue and profit equations in this activity not give marginal revenue and marginal profit?

26. Demand An apartment complex has an exercise room and sauna, and tenants will be charged a yearly fee for the use of these facilities. A survey of tenants results in these demand/price data.

Quantity demanded	Price (dollars)
5	250
15	170
25	100
35	50
45	20
55	5

a. Find a model for price as a function of demand.

b. On the basis of the price model, give the equation for revenue.

c. Find the maximum point on the revenue model. What price and what demand give the highest revenue? What is the marginal revenue at the maximum point?

27. Refuse The yearly amount of garbage (in millions of tons) taken to a landfill outside a city during selected years from 1975 through 2005 is given.

Year	Amount (millions of tons)
1975	81
1980	99
1985	117
1990	122
1995	132
2000	145
2005	180

a. Find a model for the data.

b. Give the slope formula for the model.

c. How rapidly was the amount of garbage taken to the landfill increasing in 2005?

d. Graph the derivative of your model, and determine whether your model has a relative maximum and/or minimum. Explain how you reached your conclusion.

28. Population Refer to the Kentucky population model given on page 311.

a. Align the input data in the table below as the number of years since 1980. Graph the equation given in the text along with the data. How does the behavior of the graph of the model between 1980 and 1983 compare with the behavior of the scatter plot of the data?

Year	Kentucky population (thousands)	Year	Kentucky population (thousands)
1993	3792	1997	3908
1994	3823	1998	3934
1995	3851	1999	3961
1996	3881	2000	4042

(Source: *Statistical Abstract*, 2001.)

b. Find a model for the data. Use it to write a piecewise continuous function for the population of Kentucky between 1980 and 2000.

c. Graph the piecewise continuous model in part *b* and its derivative.

d. Find all relative extrema of the population model from 1980 through 2000 and the absolute maximum and absolute minimum of the population during that time.

e. Find the most recent population value available for Kentucky. Does this value follow the trend indicated by the model in part *b*?

29. **Price**   Imagine that you have been hired as director of a performing arts center for a mid-sized community. The community orchestra gives monthly concerts in the 400-seat auditorium. To promote attendance, the former director lowered the ticket price every 2 months. The ticket prices and corresponding average attendance are given.

Price (dollars)	Average attendance
35	165
30	200
25	240
20	280
15	335
10	400

a. Find quadratic and exponential models for the data. Which model better reflects the probable attendance beyond a $35 ticket price? Explain.

b. On the basis of the model that you believe is more appropriate, give the equation for revenue.

c. Find the maximum revenue and the corresponding ticket price and average attendance.

d. What other things besides the maximum revenue should you consider in setting price?

30. If they exist, find the absolute maxima and absolute minima of $y = \dfrac{2x^2 - x + 3}{x^2 + 2}$ over all real number

inputs. If an absolute maximum or absolute minimum does not exist, explain why not.

31. If they exist, find the absolute maximum and absolute minimum of $y = (2 - 3x + x^2)(3.5 + x)^2$ over all real number inputs. If an absolute maximum or absolute minimum does not exist, explain why not.

32. **Social Security**   The table for this activity (located on the *Calculus Concepts* CD-ROM and web site) gives the number of spouses of living retired workers receiving OASI benefits. Consider the year and spouses of retired workers beneficiaries columns in this table.

a. According to the data, when between 1975 and 2001 was the number of spouses of retired workers receiving benefits greatest and when was it least?

b. Find for these data a piecewise continuous model *S* that is composed of not more than three pieces.

c. At which points does the derivative of the model in part *a* not exist? Explain.

d. Determine all relative and absolute extrema of *S* between 1975 and 2001. Consider the function out of its modeling context.

e. Interpret the results of part *c* in context.

33. **Incarceration Rates**   The table for this activity (located on the *Calculus Concepts* CD-ROM and web site) gives national and regional incarceration rates for prisoners with sentences of more than 1 year who were imprisoned under state and federal jurisdiction at year end from 1977 through 2001.

a. Find a cubic function *N* for the Northeast incarceration rate data as a function of *x*, the number of years after 1977.

b. Determine all relative and absolute extrema of *N* between $x = 0$ and $x = 24$. Consider the function out of its modeling context.

c. Interpret the results of part *b* in context.

d. Repeat parts *a* through *c* for *M*, the Midwest incarceration rate.

5.3   Inflection Points

Unlike relative extrema, which are output values, inflection points are coordinate points.

In Section 5.2, we discussed extreme points on a graph. Another important point that can occur on a graph is an **inflection point.**

Recall from our earlier work with cubic and logistic functions that an inflection point is a point where a graph changes concavity. On a smooth, continuous graph, the inflection point can also be thought of as the point of greatest or least slope in a region around the inflection point. In real-life applications, this point is interpreted as *the point of most rapid change or least rapid change.* (See Figure 5.15.)

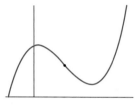

(a) Inflection point:
point of least slope, point
of most rapid decrease

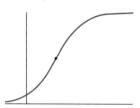

(b) Inflection point: point
of greatest slope, point
of most rapid increase

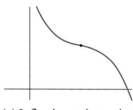

(c) Inflection point: point
of greatest slope, point
of least rapid decrease

(d) Inflection point:
point of least slope, point
of least rapid increase

FIGURE 5.15

Relative maxima and minima on a smooth, continuous graph can be found by locating the points at which the derivative graph crosses the horizontal axis. These points are among those where the original graph has horizontal tangent lines. Inflection points also can be found by examining the derivative graph and its relation to the function graph. To find the inflection point on a smooth, continuous graph, we must find the point where the slope (derivative) graph has a relative maximum or minimum. That is, we apply the method for finding maxima and minima to the *derivative graph.*

The Second Derivative

Consider, from the discussion in the previous section, the model for the population of Kentucky from 1980 through 1993:

$$p(x) = 0.395x^3 - 6.674x^2 + 30.257x + 3661.147 \text{ thousand people}$$

where x is the number of years since the end of 1980. Graphs of the function and its derivative are the first two graphs shown in Figure 5.16.

We wish to determine where the inflection point occurs—that is, where the population was declining most rapidly. It appears that p' has a minimum when p has an inflection point. In fact, this is exactly the case, so we can find the inflection point of p by finding the minimum of p'. To find the minimum of p' for this smooth, continuous function p, we must find where *its* derivative crosses the x-axis. The derivative

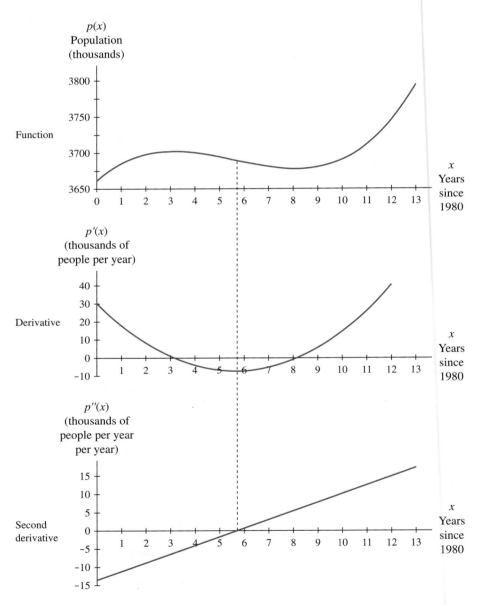

FIGURE 5.16

Other notations for the second derivative of p with respect to x include $\frac{d^2p}{dx^2}$ and $\frac{d^2}{dx^2}[p(x)]$.

of p' is called the **second derivative** of p, because it is the derivative of a derivative. The second derivative of p is denoted p''. In this case, the second derivative is given by

$$p''(x) = 2.37x - 13.348 \text{ thousand people per year per year}$$

where $x = 0$ in 1980.

Because the second derivative represents the rate of change of the first derivative, the output units of p'' are

$$\frac{\text{output units of } p'}{\text{input units of } p'}$$

The input/output diagram for this second derivative is shown in Figure 5.17.

The third graph in Figure 5.16 is a graph of the second derivative. The graph of p'' crosses the x-axis where the graph of p has an inflection point. Note that this identifies the minimum point on the graph of the derivative of p where the tangent line is horizontal.

Setting the second derivative equal to zero and solving for x gives $x \approx 5.63$. According to the model, the population was declining most rapidly in mid-1986 at a rate of approximately $p'(5.63) \approx -7.3$ thousand people per year. At that time, the population was approximately $p(5.63) \approx 3690$ thousand people.

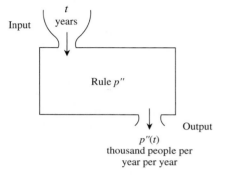

FIGURE 5.17

EXAMPLE 1   *Using the Second Derivative to Locate an Inflection Point*

5.3.1

Education   Consider a model for the percentage* of students graduating from high school in South Carolina from 1982 through 1990 who entered post-secondary institutions:

$$f(x) = -0.1057x^3 + 1.355x^2 - 3.672x + 50.792 \text{ percent}$$

where x is the number of years since 1982.

a. Find the inflection point of the function.

b. Determine the year between 1982 and 1990 in which the percentage was increasing most rapidly.

c. Determine the year between 1982 and 1990 in which the percentage was decreasing most rapidly.

Solution

a. Consider the point(s) at which the second derivative is zero. The rate-of-change formula for this function is

$$f'(x) = -0.3171x^2 + 2.71x - 3.672 \text{ percentage points per year}$$

where x is the number of years since 1982. The second derivative is

$$f''(x) = -0.6342x + 2.71 \text{ percentage points per year}$$

*Based on data in *South Carolina Statistical Abstract*, 1992.

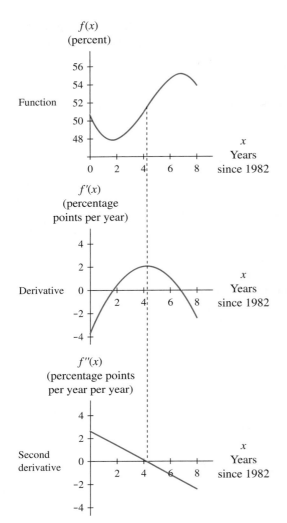

$f(x)$
(percent)

Function

$f'(x)$
(percentage
points per year)

Derivative

$f''(x)$
(percentage points
per year per year)

Second
derivative

FIGURE 5.18

where x is the number of years since 1982. The second derivative is zero when $x \approx 4.27$ years after 1982. Next, look at the graph of f shown in Figure 5.18. It does appear that $x = 4.27$ is the approximate input of the inflection point. The output is $f(4.27) \approx 51.6\%$, and the rate of change at that point is $f'(4.27) \approx 2.1$ percentage points per year.

b. Although f is a continuous function, it can be interpreted only at integer values of x because educational data such as these are reported for the fall of each year. Thus, to determine which valid input actually has the greatest rate of change, we evaluate f' at the integer values on either side of $x = 4.27$.

The rate of change of the model in 1986 is $f'(4) \approx 2.09$ percentage points per year. The rate of change in 1987 is $f'(5) \approx 1.95$ percentage points per year. Thus we can say that according to the model, the percentage of South Carolina high school graduates who enter post-secondary institutions was increasing most rapidly in 1986. The percentage of graduates going on for post-secondary education in 1986 was approximately $f(4) = 51.0\%$. The percentage was increasing by about $f'(4) = 2.1$ percentage points per year at that time.

Figure 5.18 shows the function, its derivative, and its second derivative. Note again the relationship among the points at which the second derivative crosses the x-axis, the derivative has a maximum, and the function has an inflection point.

c. Observe from the graph of f shown in Figure 5.18 that the most rapid decrease occurs at one of the endpoints. Evaluate the derivative of f at both endpoints.

$f'(0) = -3.672$ percentage points per year

$f'(8) = -2.2864$ percentage points per year

The percentage was declining most rapidly in 1982 at a rate of approximately 3.7 percentage points per year. ●

You have just seen two examples of how the second derivative of a function can be used to find an inflection point. It is important to use the second derivative whenever possible, because it gives an exact answer. Sometimes, however, finding the second derivative of a function can be tedious. In such cases, you will have to decide how important extreme accuracy is. If a close approximation will suffice (as is often the case in real-world modeling), then you may wish to find the first derivative only and use appropriate technology to estimate where its maximum (or minimum) occurs.

EXAMPLE 2 *Using Technology to Locate an Inflection Point*

5.3.2

Epidemic Consider the following model for the number of polio cases in the United States in 1949.

$$C(t) = \frac{42,183.911}{1 + 21,484.253e^{-1.248911t}} \text{ polio cases}$$

where $t = 1$ at the end of January, $t = 2$ at the end of February, and so forth. Find when the number of polio cases was increasing most rapidly, the rate of change of polio cases at the time, and the number of cases at that time.

Solution The graphs of C, C', and C'' are shown in Figure 5.19. We seek the inflection point on the graph of C that corresponds to the maximum point on the graph of C' that corresponds to the point at which the graph of C'' crosses the t-axis.

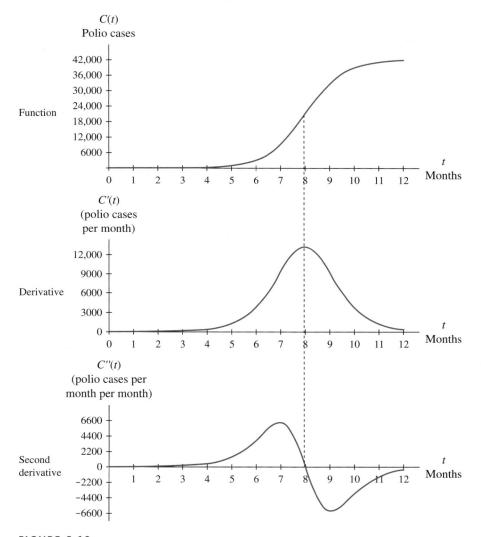

FIGURE 5.19

We choose to use technology to estimate the maximum point on the derivative graph. It occurs at $t \approx 8$ where the output is approximately 13,171. It is important to understand what these numbers represent. The t-value tells us the month in which the greatest increase occurred: $t = 8$ corresponds to the end of August. The output value is a value on the *derivative* graph. *This is the slope of the graph of C at the inflection point.* We can therefore say that polio was spreading most rapidly at the end of

August 1949 at a rate of approximately 13,171 cases per month. To find the number of polio cases at the inflection point, substitute the unrounded *t*-value of the point into the original function to obtain approximately 21,092 cases. Note that to the right of the inflection point, the *number* of polio cases was increasing whereas the *rate* at which polio cases appeared was declining.   ●

We saw in Section 5.2 that for a smooth, continuous function, a relative maximum or minimum occurs where the derivative graph *crosses* the horizontal axis, but not where the derivative graph touches the horizontal axis without crossing it. A similar statement can be made about inflection points. If the second derivative graph *crosses* the horizontal axis, then an inflection point occurs on the graph of the function. The graphs in Figure 5.20 of a function, its derivative, and its second derivative illustrate this issue. Note that the point at which the second derivative graph touches, but does not cross, the horizontal axis actually corresponds to a relative maximum on the function graph, not to an inflection point.

Two other situations that could occur are illustrated in Figure 5.21.

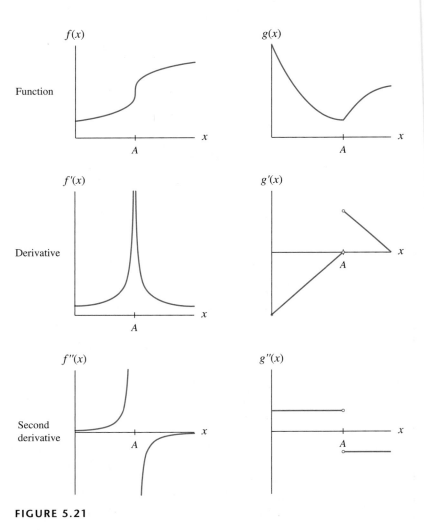

FIGURE 5.20

FIGURE 5.21

Note that the graphs of both f and g have inflection points at $x = A$ because they change concavity at that point. However, the second derivatives of f and g never cross the horizontal axis. In fact, in each case, the second derivative does not exist at $x = A$, because the first derivative does not exist there. The function f has a vertical tangent line at $x = A$, and the function g is not smooth at $x = A$. Even though such situations as this do not often occur in real-world applications, you should be aware that they could happen. Keep in mind the following result:

> At a point of inflection on the graph of a function, the second derivative is zero or does not exist. If the second derivative graph *crosses* the horizontal axis, then an inflection point of the function graph occurs at that input value.

In some applications, the inflection point can be regarded as the **point of diminishing returns.** Consider the college student who studies for 8 hours without a break before a major exam. The percentage of new material that the student will retain after studying for t hours can be modeled as

$$P(t) = \frac{45}{1 + 5.94e^{-0.969125t}} \text{ percent}$$

This function has an inflection point at $t \approx 1.8$. That is, after approximately 1 hour and 48 minutes, the rate at which the student is retaining new material begins to diminish. Studying beyond that point will improve the student's knowledge, but not as quickly. This is the idea behind diminishing returns: Beyond the inflection point, you gain fewer percentage points per hour than you gain at the inflection point; that is, output increases at a decreasing rate. The existence of this point of diminishing returns is one factor that has led many educators and counselors to suggest studying in 2-hour increments with breaks in between.

Concavity and the Second Derivative

Because the derivative of a function is simply the slope of the graph of that function, we know that a positive derivative indicates that the function output is increasing and a negative derivative indicates that the function output is decreasing. The second derivative provides similar information about where a function graph is concave up and where it is concave down.

In particular, if the second derivative is negative, it means that the first derivative graph is declining, which means that the original function graph is concave down. Similarly, a positive second derivative indicates that the first derivative is increasing, which means that the original function graph is concave up. And, as we have already seen, where the second derivative changes from positive to negative or from negative to positive, the function graph has an inflection point.

Consider, for example, the following information (see Table 5.6) about the second derivative of a function f and the information it provides about the concavity of a graph of f. We use "ccu" to indicate concave up and "ccd" to indicate concave down.

TABLE 5.6

x	0	1	2	3	4	5	6
f″(x)	15	0	-1	0	-1	0	15
Concavity of f	ccu	infl. pt.	ccd	not infl. pt.	ccd	infl. pt.	ccu

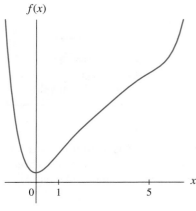

f(x)

FIGURE 5.22

We can conclude from this table that a graph of the function f has two inflection points at $x = 1$ and $x = 5$. Although the second derivative is zero at $x = 3$, the function graph does not change concavity at that value so the corresponding point is not an inflection point.

Next, consider some values of the derivative of f (see Table 5.7). We use the symbol ↗ to indicate increasing and ↘ to indicate decreasing.

TABLE 5.7

x	-1	0	1	2	3	4	5	6
f′(x)	-35.6	0	5.6	4.6	4.2	3.8	2.8	8.4
f(x)	↘	local min.	↗	↗	↗	↗	↗	↗

We conclude that the graph of the function f decreases to a local minimum at zero and then increases to the right of zero. The graph is concave down between $x = 1$ and $x = 5$ and concave up to the left of $x = 1$ and to the right of $x = 5$. A possible graph of f, based on this information, is shown in Figure 5.22.

5.3 Concept Inventory

○ *Inflection point*

○ *Second derivative*

○ *Point of diminishing returns*

○ *Conditions under which inflection points exist*

5.3 Activities

1. Production The graph shows an estimate of the ultimate crude oil production recoverable from Earth.

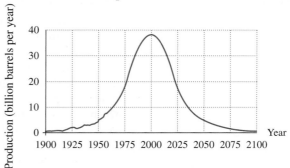

(Source: Adapted from François Ramade, *Ecology of Natural Resources.* New York: Wiley, 1984. Copyright 1984 by John Wiley & Sons, Inc. Reprinted by permission of the publishers.)

a. Estimate the two inflection points on the graph.

b. Explain the meaning of the inflection points in the context of crude oil production.

2. **Advertising** The graph shows sales (in thousands of dollars) for a business as a function of the amount spent on advertising (in hundreds of dollars).

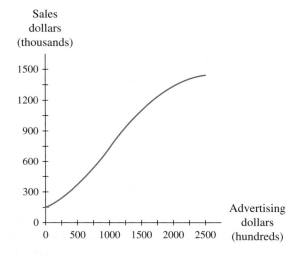

a. Mark the approximate location of the inflection point on the graph.

b. Explain the meaning of the inflection point in the context of this business.

c. Explain how knowledge of the inflection point might affect decisions made by the managers of this business.

Each of Activities 3 through 6 presents graphs of a function, its derivative, and its second derivative. (Assume that the input is on the horizontal axis.) In each case, identify each graph as the function, the derivative, or the second derivative. Give reasons for your choice.

3. a. **b.**

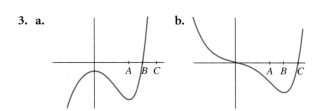

c.

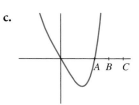

4. a. **b.**

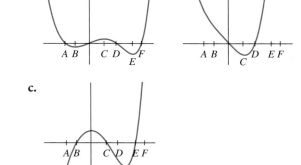

c.

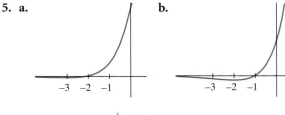

5. a. **b.**

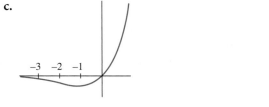

c.

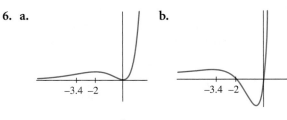

6. a. **b.**

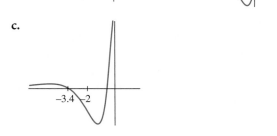

c.

7. Consider the function

$$g(x) = 0.04x^3 - 0.88x^2 + 4.81x + 12.11$$

   a. Graph g, g', and g'' between $x = 0$ and $x = 15$. Indicate the relationships among points on the three graphs that correspond to maxima, minima, and inflection points.

   b. Find the inflection point on the graph of g. Is it a point of most rapid decline or least rapid decline?

8. Consider the function

$$f(x) = \frac{20}{1 + 19e^{-0.5x}}$$

   a. Graph f, f', and f'' between $x = 0$ and $x = 15$. Indicate the points on the graphs of f' and f'' that correspond to the inflection point on the graph of f.

   b. Find the inflection point on the graph of f. Is it a point of most rapid or least rapid increase?

9. **Study Time**   The percentage of new material that a student will retain after studying t hours without a break can be modeled by

$$P(t) = \frac{45}{1 + 5.94e^{-0.969125t}} \text{ percent}$$

   a. Find the inflection point on the graph of P, and interpret the answer.

   b. Compare your answer with that given in the discussion at the end of this section.

10. **Population**   The U.S. Bureau of the Census' prediction for the percentage of the population that is 65 to 74 years old from 2000 through 2050 can be modeled by

$$p(x) = (1.619 \cdot 10^{-5})x^4 - (1.675 \cdot 10^{-3})x^3 + 0.050x^2 - 0.308x + 6.693 \text{ percent}$$

where x is the number of years after 2000.
(Source: Based on data from *Statistical Abstract*, 1998.)

   a. Determine the year between 2000 and 2050 in which the percentage is predicted to be increasing most rapidly, the percentage at that time, and the rate of change of the percentage at that time.

   b. Repeat part *a* for the most rapid decrease.

11. **Grasshoppers**   The percentage of southern Australian grasshopper eggs that hatch as a function of temperature (for temperatures between 7°C and 25°C) can be modeled by

$$P(t) = -0.00645t^4 + 0.488t^3 - 12.991t^2 + 136.560t - 395.154 \text{ percent}$$

where t is the temperature in °C.
(Source: Based on information in *Elements of Ecology*, George L. Clarke. New York: Wiley, 1954, p. 170.)

   a. Graph P, P', and P''.

   b. Find the point of most rapid decrease on the graph of P. Interpret your answer.

12. **Home Sale**   The median size of a new single-family house built in the United States between 1987 and 2001 can be modeled by the equation

$$H(x) = 0.359x^3 - 15.198x^2 + 221.738x + 826.514 \text{ square feet}$$

x years after 1980.
(Source: Based on data from the National Association of Home Builders Economics Division.)

   a. Determine the time between 1987 and 2001 when the median house size was increasing least rapidly. Find the corresponding house size and rate of change in house size.

   b. Graph H, H', and H'', indicating the relationships between the inflection point on the graph of H and the corresponding points on the graphs of H' and H''.

   c. Determine the time between 1985 and 2001 when the median house size was increasing the most rapidly.

13. **Price**   The average price (per 1000 cubic feet) of natural gas for residential use from 1994 through 2000 is given by

$$p(x) = 0.03x^4 - 0.834x^3 + 8.45x^2 - 36.7x + 63.74 \text{ dollars}$$

where x is the number of years since 1990.
(Source: Based on data from *Statistical Abstract*, 1998, 2001.)

   a. Sketch the graphs of p and its first and second derivatives. Label the vertical axes appropriately. Which points on the derivative graph correspond to the inflection points of the original function graph? Which point on the second derivative graph corresponds to the inflection point of the original function graph?

b. Find the *x*-intercepts of the second derivative graph, and interpret their meaning in context.

c. Determine when, according to the model, the average natural gas price was declining most rapidly and when it was increasing most rapidly between 1990 and 2001.

d. Repeat part *c* for the interval 1995 through 1999.

14. Cable TV The percentage of households with TVs that subscribed to cable from 1970 through 2002 can be modeled by

$$P(x) = 6 + \frac{62.7}{1 + 38.7e^{-0.258x}} \text{ percent}$$

where *x* is the number of years after 1970. When was the percentage of households with TVs that subscribed to cable increasing the most rapidly? What were the percentage and the rate of change of the percentage at that time?
(Source: Based on data from Television Bureau of Advertising.)

15. Donors The number of people who donated to an organization supporting athletics at a certain university in the southeast from 1975 through 1992 can be modeled by

$$D(t) = -10.247t^3 + 208.114t^2 - 168.805t + 9775.035 \text{ donors}$$

t years after 1975.
(Source: Based on data from IPTAY Association at Clemson University.)

a. Find any relative maxima or minima that occur on a graph of the function.

b. Find the inflection point(s).

c. How do the following events in the history of football at that college correspond with the curvature of the graph?

   i. In 1981, the team won the National Championship.

   ii. In 1988, Coach F was released and Coach H was hired.

16. Cable TV The amount spent on cable television per person per year from 1984 through 1992 can be modeled by

$$A(x) = -0.126x^3 + 1.596x^2 + 1.802x + 40.930 \text{ dollars}$$

where *x* is the number of years since 1984.
(*Source:* Based on data from *Statistical Abstract*, 1994.)

a. Find the inflection point on the graph of *A* and the corresponding points on graphs of *A′* and *A″*.

b. Find the year between 1984 and 1992 in which the average amount spent per person per year on cable television was increasing most rapidly. What was the rate of change of the amount spent per person in that year?

17. Labor A college student works for 8 hours without a break assembling mechanical components. The cumulative number of components she has assembled after *h* hours can be modeled by

$$N(h) = \frac{62}{1 + 11.49e^{-0.654h}} \text{ components}$$

a. Determine when the rate at which she was working was greatest.

b. How might her employer use the information in part *a* to increase the student's productivity?

18. Lake Level The lake level of Lake Tahoe from October 1, 1995, through July 31, 1996, can be modeled by

$$L(d) = (-5.345 \cdot 10^{-7})d^3 + (2.543 \cdot 10^{-4})d^2 - 0.0192d + 6226.192 \text{ feet above sea level}$$

d days after September 30, 1995.
(Source: Based on data from the Federal Watermaster, U.S. Department of the Interior.)

a. Determine when the lake level was rising most rapidly between October 1, 1995, and July 31, 1996.

b. What factors may have caused the inflection point to occur at the time you found in part *a*?

c. Would you expect the most rapid rise to occur at approximately the same time each year? Explain.

19. Labor The personnel manager for a construction company keeps track of the total number of labor hours spent on a construction job each week during the construction. Some of the weeks and the corresponding labor hours are given.

a. Find a logistic model for the data.

b. Find the derivative of the model. What are the units on the derivative?

Weeks after the start of a project	Cumulative labor hours
1	25
4	158
7	1254
10	5633
13	9280
16	10,010
19	10,100

c. Graph the derivative, and discuss what information it gives the manager.

d. When is the maximum number of labor hours per week needed? How many labor hours are needed in that week?

e. Find the point of most rapid increase in the number of labor hours per week. How many weeks into the job does this occur? How rapidly is the number of labor hours per week increasing at this point?

f. Find the point of most rapid decrease in the number of labor hours per week. How many weeks into the job does this occur? How rapidly is the number of labor hours per week decreasing at this point?

g. Carefully explain how the exact values for the points in parts e and f can be obtained.

h. If the company has a second job requiring the same amount of time and number of labor hours, a good manager will schedule the second job to begin so that the time when the number of labor hours per week for the first job is declining most rapidly corresponds to the time when the number of labor hours per week for the second job is increasing most rapidly. How many weeks into the first job should the second job begin?

20. **Advertising** A business owner's sole means of advertising is to put fliers on cars in a nearby shopping mall parking lot. The table shows the number of labor hours per month spent handing out fliers and the corresponding profit.

a. Find a model for profit.

b. For what number of labor hours is profit increasing most rapidly? Give the number of

labor hours, the profit, and the rate of change of profit at that number.

c. In this context, the inflection point can be thought of as the point of diminishing returns. Discuss how knowing the point of diminishing returns could help the business owner make decisions relative to employee tasks.

Labor hours each month	Profit (dollars)
0	2000
10	3500
20	8500
30	19,000
40	32,000
50	43,000
60	48,500
70	55,500
80	56,500
90	57,000

21. **Refuse** The yearly amount of garbage (in millions of tons) taken to a landfill outside a city during selected years from 1970 through 2000 is given in the table.

Year	Amount (millions of tons)
1970	81
1975	99
1980	117
1985	122
1990	132
1995	145
2000	180

a. Using the table values only, determine during which 5-year period the amount of garbage showed the slowest increase. What was the average rate of change during that 5-year period?

b. Find a model for the data.

c. Give the second derivative formula for the equation.

d. Use the second derivative to find the point of slowest increase on a graph of the equation.

e. Graph the first and second derivatives, and explain how they support your answers to part *d*.

f. In what year was the rate of change of the yearly amount of garbage the smallest? What were the rate of increase and the amount of garbage in that year?

22. Revenue   The revenue (in millions of dollars) of the Polo Ralph Lauren Corporation from 1993 through 2001 is given in the table.

Year	Revenue (millions of dollars)
1993	767.3
1994	810.7
1995	846.6
1996	1,019.9
1997	1,180.4
1998	1,470.9
1999	1,713.1
2000	1,948.7
2001	1,982.4

(Source: Hoover's Online Guide.)

a. Use the data to estimate the year in which revenue was growing most rapidly.

b. Find a model for the data.

c. Find the first and second derivatives of the model in part *b*.

d. Determine the year in which revenue was growing most rapidly. Find the revenue and the rate of change of revenue in that year.

23. Reaction   A chemical reaction begins when a certain mixture of chemicals reaches 95°C. The reaction activity is measured in Units (U) per 100 microliters (100μL) of the mixture. Measurements at 4-minute intervals during the first 18 minutes after the mixture reaches 95°C are listed in the table.
(Source: David E. Birch et al., "Simplified Hot Start PCR," *Nature*, vol. 381, May 30, 1996, p. 445.)

a. According to the table, during what time interval was the activity increasing most rapidly?

Time (minutes)	2	6	10	14	18
Activity (U/100μL)	0.10	0.60	1.40	1.75	1.95

b. Find a model for the data, and use the equation to find the inflection point. Interpret the inflection point in context.

24. Emissions   The table gives the total emissions in millions of metric tons of nitrogen oxides, NO_x, in the United States from 1940 through 1990.

Year	NO_x (millions of metric tons)
1940	6.9
1950	9.4
1960	13.0
1970	18.5
1980	20.9
1990	19.6

(Source: *Statistical Abstract*, 1992.)

a. Find a cubic model for the data.

b. Give the slope formula for the equation.

c. Determine when emissions were increasing most rapidly between 1940 and 1990. Give the year, the amount of emissions, and how rapidly they were increasing.

25. Emissions   Total yearly emissions of lead into the atmosphere between 1970 and 1995 can be modeled by the equation

$$L(x) = \frac{251.3}{1 + 0.1376e^{0.2854x}} \text{ million tons}$$

x years after 1970.
(Source: Based on data from *Statistical Abstract*, 1998.)

a. Give the slope for *L*.

b. How quickly were emissions decreasing in 1970 and 1995?

c. Find the year in which lead emissions were declining most rapidly. Give the emissions and the rate of change of the emissions in that year.

26. Gender Ratio   The table shows the number of males per 100 females in the United States based on census data. The number is referred to as the gender ratio.

a. Find a cubic model for the data.

b. Write the second derivative of the equation you found in part a.

c. In what year does the output of the model exhibit the most rapid growth? What was the gender ratio in that year, and how rapidly was it changing?

Year	Males per 100 females
1900	104.6
1910	106.2
1920	104.1
1930	102.6
1940	100.8
1950	98.7
1960	97.1
1970	94.8
1980	94.5
1990	95.1
2000	96.3

27. For a function f, $f(x) > 0$ for all real number input values. Describe the concavity of a graph of f, and sketch a function for which this condition is true.

28. Draw a graph of a function g such that $g''(x) = 0$ for all real number input values.

29. For a function f, the following statements are true:

$$f''(x) > 0 \quad \text{when } 0 \le x < 2$$
$$f''(x) = 0 \quad \text{when } x = 2$$
$$f''(x) < 0 \quad \text{when } 2 < x \le 4$$

a. Describe the concavity of a graph of f between $x = 0$ and $x = 4$.

b. Draw two completely different graphs that satisfy these second-derivative conditions.

30. For a function h, the following statements are true:

$$h''(x) > 0 \quad \text{when } 0 < x < 2$$
$$h''(x) = 0 \quad \text{when } x = 2 \text{ and } x = 0$$
$$h''(x) < 0 \quad \text{when } x < 0 \text{ and } x > 2$$

a. Describe the concavity of a graph of h.

b. Draw a graph that satisfies all three of these second-derivative conditions.

31. Which of the six basic models discussed thus far in this book could have inflection points?

32. Discuss in detail all of the options that are available for finding inflection points of a function.

33. **Incarceration Rates** The table for this activity (located on the *Calculus Concepts* CD-ROM and web site) gives national and regional incarceration rates for prisoners with sentences of more than 1 year who were imprisoned under state and federal jurisdiction at year end from 1977 through 2001.

a. Find a cubic function N for the Northeast incarceration rate data as a function of x, the number of years after 1977.

b. Write the second derivative of the equation you found in part a.

c. In what year does the output of N exhibit the most rapid growth? What was the incarceration rate in that year, and how rapidly was it changing?

d. Repeat parts a–c for M, the Midwest incarceration rate.

34. **Social Security** The table for this activity (located on the *Calculus Concepts* CD-ROM and web site) gives the number of mother and/or father beneficiaries of OASI benefits between 1975 and 2001.

a. Using the table values only, determine during which year between 1975 and 2001 the number of mother and/or father beneficiaries showed the fastest increase. What was the average rate of change during that year?

b. Find a logistic model for the number of mother and/or father beneficiaries as a function of the number of years after 1975.

c. Use the equation in part b to find the inflection point. Interpret the inflection point in context.

5.4   Derivatives in Action

By now you should have a solid understanding of the basic method of finding both relative and absolute maximum and minimum values of a function. We now extend the techniques of optimization to situations in which you must first set up an equation to optimize and then apply optimization techniques to that equation. Many of the problems presented in this section are different from most of those in this book because no data or equation is given. You must find an equation from a word description. We begin by outlining some steps that are helpful in setting up equations and optimizing their outputs. Refer to this list often until these steps become second nature to you.

Step 1: Read the question carefully, and identity the quantity to be maximized or minimized (the output variable) and the quantity or quantities on which the output quantity depends (the input variable or variables). This information is often found in the final sentence of the problem statement. Because this step is the most crucial one for correctly directing your thinking as you approach the problem, we present four questions in Example 1 to give you some practice in identifying input and output quantities.

EXAMPLE 1   *Determining Output and Input When Optimizing*

For each of the following statements, determine the quantity to be maximized or minimized and the corresponding input quantity or quantities.

a. Find the order size that will minimize cost.

b. Determine the dimensions that will result in a box of greatest volume.

c. For what size dose is the sensitivity to the drug greatest?

d. What is the most economical way to lay the pipe?

Solution

a. Output quantity to be minimized: cost
    Input quantity: order size

b. Output quantity to be maximized: volume
    Input quantity: dimensions of box (height, length, and width)

c. Output quantity to be maximized: sensitivity
    Input quantity: dosage

d. Output quantity to be minimized: cost
    Input quantity: Not enough information is given in this sentence, but we can conclude that the input is probably some measure of distance related to how the pipe is laid. ●

WEB/CD

**Strengthening
the Concepts:
Formulas for
Geometry-
Related
Problems**

Step 2: Sketch a picture if the problem situation is geometric in nature—that is, if the problem involves constructing containers, laying cable or wire, building enclosures, or the like. The picture should be labeled with appropriate variables representing distances or sizes. You should find that at least one of the variables represents the input quantity identified in Step 1.

Step 3: Begin building the equation. From the results of Step 1, you will know the quantity for which you need an equation, such as cost, volume, or drug sensitivity. If the quantity is a geometric one (such as volume of a right circular cylinder, area of a trapezoid, hypotenuse of a right triangle) you may need to look up the formula in some reference source.

Step 4: Rewrite the equation as a function of only one variable if it is not already in this form. This step may require the use of other secondary equations relating the input variables in the equation in Step 3. For example, if the equation in Step 3 is for volume of a box as a function of length and width, you need to find an equation relating length and width. You will then use this equation to solve for one variable in terms of the other and substitute for that variable in the equation in Step 3.

Step 5: Determine any limitations on the input variable. State the limitation as an interval in which the input must lie.

Step 6: Apply optimization techniques to the final equation. This step includes finding all relative extrema and determining all absolute extrema over the interval identified in Step 5. This step should also include graphing the equation or some other method of verifying that the answer you found is indeed a maximum or minimum on the interval.

Step 7: Read the problem again to be sure you have answered the question asked. Sometimes the question asks you to find the output quantity, sometimes the input quantity, sometimes both, and occasionally some other quantity that can be calculated using the input and output values that correspond to the extreme point.

Before demonstrating this multistep problem-solving method with three examples, we summarize the seven steps.

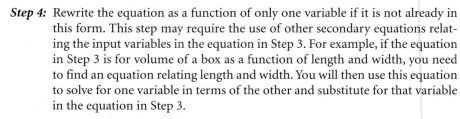

Problem-Solving Strategy for Optimization Problems

Step 1: Identify the quantity to be maximized or minimized (the output) and the quantity or quantities on which the output quantity depends (the input).

Step 2: Sketch and label a picture if the problem is geometric in nature.

Step 3: Build a model for the quantity that is to be maximized or minimized.

Step 4: Rewrite the equation in Step 3 in terms of only one input variable.

Step 5: Identify the input interval.

Step 6: Apply optimization techniques to the final equation.

Step 7: Answer the question or questions posed in the problem.

EXAMPLE 2   *Applying the Problem-Solving Strategy*

Ranching   A rancher removed 200 feet of wire fencing from a field on his ranch. He wishes to reuse the fencing to create a rectangular corral into which he will build a 6-foot-wide wooden gate. What dimensions will result in a corral with the greatest possible area? What is the greatest area?

Solution

Step 1: Output variable to maximize: area of rectangular corral
Input variables: length and width of the corral

Step 2: A sketch of the corral is shown in Figure 5.23. (Note that you can put the gate on any side of the corral.)

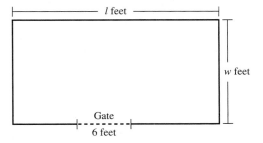

FIGURE 5.23

Step 3: Recall that the area of a rectangle is $A = lw$ square feet, where l is the length of the rectangle in feet and w is its width in feet. This equation is the one we seek to maximize.

Step 4: Because the equation in Step 3 has two input variables, we need to find a second equation relating the length and the width. A clue to this equation is found in the statement "A rancher removed 200 feet of wire fencing." If the area of the corral is to be a maximum, it makes sense that the rancher will use all of the fencing. Referring to Figure 5.23, we see that if we calculate the distance around the corral and subtract the 6-foot-wide gate, we should obtain 200 feet. This leads to the equation

$$2w + 2l - 6 = 200$$

Solving this equation for one of the variables (it doesn't matter which one) results in the equation

$$l = \frac{206 - 2w}{2} = 103 - w$$

Substituting this expression for l into the area equation gives

$$A = lw = (103 - w)w = 103w - w^2 \text{ square feet}$$

where w is the width of the corral in feet.

Step 5: Look again at Figure 5.23. If the length of the corral were 0, then the width would be 100 feet. We also know that width cannot be zero or negative. Thus we have the input interval $0 < w < 100$.

Step 6: Note that the graph of the area A is a concave-down parabola. We find the maximum point on the parabola by determining where the derivative of A is zero (that is, where its graph crosses the w-axis).

$$\frac{dA}{dw} = 103 - 2w = 0$$

$$w = 51.5$$

This value lies in the interval in Step 5. Because the interval is an open interval, we have no endpoints to check. Thus the corral with maximum area should be 51.5 feet wide.

Step 7: The statement of the problem asked for the dimensions of the corral of greatest area as well as for the area. We know the width and need to determine the length of the corral. To do so, we use the equation from Step 4 that expresses length in terms of width.

$$l = 103 - w$$

$$l = 103 - 51.5 = 51.5 \text{ feet}$$

This calculation tells us that the corral with maximum area is a square with sides that are 51.5 feet long. To determine the corresponding area, simply square 51.5 feet.

$$A = (51.5 \text{ feet})(51.5 \text{ feet}) = 2652.25 \text{ square feet} \quad \bullet$$

The next example shows the solution to a more involved geometric problem.

EXAMPLE 3   *Optimizing in a Geometric Setting*

Popcorn Tins   In an effort to be environmentally responsible, a confectionery company is rethinking the dimensions of the tins in which it packages popcorn. Each cylindrical tin is to hold 3.5 gallons. The bottom and lid are both circular, but the lid must have an additional $1\frac{1}{8}$ inch around it in order to form a lip. (Consider the amount of metal needed to create a seam on the side and to join the side to the bottom to be negligible.) What are the dimensions of a tin that meets these specifications but uses the least amount of metal possible?

Solution

Step 1: Output quantity to be minimized: amount of metal
Input quantities: dimensions of the tin

Step 2: Figure 5.24 shows the tin and its components. We have chosen to label the height of the tin h and the radius of the bottom of the tin r. (It is also possible to use the diameter, instead of the radius, of the top or bottom as the other variable.)

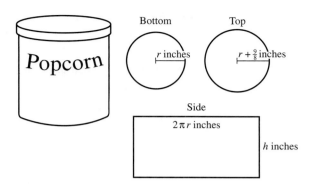

FIGURE 5.24

Step 3: The amount of metal used to construct the tin is measured by finding the combined surface area of the side, top, and bottom. Thus the total surface area is determined by the equation

$$S = \text{area of side} + \text{area of bottom} + \text{area of top}$$

The area of the side is $2\pi rh$. The area of the bottom is πr^2. The radius of the circle used to make the top is $r + \frac{9}{8}$; thus the area of the top is $\pi\left(r + \frac{9}{8}\right)^2$. Using these formulas, we have the model for the surface area of the container:

$$S = 2\pi rh + \pi r^2 + \pi\left(r + \frac{9}{8}\right)^2 \text{ square inches}$$

where r is the radius of the bottom of the container in inches and h is the height of the container in inches.

Step 4: We need to rewrite the formula for surface area in terms of either radius or height. To do so, we need an equation relating these two quantities. Recall from the statement of the problem that the volume of the container needs to be 3.5 gallons. This information allows us to set up a second equation. The volume of a cylinder is given by the equation $V = \pi r^2 h$. Because we will be using dimensions in inches, we must express the volume in cubic inches. One gallon is 231 cubic inches (the authors found this value in a dictionary), so 3.5 gallons is $3.5(231) = 808.5$ cubic inches. Thus we have the equation

$$V = \pi r^2 h = 808.5$$

We must solve for one variable in terms of the other. It makes sense to choose the variable for which this can be done most simply. In this case, we solve for h in terms of r.

$$h = \frac{808.5}{\pi r^2} \quad \text{for } r \neq 0$$

Next, we substitute this expression for h into the surface area equation in Step 3.

$$S = 2\pi r\left(\frac{808.5}{\pi r^2}\right) + \pi r^2 + \pi\left(r + \frac{9}{8}\right)^2 = \frac{1617}{r} + \pi r^2 + \pi\left(r + \frac{9}{8}\right)^2 \text{ square inches}$$

Step 5: Because we are given no indication that the dimensions are restricted (except by the volume of the container, which was taken into account in Step 4), our only restriction is that the radius must be positive: $r > 0$.

Step 6: Next we find the minimum of S by determining where a graph of $\frac{dS}{dr}$ crosses the horizontal axis. To do so, we find the derivative of S with respect to r.

$$\frac{dS}{dr} = \frac{-1617}{r^2} + 2\pi r + 2\pi\left(r + \frac{9}{8}\right)(1) = \frac{-1617}{r^2} + 4\pi r + \frac{9\pi}{4}$$

Setting this expression equal to zero and solving for r gives $r \approx 4.87$ inches, which lies in the interval in Step 5. To confirm that this value of r does indeed give a minimum surface area, we examine a graph of the surface area function. See Figure 5.25.

Step 7: Our final step is to answer the question posed. In this case, we must provide the dimensions (height and radius) of the tin with minimum surface area. We know the radius. In order to find the height, we substitute the unrounded value of the radius into the equation that we found in Step 4 for height in terms of radius.

$$h \approx \frac{808.5}{\pi(4.87)^2} \approx 10.86 \text{ inches}$$

In order to use the least amount of metal possible, the popcorn tins should be constructed with height approximately 10.86 inches and radius approximately 4.87 inches.* ●

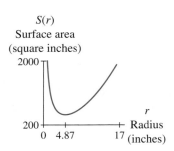

$S(r)$
Surface area
(square inches)

FIGURE 5.25

Our final example illustrates optimization in a nongeometric setting.

EXAMPLE 4   *Optimizing Revenue*

Cruise   A travel agency offers spring-break cruise packages. The agency advertises a cruise to Cancun, Mexico, for $1200 per person. In order to promote the cruise among student organizations on campus, the agency offers a discount for student groups selling the cruise to over 50 of their members. The price per student will be discounted by $10 for each student in excess of 50. For example, if an organization had 55 members go on the cruise, each of those 55 students would pay $1200 - 5(\$10) = \$1150$.

a. What size group will produce the largest revenue for the travel agency, and what is the largest possible revenue?

b. If the travel agent limits each organization to 75 tickets, what is the agent's maximum revenue for each organization?

c. If you were the travel agent, would you set a limit on the number of students per organization?

---

*Point of Interest: The authors found that the popcorn tin they purchased as a reference tool for this example did have dimensions that matched those of a tin with minimum surface area.

Solution

a. ***Step 1:*** Output quantity to be maximized: travel agency's revenue from student group

Input quantity: size of the group

Step 2: Because this scenario is not geometric, there is no need for a picture.

Step 3: We need an equation for the travel agency's revenue from the students. In this example, revenue is the number of students traveling on the cruise multiplied by the price each student pays:

Revenue = (number of students)(price per student)

In Step 1, we determined that the input quantity is the size of the group. Because the factor affecting price is the number of students in excess of 50, we choose to use a variable to represent that quantity. (It is also possible to use a variable representing the total number of students.) If s is the number of students in excess of 50, then the total number of students is $50 + s$. The price is $1200 minus $10 for each student in excess of 50. Converting this statement into math symbols, we have the price per student as $1200 - 10s$ dollars. Thus

Revenue $= R(s) = (50 + s)(1200 - 10s) = 60{,}000 + 700s - 10s^2$ dollars

where there are $50 + s$ students on the cruise.

Step 4: Because the revenue function has only one input variable, this step is not necessary.

Step 5: We are not given any restriction concerning the number of students. Our interval is $s > 0$.

Step 6: The revenue equation is a concave-down parabola, so we can be confident that it has a maximum. We simply set the derivative of the revenue function equal to zero and solve for s.

$$R'(s) = 700 - 20s = 0$$
$$s = 35$$

Revenue is maximized when $s = 35$.

Step 7: The question asks for the number of students and the corresponding revenue. Recall that the variable s represents the number of students in excess of 50. Thus the total number of students is $50 + 35 = 85$. The price per student is $1200 - 35(10) = 850. The revenue is (85 students)($850 per student) = $72{,}250$. Revenue is maximized at $72,250 for a campus group of 85 students.

b. If the number of students is limited to 75, then the interval in Step 5 is $0 < s \leq 25$. The solution to part a is no longer valid because it does not lie in this interval. In this case, the maximum revenue occurs at the endpoint of the interval (when there are 75 students). Examine a graph of R, and observe that this is true. The price each student pays is $1200 - 25(10) = 950. The travel agent's revenue in this case is (75 students)($950 per student) = $71{,}250$.

c. In part *a*, we found that when 85 students bought cruise tickets, the travel agent's revenue was a maximum. If more than 85 students bought tickets, the agent's revenue would actually decline because of the low price. (Recall that the revenue function graph was a concave-down parabola.) Therefore, it would make sense for the agent to limit the number of students per organization to 85. ●

Optimization problems such as those discussed in this section are often challenging, but as long as you approach them using the seven steps outlined here, you should be able to solve them—or at least identify the point at which you need help. As we have said before, the more problems you work, the more skilled and confident you will become.

5.4   Concept Inventory

○ *Seven-step problem-solving strategy for optimization*

5.4   Activities

1. **Popcorn Tins**   Refer to Example 3, and repeat the solution using the diameter of the bottom of the tin, instead of the radius, as an input variable. Show that the dimensions you obtain for minimum surface area are equivalent to those found in Example 3.

2. **Cruise**   Refer to Example 4, and repeat the solution using the total number of students as the input variable. Show that the maximum revenue and the corresponding number of students and price are equivalent to those found in Example 4.

3. **Garden Area**   A rectangular-shaped garden has one side along the side of a house. The other three sides are to be enclosed with 60 feet of fencing. What is the largest possible area of such a garden?

4. **Floral Frame**   A florist uses wire frames to support flower arrangements displayed at weddings. Each frame is constructed from a wire of length

9 feet that is cut into six pieces. The vertical edges of the frame consist of four of the pieces of wire that are each 12 inches long. One of the remaining pieces is bent into a square to form the base of the frame; the final piece is bent into a circle to form the top of the frame. See the figure.

a. How should the florist cut the wire of length 9 feet in order to minimize the combined area of the circular top and the square base of the frame?

b. Verify that the answer to part *a* minimizes the combined area.

c. What is the answer to part *a* if the frame must be constructed so that the area enclosed by the square is twice the area enclosed by the circle?

5. **Game Show**   You have been selected to appear on the *Race for the Money* television program. Before the program is aired on television, you and three other contestants are sent to Myrtle Beach, South Carolina. Each of you is given a piece of cardboard that measures 8 inches by 10 inches, a pair of scissors, a ruler, and a roll of tape. The contestant who carries away the most sand from the beach is eligible to win the grand prize of $250,000.

There are two restrictions in the race. The sand must be carried away in an open box that each person makes from the piece of cardboard by cutting equal squares from each corner and turning up the sides. A second restriction is that the sand cannot be piled higher than the sides of the box.

a. What length should the corner cuts be for a contestant to carry away the most sand? What is the largest volume of sand that can be carried away in such a box?

b. What length should the cuts be if there is a third restriction that the box can hold no more than 50 cubic inches of sand?

c. Do you feel you would have a better chance of winning the sand race with the two restrictions given in part *a* or with all three restrictions on the amount of sand that can be carried away?

6. **Ferrying Supplies** A portion of the shoreline of a Caribbean island is in the shape of the curve $y = 2\sqrt{x}$. A hut is located at point C, as shown in the figure. You wish to deliver a load of supplies to the shoreline and then hire someone to help you carry the supplies to your hut. The helper will charge you $10 per mile. At what point (x, y) on the shoreline should you land in order to minimize the amount you must pay to have the supplies carried to the hut? How much currency should you have on hand to pay the helper?

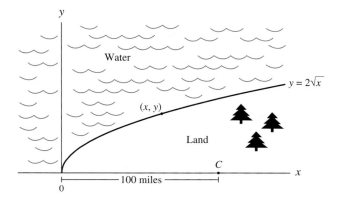

7. **Costs** During one calendar year, a year-round elementary school cafeteria uses 42,000 Styrofoam plates and packets, each containing a fork, spoon, and napkin. The smallest amount the cafeteria can order from the supplier is one case containing 1000 plates and packets. Each order costs $12, and the cost to store a case for the whole year is approximately $4. In order to find the optimal order size that will minimize costs, the cafeteria manager must balance the ordering costs incurred when many small orders are placed with the storage costs incurred when many cases are ordered at once.

a. How many cases does the cafeteria need over the course of 12 months?

b. If x is the number of cases in each order, how many times will the manager need to order during one calendar year?

c. What is the total cost of ordering for 1 calendar year?

d. Assume that the average number of cases stored throughout the year is half the number of cases in each order. What is the total storage cost for 1 year?

e. Write a model for the combined ordering and storage costs for 1 year.

f. What order size minimizes the total yearly cost? (Note that only full cases may be ordered.) How many times a year should the manager order? What will the minimum total ordering and storage costs be for the year?

8. **Costs** A situation similar to the one in Activity 7 occurs when a company uses a machine to produce different types of items. For example, the same machine may be used to produce popcorn tins and 30-gallon storage drums. The machine is set up to produce a quantity of one item and then be reconfigured to produce a quantity of another item.

The plant produces a run and then ships the tins out at a constant rate so that the warehouse is empty for storing the next run. Assume that the number of tins stored on average during 1 year is half of the number of tins produced in each run. A plant manager must take into account the cost to reset the machine (similar to the ordering cost in Activity 7) and the cost to store inventory. Although it might otherwise make sense to produce an entire year's inventory of popcorn tins at once, the cost of storing all the tins over a year's time would be prohibitive.

Suppose a company needs to produce 1.7 million popcorn tins over the course of a year. The cost to set up the machine for production is $1300, and the cost to store one tin for a year is approximately $1. What size production run will minimize set-up and storage costs? How many runs are needed during 1 year, and how often will the plant manager need to schedule a run of popcorn tins?

9. **Juice Cans**   Frozen juice cans are constructed with cardboard sides and metal top and bottom. A typical can holds 12 fluid ounces. If the metal costs twice as much per square inch as the cardboard, what dimensions will minimize the cost of a 12-fluid-ounce can?

10. **Booth Design**   As a sales representative, you are required to travel to trade shows and display your company's products. You need to design a display booth for such shows. Because your company generally must pay for the amount of square footage your booth requires, you want to limit the floor size to 300 square feet. The booth is to be 6 feet tall and three-sided, with the back of the booth a display board and the two sides of the booth made of gathered fabric. The display board for the back of the booth costs $30 per square foot. The fabric costs $2 per square foot and needs to be twice the length of the side to allow for gathering.

   a. Find the minimum cost of constructing a booth according to these specifications. What should be the dimensions of the booth?

   b. In order to accommodate your company's display, the back of the booth must be at least 15 feet wide. Does this restriction change the dimensions necessary to minimize cost? If so, what are the optimal dimensions now?

11. **Dog Run**   A dog kennel owner needs to build a dog run adjacent to one of the kennel cages. See the figure.

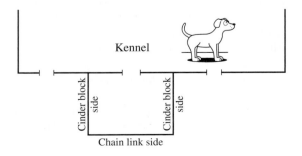

The ends of the run are to be cinder block, which costs 50 cents per square foot. The side is to be chain link, which costs $2.75 per square foot. Local regulations require that the walls and fence be at least 7 feet high and that the area measures no less than 120 square feet.

   a. Determine the dimensions of the dog run that will minimize cost but still meet the regulatory standards.

   b. Repeat part *a* for conditions where two identical dog runs are to be built side by side, sharing one cinder block wall.

   c. Do the answers to parts *a* and *b* change if there is a mandatory minimum width or length of 6 feet?

12. **Duplex Rent**   A set of 24 duplexes (48 units) was sold to a new owner. The previous owner charged $700 per month rent and had a 100% occupancy rate. The new owner, aware that this rental amount was well below other rental prices in the area, immediately raised the rent to $750. Over the next 6 months, an average of 47 units were occupied. The owner raised the rent to $800 and found that 2 of the units were unoccupied during the next 6 months.

   a. What was the monthly rental income made by the previous owner?

   b. After the first rent increase, what was the monthly income?

   c. After the second increase, how much did the new owner collect in rent each month?

   d. Find an equation for the monthly rental income as a function of the number of $50 increases in the rent.

   e. If the occupancy rate continues to decrease as the rent increases in the same manner as it did for the first two increases, what rent amount will maximize the owner's rental income from these duplexes? How much will the owner collect in rent each month at this rent amount?

   f. What other considerations besides rental income should the owner take into account when determining an optimal rent amount?

13. **Bus Trip**   A sorority plans a bus trip to the Great Mall of America during Thanksgiving break. The bus they charter seats 44 and charges a flat rate of $350 plus $35 per person. However, for every empty seat, the charge per person is increased by

$2. There is a minimum of 10 passengers. The sorority leadership decides that each person going on the trip will pay $35. The sorority itself will pay the flat rate and the additional amount above $35 per person. For example, if there are only 40 passengers, the fare per passenger is $43. Each of the 40 passengers will pay $35, and the sorority will pay $350 + 40($8) = $670.

a. Find a model for the revenue made by the bus company as a function of the number of passengers.

b. Find a model for the amount the sorority pays as a function of the number of passengers.

c. For what number of passengers will the bus company's revenue be greatest? For what number of passengers will the bus company's revenue be least?

d. For what number of passengers will the amount the sorority pays be greatest? For what number of passengers will the amount the sorority pays be least?

14. **Cable Line** A cable television company needs to run a cable line from its main line ending at point P in the figure to point H at the corner of the house. The county owns the roads shown in the figure, and it costs the cable company $25 per foot to run the line on poles along the county roads. The area bounded by the house and roads is a privately owned field, and the cable company must pay for an easement to run lines underground in the field. It is also more costly for the company to run lines underground than to run them on poles. The total cost to run the lines underground across the field is $52 per foot. The cable company has the choice of running the line along the roads or cutting across the field.

a. Calculate the cost for the company to run the line along the roads to point H.

b. Calculate the cost to run the line directly across the field from point P to point H.

c. Calculate the cost to run the line along the road for 50 feet from point P and then cut across the field.

d. Set up an equation for the cost to run the line along the road a distance of x feet from point P and then cut across the field. If $x = 0$, the line will cut directly across the field, and the value of the cost function should match your answer to part a.

e. Using calculus and your equation from part d, determine whether it is less costly for the company to cut across the field. If so, at what distance from point P should the company begin laying the line through the field?

15. **Costs** A trucking company wishes to determine the recommended highway speed for its truckers to drive in order to minimize the combined cost of driver's wages and fuel required for a trip. The average wage for the truckers is $15.50 per hour, and average fuel efficiencies for their trucks as a function of the speed at which the truck is driven are shown in the table.

Speed (mph)	50	55	60	65	70
Fuel consumption (mpg)	5.11	4.81	4.54	4.09	3.62

a. Find a model for fuel consumption as a function of the speed driven.

b. For a 400-mile trip, find formulas for the following quantities in terms of speed driven:

  i. Driving time required

  ii. Wages paid to the drive

  iii. Gallons of fuel used

  iv. Total cost of fuel (use a reasonable price per gallon based on current fuel prices)

  v. Combined cost of wages and fuel

c. Using equation v in part b, find the speed that should be driven in order to minimize cost.

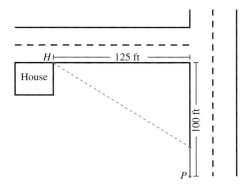

$H \vdash$———— 125 ft ————$\dashv$

House

100 ft

P

d. Repeat parts *b* and *c* for 700-mile and 2100-mile trips. What happens to the optimal speed as the trip mileage increases?

e. Repeat parts *b* and *c* for a 400-mile trip, increasing the cost per gallon of fuel by 20, 40, and 60 cents. What happens to the optimal speed as the cost of fuel increases?

f. Repeat parts *b* and *c* for a 400-mile trip, increasing the driver's wages by $2, $5, and $10 per hour. What happens to the optimal speed as the wages increase?

16. **Diskettes** A standard 3.5-inch computer diskette stores data on a circular sheet of mylar that is 3.36 inches in diameter. The disk is formatted by magnetic markers being placed on the disk to divide it into tracks (thin, circular rings). For design reasons, each track must contain the same number of bytes as the innermost track. If the innermost track is close to the center, more tracks will fit on the disk, but each track will contain a relatively small number of bytes. If the innermost track is near the edge of the disk, fewer tracks will fit on the disk, but each track will contain more bytes. For a standard double-sided, high-density 3.5-inch disk, there are 135 tracks per inch between the innermost track and the edge of the disk, and about 1400 bytes per inch of track. This type of disk holds 1,440,000 bytes of information when formatted.

a. For the upper disk in the accompanying figure, calculate the following quantities. The answers to parts *ii* and *iii* must be positive integers, and the rounded values should be used in the remaining calculations.

  i. Distance in inches around the innermost track

  ii. Number of bytes in the innermost track

  iii. Total number of tracks

  iv. Total number of bytes on one side of the disk

  v. Total number of bytes on both sides of the disk

b. Calculate the quantities in part *a* for the lower disk shown in the figure.

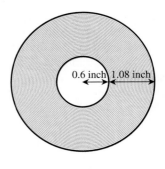

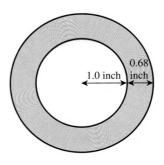

c. Repeat part *a* for a disk whose innermost track begins *r* inches from the center of the disk. Use the optimization techniques presented in this chapter to find the optimal distance the innermost track should be from the center of the disk in order to maximize the number of bytes stored on the disk. What is the corresponding number of bytes stored? How does this number compare to the total number of bytes stored by a standard double-sided, high-density 3.5-inch diskette?

d. If you are willing to destroy a used disk for the sake of curiosity, open a 3.5-inch disk and measure from the center to where the tracks begin. How does this measurement compare with the optimal distance you found in part *c*?

17. Explain how the problem-solving strategy presented in this section can be applied to situations in which data are given.

18. How can Step 2 of the problem-solving strategy help in building a model? How could it be misleading? Explain how you can keep from being misled while implementing Step 2.

5.5   Interconnected Change: Related Rates

We have seen that many situations in the world around us can be modeled mathematically and that because of this ability to model, we can use calculus to analyze changes taking place. So far we have considered how the rate of change of the output variable of a function is affected by a change in the input variable. We now consider the interaction of the rates of change of input and output variables with respect to a third variable. We shall also see that the interconnection of the input and output variables often reflects an interaction between their rates of change with respect to a third variable.

Concept Development: Interconnected-Change Equations   An equation relating the volume V of a spherical balloon to its radius r is $V = \frac{4}{3}\pi r^3$. When the balloon is inflated (over time), its volume increases and its radius increases. Note that even though V and r depend on the time, no time variable appears in the volume equation. In such applications, we refer to both volume V and radius r as **dependent variables** because their changes depend on a third variable, time t. We refer to time, the "with respect to" variable, as the **independent variable.**

In order to develop an equation relating the rate of change of the balloon's volume with respect to time and the rate of change of its radius with respect to time, we differentiate both sides of the equation with respect to time t. We use the Chain Rule to differentiate the right side of the volume equation, considering $\frac{4}{3}\pi r^3$ as the outside function and r as the inside function. The derivative of the inside function with respect to t is $\frac{dr}{dt}$. Thus we get

$$\frac{dV}{dt} = \left[\frac{4}{3}\pi(3r^2)\right]\frac{dr}{dt} = 4\pi r^2\frac{dr}{dt}$$

This equation shows how the rates of change of the volume and the radius of a sphere with respect to time are interconnected and can be used to answer questions about those rates. Such an equation is referred to as a **related-rates equation.** Example 1 demonstrates how to develop related-rates equations.

EXAMPLE 1   *Interconnecting Rates of Change*

Using each of the given equations, find an equation relating the indicated rates.

a. $p = 39x + 4$; relate $\frac{dp}{dt}$ with $\frac{dx}{dt}$.

b. $a = 4\ln t$; relate $\frac{da}{dx}$ with $\frac{dt}{dx}$.

c. $s = \pi r\sqrt{r^2 + h^2}$; relate $\frac{ds}{dx}$ with $\frac{dr}{dx}$, assuming that h is constant.

d. $v = \pi r^2 h$; relate $\frac{dr}{dt}$ with $\frac{dh}{dt}$, assuming that v is constant.

Solution

a. Differentiating the left side of $p = 39x + 4$ with respect to t gives $\frac{dp}{dt}$. We use the Chain Rule to differentiate the right side of the equation with respect to t, considering $39x + 4$ to be the outside function and x to be the inside function.

$$\frac{dp}{dt} = 39\frac{dx}{dt}$$

b. Applying the Chain Rule to the right side of $a = 4 \ln t$ yields

$$\frac{da}{dx} = 4\left(\frac{1}{t}\right)\frac{dt}{dx}$$

so

$$\frac{da}{dx} = \frac{4}{t}\frac{dt}{dx}$$

c. Differentiating the right side of $s = \pi r\sqrt{r^2 + h^2}$ requires both the Chain Rule and the Product Rule. Consider first the application of the Product Rule:

$$\frac{ds}{dt} = (\pi r)\left(\text{derivative of }\sqrt{r^2 + h^2}\text{ with respect to }x\right)$$
$$+ (\text{derivative of }\pi r\text{ with respect to }x)\left(\sqrt{r^2 + h^2}\right)$$

In order to calculate the two derivatives with respect to x, remember to apply the Chain Rule.

$$\frac{ds}{dx} = (\pi r)\left[\frac{1}{2}(r^2 + h^2)^{-1/2}(2r)\left(\frac{dr}{dx}\right)\right] + \left(\pi\frac{dr}{dx}\right)\left(\sqrt{r^2 + h^2}\right)$$

so

$$\frac{ds}{dx} = \pi\left(\frac{r}{\sqrt{r^2 + h^2}} + \sqrt{r^2 + h^2}\right)\frac{dr}{dx}$$

d. In order to develop a related-rates equation from $v = \pi r^2 h$ showing the interconnection between $\frac{dr}{dt}$ and $\frac{dh}{dt}$, we can isolate either r or h on one side and then find the derivative with respect to t, or we can find the derivative of both sides of the equation with respect to t and then solve for $\frac{dr}{dt}$. The two methods are equivalent, but sometimes one method has less involved algebra than the other method does. We choose the second method because it leads to the immediate removal of v from the equation and thus does not require substitution later.

Differentiating the left side of the equation with respect to t gives zero because v is constant. Applying the Product Rule and the Chain Rule to the right side of the equation gives

$$0 = (\pi r^2)\left(\frac{dh}{dt}\right) + 2\pi r\left(\frac{dr}{dt}\right)(h)$$

Now solve for $\frac{dr}{dt}$:

$$-2\pi rh\frac{dr}{dt} = \pi r^2\frac{dh}{dt}$$

$$-2rh\frac{dr}{dt} = r^2\frac{dh}{dt}$$

$$\frac{dr}{dt} = \frac{r^2}{-2rh}\frac{dh}{dt}$$

$$\frac{dr}{dt} = \frac{r}{-2h}\frac{dh}{dt} \quad \bullet$$

Today's news often brings us stories of environmental pollution in one form or another. One serious form of pollution is groundwater contamination. If a hazardous chemical is introduced into the ground, it can contaminate the groundwater and make the water source unusable. The contaminant will be carried downstream via the flow of the groundwater. This movement of contaminant as a consequence of the flow of the groundwater is known as *avection*. As a result of *diffusion,* the chemical will also spread out perpendicularly to the direction of flow. The region that the contamination covers is known as a chemical *plume.*

Hydrologists who study groundwater contamination are sometimes able to model the area of a plume as a function of the distance away from the source that the chemical has traveled by avection. That is, they develop an equation showing the interconnection between A, the area of the plume, and r, the distance from the source that the chemical has spread because of the flow of the groundwater.

If A is a function of r, then the rate of change $\frac{dA}{dr}$ describes how quickly the area of the plume is increasing as the chemical travels farther from the source. However, of greater interest is the rate of change of A with respect to time, $\frac{dA}{dt}$, and how it relates to the rate of change of r with respect to time, $\frac{dr}{dt}$. These relationships are examined in Example 2.

EXAMPLE 2   *Using a Given Equation in a Related-Rates Problem*

Groundwater   In a certain part of Michigan, a hazardous chemical leaked from an underground storage facility. Because of the terrain surrounding the storage facility, the groundwater was flowing almost due south at a rate of approximately 2 feet per day. Hydrologists studying this plume drilled wells in order to sample the groundwater in the area and determine the extent of the plume. They found that the shape of the plume was fairly easy to predict and that the area of the plume could be modeled as

$$A = 0.9604r^2 + 1.960r + 1.124 - \ln(0.980r + 1) \text{ square feet}$$

when the chemical had spread r feet south of the storage facility.

a. How quickly was the area of the plume growing when the chemical had traveled 3 miles south of the storage facility?

b. How much area had the plume covered when the chemical had spread 3 miles south?

Solution

a. First, note that the question posed asks for the rate of change of area with respect to time. However, the equation does not contain a variable representing time. Second, note that we are given the rate of change of r with respect to time. If we represent the time in days since the leak began as t, then we know that $\frac{dr}{dt} = 2$ feet per day and we are trying to find $\frac{dA}{dt}$. We also have an equation that shows the interconnection between the dependent variables A and r. Therefore, we differentiate with respect to t both sides of the equation that relates A and r.

$$\frac{dA}{dt} = 0.9604(2r)\frac{dr}{dt} + 1.960\frac{dr}{dt} + 0 - \left(\frac{1}{0.980r + 1}\right)\left(0.980\frac{dr}{dt}\right)$$

$$\frac{dA}{dt} = \left(1.9208r + 1.960 - \frac{0.980}{0.980r + 1}\right)\frac{dr}{dt}$$

Next, we substitute the known values $r = 3$ miles $= 15,840$ feet and $\frac{dr}{dt} = 2$ feet per day and then solve for the unknown rate.

$$\frac{dA}{dt} = \left(1.9208(15,840) + 1.960 - \frac{0.980}{0.980(15,840) + 1}\right)(2)$$

$$\approx 60,855 \text{ square feet per day}$$

When the chemical has spread 3 miles south, the plume is growing at a rate of 60,855 square feet per day.

b. The area of the plume is

$$A = 0.9604(15,840)^2 + 1.960(15,840) + 1.124 - \ln[0.980(15,840) + 1]$$

$$\approx 241,000,776 \text{ square feet}$$

$$\approx 8.6 \text{ square miles}$$

When the contamination has spread 3 miles south, the total contaminated area is approximately 8.6 square miles. ●

Note the method we used to answer the question posed in Example 2. First, we determined which variables were involved. Second, we identified an equation that connected the dependent variables. Third, we determined which rates of change were needed to relate to one another and took the derivative of each side of the equation with respect to the independent variable. Finally, we substituted given quantities and rates into the related-rates equation and solved for the unknown rate. We summarize this method:

Method of Related Rates

Step 1: Read carefully the problem and determine what variables are involved. Identify the independent variable (the "with respect to" variable) and all dependent variables.

Step 2: Use the given equation or find an equation relating the dependent variables. The independent variable may or may not appear in the equation.

Step 3: Differentiate both sides of the equation in Step 2 with respect to the independent variable to produce a related-rates equation. The Chain and/or the Product Rule(s) may be needed.

Step 4: Substitute the known quantities and rates into the related-rates equation, and solve for the unknown rate.

Step 5: Interpret in context the solution found in Step 4.

We illustrate this process in Example 3.

EXAMPLE 3   *Using Geometric Relationships in a Related-Rates Problem*

Baseball   A baseball diamond is a square with each side measuring 90 feet. A baseball team is participating in a publicity photo session, and a photographer at second base wants to photograph runners when they are halfway to first base. Suppose that the average speed at which a baseball player runs from home plate to first base is 20 feet per second. The photographer needs to set the shutter speed in terms of how fast the distance between the runner and the camera is changing. At what rate is the distance between the runner and second base changing when the runner is halfway to first base?

Solution

Step 1: The three variables involved in this problem are time, the distance between the runner and first base, and the distance between the runner and the photographer at second base. Because speed is the rate of change of distance with respect to time, the independent variable is time, and the two distances are the dependent variables.

Step 2: We need an equation that relates the distance between the runner and first base and the distance between the runner and the photographer at second base. A diagram can help us better understand the relationship between these distances. See Figure 5.26.

We consider the right triangle formed by the runner, first base, and second base. Using the Pythagorean Theorem, we know that the relationship between f and s in Figure 5.26 is

$$f^2 + 90^2 = s^2$$

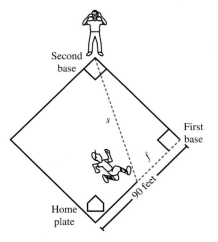

FIGURE 5.26

Step 3: Differentiating the equation in Step 2 with respect to t gives

$$2f \frac{df}{dt} + 0 = 2s \frac{ds}{dt}$$

$$\frac{ds}{dt} = \frac{f}{s} \frac{df}{dt}$$

Step 4: Note that the rate we need to find is $\frac{ds}{dt}$. We are told that $f = 45$ feet (half of the way to first base) and $\frac{df}{dt} = -20$ feet per second (negative because the distance between first base and the runner is decreasing). Substitute these values into the related-rates equation to obtain $\frac{ds}{dt} = \frac{45}{s}(-20)$. Find the value of s using the equation $f^2 + 90^2 = s^2$ and the fact that $f = 45$ feet. Thus $s = \sqrt{10{,}125} \approx 100.62$ feet, and we have

$$\frac{ds}{dt} = \frac{45 \text{ feet}}{\sqrt{10{,}125 \text{ feet}}}(-20 \text{ feet per second}) \approx -8.94 \text{ feet per second}$$

Step 5: When a runner is halfway to first base, the distance between the runner and the photographer is decreasing by about 8.9 feet per second. ●

Understanding that the change in one variable is connected to the change in another variable is important in a variety of applications. The independent variable is often—but not always—time. Being able to work with related rates can help you solve problems that occur in many real-world applications.

5.5 Concept Inventory

○ *Dependent and independent variables*
○ *Related-rates equation*
○ *Method of related rates*

5.5 Activities

For each of the equations in Activities 1 through 14, write the indicated related-rates equation.

1. $f = 3x$; relate $\frac{df}{dt}$ and $\frac{dx}{dt}$.

2. $p^2 = 5s + 2$; relate $\frac{dp}{dx}$ and $\frac{ds}{dx}$.

3. $k = 6x^2 + 7$; relate $\frac{dk}{dy}$ and $\frac{dx}{dy}$.

4. $y = 9x^3 + 12x^2 + 4x + 3$; relate $\frac{dy}{dt}$ and $\frac{dx}{dt}$.

5. $g = e^{3x}$; relate $\frac{dg}{dt}$ and $\frac{dx}{dt}$.

6. $g = e^{15x^2}$; relate $\frac{dg}{dt}$ and $\frac{dx}{dt}$.

7. $f = 62(1.02^x)$; relate $\frac{df}{dt}$ and $\frac{dx}{dt}$.

8. $p = 5 \ln(7 + s)$; relate $\frac{dp}{dx}$ and $\frac{ds}{dx}$.

9. $h = 6a \ln a$; relate $\frac{dh}{dy}$ and $\frac{da}{dy}$.

10. $v = \pi hw(x + w)$; relate $\frac{dv}{dt}$ and $\frac{dw}{dt}$, assuming that h and x are constant.

11. $s = \pi r \sqrt{r^2 + h^2}$; relate $\frac{ds}{dt}$ and $\frac{dh}{dt}$, assuming that r is constant.

12. $v = \frac{1}{3}\pi r^2 h$; relate $\frac{dh}{dt}$ and $\frac{dr}{dt}$, assuming that v is constant.

13. $s = \pi r \sqrt{r^2 + h^2}$; relate $\frac{dh}{dt}$ and $\frac{dr}{dt}$, assuming that s is constant.

14. $v = \pi h w(x + w)$; relate $\frac{dh}{dt}$ and $\frac{dw}{dt}$, assuming that v and x are constant.

15. **Trees**   Trees do a lot more than provide oxygen and shade. They also help pump water out of the ground and transpire it into the atmosphere. The amount of water an oak tree can remove from the ground is related to the tree's size. Suppose that a tree transpires

    $w = 31.54 + 12.97 \ln g$ gallons of water per day

    where g is the girth in feet of the tree trunk measured 5 feet above the ground. A tree is currently 5 feet in girth and is gaining 2 inches of girth per year.

    a. How much water does the tree currently transpire each day?

    b. If t is the time in years, find and interpret $\frac{dw}{dt}$.

16. **BMI**   The body-mass index of an individual who weighs w pounds and is h inches tall is given as

    $$B = \frac{0.45w}{0.00064516h^2} \text{ points}$$

    (Source: *New England Journal of Medicine,* September 14, 1995.)

    a. Write an equation showing the relationship between the body-mass index and weight of a woman who is 5 feet 8 inches tall.

    b. Find a related-rates equation showing the interconnection between the rates of change with respect to time of the weight and the body-mass index.

    c. Consider a woman who weighs 160 pounds and is 5 feet 8 inches tall. If $\frac{dw}{dt} = 1$ pound per month, find and interpret $\frac{dB}{dt}$.

    d. Suppose a woman who is 5 feet 8 inches tall has a body-mass index of 24 points. If her body-mass index is decreasing by 0.1 point per month, at what rate is her weight changing?

17. **BMI**   Refer to the body-mass index equation in Activity 16.

    a. Write an equation showing the relationship between the body-mass index and the height of a young teenager who weighs 100 pounds.

    b. Find a related-rates equation showing the interconnection between the rates of change with respect to time of the body-mass index and the height.

    c. If the weight of the teenager who is 5 feet 3 inches tall remains constant at 100 pounds while she is growing at a rate of $\frac{1}{2}$ inch per year, how quickly is her body-mass index changing?

18. **Heat Index**   The apparent temperature A in degrees Fahrenheit is related to the actual temperature $t°$F and the humidity $100h$% by the equation

    $$A = 2.70 + 0.885t - 78.7h + 1.20th \text{ degrees Fahrenheit}$$

    (Source: W. Bosch and L. G. Cobb, "Temperature Humidity Indices," UMAP Module 691, *UMAP Journal,* vol. 10, no. 3, Fall 1989, pp. 237–256.)

    a. If the humidity remains constant at 53% and the actual temperature is increasing from 80°F at a rate of 2°F per hour, what is the apparent temperature and how quickly is it changing with respect to time?

    b. If the actual temperature remains constant at 100°F and the relative humidity is 30% but is dropping by 2 percentage points per hour, what is the apparent temperature and how quickly is it changing with respect to time?

19. **Volume**   The lumber industry is interested in being able to calculate the volume of wood in a tree trunk. The volume of wood contained in the trunk of a certain fir has been modeled as

    $$V = 0.002198d^{1.739925}h^{1.133187} \text{ cubic feet}$$

    where d is the diameter in feet of the tree, measured 5 feet above the ground, and h is the height of the tree in feet.

    (Source: J. L. Clutter et al., *Timber Management: A Quantitative Approach.* New York: Wiley, 1983.)

    a. If the height of a tree is 32 feet and its diameter is 10 inches, how quickly is the volume of the wood changing when the tree's height is increasing by half a foot per year? (Assume that the tree's diameter remains constant.)

b. If the tree's diameter is 12 inches and its height is 34 feet, how quickly is the volume of the wood changing when the tree's diameter is increasing by 2 inches per year? (Assume that the tree's height remains constant.)

20. **Wheat Crop**   The carrying capacity of a crop is measured in the number of people for which it will provide. The carrying capacity of a certain wheat crop has been modeled as

$$K = \frac{11.56P}{D} \text{ people per hectare}$$

where P is the number of kilograms of wheat produced per hectare per year and D is the yearly energy requirement for one person in megajoules per person.
(Source: R. S. Loomis and D. J. Connor, *Crop Ecology: Productivity and Management in Agricultural Systems.* Cambridge, England: Cambridge University Press, 1982.)

a. Write an equation showing the yearly energy requirement of one person as a function of the production of the crop.

b. With time t as the independent variable, write a related-rates equation using your result from part *a*.

c. If the crop currently produces 10 kilograms of wheat per hectare per year and the yearly energy requirement for one person is increasing by 2 megajoules per year, find and interpret $\frac{dP}{dt}$.

21. **Production**   A Cobb-Douglas function for the production of mattresses is

$$M = 48.10352L^{0.6}K^{0.4} \text{ mattresses produced}$$

where L is measured in thousands of worker hours and K is the capital investment in thousands of dollars.

a. Write an equation showing labor as a function of capital.

b. Write the related-rates equation for the equation in part *a*, using time as the independent variable and assuming that mattress production remains constant.

c. If there are currently 8000 worker hours and if the capital investment is $47,000 and is increasing by $500 per year, how quickly must the number of worker hours be changing in order for mattress production to remain constant?

22. **Ladder**   A ladder 15 feet long leans against a tall stone wall. If the bottom of the ladder slides away from the building at a rate of 3 feet per second, how quickly is the ladder sliding down the wall when the top of the ladder is 6 feet from the ground? At what speed is the top of the ladder moving when it hits the ground?

23. **Height**   A hot-air balloon is taking off from the end zone of a football field. An observer is sitting at the other end of the field 100 yards away from the balloon. If the balloon is rising vertically at a rate of 2 feet per second, at what rate is the distance between the balloon and the observer changing when the balloon is 500 yards off the ground? How far is the balloon from the observer at this time?

24. **Kite**   A girl flying a kite holds the string 4 feet above ground level and lets out string at a rate of 2 feet per second as the kite moves horizontally at an altitude of 84 feet. Find the rate at which the kite is moving horizontally when 100 feet of string has been let out.

25. **Softball**   A softball diamond is a square with each side measuring 60 feet. Suppose a player is running from second base to third base at a rate of 22 feet per second. At what rate is the distance between the runner and home plate changing when the runner is halfway to third base? How far is the runner from home plate at this time?

26. **Volume**   Helium gas is being pumped into a spherical balloon at a rate of 5 cubic feet per minute. The pressure in the balloon remains constant.

a. What is the volume of the balloon when its diameter is 20 inches?

b. At what rate is the radius of the balloon changing when the diameter is 20 inches?

27. **Snowball**   A spherical snowball is melting, and its radius is decreasing at a constant rate. Its diameter decreased from 24 centimeters to 16 centimeters in 30 minutes.

a. What is the volume of the snowball when its radius is 10 centimeters?

b. How quickly is the volume of the snowball changing when its radius is 10 centimeters?

28. **Salt**   A leaking container of salt is sitting on a shelf in a kitchen cupboard. As salt leaks out of a hole in the side of the container, it forms a conical pile on

the counter below. As the salt falls onto the pile, it slides down the sides of the pile so that the pile's radius is always equal to its height. If the height of the pile is increasing at a rate of 0.2 inch per day, how quickly is the salt leaking out of the container when the pile is 2 inches tall? How much salt has leaked out of the container by this time?

29. **Yogurt** Soft-serve frozen yogurt is being dispensed into a waffle cone at a rate of 1 tablespoon per second. If the waffle cone has height $h = 15$ centimeters and radius $r = 2.5$ centimeters at the top, how quickly is the height of the yogurt in the cone rising when the height of the yogurt is 6 centimeters? (*Hint:* 1 cubic centimeter = 0.06 tablespoon and $r = \frac{h}{6}$.)

30. **Volume** Boyle's Law for gases states that when the mass of a gas remains constant, the pressure p and the volume v of the gas are related by the equation

$pv = c$, where c is a constant whose value depends on the gas. Assume that at a certain instant the volume of a gas is 75 cubic inches and its pressure is 30 pounds per square inch. Because of compression of volume, the pressure of the gas is increasing by 2 pounds per square inch every minute. At what rate is the volume changing at this instant?

31. Demonstrate that the two solution methods referred to in part d of Example 1 yield equivalent related-rates equations for the equation given in that part of the example.

32. In what fundamental aspect does the Method of Related Rates differ from the other rate-of-change applications seen so far in this text? Explain.

33. Which step of the Method of Related Rates do you consider to be most important? Support your answer.

SUMMARY

This chapter is devoted to analyzing change. The principal topics are approximating change, optimization, inflection points, and related rates.

Approximating Change

One of the most useful approximations of change in a function is to use the behavior of a tangent line to approximate the behavior of the function. Because of the Principle of Local Linearity, we know that tangent-line approximations are quite accurate over small intervals. We estimate the output $f(x + h)$ as $f(x) + f'(x) \cdot h$, where h represents the small change in input.

Optimization

The word *optimization* (as we used it in Section 5.2) refers to locating relative or absolute extreme points. A relative maximum is a point to which the graph rises and after which the graph falls. Similarly, a relative minimum is a point to which the graph falls and after which

the graph rises. There may be several relative maxima and relative minima on a graph. The highest and lowest points on a graph over an interval or over all possible input values are called the absolute maximum and absolute minimum points. These points may coincide with a relative maximum or minimum, or they may occur at the endpoints of a given interval.

Inflection Points

Inflection points are simply points where the concavity of the graph changes from concave up to concave down, or vice versa. Their importance, however, is that they identify the points of most rapid change or least rapid change in a region around the point.

Inflection points can be found where the graph of the second derivative of a function crosses the horizontal axis or, sometimes, where the second derivative fails to exist. In addition to locating input values of inflection points, the second derivative of a function can be used to determine the concavity of the function.

Related Rates

When the changes in one or more variables (called dependent variables) depend on a third variable (called the independent variable), a related-rates equation can be developed to show how the rates of change of these variables are interconnected. The Chain Rule plays an important role in the development of a related-rates equation, because the independent variable (which is often time) is not always expressed in the equation that relates the dependent variables.

CONCEPT CHECK

Can you

O   Use derivatives to approximate change?
O   Understand marginal analysis?
O   Find relative and absolute extreme points?
O   Find and interpret inflection points?
O   Understand the relationship between second derivatives and concavity?
O   Set up and solve applied optimization problems?
O   Set up and solve related-rates equations?

To practice, try

Section 5.1	Activity 17
Section 5.1	Activities 5, 19
Section 5.2	Activity 17
Section 5.3	Activities 3, 23
Section 5.3	Activity 31
Section 5.4	Activities 3, 9, 13
Section 5.5	Activities 19, 25

REVIEW TEST

1. **Tourists**   The number of tourists who visited Tahiti each year between 1988 and 1994 can be modeled by

$$T(x) = -0.4804x^4 + 6.635x^3 - 26.126x^2 + 26.981x + 134.848 \text{ thousand tourists}$$

x years after 1988.
(Source: Stephen J. Page, "The Pacific Islands," *EIU International Reports,* vol. 1, 1996, p. 91.)

   a.   Find any relative maxima and minima of $T(x)$ between $x = 0$ and $x = 6$. Explain how you found the value(s).

   b.   Find any inflection points of the graph of T between $x = 0$ and $x = 6$. Explain how you found the value(s).

   c.   Graph T, T', and T''. Clearly label on each graph the points corresponding to your answers to parts *a* and *b*.

   d.   Between 1988 and 1994, when was the number of tourists the greatest and when was it the least? What were the corresponding numbers of tourists in those years?

e. Between 1988 and 1994, when was the number of tourists increasing the most rapidly, and when was it declining the most rapidly? Give the rates of change in each of those years.

2. **Population**   Let $M(t)$ represent the population of French Polynesia (of which Tahiti is a part) at the middle of year t. If $M(2000) = 202$ thousand people, and if $M'(2000) = 4.5$ thousand people per year, estimate the following:
(Source: *Statistical Abstract*, 2001.)

   a. How much did the population of French Polynesia increase during the third quarter of 2001?

   b. What was the population at the end of 2001?

3. **Gas Pipe**   A natural gas company needs to run pipe from the point on shore marked A in the accompanying figure to the point marked B on the island. The island is 3.2 miles down shore from point A and 1.6 miles out to sea. It will cost the gas company \$27 per foot to lay pipe underground and \$143 per foot to lay pipe underwater. The company must decide the most economical distance to run the pipe underground before cutting across the water. Determine the optimal underground distance x and the corresponding total cost of laying the pipe to the island.

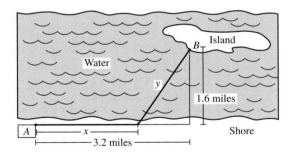

4. The graph of the derivative of a function h is shown.

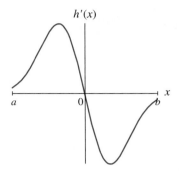

   a. Does the graph of h have a relative maximum and/or minimum between a and b? If so, mark on the derivative graph the input location of each extreme point, and classify the type of extreme point at that location. If not, explain why not.

   b. Does the graph of h have one or more inflection points between a and b? If so, mark and label on the derivative graph the input location(s) of the inflection point(s). If not, explain why not.

5. **Skid Marks**   The length of skid marks made on an asphalt road when a vehicle's brakes are applied quickly is given by

   $$S = 0.000013wv^2 \text{ feet}$$

   where the vehicle weighs w pounds and is traveling v mph when the brakes are applied. How quickly is the length of the skid marks changing when the velocity of a 4000-pound vehicle traveling at a speed of 60 mph is decreasing by 5 mph per second when the brakes are applied? (Assume that the vehicle's weight remains constant.)

Project 5.1 Hunting License Fees

Setting

In 1986, the state of California was trying to make a decision about raising the fee for a deer hunting license. Five hundred hunters were asked how much they would be willing to pay in excess of the current fee to hunt deer. The percentage of hunters to agree to a fee increase of x is given by the logistic model*

$$\text{Percentage} = \frac{1.221}{1 + 0.221e^{0.0116x}}$$

Suppose that in 1986 the license fee was $100, and 75,000 licenses were sold. Suppose that you are part of the 1986 Natural Resources Team presenting a proposed increase in the hunting license fee to the head of the California Department of Natural Resources.

Tasks

1. Illustrate how the model can be used by answering the following questions.

   a. What was the hunting license revenue in 1986?

   b. Suppose that in 1987 the fee increased to $150. What percentage of the 1986 hunters would buy another license? How many hunters is that? What would be the 1987 revenue?

   c. Repeat the questions in part *b* if the fee were increased to $300.

2. If the fee increase for 1987 were x dollars, find formulas for the following: (a) the percentage of hunters willing to pay the new fee, (b) the number of the 75,000 hunters willing to pay the new fee, (c) the new fee, and (d) the 1987 revenue.

3. Use your formulas to determine the optimal license fee. Also, find the optimal fee increase, the number of hunters who will buy licenses at the new fee, and the optimal revenue.

Reporting

1. Prepare a letter to the head of the California Department of Natural Resources. Your letter should address the fee increase and expectations for revenue. You should not make it technical but should give some support to back up your conclusions.

2. Prepare a technical written report outlining your findings as well as the mathematical methods you used to arrive at your conclusions.

*Based on information in *Journal of Environmental Economics and Management,* vol. 24, no. 1, January 1993.

Project 5.2 Fund-Raising Campaign

Setting

In order to raise funds, the mathematics department in your college or university is planning to sell T-shirts before next year's football game against the school's biggest rival. Your team has volunteered to conduct the fund raiser. Because several other students groups have also volunteered to head this project, your team is to present its proposal for the fund drive, as well as your predictions about its outcome, to a panel of mathematics faculty.

Task A

Follow the tasks for Project 2.2 on page 154. You will find a partial price listing for the T-shirt company on the *Calculus Concepts* CD-ROM and web site.

Task B

1. Review your work for Task A. If you wish to make any changes in your marketing scheme, you should do so now. If you decide to make any changes, make sure that the polling that was done is still applicable (for example, you will not be able to change your target market). Change (if necessary) any models from Task A to reflect any changes in your marketing scheme.

2. Use the models of demand, revenue, total cost, and profit developed in Task A to proceed with this section.

   Determine the selling price that generates maximum revenue. What is maximum revenue? Is the selling price that generates maximum revenue the same as the price that generates maximum profit? What is maximum profit? Which should you consider (maximum revenue or maximum profit) in order to get the best picture of the effectiveness of the drive? Re-evaluate the number of shirts you may wish to sell. Will this affect the cost you determined

above? If so, change your revenue, total cost, and profit functions to reflect this adjustment and re-analyze optimal values. Show and explain the mathematics that underlies your reasoning.

Discuss the sensitivity of the demand function to changes in price (check rates of change for $20, $14, and $8). Does the demand function have an inflection point? If so, find it. Find the rate of change of demand with respect to price at this point, and interpret its meaning and impact in this context. How would the sensitivity of the demand curve affect your decisions about raising or lowering your selling price?

On the basis of your findings, predict the optimal selling price, the number of T-shirts you intend to print, the costs involved, the number of T-shirts you expect to sell before realizing a profit (that is, the break-even point), and the expected profit.

Reporting

1. Write a report for the mathematics department concerning your proposed campaign. They will be interested in the business interpretation as well as in an accurate description of the mathematics involved. Make sure that you include graphical as well as mathematical representations of your demand, revenue, cost, and profit functions. (Include graphs of any functions and derivatives that you use. Include your calculations and your survey as appendices.) Do not forget to cover Task A in this report as well.

2. Make your proposal and present your findings to a panel of mathematics professors in a 15-minute presentation. Your presentation should be restricted to the business interpretation, and you should use overhead transparencies of graphs and equations of all models and derivatives as well as any other visual aids that you consider appropriate.

CHAPTER

6

Accumulating Change: Limits of Sums and the Definite Integral

Chapters 1 through 5 focused on the derivative, one of the two fundamental concepts of calculus. Now we begin a study of the second fundamental concept in calculus, the integral. As before, our approach is through the mathematics of change.

We start by analyzing the accumulated change in a quantity and how it is related to areas of regions between the graph of the rate-of-change function for that quantity and the horizontal axis. As we refine our thinking about area, we are led to consider limits of sums, which, in turn, show us how to account for the results of change in terms of integrals. Integrals, as we shall see, are intimately connected to derivatives by the Fundamental Theorem of Calculus.

We conclude by considering the difference of two accumulated changes and use integrals to calculate averages.

Concept Objectives

This chapter will help you understand the concepts of

O Accumulated change and area under a curve
O Area approximation techniques
O Definite integral as a limiting value
O Accumulation function
O Fundamental Theorem of Calculus
O Antiderivative
O Differences in accumulated change
O Average value

and you will learn to

O Approximate area using rectangles
O Interpret the area between a graph and the horizontal axis
O Approximate area using a limiting value of sums of areas of rectangles
O Sketch and interpret accumulation graphs
O Find simple general antiderivatives
O Find and interpret specific antiderivatives
O Recover a function from its rate-of-change equation
O Evaluate and interpret definite integrals
O Calculate and interpret the area between two curves
O Calculate and interpret the average value of a function and average rates of change

Getty Images

Concept Application If crude oil is flowing into a holding tank through a pipe, the rate at which the oil is flowing determines how quickly the holding tank will fill. It is possible that the rate varies with time and can be mathematically modeled. Such a model could then be used to answer questions such as

○ What is the change in the amount of oil in the tank during the first 10 minutes the oil is flowing into the tank?

○ If there were 5000 cubic feet of oil in the tank before the oil began flowing into the tank, how much oil is in the tank after 10 minutes?

○ How long can oil flow into the tank before the tank is full?

Questions such as these can be answered using definite integrals. Examples of this type of problem appear in Questions 1 and 4 of the Chapter 6 Review Test.

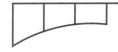

6.1   Results of Change and Area Approximations

In our study of calculus so far, we have concentrated on finding rates of change. We now consider the results of change.

Accumulated Change

Suppose that you have been driving on an interstate highway for 2 hours at a constant speed of 60 mph. Because velocity, v, is the rate of change of distance traveled, s, with respect to time, we write a function for velocity mathematically as $v(t) = s'(t) = 60$ mph, where t is the time in hours ($0 \leq t \leq 2$). A graph of this rate function over a 2-hour period of time appears in Figure 6.1.

At this constant rate, the distance traveled during a time period of t hours is (rate)(time) $= (60 \text{ mph})(t \text{ hours}) = 60t$ miles. Geometrically, we view this multiplication as giving the area of the region between the rate-of-change graph and the horizontal axis over any time period of length t hours. Figure 6.2 illustrates this fact for the 1-hour time period between 0.5 hour and 1.5 hours, the 15-minute time period between 1 hour and 45 minutes and 2 hours, and the first t hours of the trip.

FIGURE 6.1

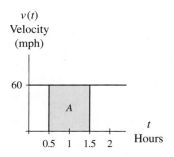

(a) Area $A = (60 \text{ mph})(1 \text{ hour})$ $= 60$ miles is the distance traveled during 1 hour

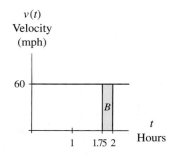

(b) Area $B = (60 \text{ mph})(0.25$ hour$) = 15$ miles is the distance traveled during 15 minutes

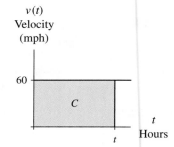

(c) Area $C = (60 \text{ mph})(t \text{ hours})$ $= 60t$ miles is the distance traveled during t hours

FIGURE 6.2

Now, imagine that after a 2-hour drive at a constant speed of 60 mph, you increase your speed at a constant rate to 75 mph over a 10-minute interval and then maintain that constant 75-mph speed during the next half hour. A graph of your speed appears in Figure 6.3a.

We know that the distance traveled during the times when speed is constant is the speed multiplied by the amount of time driven at that speed (represented by regions R_1 and R_3 in Figure 6.3b). But how can we calculate the distance driven between 2 hours and 2 hours 10 minutes when the speed is increasing linearly? If we knew the average speed over the 10-minute interval, then we could multiply that average speed by $\frac{1}{6}$ of an hour to obtain the distance traveled.

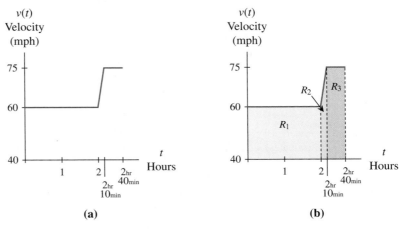

(a)   **(b)**

FIGURE 6.3

In this case, the average speed is simply the average of the beginning and ending speeds during the 10-minute interval. Thus we have

$$\begin{array}{l} \text{Distance traveled} \\ \text{between 2 hours} \\ \text{and 2 hours} \\ \text{10 minutes} \end{array} = \left(\begin{array}{c}\text{average}\\\text{speed}\end{array}\right)(\text{time}) = \left(\frac{60\text{ mph} + 75\text{ mph}}{2}\right)\left(\frac{1}{6}\text{ hour}\right)$$

This distance is, in fact, the area of the trapezoid $\left[\text{Area} = \left(\frac{\text{side 1} + \text{side 2}}{2}\right)(\text{base})\right]$ labeled R_2 in Figure 6.3b. Thus, the distance traveled during the 2-hour 40-minute interval is the area of the region beneath the velocity graph and above the time axis calculated as

$$\begin{aligned} \begin{array}{c}\text{Distance}\\\text{traveled}\end{array} &= \text{area of region } R_1 + \text{area of region } R_2 + \text{area of region } R_3 \\ &= (60\text{ mph})(2\text{ hr}) + \left(\frac{60\text{ mph} + 75\text{ mph}}{2}\right)\left(\frac{1}{6}\text{ hr}\right) + (75\text{ mph})\left(\frac{1}{2}\text{ hr}\right) \\ &= 120\text{ miles} + 11.25\text{ miles} + 37.5\text{ miles} \\ &= 168.75\text{ miles} \end{aligned}$$

Once more, the change in the distance traveled is given by the area of the region between the rate-of-change graph and the horizontal axis.

EXAMPLE 1   *Relating Area to Accumulated Change*

Draining Water   A water tank drains at a rate of $r(t) = -2t$ gallons per minute t minutes after the water began draining. The graph of the rate-of-change function is shown in Figure 6.4.

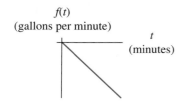

FIGURE 6.4

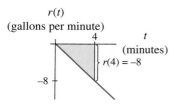

FIGURE 6.5

By *signed area* we mean the area of the region with a negative sign in front of the area value to indicate that the region lies below the input axis.

a. What are the units on height, width, and area of the region between the time axis and the rate graph?

b. Determine the change in the volume of water in the tank during the first 4 minutes the tank was draining.

Solution

a. The height corresponds to the output units, which are gallons per minute. The width is time (in minutes). In the calculation of area, the height and width are multiplied, giving area in gallons.

b. To find the change in the volume of the water, we find the area of the region between the time axis and the rate graph. This area is shaded in Figure 6.5. The region is a triangle with base of length 4. The height is determined by the function value at $t = 4$. Note that the function value is negative: $r(4) = -8$. The negative sign is simply an indication that the rate-of-change graph lies below the horizontal axis. Although we know that height is a positive measure, we use this signed value for height to remind us that the region lies beneath the horizontal axis and represents a decrease in the amount of water in the tank. The signed area of the region is

$$\frac{1}{2} (4 \text{ minutes})(-8 \text{ gallons per minute}) = -16 \text{ gallons}$$

Thus, during the first 4 minutes the tank was draining, the volume changed by -16 gallons. In other words, the tank lost 16 gallons of water. ●

The previous velocity examples illustrate the following principle:

Results of Change

The accumulated change in a quantity is represented as the area or signed area of a region between the rate-of-change function for that quantity and the horizontal axis provided the function does not cross the horizontal axis.

Left- and Right-Rectangle Approximations

The preceding velocity examples were carefully chosen so that their rate-of-change graphs were easy to obtain and the areas of the desired regions were easy to calculate. But most real-life situations are not so simple. Indeed, there are two issues that we must face:

1. Obtaining the rate-of-change function for the quantity of interest

2. Calculating the area of the desired region between the rate-of-change function and the horizontal axis

In many cases, we must resort to approximating both the rate-of-change function and the desired area. Consider the example of a store manager of a large department store who wishes to estimate the number of customers who came to a Saturday sale from 9 A.M. to 9 P.M. The manager stands by the entrance for 1-minute intervals at different times throughout the day and counts the number of people entering the store. He uses these data as an estimate of the number of customers who enter the store each minute. The manager's data may look something like Table 6.1.

TABLE 6.1

Time	Customers per minute	Time	Customers per minute
9:00 A.M.	1	2:30 P.M.	5
9:45 A.M.	2	3:10 P.M.	5
10:15 A.M.	3	4:00 P.M.	4
11:00 A.M.	4	5:15 P.M.	4
11:45 A.M.	4	6:30 P.M.	3
12:15 P.M.	5	7:30 P.M.	2
1:15 P.M.	5	8:15 P.M.	2

The number of customers who attended the sale can be calculated by summing the number of customers who entered the store during each hour for every hour of the 12-hour sale. We do not have enough information to determine the exact number of customers who entered the store each hour; however, we can estimate the number by using a continuous model for the customers-per-minute data. To build a model for the rate-of-change data, we choose to convert each of the above observation times to minutes after 9:00 A.M. Thus $m = 0$ at 9:00 A.M. and $m = 675$ at 8:15 P.M. A model for the customers-per-minute data is

$$c(m) = (4.58904 \cdot 10^{-8})m^3 - (7.78127 \cdot 10^{-5})m^2 + 0.03303m + 0.88763$$
customers per minute

where m is the number of minutes after 9:00 A.M. The graph of this model is shown in Figure 6.6. Note that c is a continuous function modeling a discrete situation.

Concept Development: Left-Rectangle Approximation
To estimate the number of customers who attended the sale, we use the model to estimate the number of customers per minute entering the store at the beginning of each hour and multiply by 60 minutes to estimate the number of customers entering the store during that hour. Summing the estimates for each of the 12 hours results in the estimate we desire for the total number of customers. This process is the same as drawing a set of 12 rectangles under the graph of c, one for each hour, and using the sum of their areas to estimate the total number of customers who came to the sale. (See Figure 6.7.)

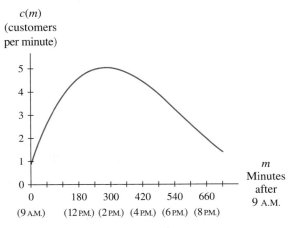

FIGURE 6.6

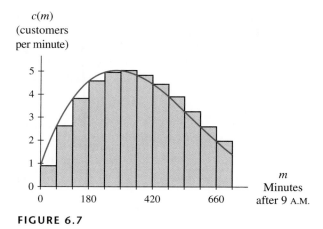

FIGURE 6.7

6.1.1

In Figure 6.7, the height of each rectangle is the function *c* evaluated at the left endpoint of the base of the rectangle. It is for this reason that we call these rectangles **left rectangles.** Because we are using 12 rectangles of equal width to span the 12-hour (720-minute) sale, the width of each rectangle is 720 ÷ 12 = 60 minutes. We use Table 6.2 to keep track of the areas that we are summing.

TABLE 6.2

Left endpoint of rectangle *m*	Height of rectangle *c(m)* (customers per minute)	Width of rectangle (minutes)	Area of rectangle* = height·width [(customers/min)(min) → customers]
9 A.M.	$c(0) \approx 0.9$	60	53.3
10 A.M.	$c(60) \approx 2.6$	60	156.0
11 A.M.	$c(120) \approx 3.8$	60	228.6
noon	$c(180) \approx 4.6$	60	274.8
1 P.M.	$c(240) \approx 5.0$	60	298.1
2 P.M.	$c(300) \approx 5.0$	60	302.0
3 P.M.	$c(360) \approx 4.8$	60	290.2
4 P.M.	$c(420) \approx 4.4$	60	266.1
5 P.M.	$c(480) \approx 3.9$	60	233.4
6 P.M.	$c(540) \approx 3.3$	60	195.7
7 P.M.	$c(600) \approx 2.6$	60	156.4
8 P.M.	$c(660) \approx 2.0$	60	119.2
		Total area of rectangles ≈ 2574 customers	

*These values were obtained using the unrounded model.

Thus, using 12 rectangles, we estimate that 2574 customers came to the Saturday sale.

Note the importance in this example of measuring time in minutes. If time is measured in hours, then the area calculated by multiplying height (measured in customers per minute) by width (measured in hours) is not the number of customers. Make sure the units correspond so that the result of their multiplication gives the desired units. Also note that our choice of using time intervals of 1 hour was arbitrary.

The previous discussion illustrates a way to approximate accumulation using left-rectangle areas. In some situations, choosing rectangles whose heights are measured at the right endpoint of the base of each rectangle may give more reasonable area approximations. Such rectangles are called **right rectangles.** The use of such rectangles is illustrated in the following example.

EXAMPLE 2   *Using Right Rectangles to Approximate Change*

6.1.2

Medicine   A pharmaceutical company has tested the absorption rate of a drug that is given in 20-milligram doses for 20 days. Researchers have modeled the rate of change of the concentration of the drug, measured in micrograms per milliliter per day (μg/mL/day), in the bloodstream as

$$r(x) = \begin{cases} 1.708(0.845^x) \ \mu\text{g/mL/day} & \text{when } 0 \le x \le 20 \\ -10.058 + 2.94 \ln x \ \mu\text{g/mL/day} & \text{when } 20 < x \le 30 \end{cases}$$

where x is the number of days after the drug is first administered. Figure 6.8 illustrates a graph of this rate-of-change function.

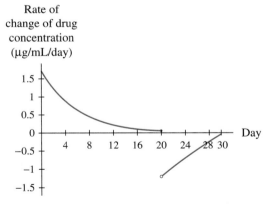

FIGURE 6.8

a.   Use the model and right rectangles of width 2 days to estimate the change in the drug concentration from the beginning of day 1 through the end of day 20.

b.   Use the model and right rectangles of width 2 days to estimate the change in the drug concentration from the beginning of day 21 through the end of day 30.

c.   Combine your answers to parts *c* and *d* to estimate the change in the drug concentration from the beginning of day 1 through the end of day 30.

Solution

a. To determine the change in the drug concentration from the beginning of day 1 ($x = 0$) through the end of day 20 ($x = 20$), we use the exponential portion of the model and 10 right rectangles as shown in Figure 6.9 and Table 6.3.

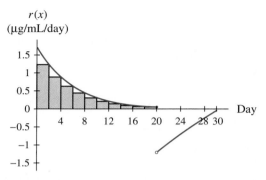

FIGURE 6.9

TABLE 6.3

Right endpoint of rectangle x	Height of rectangle $r(x)$ (μg/mL/day)	Area of rectangle = height·width [(μg/mL/day)(days) → μg/mL]
2	1.220	2.439
4	0.871	1.742
6	0.622	1.244
8	0.444	0.888
10	0.317	0.634
12	0.226	0.453
14	0.162	0.323
16	0.115	0.231
18	0.082	0.165
20	0.059	0.118
Sum of areas of rectangles ≈ 8.235 μg/mL Change in the drug concentration ≈ 8.235 μg/mL		

From the beginning of day 1 through the end of day 20, the drug concentration increased by approximately 8.235 micrograms per milliliter.

b. To determine the change in the concentration from the beginning of day 21 through the end of day 30, we use the log portion of the model and five right rectangles as shown in Figure 6.10 and Table 6.4.

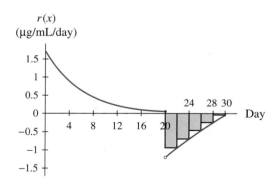

FIGURE 6.10

TABLE 6.4

Right endpoint of rectangle x	Signed height of rectangle $r(x)$ (μg/mL/day)	Signed area of rectangles = height·width [(μg/mL/day)(days) → μg/mL]
22	-0.970	-1.941
24	-0.715	-1.429
26	-0.479	-0.958
28	-0.261	-0.523
30	-0.059	-0.117
Sum of signed areas of rectangles ≈ -4.968 μg/mL Change in the drug concentration ≈ -4.968 μg/mL		

From the beginning of day 21 through the end of day 30, the drug concentration decreased by approximately 4.968 μg/mL.

c. To determine the change in concentration from the beginning of day 1 through the end of day 30, we need only subtract the amount of decline from the amount of increase.

   8.235 μg/mL − 4.968 μg/mL = 3.267 μg/mL

The drug concentration increased by approximately 3.267 μg/mL from the beginning of day 1 through the end of day 30.   ●

Part c of Example 2 illustrates the results of change using signed areas. In general, if a function f gives the rate of change of a function F and if the function f is sometimes positive and sometimes negative between inputs a and b, then the accumulated (or net) change in F is equal to the area of the region lying under the graph of f and above the x-axis minus the area of the region lying above the graph of f and below the x-axis. (See Figure 6.11.) In other words, the accumulated change in a quantity is equal to the sum of the signed areas of the regions between the rate-of-change function for that quantity and the horizontal axis over the interval from a to b.

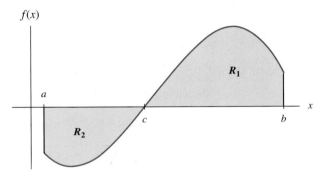

FIGURE 6.11   Accumulated change = area of R_1 − area of R_2 = sum of signed areas

Midpoint-Rectangle Approximation

It is often true that if a left-rectangle approximation is an overapproximation, then a right-rectangle area will be an underapproximation, and vice versa. For example, consider using four left and four right rectangles to approximate the area of the region between the function $f(x) = \sqrt{1 - x^2}$ and the x-axis between $x = 0$ and $x = 1$. Figure 6.12 shows the rectangles and the approximate areas.

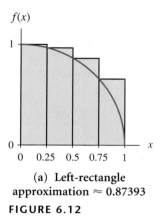

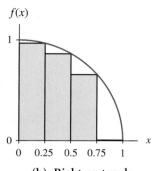

(a) **Left-rectangle**
approximation ≈ 0.87393

(b) **Right-rectangle**
approximation ≈ 0.62393

FIGURE 6.12

Note from Figure 6.12 that the left-rectangle approximation is an overestimate and the right-rectangle approximation is an underestimate.

We next consider approximating area using a third type of rectangle. The **midpoint-rectangle approximation** uses rectangles whose heights are calculated at the midpoints of the subintervals. Midpoint-rectangle approximation is illustrated in Example 3.

EXAMPLE 3   *Using Midpoint Rectangles to Approximate Change*

6.1.3a, b

Consider again the region between the function $f(x) = \sqrt{1 - x^2}$ and the x-axis from $x = 0$ to $x = 1$. Use four midpoint rectangles to approximate the area of this region.

Solution   Table 6.5 shows the calculations for the areas of the midpoint rectangles shown in Figure 6.13.

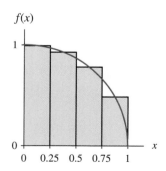

FIGURE 6.13

TABLE 6.5

Rectangle number	Midpoint of interval	Height of rectangle	Width of rectangle	Area = height·width
1	0.125	$f(0.125) \approx 0.99216$	0.25	0.24804
2	0.375	$f(0.375) \approx 0.92702$	0.25	0.23176
3	0.625	$f(0.625) \approx 0.78062$	0.25	0.19516
4	0.875	$f(0.875) \approx 0.48412$	0.25	0.12103
			Total midpoint area ≈ 0.79598	

The area of the region is approximately 0.79598 (using four midpoint rectangles).   ●

The region whose area we are approximating is the interior of a quarter-circle with radius 1, so the true area is $\frac{1}{4}(\pi \cdot \text{radius}^2) = \frac{\pi}{4}(1)^2 \approx 0.78540$. The midpoint-rectangle approximation is much closer to the actual area than are the two other approximations. This is often the case.

Working with Count Data

Frequently, rate-of-change data are not available for a quantity; instead, we have **count data** (that is, totals reported at the end of a period of time) pertaining to that quantity. Table 6.6 shows the number of aluminum cans that were recycled each year from 1978 through 1988.

TABLE 6.6

Year	Cans each year (billions)	Year	Cans each year (billions)
1978	8.0	1984	31.9
1979	8.5	1985	33.1
1980	14.8	1986	33.3
1981	24.9	1987	36.6
1982	28.3	1988	42.0
1983	29.4		

(Source: Data from the Aluminum Association, Inc.)

The number of cans recycled during 1980 can be interpreted geometrically as the area of a rectangle. We use a right rectangle because the data are reported at the end of each year. (See Figure 6.14.)

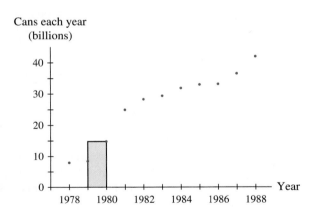

FIGURE 6.14

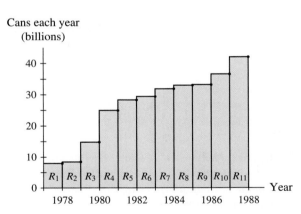

FIGURE 6.15

In other words, the number of cans recycled during 1980 is (height)(width) = (14.8 billion cans each year)(1 year) = 14.8 billion cans. If we consider the number of cans recycled during a given year as the area of a rectangle, then the number of recycled aluminum cans from the beginning of 1978 through the end of 1988 is the sum of the areas of the rectangles shown in Figure 6.15.

Number of cans recycled = the sum of the areas of the rectangles

$$= 8.0 + 8.5 + 14.8 + 24.9 + 28.3 + 29.4 + 31.9$$
$$+ 33.1 + 33.3 + 36.6 + 42.0$$

$$= 290.8 \text{ billion cans}$$

Note that because each rectangle has width 1 year, the area is simply the sum of the output data.

If some count data were missing or unequally spaced, we might be able to adjust the equal rectangle width or find a model for the data and then estimate the accumulated change by using values determined from the equation and rectangles with width 1 year.

Much of the data recorded in statistical tables and abstracts is count data. As the aluminum can illustration indicates, calculus is not necessary to find the accumulated change in the recorded quantity. In fact, although calculus could be used to approximate an accumulated change, it is a sum of discrete data values that gives the actual change.

Count Data

Do not confuse count data with rate-of-change data. When working with rate-of-change data, accumulated change is calculated using areas. When working with count data, accumulated change is calculated by summing the output data.

The most important concept in this section is that accumulated change in a quantity can be found by finding areas between rate-of-change graphs and the horizontal axis. Sometimes we can calculate areas using geometric formulas. At other times, we approximate area using sums of areas of rectangles. The numerical calculations are often tedious and result in only approximations, but later in this chapter we will discover a valuable calculus tool that enables us to find area and determine the accumulated change in a quantity accurately and quickly.

6.1 Concept Inventory

○ *Area or signed area of a region between a rate-of-change function and the horizontal axis between a and b = accumulated change in the amount function between a and b*

○ *Left-rectangle approximation*

○ *Right-rectangle approximation*

○ *Midpoint-rectangle approximation*

○ *Count data*

6.1 Activities

1. **Bacteria**   The growth rate of bacteria (in thousands per hour) in milk at room temperature is $B(t)$, where t is the number of hours that the milk has been at room temperature. We wish to use rectangles to estimate the area of the region between a graph of B and the t-axis. What are the units on

   a. the heights of the rectangles?

   b. the widths of the rectangles?

   c. the areas of the rectangles?

   d. the area of the region between the graph of B and the t-axis?

   e. the accumulated change in the number of bacteria in the milk during the first hour that the milk is at room temperature?

2. **Road Test**   The acceleration of a car (in feet per second per second) during a test conducted by a car manufacturer is given by $A(t)$, where t is the number of seconds since the beginning of the test.

   a. What does the area of the region between the portion of the graph of A lying above the t-axis and the t-axis tell us about the car?

   b. What are the units on

      i. the heights and widths of rectangles used to estimate area?

      ii. the area of the region between the graph of A and the t-axis?

3. **Braking**   The distance required for a car to stop is a function of the speed of the car when the brakes are applied. The rate of change of the stopping distance could be expressed in feet per mile per hour, where the input is the speed of the car, in miles per hour, when the brakes are applied.

   a. What does the area of the region between the rate-of-change graph and the input axis from 40 mph to 60 mph tell us about the car?

   b. What are the units on

      i. the heights and widths of rectangles used to estimate the area in part a?

      ii. the area in part a?

4. **Emissions**   The atmospheric concentration of CO_2 is growing exponentially. If the growth rate in ppm per year is $C(t)$, where t is the number of years since 1980, what are the units on

   a. the area of the region between the graph of C and the t-axis from $t = 0$ to $t = 20$?

   b. the heights and widths of rectangles used to estimate the area in part a?

   c. the change in the CO_2 concentration from 1980 through 2000?

5. The graph of a function g is shown.

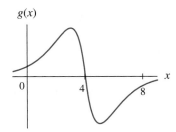

   a. Discuss how to approximate the area of the region between the graph of g and the x-axis between $x = 0$ and $x = 8$ with eight right rectangles that have the same width. Copy the graph, and draw the rectangles on the figure.

   b. Repeat part a using eight left rectangles.

6. **Stock Value**   On January 4, 2000, DuPont stock was worth $65 per share.

   a. Write a function for the value of x shares of stock. Graph this function. Note that this is a continuous model representing a discrete situation.

   b. Write the function for the rate of change of the continuous model for the value of DuPont stock with respect to the number of shares held. Graph this rate-of-change function.

   c. Find the change in the value of stock held if the number of shares held is increased from 250 to 300 shares. Depict this change as the area of a region on the rate-of-change graph.

7. The graph of a function f is shown.

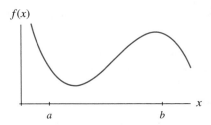

   Discuss how to approximate the area of the region beneath the graph of f from $x = a$ to $x = b$ with four midpoint rectangles that have the same width. Draw the rectangles.

8. Approximate the area of the region beneath the graph of $f(x) = e^{-x^2}$ from $x = -1$ to $x = 1$ using four left rectangles, right rectangles, and midpoint rectangles.

   a. In each case,

      i. sketch the graph of f from $x = -1$ to $x = 1$.

      ii. label the points on the x-axis, and draw the rectangles.

      iii. calculate the approximating areas.

   b. Proceed as in part a to approximate the area of the region beneath the graph of f from $x = -1$ to

$x = 1$ using eight left rectangles, eight right rectangles, and eight midpoint rectangles.

c. The area, to nine decimal places, of the region beneath the graph of $f(x) = e^{-x^2}$ is 1.493648266. Which approximation is the most accurate?

9. **Production** The graph shows two estimates, labeled A and B, of oil production rates (in billions of barrels per year).

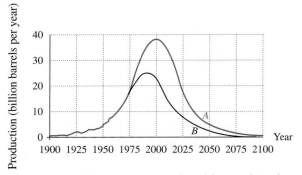

(Source: The figure for Activity 9 was adapted from *Ecology of Natural Resources* by François Ramade. Copyright 1984 by John Wiley & Sons, Inc. Reprinted by permission of the publisher.)

a. Use midpoint rectangles of width 25 years to estimate the total amount of oil produced from 1900 through 2100 using graph A.

b. Repeat part a for graph B.

c. On page 31 of *Ecology of Natural Resources,* the total oil production is estimated from graph A to be 2100 billion barrels and from graph B to be 1350 billion barrels. How close were your estimates?

d. What can you conclude from the graph about the future of oil production?

10. **Climate** Scientists have long been interested in studying global climatological changes and the effect of such changes on many aspects of the environment. From carefully controlled experiments, two scientists constructed a model to simulate daily snow depth in a region of the Northwest Territories in Canada. Rates of change (in equivalent centimeters of water per day) estimated from their model are shown as a scatter plot in the accompanying figure. Appropriate models have been sketched on the scatter plot. Note that both vertical and horizontal scales change after June 9.

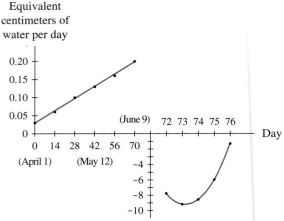

(Source: R. G. Gallimore and J. E. Kutzbach, "Role of Orbitally Induced Changes in Tundra Area in the Onset of Glaciation," *Nature,* vol. 381, June 6, 1996, pp. 503–505)

a. What does the figure indicate occurred between June 9 and June 11?

b. Estimate the area of the region beneath the curve from April 1 through June 9. Interpret your answer.

c. Use four midpoint rectangles to estimate the area of the region from June 11 through June 15. Interpret your answer.

11. **Robot Speed** A mechanical engineering graduate student designed a robot and is testing the ability of the robot to accelerate, decelerate, and maintain speed. The robot takes 1 minute to accelerate to 10 mph (880 feet per minute). The robot maintains that speed for 2 minutes and then takes half a minute to come to a complete stop. Assume that this robot's acceleration and deceleration are constant.

a. Draw a graph of the robot's speed during the experiment.

b. Find the area of the region between the graph in part a and the horizontal axis.

c. What is the practical interpretation of the area found in part b?

12. **Expanding Gas** A certain gas expands as it is heated. The accompanying figure shows the rate of expansion of the gas measured at several temperatures.

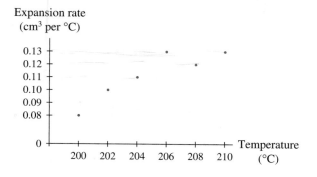

Expansion rate
(cm³ per °C)

Temperature
(°C)

a. Use only the data and left rectangles to estimate by how much the gas expanded as it was heated from 200°C to 210°C. Sketch the rectangles on the scatter plot. Do you believe your approximation is an overestimate or an underestimate? Explain.

b. Repeat part *a* using right rectangles.

13. **Bike Race**   An athlete is training for a mountain bike race. Her coach clocks her speed every 10 minutes during an hour-long ride. The speeds are shown.

Time (minutes)	0	10	20	30	40	50	60
Speed (mph)	22	18	20	23	15	17	12

a. Use only the data and six left rectangles to estimate the distance traveled by the cyclist during the 1-hour ride. Sketch a scatter plot of the data, and draw the rectangles on the scatter plot.

b. Repeat part *a* using right rectangles.

14. **Energy Use**   The graph shows the energy usage in megawatts for one day for a large university campus. The daily energy consumption for the campus is measured in megawatt-hours and is found by calculating the area of the region between the graph and the horizontal axis.

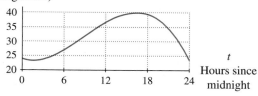

m(t)
Energy power usage
(megawatts)

t
Hours since
midnight

a. Estimate the daily energy consumption using eight left rectangles.

b. Estimate the daily energy consumption using eight right rectangles.

c. Discuss whether the estimates in parts *b* and *c* are overestimates or underestimates of the actual daily energy consumption.

15. **Population**   The rate of change of the population of North Dakota from 1970 through 1990 can be modeled as

$$p(t) = \begin{cases} 3.87 \text{ thousand} \\ \quad \text{people per year} \quad \text{when } 0 \le t < 15 \\ -7.39 \text{ thousand} \\ \quad \text{people per year} \quad \text{when } 15 \le t \le 20 \end{cases}$$

where *t* represents the number of years since 1970.
(Source: Based on data from *Statistical Abstract*, 1994.)

a. Sketch a graph of the rate-of-change function.

b. Find the area of the region between the graph of *p* and the horizontal axis from 0 to 15. Interpret your answer.

c. Find the area of the region between the graph of *p* and the horizontal axis from 15 to 20. Interpret your answer.

d. Was the population of North Dakota in 1990 greater or less than the population in 1970? By how much did the population change between 1970 and 1990?

e. What information would you need to determine the population of North Dakota in 1990?

16. **Cottage Cheese**   The rate of change of the per capita consumption of cottage cheese in the United States between 1980 and 1996 can be modeled by the function

$$c(t) = \begin{cases} -0.01t - 0.058 \text{ pounds} \\ \quad \text{per person per year} \quad \text{when } 0 \le t < 13 \\ -0.1 \text{ pound per person} \\ \quad \text{per year} \quad \text{when } 13 \le t \le 15 \\ 0 \text{ pound per person per year} \quad \text{when } 15 < t \le 19 \end{cases}$$

where *t* is the number of years since 1980.
(Source: Based on data from *Statistical Abstract*, 2001.)

a. Sketch a graph of the rate-of-change function.

b. Find the area of the region between the graph and the horizontal axis between *t* = 0 and *t* = 19.

c. Interpret the area in part *b* in the context of cottage cheese consumption.

d. Can you determine the per capita cottage cheese consumption in 1996? In 1999? Why or why not?

17. **Hospital Stay** The rate of change of the length of the average hospital stay between 1993 and 2000 can be modeled by the equation

$$s(t) = \begin{cases} 0.082t - 0.39 \text{ days per year} \\ \qquad\qquad\qquad\qquad \text{when } 0 \le t < 5 \\ -0.1t \text{ days per year} \qquad \text{when } 5 \le t \le 7 \end{cases}$$

where *t* is the number of years since 1993.
(Source: Based on data from *Statistical Abstract,* 2001.)

a. Graph *s* for the years between 1993 and 2000.

b. Judging on the basis of the graph, was the length of the average hospital stay increasing or decreasing between 1993 and 2000?

c. Find the area of the region lying above the axis between the graph and the *t*-axis.

d. Find the area of the region lying below the axis between the graph and the *t*-axis.

e. Judging on the basis of your answers to parts *c* and *d*, by how much did the average hospital stay change between 1993 and 2000? Can you determine the average stay in 2000? Why or why not?

18. **Marginal Cost** The data show the marginal cost for compact disc production at the indicated hourly production levels.

Production (CDs per hour)	Marginal cost
100	$5
150	$3.50
200	$2.50
250	$2
300	$1.60

a. What are the units on the marginal cost?

b. Use four left rectangles to approximate the change in cost when production is increased from 100 to 300 CDs per hour.

c. Find a model for the data, and sketch a graph of the model for production levels from 100 to 300 CDs per hour.

d. Repeat part *b* using eight left rectangles. Why is a model necessary when you are using eight rectangles?

e. Sketch the rectangles used in part *d* on a graph of the model. Is the approximation larger or smaller than the area of the region between the model and the horizontal axis from 100 to 300?

19. **Births** The number of live births each year in the United States from 1950 through 2000 to women 45 years of age and older is given.

Year	Births	Year	Births
1950	5322	1980	1200
1955	5430	1985	1162
1960	5182	1990	1638
1965	4614	1995	2727
1970	3146	2000	4604
1975	1628		

(Source: www.infoplease.com. Accessed 9/24/02.)

a. Draw a scatter plot of the data, and sketch right rectangles depicting the total number of live births from 1970 through 1995. Use the data to find the total number of live births for this period. Is this number an approximation or exact? Explain.

b. Find a model for the data. Sketch a graph of this function.

c. Estimate the number of such births from the beginning of 1965 through the end of 2000 by using right rectangles with a width of 1 year. How many rectangles are needed?

d. What would be required to find the exact total number of such births from the beginning of 1965 through 2000.

20. **Temperature** During a summer thunderstorm, the temperature drops and then rises again. The rate of change of the temperature during the hour and a half after the storm began is given by

$$T(h) = 9.48h^3 - 15.49h^2 + 17.38h - 9.87 \text{ °F per hour}$$

where *h* is the number of hours since the storm began.

a. Graph the function T from $h = 0$ to $h = 1.5$. Find the point at which the graph crosses the horizontal axis.

b. Consider the portion of the graph of T lying below the horizontal axis. What does the area of the region between this portion of the graph of T and the horizontal axis represent?

c. What does the area of the region lying above the axis represent?

d. Consider the graph between 0 and 1.5 hours. Use seven right rectangles to approximate the area of the region lying below the axis from 0 to 1.5 hours.

e. Repeat part *d* for the region lying above the axis.

f. According to the model, one and a half hours after the storm began, was the temperature higher, lower, or the same as the temperature at the beginning of the storm? If it was higher or lower, by how many degrees?

21. **Air Speed**   The table shows air speed recorded from a Cessna 172. Air speed is the speed at which the air flows into a Pitot tube located on the nose or wing of an aircraft and is measured in nautical miles per hour (knots). The data were recorded for 25 seconds after the plane began to taxi for takeoff.

Time (seconds)	Air speed (knots)
0	0
10	40
15	60
20	70
25	80

a. Convert knots to miles per second, using the fact that 1 knot ≈ 1.15 mph.

b. Find a model for the converted data.

c. Estimate the area of the region between a graph of the function and the horizontal axis from 0 seconds to 18 seconds.

d. If takeoff occurred at 18 seconds, interpret your estimate in part *c* in the context of the Cessna 172 data.

22. **Baby Boomers**   As the 76 million Americans born between 1946 and 1964 (the "baby boomers") continue to age, the United States will see an increasing proportion of Americans who are at a retirement age of 65. The table shows past and projected rates of change for the number of U.S. citizens 65 years of age or older. These are instantaneous rates of change measured at the end of each year based on data from the U.S. Census Bureau.

Year	Rate of change of 65+ population (millions per year)
1940	0.21
1960	0.32
1980	0.50
2000	0.76
2010	0.94
2020	1.17
2040	1.45

a. Find a model for the rate-of-change data.

b. Use the equation and ten midpoint rectangles to estimate the change in the population 65 years of age and older from the end of 2000 through the end of 2005.

23. **Birthweight**   The rate of change of the percentage of low birthweight babies (less than 5 pounds 8 ounces) in 2000 can be modeled by

$$P(w) = -11.484(0.863408^w) \text{ percentage points per pound}$$

when the mother gains w pounds during pregnancy. The model is valid for weight gains between 18 and 43 pounds.
(Source: Based on data from *National Vital Statistics Reports*, vol. 50, no. 5, February 12, 2002.)

a. Sketch a graph of P from $w = 18$ to $w = 43$.

b. What does the fact that the graph of P lies below the *t*-axis from $w = 18$ to $w = 43$ tell you about the percentage of low birthweight babies in 2000?

c. Use five midpoint rectangles to estimate the area of the region between the graph of P and the *w*-axis from $w = 18$ to $w = 43$. Interpret your answer.

d. What percentage of babies born in 2000 had low birthweight?

24. **Trust Fund**   The rate of change of the projected total assets in the Social Security trust fund for the years 2000 through 2033 can be modeled by the equation

$$S(x) = -0.4569x^2 + 1.8528x + 108.7241$$
$$\text{billion dollars per year}$$

x years after 2000.
(Source: Based on data from the Social Security Administration.)

a. Graph S between 2000 and 2033.

b. According to the graph of S, when will the trust fund assets be growing and when will they be declining?

c. Find the point on the graph of S that corresponds to the time when the amount in the trust fund will be greatest.

d. Estimate using 10 intervals the area lying above the axis and below the graph of S. Interpret your answer.

e. Estimate using 10 intervals the area lying below the axis and above the graph of S. Interpret your answer.

f. By how much will the trust fund amount change between 2000 and 2033? What information do we need to determine how much money is projected to be in the trust fund in 2033?

25. **Life Expectancy**   Life expectancies in the United States are always rising because of advances in health care, increased education, and other factors. The rate of change (measured at the end of each year) of life expectancies for women in the United States between 1970 and 2010 are shown in the table in the next column.

a. Look at a scatter plot of the data. Does the fact that the data are declining from 1970 through 1995 contradict the statement that life expectancies are always rising? Explain.

b. Find a model for the data.

c. Use eight midpoint rectangles to estimate the change in the life expectancy for women from 1970 through 2010.

Year	Years of life expectancy per year
1970	0.36
1975	0.26
1980	0.18
1985	0.13
1990	0.09
1995	0.08
2000	0.08*
2005	0.11*
2010	0.16*

*Projected
(Source: Based on data from *Statistical Abstract*, 1998.)

26. **Road Test**   Accelerations for a vehicle during a road test are approximated in the table.

Time (seconds)	Acceleration (feet per second squared)
0	22.6
2	18.2
4	14.5
6	11.4
8	8.9
10	7.1
12	5.9

a. Find a model for the data.

b. Use 10 midpoint rectangles to estimate the area of the region between the graph of your model and the input axis from 0 to 13.5 seconds. Interpret your answer.

c. Convert your answer in part *b* to miles per hour. The actual speed of the car after 13.5 seconds was 107 mph. How close is your answer to this speed?

27. **Lake Level**   The rate of change of the level of Lake Tahoe d days after September 30, 1995, can be modeled by

$$r(d) = (-1.6035 \cdot 10^{-6})d^2 + (5.086 \cdot 10^{-4})d$$
$$- 0.0192 \text{ feet per day}$$

A graph of the model is shown.

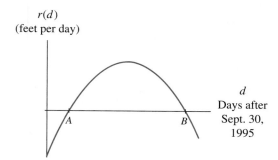

r(d)
(feet per day)

d
Days after
Sept. 30,
1995

(Source: Based on data from the Federal Watermaster,
U.S. Department of the Interior.)

a. Find the points labeled *A* and *B* on the graph.

b. Use five midpoint rectangles to estimate the area between the graph of *r* and the horizontal axis from 0 to *A*. Interpret your answer.

c. Use 10 midpoint rectangles to estimate the area between the graph of *r* and the horizontal axis from *A* to *B*. Interpret your answer.

d. Was the lake level higher or lower *B* days after September 30, 1995 than it was on that date? How much higher or lower was the lake level *B* days after September 30, 1995?

28. Grasshoppers The rate of change in the percentage of southern Australian grasshopper eggs that hatch as a function of temperature (for temperatures between 7°C and 25°C) can be modeled by the equation

$$p(t) = -0.0258t^3 + 1.464t^2 - 25.982t + 136.560$$
percentage points per degree

where *t* is the temperature in °C. A graph of *p* is shown.

a. Find the point at which the graph of *p* crosses the horizontal axis.

b. Use four midpoint rectangles to estimate the area of the region between the graph of *p* and the horizontal axis from *t* = 7 to the point at which the curve crosses the horizontal axis. Interpret your answer.

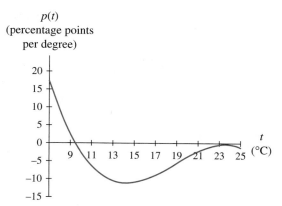

p(t)
(percentage points
per degree)

t
(°C)

(Source: Based on information in *Elements of Ecology*, George L. Clark. New York: Wiley, 1954, p. 170.)

c. Use eight midpoint rectangles to estimate the area of the region between the graph of *p* and the horizontal axis from the point at which the curve crosses the horizontal axis to *t* = 25. Interpret your answer.

d. Estimate the difference between the percentage of grasshopper eggs that hatch at 25°C and the percentage that hatch at 7°C.

29. Female Ph.D.s The table shows the number of new female Ph.D.s in computer science in selected years from 1970 through 1993.

Year	New female Ph.D.s	Year	New female Ph.D.s
1970	1	1987	51
1973	7	1988	60
1976	14	1989	87
1979	24	1990	97
1982	27	1991	113
1985	32	1992	108
1986	50	1993	126

(Source: *Computing Research News*, January 1994.)

a. Draw a scatter plot of the data, and sketch rectangles depicting the number of new female Ph.D.s in computer science from the beginning of 1985 through 1993. Find the total number of new female Ph.D.s from the beginning of 1985 through 1993. Is this number an approximation or exact?

b. Find a model for the data.

c. Use the model to estimate the total number of new female Ph.D.s in computer science from the beginning of 1970 through 1993.

d. The total number of new female Ph.D.s in computer science from the beginning of 1970 through 1993 was 987. How does your estimate from part c compare to this number?

30. **Phone Calls**   The Federal Communications Commission gives the number of international telephone calls (in millions) billed in the United States as shown.

Year	Calls (millions)	Year	Calls (millions)
1980	200	1993	1926
1985	411	1994	2313
1989	835	1995	2821
1990	984	1996	3485
1991	1371	1998	4439
1992	1643	2000	7297

a. Draw a scatter plot of the data, and sketch rectangles of width 1 year depicting the number of international calls from the beginning of 1989 through 1996. Find the total number of international calls from the beginning of 1989 through 1996. Is this an approximation or the exact number?

b. Find a model for the data.

c. Use your model to estimate the number of international calls from the beginning of 1998 through 2000.

31. In essay form (using pictures as necessary), explain why the approximation to the area of a region under a curve and above the horizontal axis generally improves when twice as many rectangles are used.

32. **Poverty**   The table for this activity (located on the *Calculus Concepts* CD-ROM and web site) lists the rates of change in the percent of people in the United States below poverty level by age for all income levels and both sexes in 2001.

a. Use only the data and right rectangles of width 17 years to estimate by how much the percent below poverty changed between birth and age 84. Sketch the rectangles on a scatter plot of the data. Do you believe your approximation is an overestimate or an underestimate? Explain.

b. Repeat part a using left rectangles.

c. Repeat part a using midpoint rectangles.

d. Which estimate do you believe is closer to the actual change in the percent of people below poverty between birth and age 84? Explain.

33. **Victims**   The table for this activity (located on the *Calculus Concepts* CD-ROM and web site) gives the rates of change in the violent crime rate per 1,000 persons in the United States for victims who are between 12 and 15 years old for 1973 to 2001.

a. Use only the data and left rectangles of width 2 years to estimate by how much the violent victimization rate changed between 1973 and 1987. Sketch the rectangles on a scatter plot of the data. Do you believe your approximation is an overestimate or an underestimate? Explain.

b. Repeat part a using right rectangles of width 2 years for 1987 through 2001.

c. Repeat part a using midpoint rectangles whose midpoints are at even-numbered years between 1973 through 2001.

6.2   Limit of Sums, Accumulated Change, and the Definite Integral

In Section 6.1 we saw that the accumulated change in a quantity can be interpreted in terms of areas of regions between the graph of a rate-of-change function for that quantity and the horizontal axis. We approximated areas of regions between a curve and the horizontal axis by using rectangles. In the Activities, you should have noticed that the approximations became closer to the actual area of the region when twice as many intervals were used. What would you expect to happen if you were to use four, eight, or even one hundred times as many intervals?

We return to Example 2 from Section 6.1, where we considered the rate of change of the concentration of a drug in the bloodstream. Recall that the rate of change of the drug concentration for the first 20 days is

$$r(x) = 1.708(0.845^x) \ \mu g/mL/day$$

where x is the number of days after the drug was first administered. We saw that we could estimate the change in the concentration of the drug between day 0 and day 20 as the area between the graph of r and the horizontal axis from $x = 0$ to $x = 20$. We then used ten right rectangles to estimate this area (see Figure 6.16) to obtain an estimate of 8.24 $\mu g/mL$.

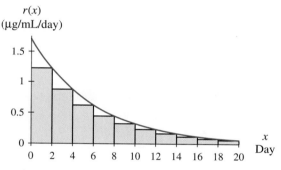

FIGURE 6.16   **FIGURE 6.17**

We know that we obtain a better estimate of this area by using midpoint rectangle, as shown in Figure 6.17.

> From now on, whenever we speak of approximating areas using rectangles, we will use midpoint rectangles because they generally give the best approximations.

An estimate of the area between the graph of r and the horizontal axis from $x = 0$ to $x = 20$ derived by using 10 midpoint rectangles is approximately 9.746 $\mu g/mL$. From this estimate of the area we can say that the drug concentration increased by approximately 9.75 $\mu g/mL$ during the first 20 days of the test.

Finding a Limit of Area Estimates

We can improve this estimate of area by using 20 rectangles instead of 10. What do you think will happen to the accuracy of the approximations if we were to use even more rectangles? Figures 6.18 through 6.21 show approximations with 10, 20, 40, and 80 rectangles, respectively.

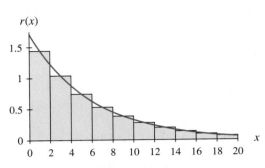

FIGURE 6.18   Ten rectangles

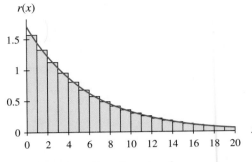

FIGURE 6.19   Twenty rectangles

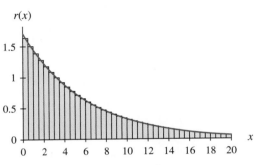

FIGURE 6.20   Forty rectangles

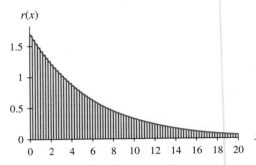

FIGURE 6.21   Eighty rectangles

By carefully inspecting the graphs in these figures, you can see that as we use more rectangles, the shaded region accounted for by the rectangles more closely approximates the region of interest. This fact will help us approximate the area. If we make a table of the approximations, we may be able to recognize a *limit*—that is, a value to which the approximations seem to be getting closer and closer as the number of rectangles becomes larger and larger.

TABLE 6.7

Number of rectangles	Approximation of area
10	9.7459
20	9.7805
40	9.7892
80	9.7913
160	9.7919
320	9.7920
640	9.7921
1280	9.7921
Limit ≈ 9.79	

The limit evident from Table 6.7 indicates that the area under consideration is approximately 9.79. Interpreting this area in context tells us that the drug concentration increased by approximately 9.79 μg/mL during the first 20 days of the test. It is important to note that we cannot conclude that the drug concentration after 20 days was 9.79 μg/mL. This is true only if the initial drug concentration was zero.

EXAMPLE 1   *Using a Limit of Sums to Obtain a More Accurate Area Approximation*

6.2.1

Consider the function $f(x) = \sqrt{1 - x^2}$ from $x = 0$ to $x = 1$. Approximate to three decimal places the area between this curve and the horizontal axis between $x = 0$ and $x = 1$, beginning with four midpoint rectangles and doubling the number each time until a trend is observed.

Solution   Numerically, we observe in Table 6.8 the midpoint-rectangle approximations approaching a limit. We can be fairly confident that the limit to three decimal places is 0.785, because the value through the fourth decimal position has remained constant in several approximations.

TABLE 6.8

Number of rectangles	Approximation of area
4	0.79598
8	0.78917
16	0.78674
32	0.78587
64	0.78557
128	0.78546
256	0.78542
512	0.78541
Limit ≈ 0.785	

Because the region is a quarter-circle with radius 1, the exact area of this region is $\frac{\pi}{4}$, which is approximately 0.7853982. When $n = 512$, the difference in the true and estimated values is approximately 0.0000074.   ●

Accumulated Change and the Definite Integral

Because the more rectangles we use to approximate area, the better we expect the approximation to be, we are led to consider area as the limiting value of the sums of areas of approximating midpoint rectangles as the number of rectangles increases without bound. Let f be a function that is continuous and non-negative over the interval from a to b. (See Figure 6.22.) Partition the interval from a to b into n subintervals of equal length $\Delta x = \frac{b - a}{n}$, and on each subinterval construct a rectangle of width Δx whose height is given by the value of f at the midpoint of the subinterval. Figures 6.23 through 6.26 show the rectangles when $n = 4, 8, 16$, and 32.

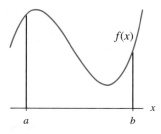

FIGURE 6.22

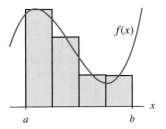

FIGURE 6.23 Four rectangles

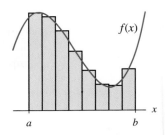

FIGURE 6.24 Eight rectangles

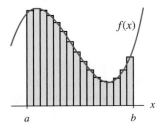

FIGURE 6.25 Sixteen rectangles

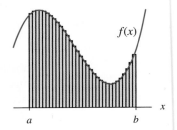

FIGURE 6.26 Thirty-two rectangles

The heights of the rectangles are given by the values

$$f(x_1), f(x_2), \ldots, f(x_n)$$

where $x_1, x_2, \ldots, x_n$ are the midpoints of the subintervals. Each rectangle has width Δx, so the areas of the rectangles are given by the values

$$f(x_1)\Delta x, f(x_2)\Delta x, \ldots, f(x_n)\Delta x$$

and the sum $[f(x_1) + f(x_2) + \cdots + f(x_n)]\Delta x$ is an approximation to the area of the region between the the graph of f and the x-axis from a to b. As our examples have shown, the approximations generally improve as n increases. In mathematical terms, the area of the region between the graph of f and the x-axis from a to b is given by a limit of sums as n gets larger and larger:

$$\text{Area} = \lim_{n \to \infty} [f(x_1) + f(x_2) + \cdots + f(x_n)]\Delta x$$

Area Beneath a Curve

Let f be a continuous or a piecewise continuous non-negative bounded function from a to b. The area of the region between the graph of f and the x-axis from a to b is given by the limit

$$\text{Area} = \lim_{n \to \infty} [f(x_1) + f(x_2) + \cdots + f(x_n)]\Delta x$$

where $x_1, x_2, \ldots, x_n$ are the midpoints of n subintervals of length $\Delta x = \dfrac{b - a}{n}$ between a and b.

EXAMPLE 2   *Relating Accumulated Change to Signed Area*

Wine Consumption   The rate of change of the per capita consumption of wine in the United States from 1970 through 1990 can be modeled* as

$$W(x) = (1.243 \cdot 10^{-4})x^3 - 0.0314x^2 + 2.6174x - 71.977 \text{ gallons per person per year}$$

where x is the number of years since the end of 1900. A graph of the function is shown in Figure 6.27.

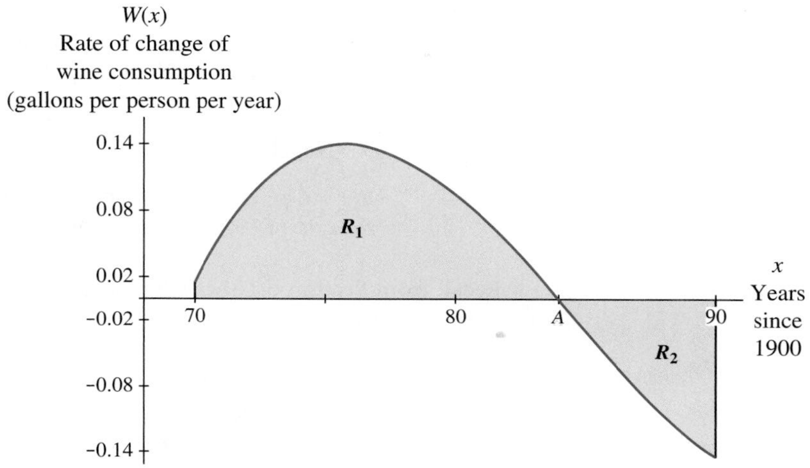

FIGURE 6.27

a. Find the input value of the point labeled A.

b. From 1970 through 1990, according to the model, when was wine consumption increasing and when was it decreasing?

c. Use a limiting value of sums to estimate the areas, to two decimal places, of the regions labeled R_1 and R_2. Interpret your answers.

d. According to the model, what was the change in the per capita consumption of wine from the end of 1970 through 1990?

TABLE 6.9

Number of rectangles	Approximation of area of region R_1
5	1.35966
10	1.34130
20	1.33671
40	1.33556
80	1.33527
160	1.33520
Limit $\approx$ 1.34	

Solution

a. Solving $W(x) = 0$ gives $A \approx 83.97$, corresponding to the end of 1984.

b. Wine consumption was increasing where the rate-of-change graph is positive—from 1970 through 1984 ($x \approx 83.97$)—and was decreasing from 1984 through 1990, where the rate-of-change graph is negative.

c. In order to find the areas of the regions R_1 and R_2, we must know the lower limit and upper limit (that is, the endpoints) of each region. Because $W(x) = 0$ when $x \approx 83.97$, we use lower limit $x = 70$ and upper limit $x \approx 83.97$ for R_1 and lower limit $x \approx 83.97$ and upper limit $x = 90$ for R_2.

*Based on data from *Statistical Abstract*, 1994.

TABLE 6.10

Number of rectangles	Approximation of signed area of region R_2
5	-0.44927
10	-0.44870
20	-0.44856
40	-0.44852
Limit ≈ -0.45	

The area of region R_1 is determined by examining sums of areas of midpoint rectangles for an increasing number of subintervals until a trend is observed, as shown in Table 6.9. The area of region R_1 is approximately 1.34 gallons per person. This indicates that wine consumption increased by approximately 1.34 gallons per person from the end of 1970 through 1984 ($x \approx 83.97$).

We calculate the signed area of region R_2 in a similar way as shown in Table 6.10. The area of region R_2 is approximately 0.45 gallon per person. This is an estimate for the decrease in per capita wine consumption from 1984 ($x \approx 83.97$) through 1990.

d. To determine the net change in the per capita consumption of wine from the end of 1970 through 1990, we subtract the decrease (area of R_2) from the increase (area of R_1).

Net change = $1.34 - 0.45 = 0.89$ gallon per person

We see an approximate net increase of 0.89 gallon per person from the end of 1970 through 1990. ●

Recall from Section 6.1 that when a rate-of-change function has both positive and negative outputs over a given input interval that the accumulated change in the quantity function is equal to the sum of the signed areas of the regions between the graph of the rate-of-change function and the horizontal axis. That is, we can sum the signed areas of rectangles over the entire interval to calculate accumulated change. We illustrate this type of calculation using a limit of sums and the rate-of-change function in Example 2 to estimate the net change in the per capita consumption of wine from the end of 1970 through 1990. Refer to Figure 6.27. Ignoring the horizontal-axis intercept, partition the interval from $x = 70$ to $x = 90$ into equal subintervals and use the formula for W and midpoint rectangles to calculate the sums of signed-rectangle areas in Table 6.11.

TABLE 6.11

Number of rectangles	Approximate accumulated change
5	0.92848
10	0.89712
20	0.88928
40	0.88732
80	0.88683
160	0.88671
320	0.88668
Limit ≈ 0.89 gallon per person	

The limit of sums $\lim_{n \to \infty} [W(x_1) + W(x_2) + \cdots + W(x_n)] \Delta x$ from $x = 70$ to $x = 90$ yields the same net change as the subtraction of non-signed areas in part d of Example 2.

It is tedious to write $\lim_{n\to\infty} [W(x_1) + W(x_2) + \cdots + W(x_n)]\Delta x$ from $x = a$ to $x = b$ every time we want to denote the limit of sums. Mathematically, we use the shorthand notation $\int_a^b f(x)\,dx$. The sign $\int$ is called an *integral sign* and resembles an elongated S to remind us that we are taking a limit of sums. The values a and b identify the input interval, f is the function, and the symbol dx reminds us of the width Δx of each subinterval. When a and b are specific numbers, $\int_a^b f(x)\,dx$ is known as a **definite integral.** We now formally define the definite integral and accumulated change.

Accumulated Change and the Definite Integral

Let f be a continuous or piecewise continuous bounded function from a to b. The accumulated change in f from a to b is given by the limit

$$\lim_{n\to\infty} [f(x_1) + f(x_2) + \cdots + f(x_n)]\Delta x = \int_a^b f(x)\,dx$$

where $x_1, x_2, \ldots, x_n$ are the midpoints of n subintervals of length $\Delta x = \dfrac{b-a}{n}$ between a and b.

$\int_a^b f(x)\,dx$ is called the definite integral of f from a to b.

EXAMPLE 3   *Relating Accumulated Change and the Definite Integral*

Wine Consumption   Recall from Example 2 the function W giving the rate of change of per capita consumption of wine in gallons per person per year from 1970 through 1990 where x is the number of years since 1900.

a.   Find the values of $\int_{70}^{83.97} W(x)\,dx$, $\int_{83.97}^{90} W(x)\,dx$, and $\int_{70}^{90} W(x)\,dx$.

b.   What was the per capita wine consumption in 1990?

Solution

a.   Refer to Figure 6.27. The area of region R_1, calculated in Example 2, is the value of the definite integral of W from 70 to 83.97.

$$\int_{70}^{83.97} W(x)\,dx \approx 1.34 \text{ gallons per person}$$

The signed area of region R_2, calculated in Example 2, is the value of the definite integral from 83.97 to 90.

$$\int_{83.97}^{90} W(x)\,dx \approx -0.45 \text{ gallon per person}$$

The limit of sums calculated using Table 6.11 gives the accumulated change in per capita wine consumption of definite integral of W from 70 to 90.

$$\int_{70}^{90} W(x)\,dx \approx 0.89 \text{ gallon per person}$$

b. The definite integral tells us the change in a quantity over an interval, not the value of the quantity at the endpoint of the interval. Thus we do not know the per capita wine consumption in 1990. ●

Although finding the limiting value of sums of areas of rectangles is an invaluable tool for finding accumulated change in a quantity, it is limited in that it must be calculated over an interval with specific numerical endpoints. In many cases, it is possible to find accumulation functions that will give as output the accumulated change in a rate-of-change function. We explore these accumulation functions graphically in Section 6.3 and algebraically in Section 6.4.

6.2   Concept Inventory

○ *Area = limiting value of sums of areas of midpoint rectangles*

○ $\int_{a}^{b} f(x)\,dx = $ *accumulated change in F where F′ = f for x between a and b*

○ $\int_{a}^{b} f(x)\,dx = $ *definite integral*

6.2   Activities

1. **Population**   The rate of change of the population of a country, in thousands of people per year, is modeled by the function P with input t, where t is the number of years since 1995. What are the units on

    a. the area of the region between the graph of P and the t-axis from $t = 0$ to $t = 10$?

    b. $\int_{10}^{20} P(t)\,dt$?

    c. the change in the population from 1995 through 2000?

2. **Lake Level**   During the spring thaw a mountain lake rises by $L(d)$ feet per day, where d is the number of days since April 15. What are the units on

    a. the area of the region between the graph of L and the d-axis from $d = 0$ to $d = 15$?

    b. $\int_{16}^{31} L(d)\,dd$?

    c. the amount by which the lake rose from May 15 to May 31?

3. **Algae Growth**   When warm water is released into a river from a source such as a power plant, the increased temperature of the water causes some algae to grow and other algae to die. In particular, blue-green algae that can be toxic to some aquatic life thrive. If $A(c)$ is the growth rate of blue-green algae (in organisms per °C) and c is the temperature of the water in °C, interpret the following in context:

    a. $\int_{25}^{35} A(c)\,dc$

    b. The area of the region between the graph of A and the c-axis from $c = 30°C$ to $c = 40°C$

4. Stock Value   The value of a stock portfolio is growing by $V(t)$ dollars per day, where t is the number of days since the beginning of the year. Interpret the following in context:

a. The area of the region between the graph of V and the t-axis from $t = 0$ to $t = 120$

b. $\int_{120}^{240} V(t)\,dt$

5. Production   The graph in the figure shows the rate of change of profit at various production levels for a pencil manufacturer. Fill in the blanks in the following discussion of the profit. If it is not possible to determine a value, write NA in the corresponding blank.

Rate of change
of profit
(dollars per box)

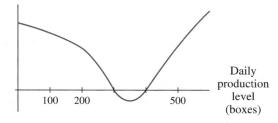

Daily production level (boxes)

Profit is increasing when between (a) _____ and (b) _____ boxes of pencils are produced each day. The profit when 500 boxes of pencils are produced each day is (c) _____ dollars. Profit is higher than nearby profits at a production level of (d) _____ boxes each day, and it is lower than nearby profits at a production level of (e) _____ boxes of pencils each day. The profit is decreasing most rapidly when (f) _____ boxes are produced each day. The area between the rate-of-change-of-profit function and the production-level axis between production levels of 100 and 200 boxes each day has units (g) _____. If $p'(b)$ represents the rate of change of profit (in dollars per box) at a daily production level of b boxes, would $\int_{300}^{400} p'(b)\,db$ be more than, less than, or the same value as $\int_{100}^{200} p'(b)\,db$? (h) _____

6. Cost   The graph shows the rate of change of cost for an orchard in Florida at various production levels during grapefruit season. Fill in the blanks in the following cost function discussion. If it is not possible to determine a value, write NA in the corresponding blank.

Rate of
change of cost
(dollars per carton)

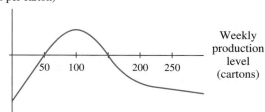

Weekly production level (cartons)

Cost is increasing when between (a) _____ and (b) _____ cartons of grapefruit are harvested each week. The cost to produce 100 cartons of grapefruit each week is (c) _____ dollars. The cost is lower than nearby costs at a production level of (d) _____ cartons, and it is higher than nearby costs at a production level of (e) _____ cartons of grapefruit each week. The cost is increasing most rapidly when (f) _____ cartons are produced each week. The area between the rate-of-change-of-cost function and the production-level axis between production levels of 50 and 150 cartons each week has units (g) _____. If $c'(p)$ represents the rate of change of cost (in dollars per carton) at weekly production level of p cartons of grapefruit, would $\int_{200}^{250} c'(p)\,dp$ be greater than, less than, or the same value as $\int_{50}^{100} c'(p)\,dp$? (h) _____

7. Weight   The rate of change of the weight of a laboratory mouse can be modeled by the equation

$$w(t) = \frac{13.785}{t} \text{ grams per week}$$

where t is the age of the mouse in weeks and $1 \le t \le 15$.

a. Use the idea of a limit of sums to estimate the value of $\int_{3}^{11} w(t)\,dt$.

b. Label units on the answer to part *a*. Interpret your answer.

c. If the mouse weighed 4 grams at 3 weeks, what was its weight at 11 weeks of age?

8. Sales   The rate of change of annual U.S. factory sales (in billions of dollars per year) of consumer electronic goods to dealers from 1990 through 2001 can be modeled by the equation

$$s(x) = 0.1164x^2 - 0.99x + 5.698$$
$$\text{billion dollars per year}$$

where *x* is the number of years since 1990.
(Sources: Based on data from *Statistical Abstract*, 2001, and Consumer Electronics Association.)

a. Use the idea of a limit of sums to estimate the change in factory sales from 1990 through 2001.

b. Write the definite integral symbol for this limit of sums.

c. If factory sales were $43.0 billion in 1990, what were they in 2001?

9. Production   On the basis of data obtained from a preliminary report by a geological survey team, it is estimated that for the first ten years of production, a certain oil well can be expected to produce oil at the rate of $r(t) = 3.93546t^{3.55}e^{-1.35135t}$ thousand barrels per year *t* years after production begins.

a. Use the idea of a limit of sums to estimate the yield from this oil field during the first 5 years of production.

b. Use the idea of a limit of sums to estimate the yield during the first 10 years of production.

c. Write the definite integral symbols representing the limits of sums in parts *a* and *b*.

d. Estimate the percentage of the first 10 years' production that your answer to part *a* represents.

10. Temperature   The rate of change of the temperature during the hour and a half after a thunderstorm began is modeled by the equation

$$T(h) = 9.48h^3 - 15.49h^2 + 17.38h - 9.87 \text{ °F per hour}$$

where *h* is the number of hours since the storm began. A graph of *T* is shown in the figure.

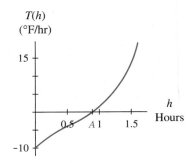

a. Determine the value of *A*.

b. Use a limit of sums to estimate $\int_0^A T(h)\,dh$. Interpret your answer.

c. Use a limit of sums to estimate $\int_A^{1.5} T(h)\,dh$. Interpret your answer.

d. Estimate $\int_0^{1.5} T(h)\,dh$. Interpret your answer.

e. What information is needed to determine the temperature when $h = 1.5$?

11. Road Test   The acceleration of a race car during the first 35 seconds of a road test is modeled by the equation

$$a(t) = 0.024t^2 - 1.72t + 22.58 \text{ ft/sec}^2$$

where *t* is the number of seconds since the test began. A graph of this acceleration function is shown in the figure.

a. Find the time at which the acceleration curve becomes negative. This time is denoted *A* in the figure.

b. Use a limit of sums to estimate $\int_0^A a(t)\,dt$. Interpret your answer.

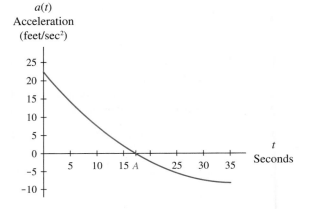

c. Use a limit of sums to estimate $\int_A^{35} a(t)\,dt$. Interpret your answer.

d. Estimate the change that occurred in the velocity of the car during the first 35 seconds of the road test.

12. Temperature Museums carefully monitor temperature and humidity. Suppose a museum has a gauge that monitors temperature as well as how rapidly temperature is changing. During a junior high school field trip to the museum, the following rates of change of temperature are recorded.

Time of day	Rate of change (°F/hr)
8:30 A.M.	0
8:45 A.M.	2.2
9:00 A.M.	2.4
9:15 A.M.	1.1
9:30 A.M.	-0.8
9:45 A.M.	-2.3
10:00 A.M.	-2.7
10:15 A.M.	-1.0

a. Basing your explanation on the rate-of-change data, discuss what happens to the temperature from 8:30 A.M. to 10:15 A.M.

b. According to the data, when is the temperature greatest?

c. Find a model for the data.

d. Use the model and a limit of sums to estimate the change in temperature between 8:30 A.M. and 9:30 A.M.

e. Use a limit of sums to estimate the change in temperature between 9:30 A.M. and 10:15 A.M.

f. What are the definite integral notations for your answers to parts *d* and *e*?

g. What information is needed to find the temperature at 10:15 A.M.?

13. Velocity An article in the May 23, 1996, issue of *Nature* addresses the interest some physicists have in studying cracks in order to answer the question

"How fast do things break, and why?" As stated in *Nature*, "In 1991, Fineberg, Gross, Swinney and others developed a method of looking at the motion of cracks, . . . [making it] possible to measure the velocity of a crack on timescales much shorter than a millionth of a second, hundreds of thousands of times in succession."

Data estimated from a graph in this article, showing the velocity of a crack during a 60-microsecond experiment, are given in the table.

Time (microseconds)	Velocity (meters per second)
10	148.2
20	159.3
30	169.5
40	180.7
50	189.8
60	200

a. In order to determine the distance the crack traveled during the 60-microsecond experiment, we wish to determine the area beneath the velocity curve from 0 to 60. What are the units on the heights and widths of rectangles under this curve?

b. Convert the data to millimeters per microsecond (there are 1000 millimeters in a meter and 1,000,000 microseconds in a second). Using the converted data, identify the area units.

c. Find a quadratic model for the converted data.

d. Use a limit of sums to determine how far the crack traveled during the experiment.

e. Give the definite integral notation for your answer to part *d*.

14. Sales The table records the volume of sales (in thousands) of a popular movie for selected months the first 18 months after it was released on video cassette.

a. Find a logistic model for the data.

b. Use 5, 10, and 15 right rectangles to estimate the number of cassettes sold during the first 15 months after release.

Months after release	Number of cassettes sold each month (thousands)
2	565
4	467
5	321
7	204
10	61
11	31
12	17
16	3
18	2

c. Which of the following would give the most accurate value of the number of cassettes sold during the first 15 months after release?

   i. The answer to part *b* for 15 rectangles.

   ii. The limiting value of the sums of midpoint rectangles using the model in part *a*.

   iii. The sum of actual sales figures for the first 15 months.

15. **Labor**   The personnel manager for a large construction company keeps records of the number of labor hours per week spent on typical construction jobs handled by the company. He has developed the following model for a labor-power curve:

$$m(x) = \frac{6,608,830e^{-0.706x}}{(1 + 925e^{-0.706x})^2} \text{ labor hours per week}$$

after the *x*th week of the construction job.

a. Use 5, 10, and 20 right rectangles to approximate the number of labor hours spent during the first 20 weeks of a typical construction job.

b. If the number of labor hours spent on a particular job exactly coincides with the model, which of the following would give the most accurate value of the number of labor hours spent during the first 20 weeks of the job?

   i. The 20-right-rectangle sum found in part *a*

   ii. The sum of 20 midpoint rectangles

   iii. The limiting value of the sums of midpoint rectangles

16. **Revenue**   Though all companies receive revenue in a discrete fashion, if a company is large enough, then its revenue can be thought of as flowing in at a continuous rate. Suppose that it is possible to measure the rate of flow of revenue for a company and that the flow rates measured at the end of the year can be modeled by the equation

$$r(x) = 9.907x^2 - 40.769x + 58.492$$
$$\text{million dollars per year}$$

x years after the end of 1987.

(Source: Based on data for Timberland in *Allen Financial Advisors*.)

a. Use the idea of a limit of sums to estimate the value of $\int_0^6 r(x)\,dx$.

b. Interpret your answer to part *a*.

c. The company's revenue from 1987 through 1993 was reported to be 543 million dollars. How close is this value to your estimate from part *a*?

17. **Customers**   In Section 6.1 we considered a department store manager's effort to estimate the number of people who attended a Saturday sale. Recall that after taking samples of the number of customers who arrived during certain 1-minute times throughout the sale, the manager estimated the number of customers who entered the store each minute with the model

$$c(m) = (4.58904 \cdot 10^{-8})m^3 - (7.78127 \cdot 10^{-5})m^2$$
$$+ 0.03303m + 0.88763 \text{ customers per minute}$$

where *m* is the number of minutes after 9:00 A.M.

a. Use a limiting value of sums to estimate the value of $\int_0^{720} c(m)\,dm$.

b. Interpret your answer to part *a* in context.

18. **Bank Account**   The table gives rates of change of the amount in an interest-bearing account for which interest is compounded continuously.

At end of year	Rate of change (dollars per day)
1	2.06
3	2.37
5	2.72
7	3.13
9	3.60

a. Convert the input to days. Disregard leap years. Why is this conversion important for a definite integral calculation?

b. Find an exponential model for the converted data.

c. Use a limiting value of sums to estimate the change in the balance of the account from the day the money was invested to the last day of the ninth year after the investment was made. Again, disregard leap years.

d. Give the definite integral notation for your answer to part c.

e. What is the balance in the account at the end of 9 years?

19. Blood Pressure   Blood pressure (BP) varies for individuals throughout the course of a day, typically being lowest at night and highest from late morning to early afternoon. The estimated rate of change in diastolic blood pressure for a patient with untreated hypertension is shown.

Time	Rate of change of diastolic BP (mm Hg per hour)
8 A.M.	3.0
10 A.M.	1.8
12 P.M.	0.7
2 P.M.	-0.1
4 P.M.	-0.7
6 P.M.	-1.1
8 P.M.	-1.3
10 P.M.	-1.1
12 A.M.	-0.7
2 A.M.	0.1
4 A.M.	0.8
6 A.M.	1.9

a. During which time intervals was the patient's diastolic blood pressure rising? falling?

b. Estimate the times when diastolic blood pressure was rising and falling most rapidly.

c. Find a model for the data.

d. Find the times at which the output of the model is zero. Of what significance are these times in the context of blood pressure?

e. Use the idea of a limiting value of sums to estimate by how much the diastolic blood pressure changed from 8 A.M. to 8 P.M.

f. Write the definite integral notation for your answer to part e.

g. What was this patient's blood pressure at 8 A.M.?

20. Income   The table for this activity (located on the *Calculus Concepts* CD-ROM and web site) gives the rates of change of the number of females aged 15 to 24 years with income at the U.S. median level between 1974 and 2001.

a. Find a model F with input t, the number of years after 1974, for the data.

b. Find the times at which the output of F is zero. Of what significance are these times in context?

c. Use the idea of a limit of sums to estimate the area that is above the t-axis and below F. Write the definite integral notation for your answer.

d. Use the idea of a limit of sums to estimate the area that is below the t-axis and above F. Write the definite integral notation for your answer.

e. Estimate by how much the number of females in this age group whose income was at the median level changed between 1974 and 2001.

21. Income   The table for this activity (located on the *Calculus Concepts* CD-ROM and web site) gives the rates of change of the number of males aged 15 to 24 years with income at the U.S. median level between 1974 and 2001.

a. Find a model M with input t, the number of years after 1974, for the data.

b. Determine the two t-intercepts, A and B $(A < B)$, for the function in part a.

c. Use the idea of a limit of sums to estimate
$$\int_0^A M(x)\,dx, \quad \int_A^B M(x)\,dx, \quad \text{and} \quad \int_B^{27} M(x)\,dx.$$
Interpret each answer.

d. Estimate $\int_0^{27} M(x)\,dx$. Interpret your answer.

22. Explain how area, accumulated change, and the definite integral are related and how they differ.

23. Why is it important to know if and where a function has horizontal-axis intercepts before using a definite integral (limit of sums) to determine the area of the region(s) bounded by the function and the horizontal axis?

6.3   Accumulation Functions

In the previous section, we saw that when we have a rate-of-change function for a certain quantity, we approximate the accumulation of change in that quantity between two values of the input variable using the area between the rate-of-change curve and the horizontal axis. Area approximation methods are valuable, but in some situations it would be helpful to have a formula that would answer the question "What was the accumulated change in the quantity from a to t for any value of t?"

For instance, it is important for hydrogeologists studying a watershed to know how much water flowed through a river since a specific starting time. Typically, data on the rate of flow is measured and used to create a flow rate model from which the accumulation of flow can be calculated.

EXAMPLE 1   *Estimating Accumulated Change*

Rising River   The flow rate past a sensor in the west fork of the Carson River in Nevada is measured periodically. Suppose the flow rates for 8 hours prior to and 20 hours following a heavy rainstorm that began at 11:45 A.M. on a Wednesday in the spring of 1996 can be modeled by the equation

$$f(t) = 0.018225t^2 - 0.1353t + 2.88154 \text{ million ft}^3/\text{hr}$$

where t is the number of hours after 11:45 A.M. Wednesday. Figure 6.28 shows a graph for the 28 hours between 3:45 A.M. Wednesday and 7:45 A.M. Thursday. Estimate to the nearest 10,000 cubic feet the amount of water that flowed past the sensor from 3:45 A.M. Wednesday to 7:45 A.M., 11:45 A.M., 3:45 P.M., 7:45 P.M., and 11:45 P.M. Wednesday and 3:45 A.M. and 7:45 A.M. Thursday, respectively.

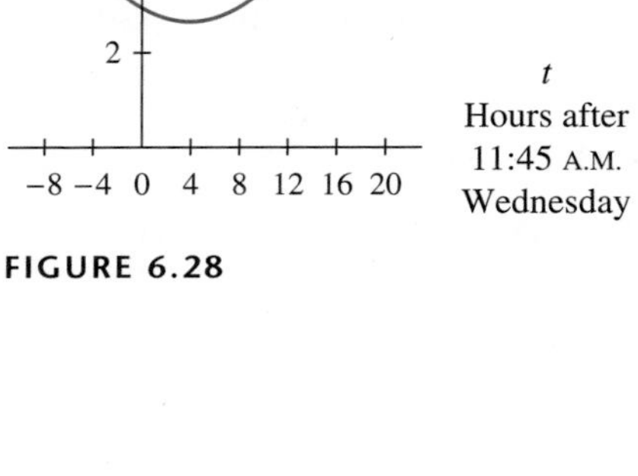

$f(t)$
(million ft³/hr)

t
Hours after
11:45 A.M.
Wednesday

FIGURE 6.28

Solution   The amount of water that flowed through the Carson River from 3:45 A.M. Wednesday ($t = -8$) and some ending time ($t = b$) can be estimated by the area between the graph of f and the horizontal axis from $t = -8$ to $t = b$. Table 6.12 shows the results of limits of sums for the various ending times.

TABLE 6.12

Ending time (input)	Area* Approximation (output)	Ending time (input)	Area* Approximation (output)
7:45 A.M. ($t = -4$)	$\int_{-8}^{-4} f(t)\,dt \approx 17.49$	11:45 P.M. ($t = 12$)	$\int_{-8}^{12} f(t)\,dt \approx 65.83$
11:45 A.M. ($t = 0$)	$\int_{-8}^{0} f(t)\,dt \approx 30.49$	3:45 A.M. ($t = 16$)	$\int_{-8}^{16} f(t)\,dt \approx 84.16$
3:45 P.M. ($t = 4$)	$\int_{-8}^{4} f(t)\,dt \approx 41.32$	7:45 A.M. ($t = 20$)	$\int_{-8}^{20} f(t)\,dt \approx 109.66$
7:45 P.M. ($t = 8$)	$\int_{-8}^{8} f(t)\,dt \approx 52.32$		

*All area units are million cubic feet.

Notice in Example 1 that the change in ending point of the time interval caused a change in the accumulation of area. As long as the starting time stays constant at some number a, we can represent the accumulated change in f from a to a variable ending time x as $\int_{a}^{x} f(t)\,dt$. This integral is called the **accumulation function of f from a to x.**

Accumulation Function

The accumulation function of a function f, denoted by $A(x) = \int_{a}^{x} f(t)\,dt$, gives the accumulation of the area between the horizontal axis and the graph of f from a to x. The constant a is referred to as the *starting value* of the accumulation.

Observe that a definite integral of f between two specified inputs simply represents an output of an accumulation function of f.

Using Estimated Areas to Sketch Accumulation Graphs

When a function is continuous and bounded over an interval, then an accumulation function for it will also be continuous over that interval. For the rising river example, we can sketch an accumulation graph (a graph showing the accumulation of flow rates since 3:45 A.M. Wednesday) by starting with the basic assumption that 0 cubic feet of water flowed past the sensor between 3:45 A.M. and 3:45 A.M. (that is, *at* 3:45 A.M.) on Wednesday. We then use a graph of f sketched on a grid to estimate the accumulated flow over some interval, say each 2 hours. Shade the region under the graph of f and above the horizontal axis for the first 2 hours after 3:45 A.M. Wednesday (see Figure 6.29a). There appear to be approximately 4.8 grid boxes included in this region. The height of a box is 1 million cubic feet per hour and the width of a box is 2 hours, so the accumulation represented by this region is (4.8)(1 million cubic feet per hour) (2 hours) = 9.6 million cubic feet. Repeating the process for the second 2-hour interval results in an estimate of 8.0 million cubic feet and a total accumulation since 3:45 A.M. Wednesday of 17.6 million cubic feet. (See the accumulation graph, Figure 6.29b.)

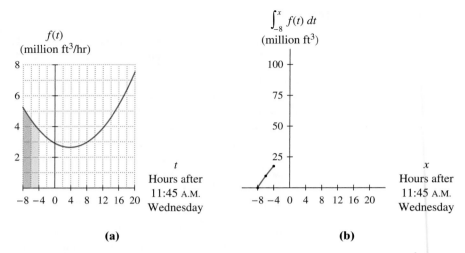

FIGURE 6.29

We repeat this graphical area approximation technique for each successive two-hour interval from 3:45 A.M. Wednesday to 7:45 A.M. Thursday, each time adding the additional area to the total accumulation. The area estimates and total accumulations are shown in Figures 6.30a through f and noted in Table 6.13.

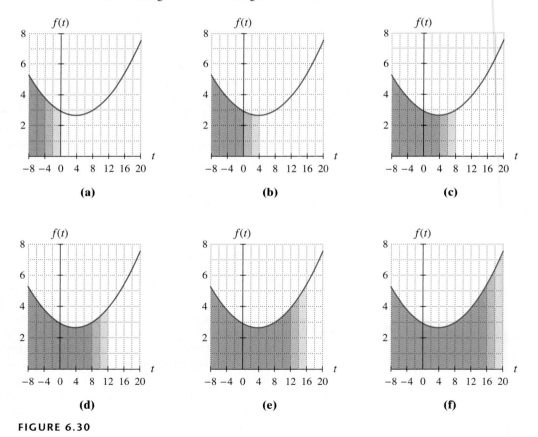

FIGURE 6.30

TABLE 6.13

Time (hours)	-6	-4	-2	0	2	4	6	8	10	12	14	16	18	20
Additional number of boxes	4.8	4.0	3.5	3.0	2.8	2.6	2.7	2.8	3.1	3.6	4.2	5.0	5.9	6.9
Accumulated number of boxes	4.8	8.8	12.3	15.3	18.1	20.7	23.4	26.2	29.3	32.9	37.1	42.1	48.0	54.9
Accumulated change (mill. ft³)	9.6	17.6	24.6	30.6	36.2	41.4	46.8	52.4	58.6	65.8	74.2	84.2	96.0	109.8

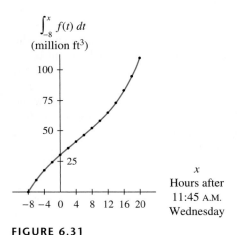

$\int_{-8}^{x} f(t)\,dt$
(million ft³)

x
Hours after
11:45 A.M.
Wednesday

FIGURE 6.31

Accumulation is occurring continuously, so a smooth curve can be used to represent the accumulation function. (See Figure 6.31.)

It is interesting to note that even though the graph of the flow rate function is decreasing between 3:45 A.M. and about 3:30 P.M. on Wednesday, the accumulation graph is increasing, although it is growing at a slower and slower rate. Then, after 3:30 P.M. Wednesday, the flow rate increases again, and the accumulation graph (still increasing) grows at a faster and faster rate. The accumulation graph seems to exhibit an inflection point near 3:30 P.M. Wednesday when the flow rate graph exhibits a minimum.

Notice that we could have used the numerical estimates found with the limits of sums instead of graphically estimating the accumulation because we had a function for the flow rate of the Carson River. Compare the accumulated changes found in Example 1 with the graph sketched in Figure 6.31.

When a portion of a graph is negative, the area below the horizontal axis indicates a decrease in the accumulation. Example 2 illustrates how to sketch accumulation graphs for graphs that go below the horizontal axis.

EXAMPLE 2   *Using Estimated Grid Areas to Sketch Accumulation Functions*

Consider the graph of f shown in Figure 6.32.

a. Construct a table of accumulation function values for $x = -3, -2.5, \ldots, 2.5, 3$.

b. Sketch a scatter plot and continuous graph of the accumulation function $A(x) = \int_{-3}^{x} f(t)\,dt$.

Solution

a. Begin estimating accumulated area from the far left side of the graph by counting the boxes between the graph of f and the horizontal axis from -3 to x. You should obtain values similar to those in the second column of Table 6.14. Note that the boxes have height 1 unit and width 0.5 unit; thus the area of each box is $(1)(0.5) = 0.5$ unit². The third column of Table 6.14 is the number of boxes multiplied by 0.5 to obtain the accumulated area value.

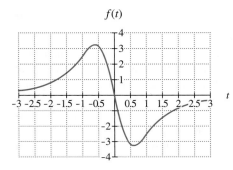

$f(t)$

t

FIGURE 6.32

TABLE 6.14

x	Accumulated number of boxes	Accumulation function value	x	Accumulated number of boxes	Accumulation function value
-3	0	0	0.5	$9 - 2 = 7$	3.5
-2.5	0.5	0.25	1	$7 - 3 = 4$	2
-2	1	0.5	1.5	$4 - 2 = 2$	1
-1.5	2	1	2	$2 - 1 = 1$	0.5
-1	4	2	2.5	$1 - 0.5 = 0.5$	0.25
-0.5	7	3.5	3	$0.5 - 0.5 = 0$	0
0	9	4.5			

For values of x greater than zero, the boxes lie below the horizontal axis. The areas of these boxes should be subtracted from the accumulated area of boxes above the horizontal axis in order to obtain the net number of accumulated boxes. For example, the number of boxes from -3 to 0 is approximately 9. The number of boxes from 0 to 0.5 is approximately 2, and these boxes lie below the horizontal axis. To determine the accumulated number of boxes from -3 to 0.5, we subtract the number of boxes lying below the axis from the number lying above the axis: $9 - 2 = 7$. Because the number of boxes from -3 to 0 is the same as the number of boxes from 0 to 3, the net result of accumulated area from -3 to 3 is zero.

b. A scatter plot and continuous graph for $A(x) = \int_{-3}^{x} f(t)\,dt$ are shown in Figure 6.33.

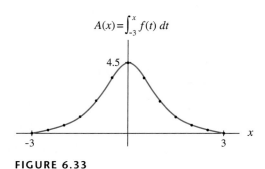

$$A(x) = \int_{-3}^{x} f(t)\,dt$$

FIGURE 6.33

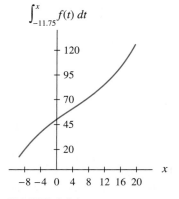

$$\int_{-11.75}^{x} f(t)\,dt$$

FIGURE 6.34

What happens to an accumulation graph if the starting point of accumulation is changed? Suppose that a hydrogeologist determined that 20 million cubic feet of water flowed past the sensor in the Carson River between midnight Tuesday and 3:45 A.M. Wednesday. The volume of water that flowed in the river since 12 A.M. Wednesday (instead of since 3:45 A.M. Wednesday) is found by adding 20 to all of the total accumulation estimates between 3:45 A.M. Wednesday and 7:45 A.M. Thursday in Table 6.13. This addition would have the effect of shifting the graph up but not changing its shape. (See Figure 6.34.)

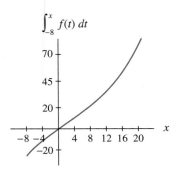

$\int_{-8}^{x} f(t)\, dt$

FIGURE 6.35

On the other hand, if the volume being considered is the accumulation since 11:45 A.M. on Wednesday, 30.6 million cubic feet would have to be subtracted from each of the total accumulation estimates in Table 6.13. (30.6 million cubic feet is the approximate accumulated area between $t = -8$ and $t = 0$.) This subtraction would have the effect of shifting the graph down but not changing its shape. (See Figure 6.35.)

The following box gives a general rule for sketching accumulation functions, regardless of their starting points.

Sketching Accumulation Graphs

Given the graph of a function, consider accumulation beginning at the far left, regardless of the location of the specified starting point. Sketch the accumulation graph starting at the far left, and then shift the graph up or down so that the output value at the specified starting point is zero.

EXAMPLE 3   *Sketching Accumulation Functions with Different Starting Points*

Consider again the function graph and accumulation graph from Example 2, redrawn here as Figures 6.36 and 6.37.

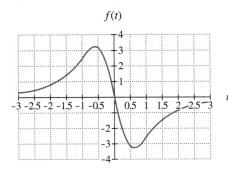

FIGURE 6.36

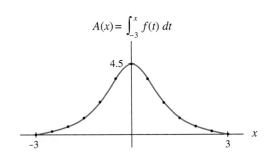

$A(x) = \int_{-3}^{x} f(t)\, dt$

FIGURE 6.37

Use the graph in Figure 6.37 to sketch a graph of the accumulation function
$$B(x) = \int_{0}^{x} f(t)\, dt.$$

Solution   In order to sketch the accumulation graph with 0 as the starting point, we vertically shift the graph in Figure 6.37 so that the function value at the starting point is zero. In this example, we shift the graph down 4.5 units so that the peak of the graph is the point $(0, 0)$. (See Figure 6.38.)

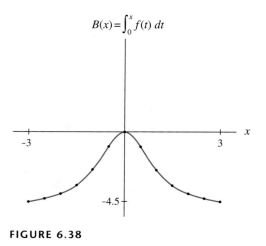

FIGURE 6.38

By now, you should have a good idea of what we mean by an accumulation function. The following information gives a procedure for sketching an accumulation function graph without making use of a grid.

Sketching Accumulation Functions

To sketch an accumulation function with starting value *a* given a graph of *f*,

Step 1: Choose as an initial starting value the input value at the left end of the horizontal axis shown in the graph.

Step 2: Sketch the accumulation of area from the far left such that

○ Regions between the graph of *f* and the horizontal axis that lie above the axis contribute positive accumulation equal to the area of the region.

○ Regions that lie beneath the horizontal axis contribute negative accumulation equal to the negative of the area of the region.

Step 3: Vertically shift the accumulation graph in Step 2 so that the graph is zero at the actual starting value *a*.

In order to more accurately sketch the graph of an accumulation function, an understanding of how rapidly the accumulated area grows or declines is necessary.

Concavity and Accumulation

An important tool for accurately sketching accumulation function graphs is an understanding of how the increasing or decreasing nature, as well as the concavity, of the graph of an accumulation function describe the accumulation of area. Increasing accumulation graphs describe rising accumulation associated with accumulation of area above the horizontal axis, whereas decreasing accumulation graphs describe declining accumulation associated with accumulation of area below the horizontal axis. Along with the increasing or decreasing nature of an accumulation graph, we also consider how the concavity of an accumulation graph describes the accumulation of area.

Concavity and Accumulation

A concave-up, increasing shape ⟋ describes faster and faster positive accumulation. A concave-down, increasing shape ⌒ indicates slower and slower positive accumulation. A concave-down, decreasing shape ⌢ indicates faster and faster decline (negative accumulation). A concave-up, decreasing shape ⌣ indicates slower and slower decline.

Note that the direction of the graph of the accumulation function A depends on whether the graph of f is positive or negative, not on whether the graph of f is increasing or decreasing.

Example 4 shows how we can apply these concepts to sketching a general accumulation function graph.

EXAMPLE 4   *Sketching a General Accumulation Function Graph*

Consider the graph shown in Figure 6.39.

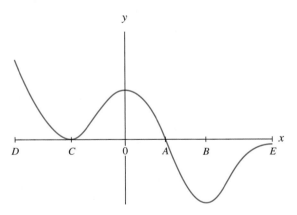

FIGURE 6.39

a.   Consider each portion of the graph between the points marked on the horizontal axis. Label these regions as contributing positive or negative accumulated area, and indicate whether each area is accumulated faster and faster or slower and slower.

b.   On the basis of the labels in part *a*, sketch the accumulation function with zero as the starting point.

Solution

a. Figure 6.40 shows the correct labels.

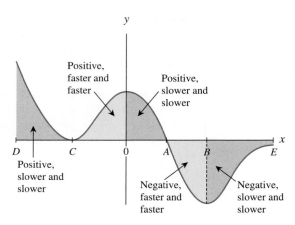

FIGURE 6.40

b. We begin at the far left with the accumulation function output at zero. On the basis of the labels in part *a*, we conclude that the accumulation function graph is increasing and concave down until *C* and is increasing and concave up between *C* and 0. Between 0 and *A*, the accumulation function graph continues to increase but is concave down. Once the graph goes below the horizontal axis, the accumulated area begins decreasing. The accumulation function graph is decreasing and concave down between *A* and *B* and is decreasing and concave up to the right of *B*.

A graph illustrating these properties is shown in Figure 6.41.

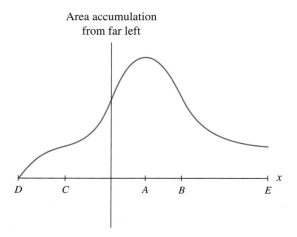

FIGURE 6.41

To obtain the accumulation function graph with zero as the starting point, shift the graph down so that it passes through $(0, 0)$. See Figure 6.42.

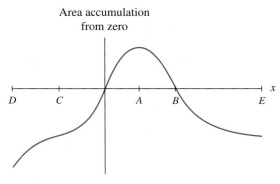

Area accumulation
from zero

FIGURE 6.42

Recovering a Function

Recovering a function is the phrase we use for the process of beginning with a rate-of-change function for a quantity and obtaining a function for the quantity. An important part of recovering a quantity function from its rate of change is recovering the units of the quantity function from the units of its rate of change. If $\frac{dM}{dt}$ is the rate-of-change function of the amount of insulin in a patient's body t hours after an injection, and if it is reported in milliliters per hour, then we can recover the units of the amount function M by recalling that the rate of change of a function is a slope of a tangent line. Slope is calculated as $\frac{\text{rise}}{\text{run}}$, where the units are $\frac{\text{output units}}{\text{input units}}$, so we know that the units of a rate-of-change function are the output units of its quantity function divided by the input units of its quantity function. From the geometric viewpoint of area, we know that the output units of the quantity function are

(Rate-of-change function units) · (input units of quantity function)

$$= \left(\frac{\text{output units of quantity function}}{\text{input units of quantity function}} \right) \cdot (\text{input units of quantity function})$$

In the case of $\frac{dM}{dt}$, the units milliliters per hour can be rewritten as $\frac{\text{milliliters}}{\text{hour}}$. Now we can see that the output units of M are milliliters and the input units are hours. Figure 6.43 shows input/output diagrams for $\frac{dM}{dt}$ and M.

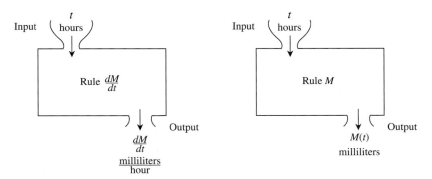

FIGURE 6.43

Occasionally, units of a rate of change are expressed in terms of a squared unit. For example, acceleration is often expressed in feet per second squared. Suppose $A(t)$ is acceleration of a vehicle in feet per second squared, where t is the number of seconds since the vehicle began accelerating. Again, this is obviously the rate of change of some function. In fact, it is the rate of change of velocity. However, the input and output units of that velocity function may not be immediately apparent because of the use of the word *squared*. Rewrite feet per second squared as

$$\frac{\text{feet}}{(\text{second})^2} = \frac{\text{feet}}{(\text{second})(\text{second})} = \frac{\text{feet}}{\text{second}} \div (\text{second})$$

$$= (\text{feet per second}) \text{ per second}$$

The output units of the velocity function are now identifiable as feet per second. Input/output diagrams for these functions are shown in Figure 6.44.

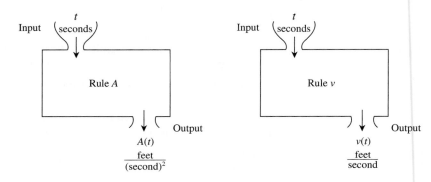

FIGURE 6.44

We have seen how to sketch an accumulated quantity function from a graph of its rate-of-change function and how to recover the quantity units from rate-of-change units. In the next section we will see that the Fundamental Theorem of Calculus implies that we have a powerful algebraic tool for recovering the accumulated quantity formula from its rate-of-change formula.

6.3 Concept Inventory

- ○ *Accumulation and area*
- ○ *Accumulation function:*
  - Sketching graphs
  - Interpreting
  - Finding simple formulas
- ○ *Accumulation and concavity*

6.3 Activities

1. Velocity Refer to the velocity graph.

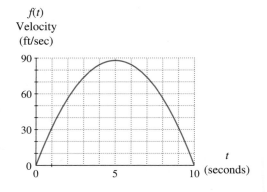

a. Sketch the accumulation function with 5 seconds as the starting point.

b. Give the mathematical notation for the function you sketched in part *a*.

c. In the context of the moving vehicle, what is the interpretation of the output values of the accumulation function in part *a*?

2. **Plant Growth**   Refer to the plant growth rate function graph.

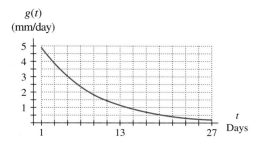

g(t)
(mm/day)

a. Sketch the accumulation function with day 13 as the starting point.

b. Give the mathematical notation for the function you sketched in part *a*.

c. In the context of plant growth, what is the interpretation of the output values of the accumulation function in part *a*?

3. **Stock Value**   Consider the graph of the rate of change in the price of a certain technology stock during the first 55 trading days of 2003.

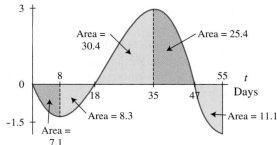

r(t)
Rate of change in price
(dollars per day)

a. What does the area of the region between the graph and the horizontal axis between days 0 and 18 represent?

b. What does the area of the region between the graph and the horizontal axis between days 18 and 47 represent?

c. Was the stock price higher or lower on day 47 than it was on day 0? How much higher or lower?

d. Was the stock price higher or lower on day 55 than it was on day 47? How much higher or lower?

e. Using the information presented in the graph, fill in the accumulation function values in the table.

x	0	8	18	35	47	55
$\int_0^x r(t)\,dt$						

f. Graph the function $R(x) = \int_0^x r(t)\,dt$ for values of x between 0 and 55, labeling the vertical axis as accurately as possible.

g. If the stock price was $127 on day 0, what was the price on day 55?

4. **Stock Value**   Refer again to the figure in Activity 3.

a. Label each of the shaded areas as representing positive or negative change in price. Also label each region as describing faster and faster or slower and slower change in stock price.

b. On the basis of your answers to part *a*, sketch a graph of the accumulation function
$$P(x) = \int_{18}^x r(t)\,dt.$$

c. Sketch a graph of the accumulation function
$$Q(x) = \int_{35}^x r(t)\,dt.$$

d. What differences do you notice among the three accumulation functions in part *f* of Activity 3 and parts *b* and *c* of this activity?

5. **Subscribers**   The accompanying graph shows the rate of change of the number of subscribers to an Internet service provider during its first year of business.

a. According to the graph, did the number of subscribers ever decline during the first year?

b. What is the significance of the peak in the graph at 20 weeks?

Subscribers
per day

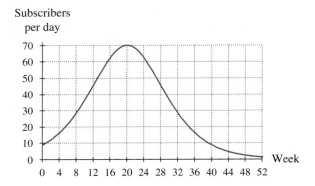

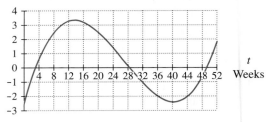

c. If $n(x)$ is the number of subscribers per day at the end of the xth day of the year, what does the accumulation function $N(t) = \int_{140}^{t} n(x)\,dx$ describe?

d. How many subscribers does each box in the figure represent?

e. Use the grid and graph in the figure to estimate the accumulation function values in the table below.

Week	t (days)	$\int_0^t n(x)\,dx$	Week	t (days)	$\int_0^t n(x)\,dx$
4	28		28	196	
8	56		36	252	
12	84		44	308	
16	112		52	364	
20	140				

f. Sketch a graph of the accumulation function with 140 days as the starting value.

6. Profit The graph shown is a model of the rate of change of profit for a new business during its first year. The input is the number of weeks since the business opened, and the output units are thousands of dollars per week.

a. What does the area of each box in the grid represent?

b. What is the interpretation, in context, of the accumulation function $P(x) = \int_0^x p(t)\,dt$?

c. Count boxes to estimate accumulation function values from 0 to x for the values of x given in the table below.

x	Accumulation function value	x	Accumulation function value
0		28	
4		32	
8		36	
12		40	
16		44	
20		48	
24		52	

d. Use the data in part c to sketch an accurate graph of the accumulation function $P(x) = \int_0^x p(t)\,dt$. Label units and values on the horizontal and vertical axes.

7. Rainfall The graph of $r(t)$ represents the rate of change of rainfall in Florida during a severe thunderstorm t hours after the rain began falling. Draw a graph of the total amount of rain that fell during this storm, using the facts that

a. The rain first started falling at noon and did not stop until 6 P.M.

b. Three inches of rain fell between noon and 3 P.M.

c. The total amount of rain that fell during the storm was 5.5 inches.

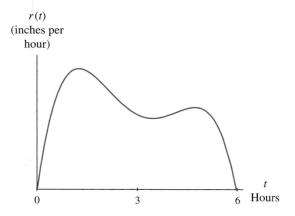

$r(t)$
(inches per
hour)

0   3   6   Hours

8. Population The Brazilian government has established a program to protect a certain species of endangered bird that lives in the Amazon rain forest. The program is to be phased out gradually by the year 2020. An environmental group believes that the government's program is destined to fail and has projected that the rate of change in the bird population between 2000 and 2050 will be as shown in the figure.

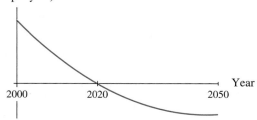

Rate of change
(birds per year)

2000   2020   2050   Year

Draw a graph of the bird population between 2000 and 2050, using the facts that (1) at the beginning of 2000 there were 1.3 million birds in existence, and (2) the birds will be extinct by 2050.

In Activities 9 through 14, sketch the indicated accumulation function graphs.

9. **a.** $\int_A^x f(t)\,dt$

   b. $\int_B^x f(t)\,dt$

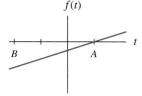

$f(t)$

B A t

10. **a.** $\int_0^x f(t)\,dt$

   b. $\int_A^x f(t)\,dt$

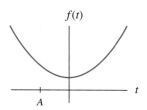

$f(t)$

A t

11. **a.** $\int_0^x f(t)\,dt$

   b. $\int_A^x f(t)\,dt$

   c. $\int_B^x f(t)\,dt$

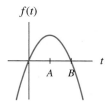

$f(t)$

A B t

12. $\int_0^x f(t)\,dt$

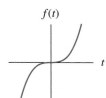

$f(t)$

t

13. $\int_0^x f(t)\,dt$

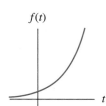

$f(t)$

t

14. $\int_A^x f(t)\,dt$

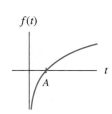

$f(t)$

A t

In each of Activities 15 through 18, a graph is given. Identify, from graphs a through f on page 414, the derivative graph and the accumulation graph (with 0 as the starting point) of the given graph. Graphs a through f may be used more than once.

15.

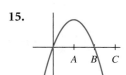

A B C

16.

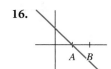

A B

17.

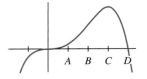

18.

a.

b.

c.

d.

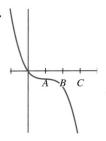

e.

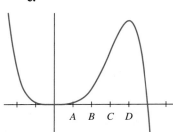

f.

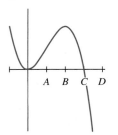

In each of Activities 19 and 20, a table of selected values for a function is given. Also shown are tables of values for the derivative and the accumulation function with 0 as the starting point. Determine which table contains values of the derivative and which contains values of the accumulation function. Justify your choice.

19.

t	f(t)	Input	Output	Input	Output
0	4	0	0	0	0
1	3	1	-2	1	3.667
2	0	2	-4	2	5.333
3	-5	3	-6	3	3
4	-12	4	-8	4	-5.333

20.

m	p(m)	Input	Output	Input	Output
0	0	0	0	0	0
1	-8	1	-12	1	-3
2	-16	2	0	2	-16
3	0	3	36	3	-27
4	64	4	96	4	0
5	200	5	180	5	125

For each of the rates of change in Activities 21 through 24,

    a. Write the units of the rate of change as a fraction.

    b. Draw an input/output diagram for the recovered function.

    c. Interpret the recovered function in a sentence.

21. When m thousand dollars are being spent on advertising, the annual revenue of a corporation is changing by $\frac{dR}{dm}$ million dollars per thousand dollars spent on advertising.

22. The percentage of households with washing machines was changing by $\frac{dW}{dt}$ percentage points per year, where t is the number of years since 1950.

23. The concentration of a drug in the bloodstream of a patient is changing by $\frac{dc}{dh}$ milligrams per liter per hour h hours after the drug was given.

24. The level of production at a tire manufacturer h hours after production began is increasing by $P'(h)$ tires per hour squared.

25. Consider a rate-of-change graph that is increasing but negative over an interval. Explain why the accumulation graph decreases over this interval.

26. What behavior in a rate-of-change graph causes the following to occur in the accumulation graph: a minimum? a maximum? an inflection point? Explain.

6.4   The Fundamental Theorem

In Section 6.3 we graphed accumulation functions from graphs of rate-of-change functions. We saw that we could use numerical estimates of definite integrals to help us sketch accumulation graphs. In this section, we develop some algebraic tools to help us write formulas for accumulation functions.

The Slope Graph of an Accumulation Graph

We began Section 6.3 with the function in Figure 6.45a and drew the accumulation function graph shown in Figure 6.45b.

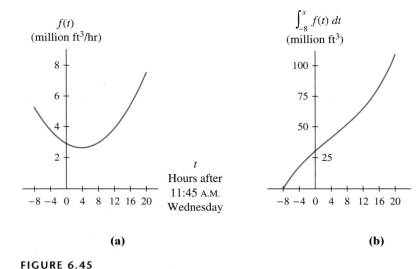

(a)   (b)

FIGURE 6.45

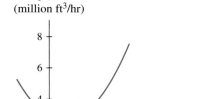

FIGURE 6.46

Now we sketch the slope graph (derivative graph) of the accumulation function in Figure 6.45b. Note that the slopes are positive everywhere but become smaller and smaller as t goes from -8 to approximately 3.7. There appears to be a point of least slope near $t \approx 3.7$ after which the slopes become more and more positive. The slope graph appears in Figure 6.46. The slope graph is exactly the graph with which we began in Figure 6.45a (with the input variable labeled x instead of t).

In Examples 2 and 3 in Section 6.3, we began with the graph in Figure 6.47a and sketched the accumulation function shown in Figure 6.47b.

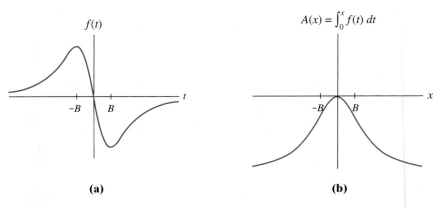

(a)   **(b)**

FIGURE 6.47

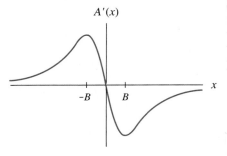

FIGURE 6.48

Again, let us sketch the slope graph of the accumulation function. To the left of zero, the graph has positive slopes. The slopes are near zero to the far left, and the graph becomes steeper until $-B$. Between $-B$ and 0, the slopes are still positive but are approaching 0 as the accumulation function approaches its maximum. At $x = 0$, the slope is zero. To the right of $x = 0$, the slopes are negative. They become more and more negative until B. To the right of B, the slopes are still negative but are getting closer and closer to zero as the graph levels off. The slope graph is shown in Figure 6.48. Again, this is exactly the graph with which we began in Figure 6.47a.

You are probably beginning to see that if we begin with a function f with input t, graph or find a formula for the accumulation function

$$A(x) = \int_a^x f(t)\,dt,$$ and then take the derivative or draw the slope graph, we

get f, the function with which we began but in terms of x. In order to explore this connection between accumulation functions and derivatives, consider the following argument:

Let A be the accumulation function of f from a to x. The graph in Figure 6.49a shows the function f and the area representing the accumulation function value from a to x. Figure 6.49b shows the region whose area is the accumulation value from a to $x + h$.

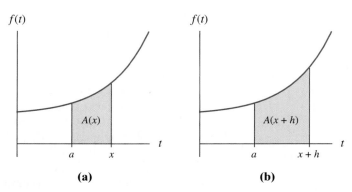

(a)   **(b)**

FIGURE 6.49

Next, consider the difference between the two areas. The small region with this area is shown in Figure 6.50a.

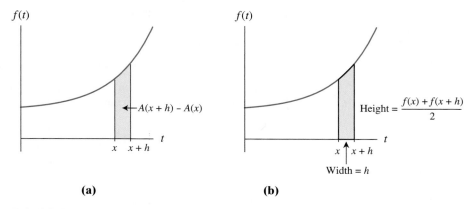

(a) **(b)**

FIGURE 6.50

We can approximate the area of this region by using a trapezoid. Recall that the height of a trapezoid constructed over an interval is the average of the function value at the two endpoints. Figure 6.50b shows this trapezoid and its height and width.

The true area of the shaded region in Figure 6.50a is $A(x + h) - A(x)$, and it can be approximated by the area of the trapezoid. Thus we have

$$A(x + h) - A(x) \approx \left[\frac{f(x) + f(x + h)}{2}\right]h$$

We now divide each side of the expression by h. The reason for this division will be evident later.

$$\frac{A(x + h) - A(x)}{h} \approx \frac{f(x) + f(x + h)}{2}$$

Consider what happens as h becomes smaller and smaller. In other words, what happens when we take the limit of the above expression as h approaches zero?

$$\lim_{h \to 0} \frac{A(x + h) - A(x)}{h} \approx \lim_{h \to 0} \frac{f(x) + f(x + h)}{2}$$

You should recognize the term on the left as the derivative of A. As h approaches 0, $f(x + h)$ gets closer and closer to $f(x)$, so the term on the right approaches $\frac{f(x) + f(x)}{2} = \frac{2f(x)}{2} = f(x)$. Thus we have

$$\frac{dA}{dx} \approx f(x)$$

In fact, we can replace the approximation with an equality (although a discussion of why this is the case is beyond the scope of this book):

$$\frac{dA}{dx} = f(x)$$

6.4.1a, b

This expression says that the derivative of the accumulation function is the original function. This is actually a surprising result—accumulation functions and areas are related to derivatives! In fact, this is such an important result that it is called the **Fundamental Theorem of Calculus.**

The Fundamental Theorem of Calculus

For any continuous function f with input t, the derivative of an accumulation function of f is the function f in terms of x. In symbols, we write

$$\frac{d}{dx}\left(\int_a^x f(t)\,dt \right) = f(x)$$

We can infer from the Fundamental Theorem that to find an accumulation formula, we need only reverse the process of finding a derivative. For this reason, we call the accumulation function $F(x) = \displaystyle\int_a^x f(t)\,dt$ an **antiderivative** of the function f.

Antiderivative

Let f be a function of x. A function F is called an antiderivative of f if $F'(x) = f(x)$; that is, the derivative of F is f.

Our motivation for developing accumulation functions (antiderivatives) is not only to have a formula for accumulated change but also (and more importantly) to develop a function for a quantity if we know a function for that quantity's rate of change.

Antiderivative Formulas

Given a function, how do we find an antiderivative? As we have seen, we must reverse the process of differentiation. Antidifferentiation starts with the known derivative and finds the unknown function. For example, consider the constant function $f(x) = 3$. To find an antiderivative of f, we need to find a function of x whose derivative is 3. One such function is $F(x) = 3x$. Other functions whose derivatives are 3 include $F(x) = 3x + 7$ and $F(x) = 3x - 24.9$.

In fact, having found one antiderivative F for a given function f, we can obtain infinitely many antiderivatives for that function by adding an arbitrary constant C to F. Thus we call $y = F(x) + C$ the **general antiderivative of f.** We use the notation

$$\int f(x)\,dx = F(x) + C$$

6.4.2

for the general antiderivative. The general antiderivative is a group of infinitely many functions. (A particular accumulation function is one specific function from that group.) Note that the integral sign has no upper and lower limits for general antiderivative notation. The dx in this notation is to remind us that we are finding the general antiderivative with respect to x, so our antiderivative formula will be in terms of x. For example, we say that the general antiderivative for $f(x) = 3$ is $F(x) = 3x + C$, and we write $\int 3dx = 3x + C$.

An antiderivative (accumulation formula) of any constant function will be a line because lines have constant derivatives. We can write this general rule as

$$\int kdx = kx + C$$

where k and C are any constants.

Now consider finding the general antiderivative of $f(x) = 2x$. We are seeking a function whose derivative is $2x$. The function is $F(x) = x^2 + C$, and we write

$$\int 2xdx = x^2 + C$$

It is more difficult to reverse the derivative process for $f(x) = x^2$. Recall that the power rule for derivatives $\left[\frac{d}{dx}(x^n) = nx^{n-1}\right]$ instructs us to

- Multiply by the power.
- Subtract 1 from the power to get the new power.

To reverse the process for antiderivatives, we

- *Add* 1 to the power to get the new power.
- *Divide* by that new power.

This formula is known as the **Simple Power Rule** for antiderivatives.

Simple Power Rule for Antiderivatives

$$\int x^n dx = \frac{x^{n+1}}{n+1} + C \quad \text{for any } n \neq -1$$

This rule requires that $n \neq -1$, because otherwise, we would be dividing by zero.

In the case of $f(x) = x^2$, the general antiderivative is $F(x) = \frac{x^{2+1}}{2+1} + C = \frac{x^3}{3} + C$:

$$\int x^2 dx = \frac{x^3}{3} + C$$

EXAMPLE 1   *Using the Simple Power Rule to Find Antiderivatives*

Find the following general antiderivatives and their appropriate units.

a. $\int -7dx$ degrees per hour where x is in hours

b. $\int h^{0.5} dh$ parts per million per day where h is in days

Solution

a. $\int -7dx = -7x + C$ degrees

b. $\int h^{0.5} dh = \frac{h^{1.5}}{1.5} + C$ parts per million   ●

Recall the Constant Multiplier Rule for derivatives:

If $g(x) = kf(x)$, then $g'(x) = kf'(x)$, where k is a constant.

A similar rule applies for antiderivatives:

> ### Constant Multiplier Rule for Antiderivatives
> $$\int kf(x)\,dx = k\int f(x)\,dx$$

Thus $\int 12x^6\,dx = 12\int x^6\,dx = 12\left(\dfrac{x^7}{7}\right) + C = \dfrac{12x^7}{7} + C.$

Another property of antiderivatives that can be easily deduced from a similar property for derivatives is the Sum Rule.

> ### Sum Rule
> $$\int [f(x) \pm g(x)]\,dx = \int f(x)\,dx \pm \int g(x)\,dx$$

The Sum Rule lets us find an antiderivative for a sum (or difference) of functions by operating on each function independently. For example,

$$\int (7x^3 + x)\,dx = \int 7x^3\,dx + \int x\,dx$$

$$= \left(\frac{7x^4}{4} + C_1\right) + \left(\frac{x^2}{2} + C_2\right)$$

$$= \frac{7x^4}{4} + \frac{x^2}{2} + C \quad \text{(Combine } C_1 \text{ and } C_2 \text{ into one constant } C.)$$

Repeated applications of the Simple Power Rule, the Constant Multiplier Rule, and the Sum Rule enable us to find an antiderivative of any polynomial function. We now have the tools we need to begin with a simple polynomial rate-of-change function for a quantity and recover an amount function for that quantity.

EXAMPLE 2   *Using Given Information to Find an Antiderivative*

Birth Rate   An African country has an increasing population but a declining birth rate, a situation that results in the number of babies born each year increasing but at a slower rate. The rate of change in the number of babies born each year is given by

$$b(t) = 87{,}000 - 1600t \text{ births per year}$$

t years from the end of this year. Also, the number of babies born in the current year is 1,185,800.

a.   Find a function describing the number of births each year t years from now.

b.   Use the function in part *a* to estimate the number of babies born next year.

Solution

a. A function B describing the number of births each year is found as

$$B(t) = \int b(t)dt = 87{,}000t - \frac{1600}{2}t^2 + C = 87{,}000t - 800t^2 + C \text{ births}$$

t years from now.

    We also know that $B(0) = 1{,}185{,}800$, so $1{,}185{,}800 = 87{,}000(0) - 800(0)^2 + C$, which gives $C = 1{,}185{,}800$. Thus we have

$$B(t) = 87{,}000t - 800t^2 + 1{,}185{,}800 \text{ births each year}$$

t years from now.

b. The number of babies born next year is estimated as $B(1) = 1{,}272{,}000$ babies.   ●

We have just presented and applied three antiderivative rules: the Simple Power Rule, the Constant Multiplier Rule, and the Sum Rule. Now let us look at four more rules for finding antiderivatives.

Refer to the Simple Power Rule, and note that it did not apply for $n = -1$. The case where $n = -1$ is special. This results in the antiderivative $\int x^{-1}dx = \int \frac{1}{x}\,dx$. Recall that $\frac{d}{dx}(\ln x) = \frac{1}{x}$. This is valid only for $x > 0$, because $\ln x$ is not defined for $x < 0$. When $x < 0$, we use $\ln(-x)$ because $\frac{d}{dx}[\ln(-x)] = \frac{-1}{-x} = \frac{1}{x}$. Both cases ($x > 0$ and $x < 0$) can be handled very simply by using $\ln |x|$.

Natural Log Rule

$$\int \frac{1}{x}\,dx = \ln |x| + C$$

The two final antiderivative formulas that we consider are for exponential functions. Recall that the derivative of $f(x) = e^x$ is e^x. Similarly, the general antiderivative of $f(x) = e^x$ is also e^x plus a constant.

---

e^x Rule

$$\int e^x dx = e^x + C$$

The other exponential function that we have encountered is $f(x) = b^x$. Its derivative was found by multiplying b^x by $\ln b$. To find the general antiderivative, we divide b^x by $\ln b$ and add a constant:

Exponential Rule

$$\int b^x dx = \frac{b^x}{\ln b} + C$$

Note that the e^x Rule is a special case of this Exponential Rule with $b = e$.

Because we often encounter functions of the form $f(x) = e^{ax}$, it is helpful to know this function's derivative formula. Recall that the Chain Rule for derivatives applied to $f(x) = e^{ax}$ gives $f'(x) = ae^{ax}$. Thus, to reverse the derivative process, we leave the e^{ax}-term and divide, rather than multiply, by a.

---

e^{ax} Rule

$$\int e^{ax}\,dx = \frac{1}{a}e^{ax} + C \quad \text{for any } a \neq 0$$

In summary, we now have the following antiderivative formulas, where k, n, b, and a are constants.

Antiderivative Formulas

	Function: f	General antiderivative: $\int f(x)\,dx$
Constant Rule	k	$kx + C$
Simple Power Rule	$x^n, n \neq -1$	$\left(\frac{1}{n+1}\right)x^{n+1} + C$
Natural Log Rule	$\frac{1}{x}$	$\ln|x| + C$
Exponential Rule	$b^x, b > 0$	$\left(\frac{1}{\ln b}\right)b^x + C$
e^x Rule	e^x	$e^x + C$
e^{ax} Rule	$e^{ax}, a \neq 0$	$\frac{1}{a}e^{ax} + C$
Constant Multiplier Rule	$kg(x)$	$k\int g(x)\,dx$
Sum Rule	$g(x) \pm h(x)$	$\int g(x)\,dx \pm \int h(x)\,dx$

EXAMPLE 3   *Using Antiderivative Formulas to Find General Antiderivatives*

Find the following general antiderivatives.

a. $\int\left(3^x - 7e^x + \frac{5}{x}\right)dx$ quarts per hour where x is measured in hours

b. $\int(4\sqrt{x} + 100e^{0.06x} + 0.46)\,dx$ mpg per mph where x is measured in mph

Solution

a. $\int \left(3^x - 7e^x + \dfrac{5}{x}\right) dx = \int 3^x dx - 7 \int e^x dx + 5 \int \dfrac{1}{x} dx$

$$= \dfrac{3^x}{\ln 3} - 7e^x + 5 \ln |x| + C \text{ quarts}$$

b. We must first rewrite $\sqrt{x}$ as $x^{1/2}$ and then apply the appropriate rules:

$$\int \left(4\sqrt{x} + 100e^{0.06x} + 0.46\right) dx = \int [4x^{1/2} + 100e^{0.06x} + 0.46] dx$$

$$= 4 \int x^{1/2} dx + 100 \int e^{0.06x} dx + \int 0.46 \, dx$$

$$= 4 \dfrac{x^{3/2}}{3/2} + 100 \dfrac{e^{0.06x}}{0.06} + 0.46x + C$$

$$= \dfrac{8}{3} x^{3/2} + \dfrac{100}{0.06} e^{0.06x} + 0.46x + C \text{ mpg} \quad \bullet$$

Specific Antiderivatives

We have seen that any function has infinitely many antiderivatives, each differing by a constant. When we seek an antiderivative with a particular constant, we call the resulting function a **specific antiderivative.** An example of a specific antiderivative is the function in Example 2 for the number of births in an African country. Using the information that in the current year the number of births is 1,185,800, we were able to solve for the constant to obtain the specific antiderivative $B(t) = 87,000t - 800t^2 + 1,185,800$ births each year t years from now.

In general, to find a specific antiderivative, you must be given additional information about the output quantity that the antiderivative describes. After finding a general antiderivative, you simply substitute the given input and corresponding output into the general antiderivative and solve for the constant. Then you replace the constant in the general antiderivative formula with the value you found to obtain the specific antiderivative. Example 4 illustrates this process.

EXAMPLE 4   *Finding a Specific Antiderivative*

Marginal Cost   Suppose that a manufacturer of small toaster ovens has collected the data given in Table 6.15, which shows, at various production levels, the approximate cost to produce one more oven. Recall from Section 5.1 that this is marginal cost and can be interpreted as the rate of change of cost.

TABLE 6.15

Production level (ovens per day)	200	300	400	500	600	700
Cost to produce an additional oven	$29	$20	$15	$11	$9	$7

The manufacturer also knows that the total cost to produce 250 ovens is $12,000.

a. Find a model for the marginal cost data.

b. Recover a cost model from the model you found in part *a*.

c.   Estimate the cost of producing 500 ovens.

Solution

a.   Either a quadratic or an exponential model is a good fit to the data. We choose an exponential model:

$$C'(x) = 47.638(0.9972^x) \text{ dollars per oven}$$

where x is the number of ovens produced each day.

b.   To recover a model for cost, we need an antiderivative of C' satisfying the known condition that the cost to produce 250 ovens is $12,000. The general antiderivative is

$$C(x) = \frac{47.638(0.9972^x)}{\ln 0.9972} + K \approx -16{,}991.852(0.9972^x) + K$$

where K is a constant.
   Using the fact that $C(250) = \$12{,}000$, we substitute 250 for x, set the antiderivative equal to 12,000, and solve for K.

$$-16{,}991.852(0.9972^{250}) + K = 12{,}000$$

$$-8430.309 + K = 12{,}000$$

$$K \approx 20{,}430.309$$

Thus the specific antiderivative giving the approximate cost of producing x toaster ovens is

$$C(x) = -16{,}991.852(0.9972^x) + 20{,}430.309 \text{ dollars}$$

c.   You can readily verify that the cost of producing 500 toaster ovens is estimated by $C(500) \approx \$16{,}248$.   ●

It is sometimes necessary to find an antiderivative twice in order to obtain the appropriate accumulation formula. For example, to obtain distance from acceleration, you must determine the specific antiderivative of the acceleration function to obtain a velocity function and then determine the specific antiderivative of the velocity function to obtain a function for distance traveled.

EXAMPLE 5   *Recovering Distance from Acceleration*

Falling Pianos   A mathematically inclined cartoonist wants to make sure his animated cartoons accurately portray the laws of physics. In a particular cartoon he is creating, a grand piano falls from the top of a 10-story building. How many seconds should he allow the piano to fall before it hits the ground? Assume that one story equals 12 feet and that acceleration due to gravity is −32 feet per second squared.

Solution   We begin with the equation for acceleration, $a(t) = -32$ feet per second squared t seconds after the piano falls. The antiderivative of acceleration gives an equation for velocity.

$$\int a(t)\,dt = v(t) = -32t + C \text{ feet per second after } t \text{ seconds of fall}$$

To find the specific velocity equation for this example, we need information about the velocity at a specific time. The fact that the piano falls (rather than being pushed with an initial force) tells us that the velocity is zero when the time is zero. Substituting zero for both t and v and solving for C give $C = 0$. Thus the specific antiderivative describing velocity is $v(t) = -32t$ feet per second after t seconds of fall.

From this velocity equation, we can derive an equation for position (the distance the piano is above the ground) by finding the antiderivative. In this case, the general antiderivative of v is

$$\int v(t)\,dt = s(t) = -16t^2 + K \text{ feet after } t \text{ seconds of fall}$$

Again, we find the specific position equation by using information given about the position of the piano at a certain time. In this example we know that when time is zero, the piano is 10 stories, or 120 feet, above the ground. We substitute this information into the position equation and solve for K, obtaining $K = 120$ feet. Substituting this value for K in the position equation yields the specific position equation

$$s(t) = -16t^2 + 120 \text{ feet after } t \text{ seconds of fall}$$

Now that we have equations for acceleration, velocity, and position, we can answer the question posed: How long until the piano hits the ground? Let us phrase this question mathematically: When position is zero, what is time? To answer this question, we set the position equation equal to zero and solve for t.

$$0 = -16t^2 + 120$$
$$16t^2 = 120$$
$$t^2 = \frac{120}{16}$$
$$t \approx \pm\, 2.7 \text{ seconds}$$

Because negative time doesn't make sense in the context of the question, we conclude that the cartoonist should allow approximately 2.7 seconds for the piano to fall. ●

Let us summarize what we have learned thus far about integrals. The definite integral $\int_a^b f(x)\,dx$ is a limiting value of sums of areas of rectangles and gives us the area of the region between the graph of f and the x-axis if the graph lies above the horizontal axis from a to b. When the graph of f lies below the horizontal axis from a to b, the definite integral is the negative of the area inscribed. If $f(x)$ is a rate of change of some quantity, then $\int_a^b f(x)\,dx$ is the change in the quantity from a to b.

The accumulation function $A(x) = \int_a^x f(t)\,dt$ is a formula in terms of x for the accumulated change in $f(t)$ from a to x. We use the integral symbol without the upper and lower limits, $\int f(x)\,dx$, to represent the general antiderivative of f.

Although these three symbols are similar, it is important that you have a clear understanding of what each one represents. Their interpretations are summarized in Table 6.16.

TABLE 6.16

Symbol	Name	Interpretation
$\int_a^b f(x)\,dx$	definite integral	a number that can be thought of in terms of area
$\int_a^x f(t)\,dt$	accumulation function, specific antiderivative	a formula for an accumulated amount
$\int f(x)\,dx$	general antiderivative	a formula whose derivative is f

The Fundamental Theorem of Calculus tells us that accumulation functions are specific antiderivatives. As we shall see in Section 6.5, antiderivatives enable us to find areas algebraically by using accumulation formulas rather than numerically as limiting values of sums of areas of rectangles.

Apart from helping us find areas, antiderivatives are useful in allowing us to recover functions from rates of change. We have seen several examples of that in this section. It may seem difficult to reverse your thinking from finding derivatives to finding antiderivatives, but with practice you will soon be proficient at both.

6.4 Concept Inventory

○ *Fundamental Theorem of Calculus*
○ *Antiderivative*
○ *Recovering a function from its rate of change*
○ $\int f(x)\,dx$ *= general antiderivative*
○ *Antiderivative formulas*
○ *Specific antiderivative*

6.4 Activities

In Activities 1 through 4, *a* and *b* are constants and *x* and *t* are variables. In these activities, label each notation as always representing

   a. A function of *x*

   b. A function of *t*

   c. A number

1. **a.** $f'(t)$ **b.** $\dfrac{df}{dx}$ **c.** $f'(3)$

2. **a.** $\int f(t)\,dt$ **b.** $\int f(x)\,dx$ **c.** $\int_a^b f(t)\,dt$

3. **a.** $\int_a^b f(x)\,dx$ **b.** $\int_a^x f(t)\,dt$ **c.** $\int_b^t f(x)\,dx$

4. **a.** $\dfrac{d}{dx}\int_a^x f(t)\,dt$ **b.** $\dfrac{d}{dt}\int_a^t f(x)\,dx$ **c.** $\dfrac{d}{dx}\int_a^a f(t)\,dt$

5. Illustrate and explain the Fundamental Theorem of Calculus from a numerical viewpoint.

6. Write the Fundamental Theorem of Calculus from an algebraic viewpoint.

7. Write the Fundamental Theorem of Calculus from a verbal viewpoint. Do not include mathematical symbols or graphs.

8. Illustrate and explain the Fundamental Theorem of Calculus from a graphical viewpoint.

Find the general antiderivative as indicated in Activities 9 through 14. Check each of your antiderivatives by taking its derivative.

WEB/CD

Skills

9. $\int 19.436(1.07^x)\,dx$

10. $\int 39.24e^{3.9x}\,dx$

11. $\int [6e^x + 4(2^x)]\,dx$

12. $\int (32.685x^3 + 3.296x - 15.067)\,dx$

13. $\int (10^x + 4\sqrt{x} + 8)\,dx$

14. $\int \left[\dfrac{1}{2}x + \dfrac{1}{2x} + \left(\dfrac{1}{2}\right)^x\right]\,dx$

For each of the rate-of-change functions in Activities 15 through 18, find the general antiderivative, and label the units on the antiderivative.

15. $s(m) = 6250(0.92985^m)$ CDs per month m months since the beginning of the year

16. $p(x) = 0.03731x^2 - 0.4841x + 1.4069$ dollars per 1000 cubic feet per year x years since 1989

17. $c(x) = \dfrac{0.7925}{x} + 0.3292(0.009324^x)$ dollars per unit squared when x units are produced

18. $p(t) = 1.724928e^{0.0256t}$ millions of people per year t years after 1990

In Activities 19 through 21, find F, the specific antiderivative of f.

19. $f(t) = t^2 + 2t$; $F(12) = 700$

20. $f(u) = \dfrac{2}{u} + u$; $F(1) = 5$

21. $f(z) = \dfrac{1}{z^2} + e^z$; $F(2) = 1$

22. Bond Yields The rate of change of the average yield of short-term German bonds can be described by the equation $G(t) = \dfrac{0.5696}{t}$ percentage points per year for a bond with a maturity time of t years. The average 10-year bond has a yield of 4.95%. Find the specific antiderivative describing the average yield of short-term German bonds. How is this specific antiderivative related to an accumulation function of G?

23. Weight The rate of change of the weight of a laboratory mouse can be modeled by the equation $W(t) = \dfrac{7.372}{t}$ grams per week, where t is the age of the mouse, in weeks, beyond 2 weeks. At an age of 9 weeks, the mouse weighed 26 grams. Find the specific antiderivative describing the weight of the mouse. How is this specific antiderivative related to an accumulation function of W?

24. Fuel Use The rate of change of the average annual fuel consumption of passenger vehicles, buses, and trucks from 1970 through 2000 can be modeled by the equation $f(t) = 0.798t - 15.886$ gallons per vehicle per year t years after 1970. The average annual fuel consumption was 712 gallons per vehicle in 1980. (Source: Based on data from Bureau of Transportation Statistics.)

   a. Find the specific antiderivative giving the average annual fuel consumption.

   b. How is this specific antiderivative related to an accumulation function of f?

25. Gender Ratio The rate of change of the gender ratio for the United States during the twentieth century can be modeled as $g(t) = (1.667 \cdot 10^{-4})t^2 - 0.01472t - 0.103$ males per 100 females per year t years after 1900. In 1970 the gender ratio was 94.8 males per 100 females. (Source: Based on data from U.S. Census Bureau.)

   a. Find a specific antiderivative giving the gender ratio.

   b. How is this specific antiderivative related to an accumulation function of g?

26. Labor Force During the 1990s the civilian labor force in the Trident region of South Carolina was expanding at a rate of $p(x) = -2915x + 19{,}433$ people each year x years after 1989. The civilian labor force was 227,120 people in 1989. (Source: Based on information in 1993 Trident Economic Forecast.)

   a. Find a model for the size of the civilian labor force in the 1990s.

   b. Determine, according to the model, how many people were in the labor force by 1992.

   c. According to the model, how many people were added to the labor force between 1989 and 1991?

27. Subscribers On the basis of data from the Cellular Telecommunications Industry Association, the rate of change in the number of cellular telephone subscribers in the United States from 1988 through 1995 can be modeled by the equation $p(x) = 891.6(1.5^x)$ thousand subscribers per year x years after 1988.

a. Find a model for the number of cellular phone subscribers in the United States. Use the fact that there were 2.069 million subscribers in 1988.

b. Find the change in the number of cellular phone subscribers in the United States between 1990 and 1994.

28. **Investment**   An investment worth $1 million in 2000 has been growing at a rate of $f(t) = 0.1397619(1.15^t)$ million dollars per year t years after 2000.

a. Determine how much the investment has grown since 2000 and how much it is projected to grow over the next year.

b. Recover the amount function, and determine the current value of the investment and its projected value next year.

29. **Dropped Coin**   The Washington Monument, located at one end of the Federal Mall in Washington, D.C., is the world's tallest obelisk at 555 feet. Suppose that a tourist drops a penny from the observation deck atop the monument. Let us assume that the penny falls from a height of 540 feet.

a. Recover the velocity function for the penny using the facts that

  i. Acceleration due to gravity near the surface of the earth is -32 feet per second squared.

  ii. Because the penny is dropped, velocity is 0 when time is 0.

b. Recover the distance function for the penny using the velocity function from part a and the fact that distance is 540 feet when time is 0.

c. When will the penny hit the ground?

d. What is the impact velocity of the penny in miles per hour?

30. **High Dive**   According to the *Guinness Book of Records*, the world's record high dive from a diving board is 176 feet, 10 inches. It was made by Olivier Favre (Switzerland) in 1987. Ignoring air resistance, approximate Favre's impact velocity in miles per hour from a height of 176 feet, 10 inches.

31.* **Velocity**   In the 1960s, Donald McDonald claimed in an article in the *New Scientist* that plummeting cats never fall faster than 40 mph.

a. What is the impact velocity (in feet per second and miles per hour) of a cat that accidentally falls off a building from a height of 66 feet $\left(5\frac{1}{2}\text{ stories}\right)$?

b. What accounts for the difference between your answer to part a and McDonald's claim (assuming McDonald's claim is accurate)?

32. **Donors**   The table gives the increase or decrease in the number of donors to a college athletics support organization for selected years.

Year	Rate of change in donors (donors per year)
1985	-169
1988	803
1991	1222
1994	1087
1997	399
2000	-842

a. Find a model for the rate of change in the number of donors.

b. Find a model for the number of donors. Use the fact that in 1990 there were 10,706 donors.

c. Estimate the number of donors in 2002.

33. **Employees**   From 1997 through 2002, an Internet company was hiring new employees at a rate of $n(x) = \frac{593}{x} + 138$ new employees per year, where x represents the number of years since 1996. By 2001 the company had hired 896 employees.

a. Write the function that gives the number of employees who had been hired by the xth year after 1996.

b. For what years will the function in part a apply?

c. Find the total number of employees the company had hired between 1997 and 2002. Would this figure necessarily be the same as the number of employees the company had at the end of 2002? Explain.

*McDonald's study referred to in Activity 31 was based on observations of veterinarians who treated cats that had fallen from buildings in New York City. None of the cats' falls were deliberately caused by the researchers.

6.5   The Definite Integral

So far, we have been finding general and specific antiderivatives to recover functions from their rate-of-change functions. The Fundamental Theorem of Calculus tells us that accumulation functions are antiderivatives and gives us a method for finding them. In general, we know that $\int_a^x f(t)\,dt = F(x) + C$, where F is an antiderivative of f. We also know that when $x = a$, the accumulation function is zero.

$$\int_a^a f(t)\,dt = F(a) + C = 0$$

This tells us that $C = -F(a)$. Thus we have

$$\int_a^x f(t)\,dt = F(x) - F(a)$$

To find the value of the accumulation function from a to b, we simply substitute b for x.

$$\int_a^b f(t)\,dt = F(b) - F(a)$$

We now have an efficient algebraic method for evaluating definite integrals:

Evaluating a Definite Integral

If f is a continuous function from a to b and F is any antiderivative of f, then

$$\int_a^b f(x)\,dx = F(b) - F(a)$$

Antiderivatives and Definite Integrals

Recall from Section 6.2 that we define the definite integral, $\int_a^b f(x)\,dx$, as the limiting value of sums of signed areas of rectangles. That is,

$$\int_a^b f(x)\,dx = \lim_{n \to \infty} [f(x_1) + f(x_2) + \cdots + f(x_n)]\Delta x$$

The antiderivative definition for a definite integral, $\int_a^b f(x)\,dx = F(b) - F(a)$, gives us a second, less tedious method for evaluating a definite integral for many functions. For example, we can calculate the area of the region between the graph of $f(x) = x^2 + 2$ and the x-axis from -2 to 4. (See Figure 6.51.)

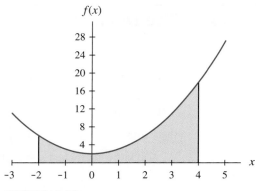

FIGURE 6.51

To find the area of this region in Sections 6.1 and 6.2, we would have used sums of areas of rectangles. Now all we need to do is calculate the value of the definite integral, $\int_{-2}^{4}(x^2 + 2)\,dx$, by simply finding an antiderivative and subtracting the value of the antiderivative at -2 from the value at 4.

$$\int_{-2}^{4}(x^2 + 2)\,dx = F(4) - F(-2)$$

where F is an antiderivative of $x^2 + 2$. Here are the details of this process:

1. Find an antiderivative:

$$\int_{-2}^{4}(x^2 + 2)\,dx = \left(\frac{x^3}{3} + 2x\right)\Big|_{-2}^{4}$$

← This notation is used to indicate that we have found an antiderivative and now must evaluate it at 4 and -2 and then subtract the results.

2. Evaluate at upper and lower limits and then subtract:

$$= \left[\frac{4^3}{3} + 2(4)\right] - \left[\frac{(-2)^3}{3} + 2(-2)\right]$$

$$= 29\frac{1}{3} - \left(-6\frac{2}{3}\right) \approx 29.333 - (-6.667)$$

$$= 36$$

Thus the area of the region depicted in Figure 6.51 is exactly 36.

EXAMPLE 1   *Evaluating and Interpreting a Definitive Integral*

a. Find a formula for $\int 69.7966(1.07229^t)\,dt$.

b. Find a formula for the accumulation function $A(x) = \int_{1}^{x} 69.7966(1.07229^t)\,dt$.

c. Determine $\int_{1}^{4} 69.7966(1.07229^t)\,dt$. Interpret your answer graphically.

d. If $f(t) = 69.7966(1.07229^t)$ is the rate of change of the balance in a savings account given in dollars per year and t is the number of years since the savings account was opened, what does your answer in part c represent?

Solution

a. $\displaystyle\int 69.7966(1.07229^t)\,dt = \frac{69.7966(1.07229)^t}{\ln 1.07229} + C$

b. The accumulation function is the specific antiderivative (in terms of x) for which the antiderivative is zero when $x = 1$.

$$A(1) = \frac{69.7966(1.07229^1)}{\ln 1.07229} + C = 0$$

$$C \approx -1072.2908$$

Thus we have

$$A(x) = \int_1^x 69.7966(1.07229^t)\,dt \approx \frac{69.7966(1.07229^x)}{\ln 1.07229} - 1072.2908$$

c. $\displaystyle\int_1^4 69.7966(1.07229^t)\,dt = A(t)\Big|_1^4$

$$= \frac{69.7966(1.07229^t)}{\ln 1.07229}\Big|_1^4$$

$$= \frac{69.7966(1.07229^4)}{\ln 1.07229} - \frac{69.7966(1.07229^1)}{\ln 1.07229}$$

$$\approx 249.7637.$$

This value is the area of the region between the graph of $f(t) = 69.7966(1.07229^t)$ and the horizontal axis from 1 to 4.

d. The answer to part c represents the change in the amount in the savings account between 1 and 4 years. The amount grew by \$249.76. ●

Now consider the result of part d of Example 1 if $f(1) = \$1000$. The function giving the balance would then be $A(t) \approx \dfrac{69.7966(1.07229^t)}{\ln 1.07229} - 72.2908$ dollars t years after the savings was opened. The definite integral calculation would be

$$\int_1^4 69.7966(1.07229^t) = A(t)\Big|_1^4$$

$$= \left[\frac{69.7966(1.07229^t)}{\ln 1.07229} - 72.2908\right]\Big|_1^4$$

$$= \left[\frac{69.7966(1.07229^4)}{\ln 1.07229} - 72.2908\right]$$

$$- \left[\frac{69.7966(1.07229^1)}{\ln 1.07229} - 72.2908\right]$$

$$\approx 249.7637 \approx \$249.76$$

Note that the balance at the starting value $t = 1$ made no difference in the definite integral calculation. This is not a coincidence. If F_1 and F_2 are any two antiderivatives of f, they differ only by a constant: $F_1(x) = F_2(x) + C$. Thus the change in $F_1(x)$ from a to b is

$$F_1(b) - F_1(a) = [F_2(b) + C] - [F_2(a) + C] = F_2(b) - F_2(a)$$

which is the change in $F_2(x)$ from a to b. The constant term C always cancels out during the definite integral calculation.

> The constant term in an antiderivative does not affect definite integral calculations. If you are concerned only with finding change in a quantity, finding the constant in the antiderivative is not necessary.

EXAMPLE 2   Using a Definite Integral to Find Accumulated Change

Marginal Cost   In Example 4 of Section 6.4 we modeled the marginal cost for toaster ovens using the exponential function

$$C'(x) = 47.638(0.9972^x) \text{ dollars per oven}$$

where x is the number of ovens produced per day. Suppose that the current production level is 300 ovens per day and that the manufacturer wishes to increase production to 500 ovens per day. How will this increase affect production cost?

Solution   The definite integral $\int_{300}^{500} C'(x)\,dx = C(500) - C(300)$ gives the change in cost as a result of this increase. Finding the change requires two steps. First, find an antiderivative of C'. Then evaluate the antiderivative at the two production levels and subtract the value at the lower limit from the value at the upper limit of the integral.

In Section 6.4 (using an unrounded model), we found the general antiderivative for C' to be

$$C(x) = \int C'(x)\,dx = \text{-}16{,}991.852(0.9972^x) + K$$

The constant K will not affect our calculations of change, so we set K to be 0. (Note that if we need to answer other questions about cost, we should use the information that the cost to produce 250 ovens is \$12,000 to find the proper value of K, as was illustrated in Example 4.)

The definite integral is

$$\int_{300}^{500} C'(x)\,dx = C(x) \Big|_{300}^{500}$$

$$= C(500) - C(300)$$

$$= \text{-}16{,}991.852(0.9972^{500}) - [\text{-}16{,}991.852(0.9972^{300})]$$

$$\approx \$3145$$

We conclude that when production is increased from 300 to 500 ovens per day, cost increases by approximately \$3145.   ●

Sums of Definite Integrals

Definite integral calculations for piecewise continuous functions require special care. In Example 1 of Section 6.3, we saw flow rates for a river during heavy rains. The model in that example is based on the data in Table 6.17. A scatter plot of the data is given in Figure 6.52.

TABLE 6.17

Hours since 11:45 A.M. Wednesday	Flow rate (cubic feet per hour)
0	2,826,000
4	2,710,800
8	3,002,400
12	3,852,000
16	5,292,000
20	7,524,000
23	6,624,000
27	5,760,000

(Source: As reported in the *Reno Gazette–Journal*, May 17, 1996, p. 4a.)

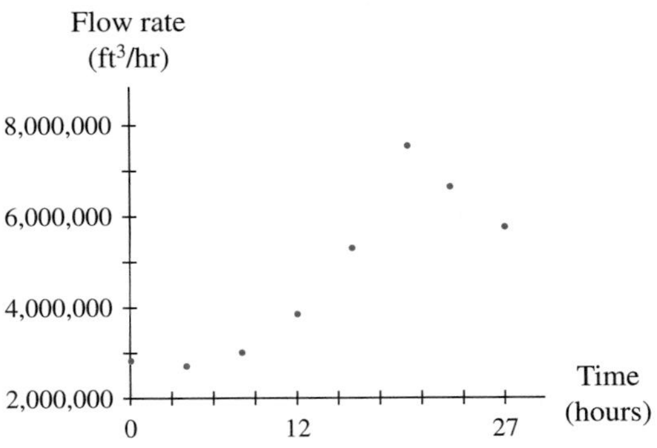

FIGURE 6.52

When the entire data set is viewed, the three points on the right appear to follow a linear pattern. However, when viewed in isolation, the points are distinctively concave up. For this reason, we chose quadratic models for both portions of the data.

A piecewise continuous function is appropriate. We choose to divide the data at 20 hours and obtain the model

$$f(h) = \begin{cases} 18{,}225h^2 - 135{,}334.3h + 2{,}881{,}542.9 \text{ ft}^3/\text{hr} & \text{when } 0 \le h \le 20 \\ 12{,}000h^2 - 816{,}000h + 19{,}044{,}000 \text{ ft}^3/\text{hr} & \text{when } 20 < h \le 27 \end{cases}$$

where h is the number of hours since 11:45 A.M. Wednesday.

To estimate the amount of water that flowed through the river from 11:45 A.M. Wednesday to 2:45 P.M. Thursday, 27 hours later, we calculate the value of the definite integral, $\int_0^{27} f(h)\,dh$. Note that the point of division for the model occurs in the interval from 0 to 27. For this reason, we cannot calculate the value of the definite integral simply by evaluating an antiderivative of f at 27 and 0 and subtracting.

Note that the area of the region from a to b shaded in Figure 6.53 is equal to the sum of the area of R_1 and the area of R_2.

This figure illustrates the following property of integrals:

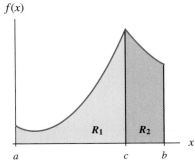

FIGURE 6.53

> ### Sum Property of Integrals
>
> $$\int_a^b f(x)\,dx = \int_a^c f(x)\,dx + \int_c^b f(x)\,dx$$
>
> where c is between a and b.

It is this property that enables us to calculate definite integrals for piecewise continuous functions.

Returning to the river flow function, in order to calculate $\int_{0}^{27} f(h)\,dh$, we divide the integral into two pieces at the point where the model changes and sum the results.

$$\int_{0}^{27} f(h)\,dh = \int_{0}^{20} f(h)\,dh + \int_{20}^{27} f(h)\,dh$$

$$= \int_{0}^{20} (18{,}225h^2 - 135{,}334.3h + 2{,}881{,}542.9)\,dh$$

$$+ \int_{20}^{27} (12{,}000h^2 - 816{,}000h + 19{,}044{,}000)\,dh$$

$$= \left(\frac{18{,}225}{3}h^3 - \frac{135{,}334.3}{2}h^2 + 2{,}881{,}542.9h \right) \Big|_{0}^{20}$$

$$+ \left(\frac{12{,}000}{3}h^3 - \frac{816{,}000}{2}h^2 + 19{,}044{,}000h \right) \Big|_{20}^{27}$$

$$= (79{,}164{,}000 - 0) + (295{,}488{,}000 - 249{,}680{,}000)$$

$$= 79{,}164{,}000 + 45{,}808{,}000$$

$$= 124{,}972{,}000 \text{ cubic feet}$$

We estimate that during the first 20 hours, 79,164,000 ft³ of water flowed through the river. Between 20 and 27 hours, the volume of water was about 45,808,000 ft³. Summing these two values, we estimate that from 11:45 A.M. Wednesday to 2:45 P.M. Thursday, 124,972,000 ft³ of water flowed through the river.

In order to calculate the change from a to b in a function whose graph is sometimes above and sometimes below the horizontal axis, it is necessary only to calculate $\int_{a}^{b} f(x)\,dx$ and not calculate the change over separate intervals. This concept is illustrated in Example 3.

EXAMPLE 3   *Illustrating the Sum Property of Integrals*

6.5.1a, b

Sea Level   Scientists believe that the average sea level is dropping and has been for some 4000 years. They also believe that was not always the case. Estimated rates of change in the average sea level in meters per year during the past 7000 years are given in Table 6.18. A quadratic model for the data is

$$r(t) = 0.14762t^2 + 0.35952t - 0.8 \text{ meters per yard}$$

t thousand years from the present (past years are represented by negative numbers).

TABLE 6.18

Time, t (thousands of years before present)	-7	-6	-5	-4	-3	-2	-1
Rate of change of average sea level, $r(t)$ (meters/year)	3.8	2.6	1.0	0.1	-0.6	-0.9	-1.0

(Source: Estimated from information in *Ecology of Natural Resources*, François Ramade. New York: Wiley, 1981.)

A graph of r is shown in Figure 6.54.

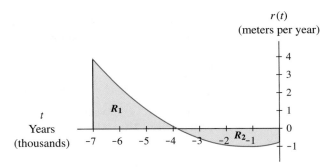

FIGURE 6.54

a. Find the areas of the regions above and below the t-axis from $t = -7$ to $t = 0$. Interpret the areas in the context of sea level.

b. Find $\displaystyle\int_{-7}^{0} r(t)\,dt$, and interpret your answer.

Solution

a. The graph in Figure 6.54 crosses the t-axis at $t \approx -3.845$ thousand years. The area of region R_1 (above the t-axis) is

$$\int_{-7}^{-3.845} (0.14762t^2 + 0.35952t - 0.8)\,dt = \left(\frac{0.14762}{3}t^3 + \frac{0.35952}{2}t^2 - 0.8t\right)\bigg|_{-7}^{-3.845}$$

$$\approx 2.9365 - (-2.4694) \approx 5.4 \text{ meters}$$

The area of region R_2 (below the t-axis) is

$$-\int_{-3.845}^{0} (0.14762t^2 + 0.35952t - 0.8)\,dt = -\left(\frac{0.14762}{3}t^3 + \frac{0.35952}{2}t^2 - 0.8t\right)\bigg|_{-3.845}^{0}$$

$$\approx -(0 - 2.9365) \approx 2.9 \text{ meters}$$

From 7000 years ago through 3845 years ago, the average sea level rose by approximately 5.4 meters. From 3845 years ago to the present, the average sea level fell by approximately 2.9 meters.

b. $\displaystyle\int_{-7}^{0} r(t)\,dt = \int_{-7}^{0} (0.14762t^2 + 0.35952t - 0.8)\,dt$

$$= \left(\frac{0.14762}{3}t^3 + \frac{0.35952}{2}t^2 - 0.8t\right)\bigg|_{-7}^{0}$$

$$\approx 0 - (-2.4694) \approx 2.5 \text{ meters}$$

From 7000 years ago to the present, the average sea level has risen approximately 2.5 meters. This result is the same as that obtained by subtracting the amount that sea level has fallen from the amount that it has risen:

$$5.4 \text{ meters} - 2.9 \text{ meters} = 2.5 \text{ meters} \quad \bullet$$

Differences of Accumulated Changes

Concept Development: Area Between Two Curves   Now we turn our attention to the difference of two accumulated changes. This difference can often be thought of as the area of a region between two curves. For example, suppose the number of patients admitted to a large inner-city hospital is changing by

$$a(h) = 0.0145h^3 - 0.549h^2 + 4.85h + 8.00 \text{ patients per hour}$$

h hours after 3 A.M. We find the approximate number of patients admitted to the hospital between 7 A.M. ($h = 4$) and 10 A.M. ($h = 7$) as

$$\int_4^7 a(h)\,dh = \int_4^7 (0.0145h^3 - 0.549h^2 + 4.85h + 8.00)\,dh$$
$$= (0.003625h^4 - 0.183h^3 + 2.425h^2 + 8.00h)\Big|_4^7$$
$$\approx 120.760 - 60.016 \approx 61 \text{ patients}$$

Graphically, this value is the area of the region between the graph of a and the horizontal axis from 4 to 7. (See Figure 6.55.)

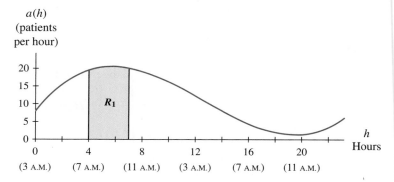

FIGURE 6.55   Area of R_1 is the number of patients admitted between 7 A.M. and 10 A.M.

Now suppose that the rate at which patients are discharged is modeled by the equation

$$y(h) = \begin{cases} 0 \text{ patients/hour} & \text{when } 0 \le h < 4 \\ -0.028h^3 + 0.528h^2 + 0.056h - 1.5 \text{ patients/hour} & \text{when } 4 \le h \le 17 \\ 0 \text{ patients/hour} & \text{when } 17 < h \le 24 \end{cases}$$

where h is the number of hours after 3 A.M. The approximate number of patients discharged between 7 A.M. and 10 A.M. is calculated as

$$\int_4^7 y(h)\,dh = \int_4^7 (-0.028h^3 + 0.528h^2 + 0.056h - 1.5)\,dh$$
$$= (-0.007h^4 + 0.176h^3 + 0.028h^2 - 1.5h)\Big|_4^7$$
$$= 34.433 - 3.92 \approx 31 \text{ patients}$$

Graphically, this value is the area of the region between the graph of y and the horizontal axis from 4 to 7. (See Figure 6.56.)

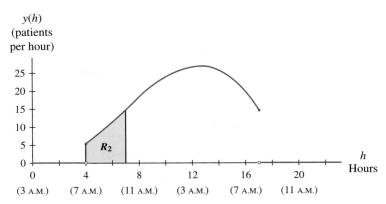

FIGURE 6.56   Area of R_2 is the number of patients discharged between 7 A.M. and 10 A.M.

The net change in the number of patients at the hospital from 7 A.M. to 10 A.M. is the difference between the number of patients admitted and the number discharged between 7 A.M. and 10 A.M. That is,

$$\text{Change in the number of patients from 7 A.M. to 10 A.M.} = \int_4^7 a(h)\,dh - \int_4^7 y(h)\,dh$$
$$\approx 60.744 - 30.513$$
$$\approx 30 \text{ patients}$$

Geometrically, we represent this value as the area of the region below the graph of a and above the graph of y from 4 to 7. (See Figure 6.57.)

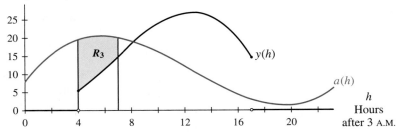

FIGURE 6.57   Area of R_3 is the net change in hospital patients between 7 A.M. and 10 A.M.

In general, when we want to find the area of a region that lies below one curve f and above another curve g from a to b (as in Figure 6.58), we calculate it as

$$\text{Area of the region between the graphs of } f \text{ and } g = \text{area beneath the graph of } f - \text{area beneath the graph of } g$$
$$= \int_a^b f(x)\,dx - \int_a^b g(x)\,dx$$

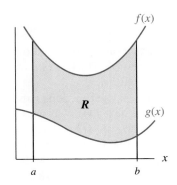

FIGURE 6.58

Using the Sum Rule for antiderivatives, we obtain

$$\text{Area of the region between the graphs of } f \text{ and } g = \int_a^b [f(x) - g(x)] \, dx$$

Note that when f and g are obtained by fitting equations to data, the input variables of the functions must represent the same quantity measured in the same units.

If while calculating the area of the region between two curves you obtain the negative of the answer you expect, then it is likely that you have interchanged the positions of the functions in the integrand.

Area of the Region Between Two Curves

If the graph of f lies above the graph of g from a to b, then the area of the region between the two curves from a to b is given by

$$\int_a^b [f(x) - g(x)] \, dx$$

EXAMPLE 4   *Determining the Area of the Region Between Two Curves*

6.5.2

Tire Manufacturers   A major European tire manufacturer has seen its sales from tires skyrocket since 1989. A model for the rate of change of sales (in U.S. dollars) accumulated since 1989 is

$$s(t) = 3.7(1.19376^t) \text{ million dollars per year}$$

where t is the number of years since the end of 1989.

At the same time, an American tire manufacturer's rate of change of sales accumulated since 1989 can be modeled by

$$a(t) = 0.04t^3 - 0.54t^2 + 2.5t + 4.47 \text{ million}$$
$$\text{dollars per year}$$

where t is the number of years since the end of 1989. These models apply through the year 2010. By how much did the amount of accumulated sales differ for these two companies from the end of 1999 through 2009?

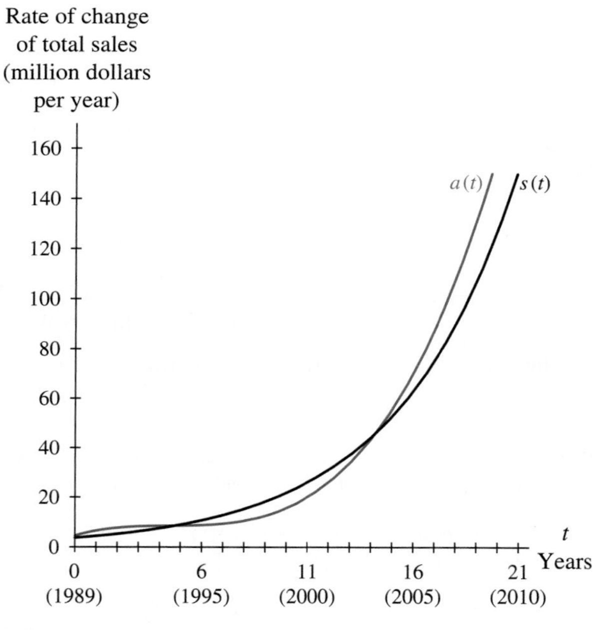

Solution   First, determine whether one graph lies above the other on the interval in the question. The two rate-of-change functions are graphed in Figure 6.59.

The graphs of the two rate-of-change functions s and a cross near $t = 5$ and cross again near $t = 14$. If we set $s(t) = 3.7(1.19376^t)$ equal to $a(t) = 0.04t^3 - 0.54t^2 + 2.5t + 4.47$ and solve for t, we find that the two functions intersect when $t \approx 4.657$ (in 1994) and when $t \approx 14.242$ (in 2004), as well as when $t \approx -0.372$ and $t \approx 28.077$. Accumulated sales were greater for the European company

Rate of change of total sales (million dollars per year)

160
140
120
100
80
60
40
20
0

$a(t)$ $s(t)$

0   6   11   16   21   Years t
(1989) (1995) (2000) (2005) (2010)

FIGURE 6.59

than for the American company from $t \approx 4.657$ to $t \approx 14.242$. Between $t \approx 14.242$ and $t = 21$, the American company saw greater accumulated sales than the European company.

From the beginning of 2000 ($t = 10$) through most of the first quarter of 2004 ($t \approx 14.242$), the European company accumulated approximately

$$\int_{10}^{14.242} [s(t) - a(t)]\,dt = 18.5383 \text{ million dollars}$$

more in sales than the American company. This is the area of region R_1 in Figure 6.60.

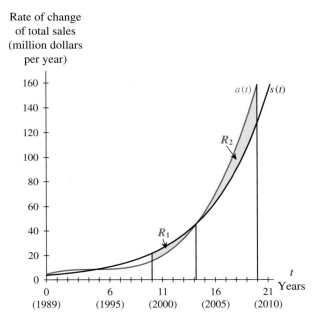

Rate of change
of total sales
(million dollars
per year)

FIGURE 6.60

From close to the end of the first quarter of 2004 ($t \approx 14.242$) through 2009 ($t = 20$), the American company accumulated approximately

$$\int_{14.242}^{20} [a(t) - s(t)]\,dt = 79.4076 \text{ million dollars}$$

more in sales than the European company. This is the area of region R_2 in Figure 6.60.

In order to calculate the estimated difference in accumulated sales between the two companies from 1999 through 2009, we subtract the portion where the American company's accumulated sales were greater from the portion where the European company's accumulated sales were greater. That is,

$$\begin{aligned}
\text{Difference in} \atop \text{accumulated sales} \;&=\; \int_{10}^{14.242} [s(t) - a(t)]\,dt - \int_{14.242}^{20} [a(t) - s(t)]\,dt \\
&\approx 18.5383 \text{ million dollars} - 79.4076 \text{ million dollars} \\
&\approx -60.869 \text{ million dollars}
\end{aligned}$$

The European company's accumulated sales were nearly 61 million dollars less than the American company's accumulated sales over the years considered.   ●

If we use the Constant Multiplier Rule and the Sum Rule for antiderivatives with the functions in Example 4, we see that we did not need to split the interval from 10 to 20 into two intervals.

$$
\begin{aligned}
\text{Difference in} \atop \text{accumulated sales}
&= \int_{10}^{14.242} [s(t) - a(t)]\,dt - \int_{14.242}^{20} [a(t) - s(t)]\,dt \\
&= \int_{10}^{14.242} [s(t) - a(t)]\,dt + \int_{14.242}^{20} [-a(t) + s(t)]\,dt \\
&= \int_{10}^{14.242} [s(t) - a(t)]\,dt + \int_{14.242}^{20} [s(t) - a(t)]\,dt \\
&= \int_{10}^{20} [s(t) - a(t)]\,dt
\end{aligned}
$$

This is true in general—when you wish to find the difference between accumulated change for two continuous rate-of-change functions, you can calculate the definite integral of the difference of the functions, regardless of where the functions intersect.

Difference of Two Accumulated Changes

If f and g are two continuous rate-of-change functions, then the difference between the accumulated change of f from a to b and the accumulated change of g from a to b is

$$
\int_a^b [f(x) - g(x)]\,dx
$$

It is important to note, however, that the integral $\int_a^b [f(x) - g(x)]\,dx$ may not represent the area between the graphs of f and g from a to b. Note that in the tire sales example, the area of the regions between the two curves from 10 to 20 is $18.538 + 79.408 = 97.946$, whereas the difference in accumulated sales is $18.538 - 79.408 = -60.869$.

If two rate-of-change functions intersect in the interval from a to b, then the difference between their accumulated changes is *not* the same as the area of the regions between the two curves.

Most practical applications of the area between two curves involve the difference between two accumulated changes. However, if total area is to be calculated, remember the distinction between these quantities.

In this section, we have seen that the Fundamental Theorem of Calculus gives us a technique for evaluating definite integrals using antiderivatives. However, there are many antiderivative formulas not covered in this text, and there are some functions to which no antiderivative rule applies. It is important for you to understand that if you are ever unable to find an algebraic formula for the antiderivative of a function, you can still estimate the value of a definite integral of that function by using the limiting value of sums of areas of rectangles.

6.5   Concept Inventory

○ $\int_a^b f(x)\,dx = F(b) - F(a)$, *where F is an antiderivative of f*

○ $\int_a^b f(x)\,dx = \int_a^c f(x)\,dx + \int_c^b f(x)\,dx$

○ *Area(s) of region(s) between two curves*

○ *Differences of accumulated changes*

6.5   Activities

For each of Activities 1 through 7, determine which of the following processes you would use when answering the question posed. Note that *a* and *b* are constants.

   a. Find a derivative.

   b. Find a general antiderivative (with unknown constant).

   c. Find a specific antiderivative (solve for the constant).

1. Given a rate-of-change function for population and the population in a given year, find the population in year *t*.

2. Given a velocity function, determine the distance traveled from time *a* to time *b*.

3. Given a function, find its accumulation function from *a* to *x*.

4. Given a velocity function, determine acceleration at time *t*.

5. Given a rate-of-change function for population, find the change in population from year *a* to year *b*.

6. Given a function, find the area of the region between the function and the horizontal axis from *a* to *b*.

7. Given a function, find the slope of the tangent line at input *a*.

In Activities 8 through 11,

   a. Graph the function *f* from *a* to *b*.

   b. Find the area of the region between the graph of *f* and the *x*-axis from *a* to *b*. Is this area equal to $\int_a^b f(x)\,dx$? Explain.

   c. Find $\int_a^b f(x)\,dx$.

8. $f(x) = -4x^{-2}$; $a = 1, b = 4$

9. $f(x) = -1.3x^3 + 0.93x^2 + 0.49$; $a = -1, b = 2$

10. $f(x) = \dfrac{9.295}{x} - 1.472$; $a = 5, b = 10$

11. $f(x) = -965.27(1.079^x)$; $a = 0.5, b = 3.5$

12. **Air Speed**   The air speed of a small airplane during the first 25 seconds of takeoff and flight can be modeled by

$$v(t) = -940{,}602t^2 + 19{,}269.3t - 0.3 \text{ mph}$$

t hours after takeoff.

   a. Find the value of $\int_0^{0.005} v(t)\,dt$.

   b. Interpret your answer in context.

13. **Phone Calls**   The rate of change of the number of international telephone calls billed in the United States between 1980 and 2000 can be described by

$$P(x) = 32.432e^{0.1826x} \text{ million calls per year}$$

where x is the number of years after 1980. Find and interpret the value of $\int_5^{15} P(x)\,dx$.

(Source: Based on data from Federal Communications Commission.)

14. **Weight**   The rate of change of the weight of a laboratory mouse t weeks (for $1 \le t \le 15$) after the beginning of an experiment can be modeled by the equation

$$w(t) = \frac{13.785}{t} \text{ grams per week}$$

Evaluate $\int_3^9 w(t)\,dt$, and interpret your answer.

15. **Revenue**   A corporation's revenue flow rate can be modeled by

$$r(x) = 9.907x^2 - 40.769x + 58.492$$
$$\text{million dollars per year}$$

x years after the end of 1987. Evaluate $\int_0^5 r(x)\,dx$, and interpret your answer.

16. **Medicine**   In Section 6.1 we saw the rate of change in the concentration of a drug modeled by

$$r(x) = \begin{cases} 1.708(0.845^x) & \text{when } 0 \le x \le 20 \\ \quad \mu g/mL/day \\ 0.11875x - 3.5854 & \text{when } 20 < x \le 29 \\ \quad \mu g/mL/day \end{cases}$$

where x is the number of days after the drug was administered. Determine the values of the following definite integrals, and interpret your answers.

a. $\int_0^{20} r(x)\,dx$ b. $\int_{20}^{29} r(x)\,dx$

c. $\int_0^{29} r(x)\,dx$

17. **Snow Pack**   The rate of change of the snow pack in an area in the Northwest Territories in Canada can be modeled by

$$s(t) = \begin{cases} 0.00241t + 0.02905 \text{ cm} \\ \quad \text{per day} & \text{when } 0 \le t \le 70 \\ 1.011t^2 - 147.971t + \\ \quad 5406.578 \text{ cm per day} & \text{when } 72 \le t \le 76 \end{cases}$$

where t is the number of days since April 1.

a. Evaluate $\int_0^{70} s(t)\,dt$, and interpret your answer.

b. Evaluate $\int_{72}^{76} s(t)\,dt$, and interpret your answer.

c. Explain why it is not possible to find the value of $\int_0^{76} s(t)\,dt$.

18. **Temperature**   The rate of change of the temperature during the hour and a half after the beginning of a thunderstorm is given by

$$T(h) = 9.48h^3 - 15.49h^2 + 17.38h - 9.87$$
$$°F \text{ per hour}$$

where h is the number of hours since the storm began.

a. Graph the function T from $h = 0$ to $h = 1.5$.

b. Calculate the value of $\int_0^{1.5} T(h)\,dh$. Interpret your answer.

19. **Temperature**   The rate of change of the temperature in a museum during a junior high school field trip can be modeled by

$$T(h) = 9.07h^3 - 24.69h^2 + 14.87h - 0.03$$
$$°F \text{ per hour}$$

h hours after 8:30 A.M.

a. Find the area of the region that lies above the axis between the graph of T and the h-axis between 8:30 A.M. and 10:15 A.M. Interpret the answer.

b. Find the area of the region that lies below the axis between the graph of T and the h-axis between 8:30 A.M. and 10:15 A.M. Interpret the answer.

c. There are items in the museum that should not be exposed to temperatures greater than 73°F. If the temperature at 8:30 A.M. was 71°F, did the temperature exceed 73°F between 8:30 A.M. and 10:15 A.M.?

20. **Road Test**   The acceleration of a race car during the first 35 seconds of a road test is modeled by

$$a(t) = 0.024t^2 - 1.72t + 22.58 \text{ ft/sec}^2$$

where t is the number of seconds since the test began.

a. Graph the function a from $t = 0$ to $t = 35$.

b. Write the definite integral notation representing the amount by which the car's speed increased during the road test. Calculate the value of the definite integral.

21. **Production**   The estimated production rate of marketed natural gas, in trillion cubic feet per year, in the United States (excluding Alaska) from 1900 through 1960 is shown in the table.

Year	Estimated production rate (trillions of cubic feet per year)
1900	0.1
1910	0.5
1920	0.8
1930	2.0
1940	2.3
1950	6.0
1960	12.7

(Source: From information in *Resources and Man*, National Academy of Sciences, 1969, p. 165.)

a. Find a model for the data in the table.

b. Use the model to estimate the total production of natural gas from 1940 through 1960.

c. Give the definite integral notation for your answer to part *b*.

22. **Advertising**   Many businesses spend money each year on advertising in order to stimulate sales of their products. The data given show the approximate increase in sales (in thousands of dollars) that an additional $100 spent on advertising, at various levels, can be expected to generate.

Advertising expenditures (hundreds of dollars)	Revenue increase due to an extra $100 advertising (thousands of dollars)
25	5
50	60
75	95
100	105
125	104
150	79
175	34

a. Find a model for these data.

b. Use the model in part *a* to determine a model for the total sales revenue $R(x)$ as a function of the amount x spent on advertising. Use the fact that revenue is approximately 877 thousand dollars when $5000 is spent on advertising.

c. Find the point where returns begin to diminish for sales revenue.

d. The managers of the business are considering an increase in advertising expenditures from the current level of $8000 to $13,000. What effect could this decision have on sales revenue?

23. **Production**   The table shows the marginal cost to produce one more compact disc, given various hourly production levels.

Production (CDs per hour)	Cost of an additional CD
100	$5
150	$3.50
200	$2.50
250	$2
300	$1.60

a. Find an appropriate model for the data.

b. Use your model from part *a* to derive an equation that specifies production cost $C(x)$ as a function of the number x of CDs produced. Use the fact that it costs approximately $750 to produce 150 CDs in a 1-hour period.

c. Calculate the value of $\int_{200}^{300} C'(x)\,dx$. Interpret your answer.

For Activities 24 and 25,

a. Sketch graphs of the functions f and g on the same axes.

b. Shade the region between the graphs of f and g from a to b.

c. Calculate the area of the shaded region.

24. $f(x) = 10(0.85^x)$ $a = 2$
    $g(x) = 6(0.75^x)$ $b = 10$

25. $f(x) = x^2 - 4x + 10$ $a = 1$
    $g(x) = 2x^2 - 12x + 14$ $b = 7$

For Activities 26 and 27,

   a. Sketch graphs of the functions f and g on the same axes.

   b. Find the input value(s) at which the graphs of f and g intersect.

   c. Shade the region(s) between the graphs of f and g from a to b.

   d. Calculate the difference in the area of the region between the graph of f and the horizontal axis and the area of the region between the graph of g and the horizontal axis from a to b.

   e. Calculate the total area of the shaded region(s).

26. $f(x) = 0.25x - 3$ $a = 15$
   $g(x) = 14(0.93^x)$ $b = 50$

27. $f(x) = e^{0.5x}$ $a = 0.5$

   $g(x) = \dfrac{2}{x}$ $b = 3$

28. Revenue/Cost The figure depicts graphs of the rate of change of total revenue $R'(x)$ (in billions of dollars per year) and the rate of change of total cost $C'(x)$ (in billions of dollars per year) of a company in year x. The area of the shaded region is 126.5.

Rates of change of
revenue and cost
(billion dollars
per year)

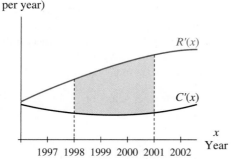

a. Interpret the area in context.

b. Write an equation for the area of the shaded region.

29. Revenue/Cost The figure shows graphs of $r'(x)$, the rate of change of revenue, and $c'(x)$, the rate of change of costs (both in thousands of dollars per thousand dollars of capital investment) associated with the production of solid wood furniture as functions of x, the amount (in thousands of dol-

lars) invested in capital. The area of the shaded region is 13.29.

Rates of change of
cost and revenue
(thousand dollars per
thousand dollars)

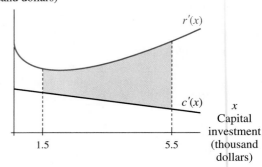

a. Interpret the area in the context of furniture manufacturing.

b. Write an equation for the area of the shaded region.

30. Epidemic The figure depicts graphs of $c(t)$, the rate at which people contract a virus during an epidemic, and $r(t)$, the rate at which people recover from the virus, where t is the number of days after the epidemic begins.

Rates of change of
contraction and recovery
(people per day)

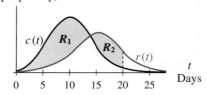

a. Interpret the area of region R_1 in the context of the epidemic.

b. Interpret the area of region R_2 in the context of the epidemic.

c. Explain how you could use a definite integral to find the number of people who contracted the virus since day 0 but have not recovered by day 20.

31. Population A country is in a state of civil war. As a consequence of deaths and people fleeing the country, its population is decreasing at a rate of $D(x)$ people per month. The rate of increase of the population as a result of births and immigration is

$I(x)$ people per month. The variable x is the number of months since the beginning of the year. Graphs of D and I are shown in the figure. Region R_1 has area 3690, and region R_2 has area 9720.

a. Interpret the area of R_1 in context.

b. Interpret the area of R_2 in context.

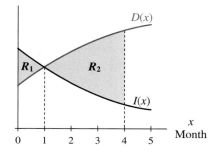

Rates of change of population (people per month)

c. Find the change in population from the beginning of January through the end of April.

d. Explain why the answer to part c is not the sum of the areas of the two regions.

32. Foreign Trade   The rate of change of the value of goods exported from the United States between 1990 and 2001 can be modeled as

$$E'(t) = -1.665t^2 + 16.475t + 7.632$$
billion dollars per year

t years after the end of 1990. Likewise, the rate of change of the value of goods imported into the United States during those years can be modeled as

$$I'(t) = 4.912t + 40.861 \text{ billion dollars per year}$$

t years after 1990.

(Source: Based on data from *World Almanac and Book of Facts,* ed. William A. McGeveran Jr. New York: World Almanac Education Group, Inc., 2003.)

a. Find the difference between the accumulated value of imports and the accumulated value of exports from the end of 1990 through 2001.

b. Is your answer from part a the same as the area of the region(s) between the graphs of E' and I'? Explain.

33. Road Test   The accompanying table shows the time it takes for a Toyota Supra and a Porsche 911 Carerra to accelerate from 0 mph to the speeds given.

Toyota Supra		Porsche 911 Carerra	
Time (seconds)	Speed (mph)	Time (seconds)	Speed (mph)
2.2	30	1.9	30
2.9	40	3.0	40
4.0	50	4.1	50
5.0	60	5.2	60
6.5	70	6.8	70
8.0	80	8.6	80
9.9	90	10.7	90
11.8	100	13.3	100

(Source: *Road and Track.*)

a. Find models for the speed of each car, given the number of seconds after starting from 0 mph. (*Hint:* Add the point (0, 0), and convert miles per hour to feet per second before modeling.)

b. How much farther than a Porsche 911 Carerra does a Toyota Supra travel during the first 10 seconds, assuming that both cars begin from a standing start?

c. How much farther than a Porsche 911 does a Toyota Supra travel between 5 seconds and 10 seconds of acceleration?

34. Postal Service   The table shows the approximate rates of change of revenue for the U.S. Postal Service (USPS), Federal Express (FedEx), and United Parcel Service (UPS) from 1993 to 2001.

Year	USPS (billions of dollars per year)	FedEx (billions of dollars per year)	UPS (billions of dollars per year)
1993	3.4	0.3	1.0
1994	3.3	0.6	1.2
1995	3.0	1.0	1.3
1996	2.6	1.5	1.5
1997	2.3	2.1	1.6
1998	2.0	2.3	1.8
1999	1.8	2.1	1.9
2000	1.7	1.4	2.1
2001	1.8	0	2.2

(Source: Based on data from Hoover's Online Guide.)

a. Using the data find and graph equations for the rates of change of revenue for USPS and UPS.

b. Find and interpret the areas of the two regions bounded by the graphs in part *a* from 1993 through 2001.

35. **Postal Service**   Refer again to the UPS, FedEx, and USPS data in Activity 34.

   a. Find and graph equations for the rates of change of revenue for FedEx and UPS.

   b. Find and interpret the areas of the three regions bounded by the graphs in part *a* between 1993 and 2001.

   c. Find the definite integral of the difference of the two equations in part *a* between 1993 and 2001. Interpret your answer.

36. **Mortality**   When modeling populations, biologists consider many factors that affect mortality. Some mortality factors (such as those that may be weather-related) are not dependent on the size of the population; that is, the proportion of the population killed by such factors remains constant regardless of the size of the population. Other mortality factors are dependent on the size of the population. In these cases, the proportion of a population killed by a certain mortality factor increases (or decreases) as the size of the population increases. Such factors are called *density-dependent mortality factors.*

   Varley and Gradwell studied the population size of a winter moth in a wooded area between 1950 and 1968. They found that predatory beetles represented the only density-dependent mortality factor in the life cycle of the winter moth. When the population of moths was small, the beetles ate few moths, searching elsewhere for food, but when the population was large, the beetles assembled in large clusters in the area of the moth population and laid eggs, thus increasing the proportion of moths eaten by the beetles.

   Suppose the annual number of winter moth larvae in Varley and Gradwell's study that survived winter kill and parasitism each year between 1961 and 1968 can be modeled by the equation

$$l(t) = -0.0505 + 1.516 \ln t \text{ hundred moths}$$
$$\text{per square meter of tree canopy}$$

and that the number of pupae surviving the predatory beetles each year during the same time period can be modeled by the equation

$$p(t) = 0.251 + 0.794 \ln t \text{ hundred moths}$$
$$\text{per square meter of tree canopy}$$

(Source: Adapted from P.J. denBoer and J. Reddingius, *Regulation and Stabilization Paradigms in Population Ecology.* London: Chapman and Hall, 1996.)

In both equations, *t* is the number of years since 1960. The area of the region below the graph of *l* and above the graph of *p* is referred to as the *accumulated density-dependent mortality* of pupae by predatory beetles. Use an integral to estimate this value between the years 1962 and 1965. Interpret your answer.

37. **Emissions**   In response to EPA regulations, a factory that produces carbon emissions plants 22 hectares of forest in 1990. The trees absorb carbon dioxide as they grow, thus reducing the carbon level in the atmosphere. The EPA requires that the trees absorb as much carbon in 20 years as the factory produces during that time. The trees absorb no carbon until they are 5 years old. Between 5 and 20 years of age, the trees absorb carbon at the rates indicated in the table.

(Source: Adapted from A. R. Ennos and S. E. R. Bailey, *Problem Solving in Environmental Biology.* Harlow, Essex, England: Longman House, 1995.)

Tree age (years)	5	10	15	20
Carbon absorption (tons per hectare per year)	0.2	6.0	14.0	22.0

   a. Find a model for the rate (in tons per year) at which carbon is absorbed by the 22 hectares of trees between 1990 and 2010.

   b. The factory produced carbon at a constant rate of 246 tons per year between 1990 and 1997. In 1997, the factory made some equipment changes that reduced the emissions to 190 tons per year. Graph the rate of emissions produced by the factory together with the model in part *a* between 1990 and 2010. Find and label the time in which the absorption rate equals the production rate.

   c. Label the regions of the graph in part *b* whose areas correspond to the following quantities:

i. The carbon emissions produced by the factory but not absorbed by the trees

ii. The carbon emissions produced by the factory and absorbed by the trees

iii. The carbon emissions absorbed by the trees from sources other than the factory

d. Determine the values of the three quantities in part *c.*

e. After 20 years, will the amount of carbon absorbed by the trees be at least as much as the amount produced by the factory during that time period, as required by the EPA?

38. Income   The table for this activity (located on the *Calculus Concepts* CD-ROM and web site) gives the rates of change of the number of males and the number of females aged 15 to 24 years with income at the U.S. median level between 1974 and 2001.

a. Find and graph (using the same axes) equations for the rates of change of the number of males and the number of females aged 15 to 24 years with income at the U.S. median level as functions of the number of years after 1974.

b. Temporarily ignoring the context, find the input *A* of the point at which the two equations in part *a* intersect.

c. Find and interpret the area between the graphs of the two equations in part *a* when the input is between 0 and *A*.

d. Find the definite integral of the difference in the two equations in part *a* when the input is between 0 and 27. Interpret your answer.

39. Poverty   The table for this activity (located on the *Calculus Concepts* CD-ROM and web site) lists the rates of change in the percent of whites and the percent of blacks in the United States who were below poverty level by age for all income levels and both sexes in 2001.

a. Find and graph (using the same axes) equations for the rates of change of the percent below poverty level for whites and blacks as functions of the person's age.

b. Find and interpret the areas of the two regions bounded by the graphs in part *a* between birth and age 84.

c. Find the definite integral of the difference in the two equations in part *a* between birth and age 84. Interpret your answer.

40. Consider the regions between *f* and *g* depicted in the figure.

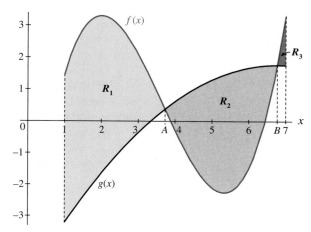

If *f* is the rate of change of the revenue of a small business and *g* is the rate of change of the costs of the business *x* years after its establishment (both quantities measured in thousands of dollars per year), interpret the areas of the regions R_1, R_2, and R_3 and the value of the definite integral $\int_1^7 [f(x) - g(x)]\,dx.$

41. How are the heights of rectangles (between two curves) determined if one or both of the graphs lie below the horizontal axis? Consider the figure when giving an example.

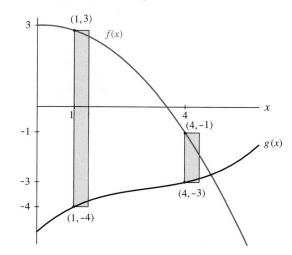

6.6   Average Value and Average Rate of Change

As a student, you are intimately acquainted with grade averages and the method of calculating averages by adding grades and dividing by the number of them. Grades are discrete data. Let us consider a situation in which discrete averaging is not practical.

Concept Development: Averages of Continuous Functions   Consider calculating a person's average heart rate over 50 minutes of moderate activity. The actual heart rate is calculated as the number of times the person's heart beats during the time period divided by the time. However, without medical monitoring devices, it is impractical to count the number of heartbeats during a 50-minute period.

We could measure heart rate every 10 minutes and use these six data points to estimate the average over the entire 50-minute period. For example, by summing the heart rates in Table 6.19 and dividing by 6, we estimate that a person's heart rate over the time period represented by the table is 100.8 beats per minute.

TABLE 6.19

Time into test (minutes)	Heart rate (beats per minute)
0	95
10	105
20	100
30	94
40	101
50	110

Obviously, if the heart rate were measured every 5 minutes instead of every 10 minutes, a more accurate estimate of the average heart rate could be obtained. Averaging the data in Table 6.20 yields an estimate of 100.73 beats per minute.

TABLE 6.20

Time into test (minutes)	Heart rate (beats per minute)	Time into test (minutes)	Heart rate (beats per minute)
0	95	30	94
5	102	35	97
10	105	40	101
15	104	45	105
20	100	50	110
25	95		

The preceding averages use a discrete number of heart rates, whereas heart rate is constantly changing. In order to compensate for this constantly changing situation, we use the data to model heart rate with a continuous function.

$$H(t) = (-4.80186 \cdot 10^{-5})t^4 + 0.00599t^3 - 0.229476t^2$$
$$+ 2.813403t + 94.370629 \text{ beats per minute}$$

where t is the number of minutes since the test began.

Integrating this function over the 50-minute interval gives a close approximation of the total number of heart beats during that period:

$$\int_0^{50} H(t)\,dt \approx 5033 \text{ beats}$$

Dividing this total number of beats by 50 minutes yields the average heart rate of 100.66 beats per minute.

Caution: Using a continuous model for a discrete situation must be done with care. In this case there are so many beats during the 50-minute interval (nearly 2 per second) that it is reasonable to model the heart rate with a continuous function.

Averaging is a balancing out of extremes. Figure 6.61 depicts a graph of the heart rate function. The average of that function's outputs, approximately 100.7 beats per minute, is shown as a dotted horizontal line on the graph.

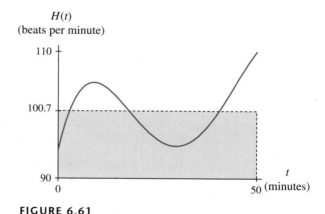

FIGURE 6.61

The average value of a function over an interval can be graphically interpreted as the height (or signed height) of a rectangle whose area equals the area between the function and the horizontal axis over the interval. (Note that we do not show the vertical axis to zero in Figure 6.61; however, it is true that the area of the rectangle and the area between the graph and the line representing the horizontal axis shown in the figure are equal.)

Average Value of a Function

If $y = f(x)$ is a continuous function, then we can approximate the average value (or average) of the function over an interval from $x = a$ to $x = b$ by dividing the interval into n equally spaced subintervals, evaluating the function at a point in each subinterval, summing the function values, and dividing by n.

$$\text{Average value} \approx \frac{f(x_1) + f(x_2) + \cdots + f(x_{n-1}) + f(x_n)}{n}$$

We denote the length of each subinterval by Δx and calculate the length as $\Delta x = \frac{b-a}{n}$. Rewrite the average value estimate by multiplying the top and bottom terms by Δx:

$$\text{Average value} \approx \frac{[f(x_1) + f(x_2) + \cdots + f(x_{n-1}) + f(x_n)]\Delta x}{n\Delta x}$$

$$= \frac{[f(x_1) + f(x_2) + \cdots + f(x_{n-1}) + f(x_n)]\Delta x}{b-a}$$

As in the heart rate example, the estimate improves as the number of intervals increases. Thus we obtain the exact average value by finding the limit of the estimate as n approaches infinity:

$$\text{Average value} = \lim_{n \to \infty} \frac{[f(x_1) + f(x_2) + \cdots + f(x_{n-1}) + f(x_n)]\Delta x}{b-a}$$

which can be written as

$$\text{Average value} = \frac{\displaystyle\int_a^b f(x)\,dx}{b-a}$$

Thus we have

Average Value

If $y = f(x)$ is a smooth, continuous function from a to b, then the average value of $f(x)$ from a to b is

$$\text{Average value of } f(x) \text{ from } a \text{ to } b = \frac{\displaystyle\int_a^b f(x)\,dx}{b-a}$$

EXAMPLE 1 *Finding Average Value and Average Rate of Change*

6.6.1a, b

Temperature Suppose that the hourly temperatures shown in Table 6.21 were recorded from 7 A.M. to 7 P.M. one day in September.

TABLE 6.21

Time	Temperature (°F)	Time	Temperature (°F)
7 A.M.	49	2 P.M.	80
8 A.M.	54	3 P.M.	80
9 A.M.	58	4 P.M.	78
10 A.M.	66	5 P.M.	74
11 A.M.	72	6 P.M.	69
noon	76	7 P.M.	62
1 P.M.	79		

a.   Find a cubic model for this set of data.

b.   Calculate the average temperature between 9 A.M. and 6 P.M.

c.   Graph the equation together with the rectangle whose upper edge is determined by the average value.

d.   Calculate the average rate of change of temperature from 9 A.M. to 6 P.M.

Solution

a.   The temperature on this particular day can be modeled as

$$t(h) = -0.03526h^3 + 0.71816h^2 + 1.584h + 13.689 \text{ degrees Fahrenheit}$$

h hours after midnight. This model applies only from $h = 7$ (7 A.M.) to $h = 19$ (7 P.M.).

b.   The average temperature between 9 A.M. ($h = 9$) and 6 P.M. ($h = 18$) is

$$\text{Average temperature} = \frac{\int_9^{18} t(h)\,dh}{18 - 9} \approx 74.4 \,°\text{F}$$

c.

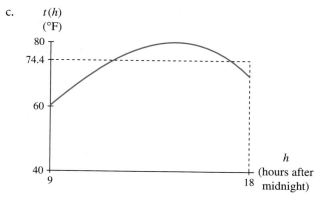

FIGURE 6.62

d. Recall from Section 3.1 that the average rate of change of a function on an interval is the change in output divided by the change in input. The average rate of change of temperature from 9 A.M. to 6 P.M. is

$$\frac{\text{Average rate of change}}{\text{of temperature}} = \frac{t(18) - t(9)}{18 - 9} \approx 0.98°\text{F per hour}$$ ●

Average Rate of Change

The preceding example asked for an average rate of change. We know from Section 3.1 that the average rate of change of a continuous function $y = f(x)$ from $x = a$ to $x = b$ is calculated as $\frac{f(b) - f(a)}{b - a}$. However, consider what happens when you have a function that describes the rate of change of a quantity (that is, you have $y = f'(x)$) and you need to find the average rate of change of the quantity $f(x)$. In this case, you do not use the average rate-of-change formula but should instead use an integral to find the average value of the rate-of-change function. Note that we use the terms *average rate of change* and *average value of the rate of change* interchangeably.

The Average Value of the Rate of Change

If $y = f'(x)$ is a smooth, continuous rate-of-change function from a to b, then the average value of $f'(x)$ from a to b is

$$\frac{\text{Average value of the rate of}}{\text{change of } f(x) \text{ from } a \text{ to } b} = \frac{\int_a^b f'(x)\,dx}{b - a}$$

$$= \frac{f(b) - f(a)}{b - a}$$

where $f(x)$ is an antiderivative of $f'(x)$.

EXAMPLE 2   *Determine Which Quantity to Average*

Population Growth   The growth rate of the population of South Carolina between 1790 and 2000 can be modeled* as

$$p'(t) = 0.1779t - 1.568 \text{ thousand people per year}$$

where t is the number of years since 1790. The population of South Carolina in 1990 was 3486 thousand people.

a. What was the average rate of change in population from 1995 through 2000?

*Based on data from *Statistical Abstract,* 2001.

b.   What was the average size of the population from 1995 through 2000?

Solution

a.   The average rate of change in population between 1995 and 2000 is calculated directly from the rate-of-change function as

$$\frac{\int_{205}^{210} p'(t)\,dt \text{ (thousand people/year)(years)}}{(210 - 205) \text{ years}} \approx 35.3 \text{ thousand people per year}$$

Keeping units of measure attached to values might help you correctly calculate and label average values.

b.   In order to calculate the average population, we must have a function for population. That is, we need an antiderivative of the rate-of-change function:

$$p(t) = \int p'(t)\,dt$$

$$= 0.08895t^2 - 1.568t + C \text{ thousand people}$$

We know that the population in 1990 was 3486 thousand people. Using this fact, we solve for C, so the function for population is

$$p(t) = 0.08895t^2 - 1.568t + 241.6 \text{ thousand people}$$

where t is the number of years since 1790. Now we calculate the average population between 1995 and 2000 as

$$\frac{\int_{205}^{210} p(t)\,dt}{210 - 205} \approx 3746 \text{ thousand people} \quad \bullet$$

Note in Example 2 that the average value of the population was found by integrating the population function, whereas the average rate of change was found by integrating the population rate-of-change function. This example illustrates an important principle:

> In using integrals to find average values, integrate the function whose output is the quantity you wish to average.

Note also that in part *a* of Example 2, we could have calculated the average rate of change by using the population function and the formula

$$\frac{p(210) - p(205)}{210 - 205} = \frac{176.731 \text{ thousand people}}{5 \text{ years}} \approx 35.3 \text{ thousand people per year}$$

We summarize this discussion as follows:

Average Values and Average Rates of Change

If $y = f(x)$ is a continuous or piecewise continuous function describing a quantity from $x = a$ to $x = b$, then the average value of the quantity from a to b is calculated by using the quantity function and the formula

$$\text{Average value of } f(x) = \frac{\int_a^b f(x)\,dx}{b - a}$$

The average value has units the same as the output of the function f.

The average rate of change of the quantity, also called the average value of the rate of change, can be calculated from the quantity function as

$$\text{Average rate of change} = \frac{f(b) - f(a)}{b - a}$$

or from the rate-of-change function as

$$\text{Average rate of change} = \frac{\int_a^b f'(x)\,dx}{b - a}$$

The average rate of change has the same units as the rate of change of f.

EXAMPLE 3   *Graphically Illustrating Average Value*

Carbon-14   Scientists estimate that 100 milligrams of the isotope ^{14}C used in carbon dating methods decay at a rate of

$$r(t) = -0.0121(0.999879^t) \text{ milligrams per year}$$

where t is the number of years since the 100 milligrams of isotope began to decay. The amount of the 100 milligrams remaining after t years of decay is

$$a(t) = 100(0.999879^t) \text{ milligrams}$$

a.   What is the average amount of the remaining isotope during the first 1000 years?

b.   What is the average rate of decay during the first 1000 years?

c.   Graphically illustrate the answers to parts a and b.

Solution

a.   We calculate the average amount remaining during the first thousand years as

$$\frac{\int_0^{1000} a(t)\,dt}{1000 - 0} \approx 94.2 \text{ milligrams}$$

b. We find the average rate of decay during the first 1000 years as

$$\frac{\int_0^{1000} r(t)\,dt}{1000 - 0} \approx -0.0114 \text{ milligram per year}$$

In other words, the amount of ^{14}C decreased by an average of 0.0114 milligram per year during the first 1000 years. Note that this average rate of change can also be calculated by using the amount function:

$$\frac{a(1000) - a(0)}{1000 - 0} \approx \frac{88.6 - 100 \text{ milligrams}}{1000 \text{ years}} = -0.0114 \text{ milligram per year}$$

c. The average amount determines the top of the rectangle shown in Figure 6.63a. The average rate of decay determines the bottom of the rectangle shown in Figure 6.63b. The average decay rate also can be graphically illustrated as the slope of a secant line through two points on the amount function.

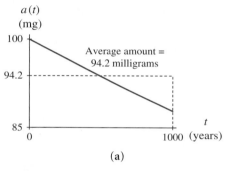

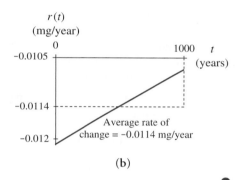

(a)   (b)

FIGURE 6.63

6.6 Concept Inventory

○ *Average value of a function*

○ *Graphical illustration of average value*

○ *Average rate of change of a function*

○ *Average value of a rate-of-change function*

6.6 Activities

1. **Traffic Speed** The Highway Department is concerned about the high speed of traffic during the weekday afternoon rush hours from 4 P.M. to 7 P.M. on a newly widened stretch of interstate highway that is just inside the city limits of a certain city. The Office of Traffic Studies has collected the data given in the table, which show typical weekday speeds during the 4 P.M. to 7 P.M. rush hours.

Time	Speed (mph)	Time	Speed (mph)
4:00	60	5:45	72.25
4:15	61	6:00	74
4:30	62.5	6:15	74.5
4:45	64	6:30	75
5:00	66.25	6:45	74.25
5:15	67.5	7:00	73
5:30	70		

a. Find a model for the data.

b. Use the equation to approximate the average weekday rush-hour speed from 4 P.M. to 7 P.M.

c. Use the equation to approximate the average weekday rush-hour speed from 5 P.M. to 7 P.M.

2. **Electronics Sales**   U.S. factory sales of electronic goods to dealers from 1990 through 2001 can be modeled by the equation

$$\text{Sales} = 0.0388x^3 - 0.495x^2 + 5.698x + 43.6$$
billion dollars

where x is the number of years since 1990.
(Sources: Based on data from *Statistical Abstract*, 2001, and Consumer Electronics Association.)

a. Use a definite integral to approximate the average annual value of U.S. factory sales of electronic goods to dealers from 1990 through 2001.

b. Sketch the graph of sales from 1990 through 2001, and draw the horizontal line representing the average value.

3. **Phone Calls**   The most expensive rates (in dollars per minute) for a 2-minute telephone call using a long-distance carrier are listed in the table in the next column.

a. Find a model for the data.

b. Use a definite integral to estimate the average of the most expensive rates from 1982 through 1990.

c. Use a definite integral to estimate the average of the most expensive rates from 1982 through 2000.

Year	Rate (dollars per minute)
1982	1.32
1984	1.24
1985	1.14
1986	1.01
1987	0.83
1988	0.77
1989	0.65
1990	0.65
1995	0.40
2000	0.20

4. **Ticket Price**   The table gives the price (in dollars) of a round-trip flight from Denver to Chicago on a certain airline and the corresponding monthly profit (in millions of dollars) for that airline for that route.

Ticket price (dollars)	Profit (millions of dollars)
200	3.08
250	3.52
300	3.76
350	3.82
400	3.70
450	3.38

a. Find a model for the data.

b. Determine the average profit for ticket prices from $325 to $450.

c. Determine the average rate of change of profit when the ticket price rises from $325 to $450.

d. Graphically illustrate the answers to parts *b* and *c*.

5. **Population**   The population of Mexico between 1921 and 2000 is given by the model

$$\text{Population} = 7.567(1.02639^t) \text{ million people}$$

where t is number of years since the end of 1900.
(Source: Based on data from **www.inegi.gob.mx.** Accessed 9/20/02.)

a. What was the average population of Mexico from the beginning of 1990 through the end of 1999?

b. In what year was the population of Mexico equal to its 1990s average?

c. What was the average rate of change of the population of Mexico during the 1990s?

6. **Veggies** The per capita utilization of commercially produced fresh vegetables in the United States from 1980 through 2000 can be modeled by

$$v(t) = 0.092t^2 + 0.720t + 149.554 \text{ pounds per person}$$

where t is the number of years after 1980.
(Sources: Based on data from *Statistical Abstract*, 2001, and **www.ers.usda.gov.** Accessed 9/25/02.)

a. Use a definite integral to estimate the average per capita utilization of commercially produced fresh vegetables in the United States between 1980 and 2000.

b. Find the average rate of change in per capita utilization between 1980 and 2000.

c. In which year was the per capita utilization closest to the average per capita utilization between 1980 and 2000?

7. **Air Accidents** The number of general-aviation aircraft accidents from 1975 through 1997 can be modeled by

$$(a)x = -100.6118x + 3967.5572 \text{ accidents}$$

where x is the number of years since 1975.
(Source: Based on data from *Statistical Abstract*, 1994 and 1998.)

a. Calculate the average rate of change in the yearly number of accidents from 1976 through 1997.

b. Use a definite integral to estimate the average number of accidents that occurred each year from 1976 through 1997.

c. Graphically illustrate the answers to parts a and b.

8. **Temperature** During a summer thunderstorm, the temperature drops and then rises again. The rate of change of the temperature during the hour and a half after the storm began is given by

$$T(h) = 9.48h^3 - 15.49h^2 + 17.38h - 9.87 \text{ °F per hour}$$

where h is the number of hours since the storm began.

a. Calculate the average rate of change of temperature from 0 to 1.5 hours after the storm began.

b. If the temperature was 85°F at the time the storm began, find the average temperature during the first 1.5 hours of the storm.

9. **Road Test** The acceleration of a race car during the first 35 seconds of a road test is modeled by

$$a(t) = 0.024t^2 - 1.72t + 22.58 \text{ ft/sec}^2$$

where t is the number of seconds since the test began. Assume that velocity and distance were both 0 at the beginning of the road test.

a. Calculate the average acceleration during the first 35 seconds of the road test.

b. Calculate the average velocity during the first 35 seconds of the road test.

c. Calculate the distance traveled during the first 35 seconds of the road test.

d. If the car had been traveling at its average velocity throughout the 35 seconds, how far would the car have traveled during that 35 seconds?

e. Graphically illustrate the answers to parts a and b. Explain how the answer to part c relates to the graphical illustration of the part b answer.

10. **Oil Production** On the basis of data obtained from a preliminary report by a geological survey team, it is estimated that for the first 10 years of production, a certain oil well in Texas can be expected to produce oil at the rate of $r(t) = 3.93546t^{3.55}e^{-1.35135t}$ thousand barrels per year, t years after production begins. Estimate the average annual yield from this oil well during the first 10 years of production.

11. **Velocity** An article in the May 23, 1996, issue of *Nature* addresses the interest some physicists have in studying cracks in order to answer the question "How fast do things break, and why?" Data estimated from a graph in this article showing

velocity of a crack during a 60-microsecond experiment are shown.

Time (microseconds)	Velocity (meters per second)
10	148.2
20	159.3
30	169.5
40	180.7
50	189.8
60	200.0

a. Find a model for the data.

b. Determine the average speed at which a crack travels between 10 and 60 microseconds.

12. **Newspapers**   The circulation (as of September 20 of each year) of daily English language newspapers in the United States between 1986 and 2000 can be modeled as

$$n(x) = 0.00792x^3 - 0.32x^2 + 3.457x + 51.588$$
million newspapers

where x is the number of years since 1980.
(Source: Based on data from *Statistical Abstract*, 1995 and 2001.)

a. Estimate the average newspaper circulation from 1986 through 2000.

b. In what year was the newspaper circulation closest to the average circulation from 1986 through 2000?

c. Graphically illustrate the answer to part *a*.

13. **Blood Pressure**   Blood pressure varies for individuals throughout the course of a day, typically being lowest at night and highest from late morning to early afternoon. The estimated rate of change in diastolic blood pressure for a patient with untreated hypertension is shown in the table.

a. Find a model for the data.

b. Estimate the average rate of change in diastolic blood pressure from 8 A.M. to 8 P.M.

c. Assuming that diastolic blood pressure was 95mm Hg at 12 P.M., estimate the average diastolic blood pressure between 8 A.M. and 8 P.M.

Time	Diastolic BP (mm Hg per hour)
8 A.M.	3.0
10 A.M.	1.8
12 P.M.	0.7
2 P.M.	-0.1
4 P.M.	-0.7
6 P.M.	-1.1
8 P.M.	-1.3
10 P.M.	-1.1
12 A.M.	-0.7
2 A.M.	0.1
4 A.M.	0.8
6 A.M.	1.9

14. **Air Speed**   The air speed of a small airplane during the first 25 seconds of takeoff and flight can be modeled by

$$v(t) = -940,602t^2 + 19,269.3t - 0.3 \text{ mph}$$

t hours after takeoff.

a. Find the average air speed during the first 25 seconds of takeoff and flight.

b. Find the average acceleration during the first 25 seconds of takeoff and flight.

15. **Swim Time**   The rate of change of the winning times for the 100-meter butterfly swimming competition at selected Summer Olympic Games between 1956 and 2000 can be described by $w(t) = 0.0106t - 1.148$ seconds per year where t is the number of years after 1900. Find the average rate of change of the winning times for the competition from 1956 through 2000.
(Source: Based on data from *Statistical Abstract*, 2001.)

16. **Emissions**   The federal government sets standards for toxic substances in the air. Often these standards are stated in the form of average pollutant levels over a period of time on the basis of the

reasoning that exposure to high levels of toxic substances is harmful, but prolonged exposure to moderate levels is equally harmful. For example, carbon monoxide (CO) levels may not exceed 35 ppm (parts per million) at any time, but they also must not exceed 9 ppm averaged over any 8-hour period.

(Source: Douglas J. Crawford-Brown, *Theoretical and Mathematical Foundations of Human Health Risk Analysis.* Boston: Kluwer Academic Publishers, 1997.)

The concentration of carbon monoxide in the air in a certain metropolitan area is measured and modeled as

$$c(h) = -0.004h^4 + 0.05h^3 - 0.27h^2 + 2.05h + 3.1 \text{ ppm}$$

h hours after 7 A.M.

a. Did the city exceed the 35-ppm maximum in the 8 hours between 7 A.M. and 3 P.M.?

b. Did the city exceed the 9-ppm maximum average between 7 A.M. and 3 P.M.?

17. **Emissions**   Refer to the discussion in Activity 16. The following table shows measured concentrations of carbon monoxide in the air of a city on a certain day between 6 A.M. and 10 P.M.

Time (hours since 6 A.M.)	CO concentration (ppm)
0	3
2	12
4	22
6	18
8	16
10	20
12	28
14	16
16	6

a. Sketch a scatter plot of the data and determine (by examination of the data) over which 8-hour period the average CO concentration was greatest.

b. Model the data. Use the equation to calculate the average CO concentration during the 8-hour period determined in part *a*.

c. Use the equation in part *b* to estimate the average CO concentration in this city between 6 A.M. and 10 P.M. The city issues air quality warnings based on the daily average CO concentration of the previous day between 6 A.M. and 10 P.M. The warnings are as follows:

Average concentration	*Warning*
$0 < \text{CO} \leq 9$	None
$9 < \text{CO} \leq 12$	Moderate pollution. People with asthma and other respiratory problems should remain indoors if possible.
$12 < \text{CO} \leq 16$	Serious pollution. Ban on all single-passenger vehicles. Everyone is encouraged to stay indoors.
$\text{CO} > 16$	Severe pollution. Mandatory school and business closures.

d. Judging on the basis of the data in the table and your answer to part *c*, which warning do you believe should be posted?

18. **Population**   Aurora, Nevada, was a mining boom town in the 1860s and 1870s. Its population can be modeled by the function

$$p(t) = \begin{cases} -7.91t^3 + 120.96t^2 + 193.92t \\ \quad - 123.21 \text{ people} & \text{when } 0.7 \leq t \leq 13 \\ 45{,}544(0.8474^t) \text{ people} & \text{when } 13 < t \leq 55 \end{cases}$$

with rate-of-change function

$$p'(t) = \begin{cases} -23.73t^2 + 241.92t + 193.92 \\ \quad \text{people per year} & \text{when } 0.7 \leq t < 13 \\ -7541.287(0.8474^t) \\ \quad \text{people per year} & \text{when } 13 < t \leq 55 \end{cases}$$

In both functions, t is the number of years since 1860.

(Source: Based on data from Don Ashbaugh, *Nevada's Turbulent Yesterday: A Study in Ghost Towns.* Los Angeles: Westernlore Press, 1963.)

a. What was the average population of Aurora between 1861 and 1871? between 1871 and 1881?

b. Demonstrate two methods for calculating the average rate of change of the population of Aurora between 1861 and 1871.

19. We know that the area of the region between the graph of a function and the horizontal axis from $x = b$ to $x = c$ is equal to the area of the rectangle whose height is the average value of the function from $x = b$ to $x = c$ and whose width is $c - b$. The graphs presented in this section show only a portion of the vertical axis. That is, in each graph, the vertical axis does not extend all the way to zero. Instead, the vertical axis is shown above (or below) a line $y = k$. However, on the interval from $x = b$ to $x = c$, the area of the region between the graph and the line $y = k$ and the area of the rectangle between the average value and $y = k$ are equal. Explain, illustrating with graphs, why this is true.

20. **Poverty**   The table for this activity (located on the *Calculus Concepts* CD-ROM and web site) lists the rates of change in the percent of whites and the percent of blacks in the United States who were below poverty level by age for all income levels and both sexes in 2001.

a. Fit functions to the data for the rates of change of the percent below poverty level for whites and blacks as functions of the person's age.

b. Use the functions in part *a* to estimate the average rate of change in the percent of whites and the percent of blacks who were below poverty level in 2001.

c. The percentage of whites below poverty level at birth in 2001 was 19.7%. Find the average percent of whites below poverty level in 2001.

d. The percentage of blacks below poverty level at age 20 in 2001 was 24.1%. Find the average percent of blacks below poverty level in 2001.

21. **Incarceration Rates**   The table for this activity (located on the *Calculus Concepts* CD-ROM and web site) gives national and regional incarceration rates for prisoners with sentences of more than 1 year who were imprisoned under state and federal jurisdiction at year-end from 1977 through 2001.

a. Find a function to fit the Northeast incarceration rate data.

b. Which was greater: the average rate of change in the Northeast incarceration rate from 1977 through 1989 or the average rate of change in the Northeast incarceration rate from 1989 through 2001?

c. Use a definite integral to estimate the average incarceration rate between 1977 and 2001.

6.7   Antiderivative Limitations

We have thus far examined situations in which it is useful to determine accumulated change in a quantity by finding or estimating the area of a region between the graph of a rate-of-change function and the horizontal axis. We estimated such areas by summing areas of rectangles until we discovered that the Fundamental Theorem of Calculus allows us to find areas using antiderivatives. However, there are limitations on our ability to find antiderivatives for functions. In this section, we further explore these limitations.

There are many relatively simple functions for which we are unable to find an antiderivative formula either because one does not exist or because the methods needed to determine a formula for the antiderivative are beyond the scope of this book. It is important for you to be able to identify such functions. In order to understand situations in which you cannot find antiderivatives, it is essential for you to keep in mind the processes involved in finding derivatives. Recall the Product Rule: Given a function that is the product of two functions, the derivative is found by summing two terms. In particular,

If $f(x) = g(x)h(x)$, then $f'(x) = g(x)h'(x) + g'(x)h(x)$.

With this rule in mind, consider the process of finding the general antiderivative $\int xe^x dx$. Seeing a product of two functions, a typical student who has forgotten the Product Rule for derivatives may reason that the antiderivative is simply the product of the antiderivatives of the two terms x and e^x and will thereby incorrectly conclude that $y = \left(\frac{x^2}{2}\right)(e^x) + C$ is the general antiderivative. A quick derivative calculation, however, reveals otherwise: $\frac{dy}{dx} = \left(\frac{x^2}{2}\right)e^x + xe^x \neq xe^x$. There is a method (known as *integration by parts*) that can be used for determining the general antiderivative of xe^x. However, we do not include it in this text. If we need to calculate a definite integral of the form $\int_a^b xe^x dx$, we choose to rely on approximation methods (using technology) to estimate the value of the integral. Example 1 illustrates antiderivatives that involve products.

EXAMPLE 1   *Finding Antiderivatives That Involve Products of Functions*

Determine whether the following general antiderivatives can be found using the techniques presented in this book. If so, find the antiderivative.

a. $\int x\sqrt{x}\,dx$ b. $\int x\sqrt{x+1}\,dx$

c. $\int e^{3x}(e^{4x} + 4)\,dx$ d. $\int e^{3x}(3x + 1)\,dx$

e. $\int e^{3x}(4x + 4)\,dx$

Solution

a. Rewrite $x\sqrt{x}$ as $x(x^{1/2}) = x^{3/2}$. Then apply the Power Rule for antiderivatives to obtain $\int x\sqrt{x}\,dx = \dfrac{2x^{5/2}}{5} + C$.

b. Unlike the expression in part *a*, it is not possible to rewrite $x\sqrt{x+1}$ in such a way that we can easily find its antiderivative.

c. Multiply the two terms to rewrite as $e^{7x} + 4e^{3x}$ and apply antiderivative rules to obtain

$$\int e^{3x}(e^{4x} + 4)\,dx = \int (e^{7x} + 4e^{3x})\,dx = \frac{e^{7x}}{7} + \frac{4e^{3x}}{3} + C$$

d. Although this question is similar to part *c*, the *x*-term complicates the solution. If we multiply the terms, we obtain $3xe^{3x} + e^{3x}$. You may recognize this expression as a result of the application of the Product Rule. The function $y = xe^{3x}$ has derivative $\frac{dy}{dx} = x(3e^{3x}) + (1)e^{3x} = 3xe^{3x} + e^{3x} = e^{3x}(3x + 1)$. Recognizing the expression as a result of the Product Rule is the key to discovering the antiderivative. Thus $\int e^{3x}(3x + 1)dx = xe^{3x} + C$. If you did not see the connection to the Product Rule and needed this antiderivative for a definite integral calculation, you could use technology to estimate the value of a definite integral involving this product.

e. This question is similar to part *d*, and a reasonable approach would be to consider $y = 4xe^{3x}$ as a candidate for the antiderivative. Finding the derivative of $y = 4xe^{3x}$ using the Product Rule, we have $\frac{dy}{dx} = 4x(3e^{3x}) + 4e^{3x} = e^{3x}(12x + 4)$. Although this answer is close to $e^{3x}(4x + 4)$, it is not the same. Therefore, the antiderivative is not $y = 4xe^{3x} + C$. This is another example of an antiderivative that can be found using methods not discussed in this text. If we needed to evaluate a definite integral of $e^{3x}(4x + 4)$, we would use technology to approximate the numerical answer. ●

In addition to keeping the Product Rule in mind when finding antiderivatives, you should also consider the Chain Rule. If a function is a composite function, you may not have the ability to find its antiderivative with the rules discussed in this book. Consider the general antiderivative $\int (x^2 + 2x)^4 dx$. A common mistake is to conclude that this general antiderivative is $y = \frac{(x^2 + 2x)^5}{5} + C$. However, if you take the derivative of y to check this answer, you find by applying the Chain Rule that $\frac{dy}{dx} = (x^2 + 2x)^4 (2x + 2)$. Therefore, the Fundamental Theorem of Calculus tells us that $y = \frac{(x^2 + 2x)^5}{5} + C$ is not the general antiderivative $\int (x^2 + 2x)^4 dx$. Note that it is possible to expand $(x^2 + 2x)^4$ by repeated multiplication to obtain $x^8 + 8x^7 + 24x^6 + 32x^5 + 16x^4$. Thus

$$\int (x^2 + 2x)^4 dx = \frac{x^9}{9} + x^8 + \frac{24x^7}{7} + \frac{16x^6}{3} + \frac{16x^5}{5} + C$$

EXAMPLE 2   *Finding Antiderivatives That Involve Composite Functions*

For the following definite integrals, find the exact answer using antiderivative formulas if possible. If finding the exact answer is not possible, use technology to estimate the answer.

a. $\int_1^6 (3x^2 - 7)^2 dx$ b. $\int_4^5 \sqrt{3x^2 - 7} dx$

c. $\int_0^2 6xe^{3x^2 - 7} dx$ d. $\int_1^2 \frac{3}{1 + 7e^{-3x}} dx$

Solution

a. Rewrite the integral as $\int_1^6 (3x^2 - 7)^2 dx = \int_1^6 (9x^4 - 42x^2 + 49)dx$. Then apply the Sum Rule, the Power Rule, and the Constant Multiplier Rule for antiderivatives:

$$\int_1^6 (9x^4 - 42x^2 + 49)dx = \left(\frac{9}{5}x^5 - 14x^3 + 49x\right)\Big|_1^6$$

$$= 11{,}266.8 - 36.8 = 11{,}230$$

b. Rewrite the integral as $\int_4^5 (3x^2 - 7)^{\frac{1}{2}} dx$. Unlike part a, there is no way to rewrite this integral in order to apply the antiderivative rules we have learned. The value of the integral can be approximated as 7.329.

c. The function in this integral is the combination of a product and a composite function. The first term, $6x$, is the derivative of the inside function, $3x^2 - 7$. This function is the result of applying the Chain Rule to the function e^{3x^2-7}. Thus the exact solution of this definite integral is $\int_0^2 6xe^{3x^2-7}dx = e^{3x^2-7}\Big|_0^2 = e^5 - e^{-7}$.

d. You should recognize this function as a logistic function. We can rewrite it as $\int_1^2 3(1 + 7e^{-3x})^{-1} dx$. Doing so makes it easy to see that $u = 1 + 7e^{-3x}$ is the inside function and $3u^{-1}$ is the outside function. Because the derivative of the inside function does not appear in the integral, the function in the integral cannot be thought of as the result of the Chain Rule, and its antiderivative cannot be found using the methods we have studied. The value of the integral can be approximated via technology as 2.718. ●

When finding antiderivatives, it is always a good idea to take the derivative of your answer in order to determine whether it is correct. Finding antiderivatives is one procedure for which there is a simple way to check the answer.

6.7 Concept Inventory

○ *Antiderivative formula limitations*

6.7 Activities

In Activities 1 through 8, find the general antiderivative if possible.

1. $\int 2e^{2x}dx$

2. $\int 2xe^{x^2}dx$

3. $\int x^2 e^{x^2} dx$

4. $\int 3(\ln 2)2^x(1 + 2^x)^3 dx$

5. $\int (1 + e^x)^2 dx$

6. $\int \sqrt{1 + e^x} dx$

7. $\int \frac{1}{2^x + 2} dx$

8. $\int \frac{e^x}{e^x + 2} dx$

WEB/CD

Skills

In Activities 9 through 20, find the exact value of the integral by using antiderivative formulas if possible. If not possible, use technology to estimate the answer. In either case, state whether your answer is exact or an approximation.

9. $\int_1^4 \ln x\,dx$

10. $\int_1^4 x \ln x\,dx$

11. $\int_2^5 \dfrac{\ln x}{x}\,dx$

12. $\int_2^5 \dfrac{5(\ln x)^4}{x}\,dx$

13. $\int_1^2 2x \ln(x^2 + 1)\,dx$

14. $\int_1^2 \ln(x^2 + 1)\,dx$

15. $\int_3^4 \dfrac{2x}{x^2 + 1}\,dx$

16. $\int_3^4 \dfrac{1}{x^2 + 1}\,dx$

17. $\int_1^6 \dfrac{2x^2}{x^2 + 1}\,dx$

18. $\int_3^4 \dfrac{x^2 + 1}{2x}\,dx$

19. $\int_3^4 \dfrac{x^2 + 1}{x^2}\,dx$

20. $\int_0^1 -x\sqrt{x^2 + 1}\,dx$

SUMMARY

Approximating Results of Change

The accumulated results of change are best understood in geometric terms: Positive accumulation is the area of a region between the graph of a positive rate-of-change function and the horizontal axis, and negative accumulation is the signed area of a region between the graph of a negative rate-of-change function and the horizontal axis. We can approximate the areas of regions of interest by summing areas of rectangular regions.

Limits of Sums and Accumulation Functions

The area of a region between the graph of a continuous, non-negative function f and the horizontal axis from a to b is given by a limit of sums:

$$\text{Area} = \lim_{n \to \infty} [f(x_1) + f(x_2) + \cdots + f(x_n)]\Delta x$$

Here, the points $x_1, x_2, \ldots, x_n$ are the midpoints of n rectangles of width $\Delta x = \dfrac{b - a}{n}$ between a and b.

More generally, we consider the limit applied to an arbitrary continuous bounded function f over the interval from a to b and call this limit the definite integral of f from a to b. In symbols, we write

$$\int_a^b f(x)\,dx = \lim_{n \to \infty} [f(x_1) + f(x_2) + \cdots + f(x_n)]\Delta x$$

An accumulation function is an integral of the form $\int_a^x f(t)\,dt$ where the upper limit x is a variable. This function gives us a formula for calculating accumulated change in a quantity.

The Fundamental Theorem of Calculus

The Fundamental Theorem sets forth the fundamental connection between the two main concepts of calculus, the derivative and the integral. It tells us that for any continuous function f,

$$\frac{d}{dx}\int_a^x f(t)\,dt = f(x)$$

In other words, the derivative of an accumulation function of $y = f(t)$ is precisely $y = f(x)$.

If we reverse the order of these two processes and begin by differentiating first, then we obtain the starting function plus a constant.

$$\int_a^x f'(t)\,dt = f(x) + C$$

A function F is an antiderivative of f if $F'(x) = f(x)$. Because the derivative of $y = \int_a^x f(t)\,dt$ is $y' = f(x)$, we see that $y = \int_a^x f(t)\,dt$ is an antiderivative of $y' = f(x)$. Each continuous bounded function has infinitely many antiderivatives, but any two differ by only a constant. The Fundamental Theorem enables us to find accumulation function formulas by finding antiderivates.

The Definite Integral

The Fundamental Theorem of Calculus enabled us to show that when f is a smooth, continuous function, the definite integral $\int_a^b f(x)\,dx$ can be evaluated by

$$\int_a^b f(x)\,dx = F(b) - F(a)$$

where F is any antiderivative of f.

The Fundamental Theorem of Calculus ensures that each continuous bounded function does indeed have an antiderivative. Thus, to the extent that we can actually obtain an algebraic expression for an antiderivative, we can easily evaluate a definite integral. In situations where an antiderivative cannot be found, we use one of the approximation techniques discussed in Sections 6.1 and 6.2, allowing technology to perform the calculations.

To compute the area between two curves, we used the fact that if the graph of f lies above the graph of g from a to b, then the integral $\int_a^b [f(x) - g(x)]\,dx$ is the area of the region between the two graphs from a to b.

Note that if the two functions intersect between a and b, then the difference between the accumulated changes of the functions is *not* the same as the total area of the regions between the two rate-of-change curves.

Average Values and Average Rates of Change

We use definite integrals to calculate the average value of a continuous function for a quantity:

$$\text{Average value of } f(x) \text{ from } a \text{ to } b = \frac{\int_a^b f(x)\,dx}{b - a}$$

When we are given a rate-of-change function $y = f'(t)$, the average rate of change of $f(t)$ from $t = a$ to $t = b$ is

$$\text{Average rate of change of } f(x) \text{ from } a \text{ to } b = \frac{\int_a^b f'(t)\,dt}{b - a}$$

CONCEPT CHECK

Can you

○ Interpret accumulated change and area?
○ Approximate areas using rectangles?
○ Interpret definite integrals?
○ Approximate area using a limiting value?
○ Sketch and interpret accumulation functions?
○ Recover the units of a quantity function?
○ Find general antiderivatives?
○ Find and interpret specific antiderivatives?
○ Recover a function from its rate-of-change equation?
○ Use the Fundamental Theorem to evaluate definite integrals?
○ Find and interpret areas between two curves?
○ Find average value and average rate of change?
○ Determine whether an approximation technique is necessary in order to estimate the value of a definite integral?

To practice, try

Section 6.1	Activity 3
Section 6.1	Activities 11, 13
Section 6.2	Activities 3, 5
Section 6.2	Activities 9, 11
Section 6.3	Activities 3, 11
Section 6.3	Activity 21
Section 6.4	Activities 9–18
Section 6.4	Activities 23, 25, 27
Section 6.4	Activity 29
Section 6.5	Activities 13, 17
Section 6.5	Activity 33
Section 6.6	Activity 7
Section 6.7	Activities 13, 15

REVIEW TEST

1. Oil Flow The rate at which crude oil flows through a pipe into a holding tank modeled by

$$r(t) = 10(-3.2t^2 + 93.3t + 50.7) \text{ ft}^3/\text{minute}$$

where t is the number of minutes the oil has been flowing into the tank.

  a. Sketch a graph of r for t between 0 and 25 minutes.

  b. Use five midpoint rectangles to estimate the area of the region between the graph of r and the t-axis from 0 to 25 minutes. Sketch the rectangles on the graph you drew in part a.

  c. Interpret your answer to part b.

2. Speed A hurricane is 300 miles off the east coast of Florida at 1 A.M. The speed at which the hurricane is moving toward Florida is measured each hour. Speeds between 1 A.M. and 5 A.M. are recorded in the table.

Time	Speed (mph)	Time	Speed (mph)
1 A.M.	15	4 A.M.	38
2 A.M.	25	5 A.M.	40
3 A.M.	35		

  a. Find a model for the data.

  b. Use a limiting value of sums of areas of midpoint rectangles to estimate how far (to the nearest tenth of a mile) the hurricane traveled between 1 A.M. and 5 A.M. Begin with five rectangles, doubling the number each time until you are confident that you know the limiting value.

3. Dieting The accompanying graph depicts the rate of change in the weight of someone who diets for 20 weeks.

  a. What does the area of the shaded region beneath the horizontal axis represent?

  b. What does the area of the shaded region above the horizontal axis represent?

Rate of change in weight
(pounds per week)

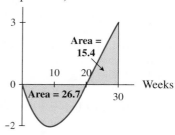

  c. Is this person's weight at 30 weeks more or less than it was at 0 weeks? How much more or less?

  d. If w is the function shown in the figure above, sketch a graph of $W(t) = \int_0^x w(t)\,dt$. Label units and values on both axes of your graph.

  e. What does the graph in part d represent?

4. Oil Flow Consider again the model for the flow rate of crude oil into a holding tank.

$$r(t) = 10(-3.2t^2 + 93.3t + 50.7) \text{ ft}^3/\text{minute}$$

after t minutes.

  a. If the holding tank contains 5000 ft^3 of oil when $t = 0$, find a model for the amount of oil in the tank after t minutes.

  b. Use your model in part a to find how much oil flowed into the tank during the first 10 minutes.

  c. If the capacity of the tank is 150,000 ft^3, according to the model, how long can the oil flow into the tank before the tank is full?

5. Investment Ten thousand dollars invested in a mutual fund is growing at a rate of

$$a(x) = 840(1.08763^x) \text{ dollars per year}$$

x years after it was invested.

  a. Determine the value of $\int_0^{2.75} a(x)\,dx$.

b. Interpret your answer to part *a.*

c. What is the average rate of growth of the investment from $x = 0$ to $x = 2.75$?

6. **Earnings**   Based on data provided by the Census Bureau for the years 1980 through 1988, the full-time average annual earnings of men and women in the United States can be modeled by the following equations:

Men: $m(t) = 0.0625t^2 - 10.38t + 466.8075$
    thousand dollars

Women: $w(t) = -0.03125t^2 + 5.695t - 234.89875$
    thousand dollars

where *t* is the number of years since 1900. From the beginning of 1980 and through the end of 1988, by how much did the 9-year earnings of a man who earned the average wage exceed those of a woman who earned the average wage?

Project 6.1 Acceleration, Velocity, and Distance

Setting

According to tests conducted by *Road and Track*, a 1993 Toyota Supra Turbo accelerates from 0 to 30 mph in 2.2 seconds and travels 1.4 miles (1320 feet) in 13.5 seconds, reaching a speed of 107 mph. *Road and Track* reported the data given in the table.

Time (seconds)	Speed reached from rest (mph)
0	0
2.2	30
2.9	40
4.0	50
5.0	60
6.5	70
8.0	80
9.0	90
11.8	100

Tasks

1. Convert the speed data to feet per second, and find a quadratic model for velocity (in feet per second) as a function of time (in seconds). Discuss how close your model comes to predicting the 107 mph reached after 13.5 seconds.

2. Add the data point for 13.5 seconds, and find a quadratic model for velocity.

3. Use four rectangles and your model from Task 2 to estimate the distance traveled during acceleration from rest to a speed of 50 mph and the distance traveled during acceleration from a speed of 50 mph to a speed of 100 mph. Repeat the estimate using twice as many rectangles.

4. Use nine rectangles to approximate the distance traveled during the first 13.5 seconds. How close is your estimate to the reported value?

5. Find the distances traveled during

   a. Acceleration from rest to a speed of 50 mph

   b. Acceleration from a speed of 50 mph to a speed of 100 mph

   c. The first 13.5 seconds of acceleration

   Compare these answers to your estimates in Tasks 3 and 4. Explain how estimating with areas of rectangles is related to calculating the definite integral.

Reporting

Prepare a written report of your work. Include scatter plots, models, graphs, and discussions of each of the above tasks.

Project 6.2 Estimating Growth

Setting

A table based on data from the Berkeley Growth Study is located on the *Calculus Concepts* CD-ROM and web site. This table lists the rate of growth of a typical male from birth to 18 years.

Tasks

1. Use the data and right rectangles to approximate the height of a typical 18-year-old male.

2. Sketch a smooth, continuous curve over a scatter plot of the data. Find a piecewise model for the data. Use no more than three pieces.

3. Use your piecewise model and limits of sums to approximate the height of a typical 18-year-old male. Convert centimeters to feet and inches, and compare your answer to the estimate you obtained using right rectangles. Which is likely to be the more accurate approximation? Why?

4. Use your piecewise model and what you know about definite integrals to find the height of a typical 18-year-old male in feet and inches. Compare your answer with the better of the approximations you obtained in Task 3.

5. Randomly choose ten 18-year-old male students, and determine their heights. (Include your data—names are not necessary, only the heights.) Discuss your selection process and why you feel that it is random. Find the average height of the 18-year-old males in your sample. Compare this average height with your answer to Task 4. Discuss your results.

6. Refer to your sketch of the rate-of-growth graph in Task 2, and draw a possible graph of the height of a typical 18-year-old male from birth to age 18.

Reporting

Prepare a report that presents your findings in Tasks 1 through 6. Explain the different methods that you used, and discuss why these methods should all give similar results. Attach your mathematical work as an appendix to your report.

Analyzing Accumulated Change: Integrals in Action

Chapter 6 established that the accumulated results of change are limiting values of approximating sums known as definite integrals. Magnitudes of accumulated change can be expressed as areas of regions between the graph of a rate-of-change function and the horizontal axis. The Fundamental Theorem of Calculus provides a simple method for evaluating definite integrals using antiderivatives.

In Chapter 7 we present several applications of integration. We use integrals to calculate perpetual accumulation, present and future values of income streams, and future values of biological streams. We conclude by discussing how integrals can be applied to economics topics and used to calculate economic quantities of interest to consumers and producers.

Concept Objectives

This chapter will help you understand the concepts of

○ Improper Integral
○ Discrete and continuous income streams
○ Present and future value
○ Supply and demand

and you will learn to

○ Evaluate and interpret improper integrals
○ Recognize when an improper integral diverges
○ Calculate and interpret present and future values for discrete and continuous income streams
○ Calculate and interpret the following economic quantities:
  consumers' willingness and ability to spend, expenditure, and surplus; suppliers' willingness and ability to receive, revenue, and surplus; market equilibrium; total social gain

Concept Application   The CEO of a large corporation must concern himself or herself with many facets of the economy and the corporation's relationship to it. For example, the CEO may be interested in answering questions as diverse as

O   What is the 5-year future value of an income stream that the corporation invests?

O   What is the market equilibrium price for a particular product produced by the corporation?

In this chapter you will learn how to answer questions such as these.

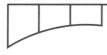

7.1   Perpetual Accumulation and Improper Integrals

Definite integrals have specific numbers for both the upper limit and the lower limit. We now consider what happens to the accumulation of change when one or both of the limits of the integral is infinite. That is, we wish to evaluate integrals of the form $\int_a^\infty f(x)\,dx$, $\int_{-\infty}^b f(x)\,dx$, or $\int_{-\infty}^\infty f(x)\,dx$. We call integrals of this form **improper integrals.** Improper integrals play a role in economics and statistics as well as in other fields of study.

Evaluating Improper Integrals

Consider evaluating the improper integral $\int_2^\infty 4.3e^{-0.06x}\,dx$. We can interpret this integral as the area of the region between the graph of $y = 4.3e^{-0.06x}$ and the x-axis from 2 to infinity. See Figure 7.1. One way to estimate this area is to consider the area between 2 and some large value. In Table 7.1 we show several area calculations for increasingly larger values.

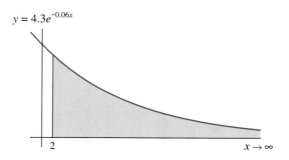

$y = 4.3e^{-0.06x}$

FIGURE 7.1

TABLE 7.1

N	$\int_2^N 4.3e^{-0.06x}\,dx$
50	59.994558
100	63.384987
200	63.562191
400	63.562631
800	63.562631
1600	63.562631
Limit ≈ 63.56263	

Note that the area between 2 and 400 is not significantly different from the area between 2 and 1600. The difference is smaller than can be shown by the technology that we used. However, the limiting value seen in the table is still just an estimate of the value of the integral $\int_2^\infty 4.3e^{-0.06x}\,dx$.

You should recognize that we are numerically investigating a limit in Table 7.1. The limit we are investigating is $\lim_{N\to\infty}\int_2^N 4.3e^{-0.06x}\,dx$. We can calculate this limit algebraically in order to obtain the exact answer. Begin by finding the general antiderivative of $4.3e^{-0.06x}$, evaluating it at 2 and N, and subtracting the results.

Before continuing, you may wish to review the discussions in Section 1.3 and Strengthening the Concepts *of limits describing end behavior.*

$$\int_2^N 4.3e^{-0.06x}\,dx = \frac{4.3}{-0.06}e^{-0.06x}\Big|_2^N$$

$$= \frac{4.3}{-0.06}e^{-0.06N} - \left(\frac{4.3}{-0.06}e^{-0.06(2)}\right) = \frac{4.3}{-0.06}e^{-0.06N} + \frac{4.3}{0.06}e^{-0.12}$$

Next find the limit of this expression as N becomes infinitely large.

$$\lim_{N\to\infty}\left(\frac{4.3}{-0.06}e^{-0.06N}+\frac{4.3}{0.06}e^{-0.12}\right)$$

$$=\lim_{N\to\infty}\frac{4.3}{-0.06}e^{-0.06N}+\lim_{N\to\infty}\frac{4.3}{0.06}e^{-0.12}$$

The first term is a decreasing exponential, so we know that as N approaches infinity, it approaches zero. The second term is a constant that is not affected by the value of N.

$$\lim_{N\to\infty}\frac{4.3}{-0.06}e^{-0.06N}+\lim_{N\to\infty}\frac{4.3}{0.06}e^{-0.12}$$

$$=0+\frac{4.3}{0.06}e^{-0.12}$$

$$=\frac{4.3}{0.06}e^{-0.12}$$

We see that this answer confirms our former numerical estimate, because

$$\frac{4.3}{0.06}e^{-0.12}\approx63.56263$$

It is important for you to understand, however, that the answer $\frac{4.3}{0.06}e^{-0.12}$ is exact, whereas the answer 63.56263 is not. Although an answer accurate to the fifth decimal place is sufficient for most applications, there are situations in which greater precision is necessary.

To summarize, an improper integral $\int_a^\infty f(x)\,dx$ is evaluated by replacing infinity with a variable, say N, and evaluating the limit of the integral $\int_a^N f(x)\,dx$ as N approaches infinity. That is, provided the limits exist,

$$\int_a^\infty f(x)\,dx=\lim_{N\to\infty}\int_a^N f(x)\,dx=[\,\lim_{N\to\infty}F(N)]-F(a)$$

$$\int_{-\infty}^b f(x)\,dx=\lim_{N\to-\infty}\int_N^b f(x)\,dx=F(b)-\lim_{N\to-\infty}F(N)$$

where F is an antiderivative of f. We now have the tools we need to apply improper integrals to some real-world problems.

EXAMPLE 1   *Using a Limit to Evaluate an Improper Integral*

Decay   Carbon-14 dating methods are sometimes used by archeologists to determine the age of an artifact. The rate at which 100 milligrams of ^{14}C is decaying can be modeled by

$$r(t)=-0.01209(0.999879^t)\text{ milligrams per year}$$

where t is the number of years since the 100 milligrams began to decay.

a.   How much of the ^{14}C will have decayed after 1000 years?

b.   How much of the ^{14}C will eventually decay?

Solution

a. The amount of ^{14}C to decay during the first 1000 years is

$$\int_0^{1000} r(t)\,dt = \int_0^{1000} -0.01209(0.999879^t)\,dt \approx -11.4 \text{ milligrams}$$

Thus approximately 11.4 milligrams will decay during the first 1000 years. Note that -11.4 milligrams is the signed area of the shaded region in Figure 7.2.

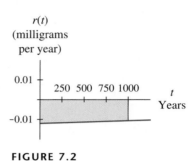

FIGURE 7.2

b. In the long run, the amount that will decay is

$$\int_0^{\infty} r(t)\,dt = \lim_{N\to\infty} \int_0^N -0.01209(0.999879^t)\,dt$$

$$= \lim_{N\to\infty} \left[\frac{-0.01209(0.999879^t)}{\ln 0.999879} \right]\Bigg|_0^N$$

$$= \lim_{N\to\infty} \left[\frac{-0.01209(0.999879^N)}{\ln 0.999879} - \frac{-0.01209(0.999879^0)}{\ln 0.999879} \right]$$

$$\approx 99.91131 \left[\lim_{N\to\infty} (0.999879^N) \right] - 99.91131$$

$$= 99.91131(0) - 99.91131 \approx -100 \text{ milligrams}$$

Eventually all of the ^{14}C will ultimately decay. In terms of the graph shown in Figure 7.3, the area of the region between the graph of the function r and the horizontal axis gets closer and closer to 99.91131 as t gets larger and larger. Because the parameters in the equation are rounded, the area is getting closer and closer to 99.91131 rather than to 100, which is the amount that must ultimately decay. ●

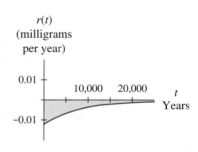

FIGURE 7.3

Divergence

It is possible that when we are evaluating an improper integral, the limit does not exist. (For example, as we numerically approximate the limit, the limit estimates become increasingly large.) In this case, we say that the improper integral **diverges.** Example 4 illustrates this situation.

EXAMPLE 2   *Recognizing That an Integral Diverges*

If possible, determine the value of $\int_1^\infty \frac{1}{x} dx$.

Solution   We begin by replacing ∞ with the variable N and finding the limit as $N \to \infty$.

$f(x) = \ln x$

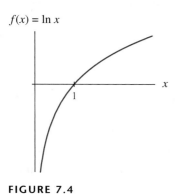

$$\int_1^\infty \frac{1}{x} dx = \lim_{N \to \infty} \ln x \Big|_1^N$$

$$= \lim_{N \to \infty} (\ln N - \ln 1)$$

$$= \lim_{N \to \infty} \ln N - \lim_{N \to \infty} \ln 1$$

To evaluate these limits, recall the shape of the graph of $y = \ln x$ (see Figure 7.4). We see from the graph that $\ln 1 = 0$ and that as the input becomes increasingly large, the output also becomes increasingly large. Thus, the limiting value does not exist. Because $\lim_{N \to \infty} \ln N \to \infty$, $\int_1^\infty \frac{1}{x} dx \to \infty$, and we say that this improper integral diverges.   ●

FIGURE 7.4

7.1   **Concept Inventory**

○ *Improper integrals*

○ *Divergence*

7.1   **Activities**

For Activities 1 through 11, evaluate the indicated improper integral.

1. $\int_0^\infty 3e^{-0.2t} dt$

2. $\int_{15}^\infty 5(0.36^t) dt$

3. $\int_{10}^{-\infty} 3x^{-2} dx$

4. $\int_{-\infty}^3 7e^{7x} dx$

5. $\int_{-\infty}^{-10} 4x^{-3} dx$

6. $\int_2^\infty \frac{1}{\sqrt{x}} dx$

7. $\int_{0.36}^\infty 9.6x^{-0.432} dx$

8. $\int_{-\infty}^{-2} \left(\frac{3}{x^2} + 1\right) dx$

9. $\int_2^\infty \frac{2x}{x^2 + 1} dx$

10. $\int_5^\infty [5(0.36^x) + 5] dx$

11. $\int_a^\infty [f(x) + k] dx$, where $\int_a^\infty f(x) dx = b$ and a, b, and k are constants

12. **Decay**   The rate at which 15 grams of ^{14}C is decaying can be modeled by

$$r(t) = -0.027205(0.998188^t) \text{ grams per year}$$

where t is the number of years since the 15 grams began decaying.

a. How much of the ^{14}C will decay during the first 1000 years? during the fourth 1000 years?

b. How much of the ^{14}C will eventually decay?

13. **Decay**   An isotope of uranium, ^{238}U, is commonly used in atomic weapons and nuclear power generators. Because of its radioactive nature, the United States government is concerned with safe ways of storing used uranium. The rate at which 100 milligrams of ^{238}U is decaying can be modeled by

$$r(t) = -1.55(0.9999999845^t) \cdot 10^{-6} \text{ milligrams per year}$$

where t is the number of years since the 100 milligrams began decaying.

a. How much of the ^{238}U will decay during the first 100 years? during the first 1000 years?

b. How much of the ^{238}U will eventually decay?

The following information is used in Activities 14 and 15:

In the study of markets, economists define *consumers' willingness and ability to spend* as the maximum amount that consumers are willing and able to spend for a specific quantity of goods or services. If some consumers will purchase a product or service regardless of its price, then the consumers' willingness and ability to spend is defined by

$$C = qp_0 + \int_{p_0}^{\infty} D(p)\,dp$$

where q is a specific quantity, p_0 is the price associated with quantity q, and $D(p)$ is the demand for the commodity when the price is p. (Section 7.3 discusses these concepts in greater detail.)

14. Demand The weekly demand for a dozen roses is given by $D(p) = 316.765(0.949^p)$ dozen roses, where $\$p$ is the price per dozen.

a. Find the price that corresponds to a weekly demand of 80 dozen roses.

b. Use the price (p_0) found in part *a* to calculate how much consumers are willing and able to spend for 80 dozen roses per week.

15. Demand The yearly demand for a certain hardback science fiction novel is

$$D(p) = 499.589(0.958^p) \text{ thousand books}$$

where $\$p$ is the price per book.

a. Find the price that corresponds to a yearly demand of 150,000 books.

b. Use the price (p_0) found in part *a* to calculate how much consumers are willing and able to spend for 150,000 books each year.

16. Work The work required to propel a 10-ton rocket an unlimited distance from the surface of Earth into space is defined in terms of force and is given by the improper integral

$$W = \int_{4000}^{\infty} \frac{160{,}000{,}000}{x^2}\,dx$$

The expression $\dfrac{160{,}000{,}000}{x^2}$ is force in tons. The variable x is the distance, measured in miles, between the rocket and the center of Earth.

a. What are the units on work in this context?

b. Calculate the work to propel this rocket infinitely into space.

The following information is used in Activities 17 and 18:

A *probability density function* is defined as a non-negative function f with the property that

$$\int_{-\infty}^{\infty} f(x)\,dx = 1.$$

17. Consider the function

$$f(x) = \begin{cases} 0.1e^{-0.1x} & \text{when } x \geq 0 \\ 0 & \text{when } x < 0 \end{cases}$$

Show that f is a probability density function.

18. Consider the function

$$g(x) = \begin{cases} \dfrac{1}{x^2} & \text{when } x \geq 1 \\ 0 & \text{when } x < 1 \end{cases}$$

Show that g is a probability density function.

7.2   Streams in Business and Biology

Picture a stream flowing into a pond. You have probably just created a mental picture of water that is flowing continuously into the pond. We can also imagine moneys that are "flowing" continuously into an investment or new individuals that are "flowing" continuously into an existing population.

It is not unreasonable to consider the income of large financial institutions and major corporations as being received continuously over time in varying amounts. For instance, consider utility companies that receive payments at varying times throughout each month. Furthermore, with electronic transfer of funds, these payments can be made at any time during the day or night. Such a flow of money is called a **continuous income stream.** When you make payments to a bank or to some other financial institution for the purpose of investing money or repaying a loan, your payments are usually for the same fixed amount and are made at regular times that are separated by a specified interval. Such a flow of money is called a **discrete income stream.** Whether continuous or discrete, an income stream is usually described as a rate $R(t)$ that varies with time t.

Determining Income Streams

Consider a business that currently posts a yearly profit of $4.3 million. The business allocates 5% of its profits in a continuous stream among several investments. There are several situations that could determine the flow rate into these investments. We consider the cases where the flow is constant or either increases or decreases by a constant or by a percentage. If the company's profits remain constant, then the function that describes the income stream flowing into the investments is $R(t) = 0.05(\$4.3$ million per year) = $0.215 million per year.

Suppose, however, that the company's profits increase by $0.2 million each year. In this case, the company's profit is linear, beginning with $4.3 million dollars the first year and increasing by a constant $0.2 million each year: $4.3 + 0.2t$ million dollars per year after t years. The investment stream is 5% of the profit, so $R(t) = 0.05(4.3 + 0.2t)$ million dollars per year t years after the company posted a profit of $4.3 million.

It is also possible that the company's profits could increase by a constant 7% each year. Recall that constant percentage change is modeled by an exponential equation. In this case, the function that describes the flow rate of the investment stream is

$$R(t) = 0.05[4.3(1.07^t)] \text{ million dollars per year}$$

t years after the company posted a profit of $4.3 million. Determining the rate at which income flows into an investment is the first step in answering questions about the present and future values of the invested income stream.

EXAMPLE 1   *Writing Flow Rate Equations*

7.2.1

Business Start-up   After you graduate from college, you start a small business that immediately becomes successful. When you establish the business, you determine that 10% of your profits will be invested each year. In the first year you post a profit of $579,000. Determine the income stream flow rate for your investments over the next several years if

a. The business's profit remains constant.

b. The profit grows by $50,000 each year.

c. The profit increases by 17% each year.

d. The profits for the first six years are as shown in Table 7.2 and are expected to follow the trend indicated by the data.

TABLE 7.2

Year	1	2	3	4	5	6
Profit (thousands of dollars)	579	600	610	618	623	627

Solution

a. If the profit remains constant, then the flow rate of the investment stream is constant, calculated as 10% of $579,000. Thus $R(t) = \$57,900$ per year.

b. Profit increasing by a constant amount each year indicates linear growth. In this case, the amount of profit that is invested is described by the linear function $R(t) = (0.10)(579,000 + 50,000t)$ dollars per year t years after the first year.

c. An increase in profit of 17% each year indicates exponential growth with a constant percentage change of 17%. In this case, the flow rate of the investment stream is described by the equation $R(t) = 0.10[579,000(1.17^t)]$ dollars per year t years after the first year of business.

d. A scatter plot of the data in Table 7.2 indicates an increasing, concave-down shape. (See Figure 7.5.) A quadratic model is a reasonably good fit, although the model indicates declining profits beyond the years given in the table. A better model for profit is the log model $P(t) = 580.117 + 26.7 \ln t$ thousand dollars per year after t years of business. Thus the investment flow rate is $R(t) = 0.10P(t) = 58.0117 + 2.67 \ln t$. Note that in parts a through c, the first year of business corresponds to an input value of 0. In this log model, the first year of business corresponds to an input value of 1. ●

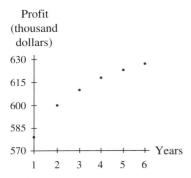

FIGURE 7.5

Future Value of a Continuous Stream

The **future value** of a continuous stream is the total accumulated value of the income stream and its earned interest. Suppose that an income stream flows continuously into an interest-bearing account at the rate of $R(t)$ dollars per year where t is measured in years and the account earns interest at the annual rate of $100r\%$ compounded continuously. What is the future value of the account at the end of T years?

To answer this question, we begin by imagining the time interval from 0 to T years as being divided into n subintervals, each of length Δt.

We regard Δt as being small—so small that over a typical subinterval $[t, t + \Delta t]$, the rate $R(t)$ can be considered constant. Then the amount paid into the account during this subinterval can be approximated by

$$\text{Amount paid in} \approx [R(t) \text{ dollars per year}](\Delta t \text{ years})$$
$$\approx R(t)\Delta t \text{ dollars}$$

We consider this amount as being paid in at t, the beginning of the interval, and earning interest continuously for $(T - t)$ years. Using the continuously compounded interest formula $(A = Pe^{rt})$, we see that the amount grows to

$$[R(t)\Delta t]e^{r(T-t)} = R(t)e^{r(T-t)}\Delta t \text{ dollars}$$

at the end of T years. Summing over the n subintervals, we have the approximation

$$\text{Future value} \approx [R(t_1)e^{r(T-t_1)} + R(t_2)e^{r(T-t_2)} + \cdots + R(t_n)e^{r(T-t_n)}]\,\Delta t \text{ dollars}$$

where $t_1, t_2, \ldots, t_n$ are the left endpoints of the n subintervals. This sum should look familiar to you. If we simplify the expression by letting $f(t) = R(t)e^{r(T-t)}$ and rewrite the sum as $[f(t_1) + f(t_2) + \cdots + f(t_n)]\Delta t$, then you should recognize it as the type of sum we used in Sections 6.1 and 6.2.

Because we are considering the income as a continuous stream and interest as being compounded continuously, we let the time interval Δt become extremely small $(\Delta t \to 0)$. That is, we use an infinite number of intervals $(n \to \infty)$. Thus

$$\text{Future value} = \lim_{n\to\infty} [f(t_1) + f(t_2) + \cdots + f(t_n)]\Delta t$$

$$= \int_0^T f(t)\,dt$$

$$= \int_0^T R(t)e^{r(T-t)}\,dt \text{ dollars}$$

Future Value of a Continuous Income Stream

Suppose that an income stream flows continuously into an interest-bearing account at the rate of $R(t)$ dollars per year where t is measured in years and the account earns interest at the annual rate of $100r\%$ compounded continuously. The future value of the account at the end of T years is

$$\text{Future value} = \int_0^T R(t)e^{r(T-t)}\,dt \text{ dollars}$$

Using the Fundamental Theorem, we can find the rate-of-change function for future value:

$$\text{Rate of change of future value} = \frac{d}{dx}\int_0^x R(t)e^{r(T-t)}\,dt \quad \text{for } 0 \le x \le T$$

$$= R(x)e^{r(T-x)} \text{ dollars per year}$$

Thus the function $f(t) = R(t)e^{r(T-t)}$ gives the rate of change after t years of the future value (in T years) of an income stream whose income is flowing continuously in at a rate of $R(t)$ dollars per year. It is the rate-of-change function $f(t) = R(t)e^{r(T-t)}$, not the flow rate of the income stream $R(t)$, that we graph when depicting future value as the area of a region beneath a rate-of-change function.

EXAMPLE 2   *Finding Future Value of a Continuous Income Stream*

Airline Expansion   The owners of a small airline are making big plans. They hope to be able to buy out a larger airline 10 years from now by investing into an account returning 9.4% APR. Assume a continuous income stream and continuous compounding of interest.

a. The owners have determined that they can afford to invest $3.3 million each year. How much will these investments be worth 10 years from now?

b. If the airline's profits increase so that the amount the owners invest each year increases by 8% per year, how much will their investments be worth in 10 years?

Solution

a. The flow rate of the income stream is $R(t) = 3.3$ million per year with $r = 0.094$ and $T = 10$ years. The value of these investments in 10 years is calculated as

$$\text{Future value} = \int_0^{10} 3.3e^{0.094(10-t)}dt$$

$$= \int_0^{10} 3.3e^{0.94}e^{-0.094t}dt$$

$$= \frac{3.3e^{0.94}}{-0.094}e^{-0.094(10)} - \frac{3.3e^{0.94}}{-0.094}e^{-0.094(0)}$$

$$\approx -35.106 + 89.872$$

$$\approx \$54.8 \text{ million}$$

b. The function modeling exponential growth of 8% per year in the investment stream is $R(t) = 3.3(1.08^t)$ million dollars per year after t years. The future value is calculated (using technology) as $\int_0^{10} 3.3(1.08^t)e^{0.094(10-t)}dt \approx \$77.7$ million.   ●

Present Value of a Continuous Stream

The **present value** of a continuous income stream is the amount P that would have to be invested now in an interest-bearing account in order for the amount to grow to a given future value. Because P dollars earning continuously compounded interest would grow to a future value of Pe^{rT} dollars in T years, we have

$$Pe^{rT} = \int_0^T R(t)e^{r(T-t)}dt = \int_0^T R(t)e^{rT}e^{-rt}dt = e^{rT}\int_0^T R(t)e^{-rt}dt$$

Solving for P, we obtain

$$\text{Present value} = P = \int_0^T R(t)e^{-rt}dt$$

Present Value of a Continuous Income Stream

Suppose that an income stream flows continuously into an interest-bearing account at the rate of $R(t)$ dollars per year where t is measured in years and that the account earns interest at the annual rate of $100r\%$ compounded continuously. The present value of the account is

$$\text{Present value} = \int_0^T R(t)e^{-rt}dt \text{ dollars}$$

It is worth noting that once you have calculated future value, it is easy to calculate the associated present value by solving for P in the equation

$$Pe^{rt} = \text{future value}$$

EXAMPLE 3   *Finding Present Value from Future Value*

Profit   Last year, profit for the HiTech Corporation was $17.2 million. Assuming that HiTech's profits increase continuously for the next 5 years at a rate of $1.3 million per year, what are the future and present values of the corporation's 5-year profits? Assume an interest rate of 12% compounded continuously.

Solution   We note that the rate of the stream is $R(t) = 17.2 + 1.3t$ million dollars per year in year t. In order to calculate the future value of this stream, we evaluate $\int_0^5 (17.2 + 1.3t)e^{0.12(5-t)}dt$. We have not developed a method for finding the antiderivative of $f(t) = (17.2 + 1.3t)e^{0.12(5-t)}$, so we numerically estimate the definite integral using a limiting value of sums or use technology to evaluate the integral.

$$\text{Future value} = \int_0^5 (17.2 + 1.3t)e^{0.12(5-t)}dt \approx \$137.9 \text{ million}$$

The invested revenue will be worth approximately $138 million in 5 years. Again, numerically estimate a limiting value of sums or use technology to find the present value.

$$\text{Present value} = \int_0^5 (17.2 + 1.3t)e^{-0.12t}dt \approx \$75.7 \text{ million}$$

This amount is the lump sum ($75.7 million) that would have to be invested at 12% compounded continuously in order to earn $137.9 million (the future value) in 5 years.

We could also use the future value ($137.9 million) to calculate the present value:

$$Pe^{(0.12)(5)} \approx \$137.9 \text{ million} \qquad \text{so} \qquad P \approx \$75.7 \text{ million}$$

The integral definition of the present value is most useful in situations in which you do not know the future value.   ●

Discrete Income Streams

The assumptions that income is flowing continuously and that interest is compounded continuously make it possible to use calculus and are often imposed by economists. Unfortunately, they do not generally hold in the real world of business. It is much more realistic to consider an income stream that flows monthly into an account with monthly compounding of interest or a stream that flows quarterly with quarterly compoundings.

The process of finding the future value for discrete income streams begins in a similar way to that for continuous streams: Determine the rate-of-flow function for the income stream, and multiply by a term that accounts for compounding interest. In the discrete case, we base our interest calculations on the formula $A = P\left(1 + \frac{r}{n}\right)^{nt}$ where A is the dollar amount accumulated after t years when P dollars are invested at an annual interest rate of $100r\%$ compounded annually n times a year. Instead of integrating the resulting function, we sum a series of values.

Consider a small business that begins investing 7.5% of its monthly profit into an account that pays 10.3% annual interest compounded monthly. When the company begins investing, the monthly profit is $12,000 and is growing by $500 each month. We wish to determine the 2-year future value of the company's investment (assuming that profit continues to grow in the manner described).

The company will make a total of 24 deposits during the 2-year period. The first deposit is $(0.075)(\$12,000) = \$900$. This deposit earns interest each month for 24 months. The future value of this first deposit is $\$900\left(1 + \frac{0.103}{12}\right)^{24} = \$1104.91$.

The second deposit is $(0.075)(\$12,000 + \$500) = \$937.50$. This deposit earns interest each month for 23 months. The future value of the second deposit is $\$937.50\left(1 + \frac{0.103}{12}\right)^{23} = \$1141.15$.

The third deposit is $(0.075)(\$12,000 + \$1000) = \$975$. This deposit earns interest each month for 22 months. The future value of the third deposit is $\$975\left(1 + \frac{0.103}{12}\right)^{22} = \$1176.60$.

By now you should be able to see a pattern in the future values of these successive monthly deposits. (See Table 7.3.)

TABLE 7.3

Time d (months after first deposit)	Future value of monthly deposit $F(d)$
0	$F(0) = \$900.00\left(1 + \frac{0.103}{12}\right)^{24} = \$1104.91$
1	$F(1) = \$937.50\left(1 + \frac{0.103}{12}\right)^{23} = \$1141.15$
2	$F(2) = \$975.00\left(1 + \frac{0.103}{12}\right)^{22} = \$1176.60$
$\vdots$	$\vdots$
22	$F(22) = \$1725.00\left(1 + \frac{0.103}{12}\right)^{2} = \$1754.74$
23	$F(23) = \$1762.50\left(1 + \frac{0.103}{12}\right)^{1} = \$1777.63$
Future value = sum of 24 values = $35,204.03	

The amount deposited each month is given by $R(d) = (0.075)(\$12,000 + \$500d)$, where d is the number of deposits after the first one. The deposit associated with input d accrues interest for $24 - d$ months. The future value of each month's deposit is given by the formula

$$F(d) = R(d)\left(1 + \frac{0.103}{12}\right)^{24-d} = (0.075)(12,000 + 500d)\left(1 + \frac{0.103}{12}\right)^{24-d}$$

To determine the 2-year future value, we add the future values of each month's deposit beginning with 0 (month 1) and ending with 23 (month 24). Using summation notation, we write

$$\text{Future value} = \sum_{d=0}^{23} F(d) = \sum_{d=0}^{23} (0.075)(12,000 + 500d)\left(1 + \frac{0.103}{12}\right)^{24-d}$$

The symbol $\displaystyle\sum_{d=0}^{23} F(d)$ is the notation used for the sum $F(0) + F(1) + F(2) + \cdots + F(23)$. In this example, the future value of the deposits is $35,204.03. We generalize the definition of future value for discrete income streams as follows:

Future Value of a Discrete Income Stream

Suppose that a deposit is made into an interest-bearing account at n equally spaced times throughout a year. The value of the dth deposit is $R(d)$ dollars per period, and it earns interest at an annual percentage rate of $100r\%$ compounded once in each deposit period. The future value of the deposits at the end of D deposit periods is

$$\text{Future value} = \sum_{d=0}^{D-1} R(d)\left(1 + \frac{r}{n}\right)^{D-d} \text{ dollars}$$

Once you have determined future value, present value can be found by solving for P in the equation

$$P\left(1 + \frac{r}{n}\right)^{D} = \text{future value}$$

EXAMPLE 4   *Determining Future Value of a Discrete Income Stream*

7.2.2

Savings   When you graduate from college (say, in 3 years), you would like to purchase a car. You have a job and can put $75 into savings each month for this purchase. You choose a money market account that offers an APR of 6.2% compounded quarterly.

a. How much money will you have deposited in 3 years?

b. What will be the value of your savings in 3 years?

c. How much money would you have to deposit now (in one lump sum) to achieve the same future value in 3 years?

d. You are considering a second money market account that pays monthly interest of 0.5%. Will this account result in a greater future value than that calculated in part *b*?

Solution

a. The total amount deposited is $(36)(\$75) = \$2700$.

b. Because interest is compounded quarterly ($n = 4$), the monthly deposits each quarter do not earn interest until the end of the quarter in which they are deposited. We therefore consider three $75 monthly deposits equivalent to one $225 quarterly deposit, and this gives $R(d) = \$75(3) = \$225$ per quarter d quarters after the first one. We assume that the first deposit is made at the beginning of a quarter. During the 3 years, there will be 12 quarters to consider ($D = 12$). Table 7.4 shows the pattern of the future values of each quarter's deposit.

The 3-year future value is given by

$$\text{Future value} = \sum_{d=0}^{12-1} R(d)\left(1 + \frac{0.062}{4}\right)^{12-d}$$

$$= \sum_{d=0}^{11} \$225\left(1 + \frac{0.062}{4}\right)^{12-d} = \$2988.10$$

TABLE 7.4

Time (quarters after first deposit)	Future value of quarterly deposit
0	$\$225\left(1 + \frac{0.062}{4}\right)^{12} = \$270.61$
1	$\$225\left(1 + \frac{0.062}{4}\right)^{11} = \$266.48$
2	$\$225\left(1 + \frac{0.062}{4}\right)^{10} = \$262.41$
⋮	⋮
10	$\$225\left(1 + \frac{0.062}{4}\right)^{2} = \$232.03$
11	$\$225\left(1 + \frac{0.062}{4}\right)^{1} = \$228.48$
Future value = sum of 12 values = $\$2988.10$	

c. Because we know the future value, we can solve $P\left(1 + \frac{0.062}{4}\right)^{12} \approx 2988.10$ for the present value P to obtain $P \approx \$2484.48$. This is the amount that you would need to deposit now to have $2988.10 in 3 years.

d. Because the account pays interest monthly, we have the values $n = 12$ and $D = 3(12) = 36$. Also, because interest is compounded monthly, we consider the deposits being made monthly, so that $R(d) = \$75$. In this case, the interest rate is given in terms of a monthly rather than a yearly rate. In other words, we are given the term $\frac{r}{n} = \frac{r}{12} = 0.005$. Thus the 3-year future value of $75 monthly deposits into this money market account is

$$\text{Future value} = \sum_{d=0}^{35} \$75(1.005)^{36-d} = \$2964.96$$

This is a smaller future value than for the account that pays 6.2% interest quarterly.   ●

Streams in Biology

Biology and other fields involve situations very similar to income streams. An example of this is the growth of populations of animals. As of 1978, there were approximately 1.5 million sperm whales* in the world's oceans. Each year approximately 0.06 million sperm whales are added to the population. Also each year, 4% of the sperm whale population either die of natural causes or are killed by hunters. Assuming that these rates (and percentage rates) have remained constant since 1978, we estimate the sperm whale population in 1998 using the same procedure as when determining future value of continuous income streams.

There are two aspects of the population that we must consider when estimating the population of sperm whales in 1998. First, we must determine the number of whales that were living in 1978 that will still be living in 1998. Because 4% of the sperm whales die each year, we calculate the number of whales that have survived the entire 20 years as $1.5(0.96^{20}) \approx 0.663$ million whales.

The second aspect that we must consider is the impact on the population made by the birth of new whales. We are told that 0.06 million whales per year are added to the population and that 96% of those survive each year. Therefore, the growth rate of the population of sperm whales associated with those that were born t years after 1978 is

$$f(t) = 0.06(0.96)^{20-t} \text{ million whales per year}$$

Thus the sperm whale population in 1998 is calculated as

$$\text{Whale population} = 1.5(0.96^{20}) + \int_0^{20} 0.06(0.96)^{20-t}\,dt$$

$$\approx 0.663 + \int_0^{20} 0.06(0.96^{20})(0.96)^{-t}\,dt$$

$$= 0.663 + 0.06(0.96^{20})\int_0^{20}(0.96^{-1})^t\,dt$$

$$= 0.663 + \frac{0.06(0.96^{20})(0.96^{-1})^t}{\ln(0.96^{-1})}\Bigg|_0^{20}$$

$$\approx 1.48 \text{ million sperm whales}$$

Functions that model such biological streams, in which new individuals are added to the population and the rate of survival of the individuals is known, are referred to as *survival* and *renewal functions.*

Future Value of a Biological Stream

The future value (in b years) of a biological stream with initial population size P, survival rate $100s\%$, and renewal rate $r(t)$, where t is the number of years, is

$$\text{Future value} \approx Ps^b + \int_0^b r(t)s^{b-t}\,dt$$

---

*Delphine Haley, *Marine Mammals* (Seattle, WA: Pacific Search Press, 1978).

In the whale example, the initial population is $P = 1.5$ million. The survival rate is 96% per year, so $s = 0.96$. The renewal rate is $r(t) = 0.06$ million whales per year.

EXAMPLE 5   *Determining Future Value of a Biological Stream*

Flea Population   An example of a stream in entomology is the growth of a flea population. In cooler areas of the country, adult fleas die before winter, but flea eggs survive and hatch the following spring when temperatures again reach 70°F. Not all the eggs hatch at the same time, so part of the growth in the flea population is due to the hatching of the original eggs. Another part of the growth in the flea population is due to propagation. Suppose fleas propagate at the rate of 134% per day and that the original set of fleas (from the dormant eggs) become reproducing adults at the rate of 600 fleas per day. What will the flea population be 10 days after the first 600 fleas begin reproducing? Assume that none of the fleas die during the 10-day period and that all fleas become reproducing adults 24 hours after hatching and propogate every day thereafter at the rate of 134% per day.

Solution   We first note that because we begin counting when the first 600 fleas have become mature adults, we consider the initial population to be $P = 600$ fleas. The renewal rate is also 600 fleas per day, so $r(t) = 600$.

Because, in this case, the renewal rate function r does not account for renewal due to propagation, we must incorporate the propagation rate of 134% into the survival rate of 100%. Thus the survival/propagation rate is $s = 2.34$.

Because the renewal rate and survival/propagation rate are given in days, we let t be the input variable measured in days. The flea population will grow over 10 days to

$$\text{Flea population} \approx Ps^{10} + \int_0^{10} r(t)s^{10-t}dt = 600(2.34^{10}) + \int_0^{10} 600(2.34)^{10-t}dt$$

$$\approx 2{,}953{,}315 + 3{,}473{,}166 \approx 6.4 \text{ million fleas} \quad \bullet$$

7.2   Concept Inventory

○ *Income streams*

○ *Flow rate of a stream*

○ *Future and present value of a continuous stream*

○ *Future and present value of a discrete stream*

○ *Biological stream*

○ *Future value of a biological stream*

7.2   Activities

1. **Savings**   Suppose that after you graduate, you are hired by a company in the San Francisco Bay area. Housing prices in that region are the highest in the nation; however, you are determined to buy a house within 5 years of beginning your new job. Your starting salary is $47,000. After talking with fellow employees, you consider three possibilities for what might happen to your salary over the next 5 years:

   i. Your salary remains at your starting level.

   ii. Your salary increases by $100 a month.

   iii. Your salary increases by 0.5% each month.

a. You have decided to save 20% of your salary each month for a down payment on a house. Give the function describing your monthly investments for each of the three salary possibilities.

b. You estimate that you will need $60,000 for a down payment, and you are unwilling to accept any investment risk. You will be investing your money monthly in a bank savings account that pays an annual interest rate of 5% compounded monthly. For which salary possibilities will the total amount saved in 5 years be at least $60,000?

2. Investment A company is hoping to expand its facilities but needs capital to do so. In an effort to position itself for expansion in 3 years, the company will direct half of its profits into investments in a continuous manner. The company's profits for the past 5 years are shown in the table.

Years ago	5	4	3	2	1
Profit (thousands of dollars)	860	890	930	990	1050

The company's current yearly profit is $1,130,000. Find the function that describes the flow of the company's investments for each of the following profit scenarios:

a. The profit for the next 3 years follows the trend shown in the table.

b. The profit increases each year for the next 3 years by the same percentage that it increased in the current year.

c. The profit remains constant at the current year's level.

d. The profit increases each year for the next 3 years by the same fixed amount that it increased this year.

e. If the company's investments can earn 16.4% annual interest compounded continuously, how much capital will it have saved after 3 years of investing for each of the profit scenarios given?

3. Revenue For the year ending June 30, 2002, the revenue of the Sara Lee Corporation was $17.628 billion. Assume that Sara Lee's revenue will increase by 5% per year and that beginning on July 1, 2002, 12.5% of the revenue was invested each year (continuously) at an APR of 7% compounded continuously. What is the future value of the investment at the end of the year 2006?
(Source: Hoover's Online Guide.)

4. Savings A high school student is trying to save money to help pay for her first-year college tuition. She plans to invest $300 each quarter for 3 years into an account that pays interest at an APR of 4% compounded quarterly. Her parents decide to lend her the money so that she can devote more time to her studies while in high school. How much should they lend her so that she can invest the loaned amount now, as one lump sum, into the account and accumulate the same amount as if she had made the quarterly deposits for 3 years?

5. Revenue The revenue of General Motors Corporation (GM) in December 2001 was $177.26 billion. Assume that GM's revenue remains constant and that 3% of the revenue is invested continuously throughout each year beginning at the end of December 2001 into an account that pays interest at a rate of 8.8% compounded continuously.
(Source: Hoover's Online Guide.)

a. Find the value of the account in December 2008.

b. How much would GM have had to invest at the end of December 2001, in one lump sum, into this account in order to build the same 7-year future value as the one found in part *a*?

6. Revenue For the year ending December 31, 2001, the General Electric Company's revenue was $125.68 billion. Assume that the revenue increases by 8% per year and that General Electric will (continuously) invest 10% of its profits each year at an APR of 8.5% compounded continuously for a period of 9 years beginning at the end of December of 2001. What is the present value of this 9-year investment?
(Source: Hoover's Online Guide.)

7. Savings To save for the purchase of your first home (in 6 years), suppose you begin investing $500 per month in an account with a fixed rate of return of 6.34%.

a. Assuming a continuous stream, what will the account be worth at the end of 6 years?

b. Assuming monthly activity (deposits and interest compounding), what will the account be worth at the end of 6 years?

c. Is the answer to part *a* or part *b* more likely to be the actual future value of the account? Explain.

8. **Investment** In preparing for your retirement (in 40 years), suppose you plan to invest 14% of your salary each month in an annuity with a fixed rate of return of 6.2%. You currently make $2800 per month and expect your income to increase by 3% per year.

   a. Assuming a continuous stream, what will the annuity be worth at the end of 40 years?

   b. Assuming monthly activity (deposits and interest compounding), what will the annuity be worth at the end of 40 years?

   c. Is the answer to part *a* or part *b* more likely to be the actual future value of the annuity? Explain.

9. **Profit** Ticketmaster is the world's largest ticket retailer. Ticketmaster's 2002 third-quarter gross profits were $82.1 million. Assume that these profits will increase by 5% per quarter and that Ticketmaster will invest 15% of its quarterly profits in an investment with a quarterly return of 9%.
   (Source: Hoover's Online Guide.)

   a. Write a function for the rate at which money flows into this investment each quarter.

   b. Write a function for the rate at which the 4-year future value of this investment is changing.

   c. Find the value of this investment at the end of the year 2006. (Assume a quarterly stream beginning on January 1, 2003 with the investment of 4th quarter 2002 profits.)

10. **Net Income** For the 2002 fiscal year, Lowe's Companies, Inc., reported an annual net income of $1,023,300,000. Assume the income can be reinvested continuously at an annual rate of return of 10% compounded continuously. Also assume that Lowe's will maintain this annual net income for the next 5 years.
    (Source: Hoover's Online Guide.)

    a. What is the future value of its 5-year net income?

    b. What is the present value of its 5-year net income?

11. **Cola Sales** In 1993, PepsiCo installed a new soccer scoreboard for Alma College in Alma, Michigan. The terms of the installation were that Pepsi would

have sole vending rights at Alma College for the next 7 years. It is estimated that in the 3 years after the scoreboard was installed, Pepsi sold 36.4 thousand liters of Pepsi products to Alma College students, faculty, staff, and visitors. Suppose that the average yearly sales and associated revenue remained constant and that the revenue from Alma College sales was reinvested at 4.5% APR. Also assume that PepsiCo makes a revenue of $0.80 per liter of Pepsi.

   a. The vending of Pepsi products on campus can be considered a continuous process. Assuming that the revenue was invested in a continuous stream and that interest on that investment was compounded continuously, how much did Pepsi make from its 7 years of sales at Alma College?

   b. Still assuming a continuous stream, find how much Pepsi would have had to invest in 1993 to create the same 7-year future value.

12. **Investment** Refer to Activity 8. How much would you have to invest now, in one lump sum instead of in a continuous stream, in order to build to the same future (40-year) value?

13. **Savings** Refer to Activity 7.

    a. How much would you have to invest now, in one lump sum instead of in a continuous stream, in order to build to the same future (6-year) value? Assume that interest is compounded continuously.

    b. How much would you have to invest now, in one lump sum instead of in a monthly stream, in order to build to the same future (6-year) value? Assume monthly compounding of interest.

    c. Is the answer to part *a* or to part *b* more likely to be the actual present value of the annuity? Explain.

14. **Revenue** Between 1995 and 2001, the revenue of Sears Roebuck and Co. can be modeled as

$$R(t) = \frac{11.24}{1 + 1.366e^{-1.55t}} + 30 \text{ billion dollars per year}$$

t years after 1995. Assume that the revenue can be reinvested at 9.5% compounded continuously.
(Source: Based on data from the Hoover's Online Guide.)

   a. How much is Sears' revenue invested since 1995 worth in 2003?

b. How much was this accumulated investment worth in 1995?

15. **Buyout** In 1956, AT&T laid its first underwater phone line. By 1996, AT&T Submarine Systems, the division of AT&T that installs and maintains undersea communication lines, had seven cable ships and 1000 workers. On October 5, 1996, AT&T announced that it was seeking a buyer for its Submarine Systems division. The Submarine Systems division of AT&T was posting a profit of $850 million per year.
(Source: "AT&T Seeking a Buyer for Cable-Ship Business," *Wall Street Journal,* October 5, 1996.)

a. If AT&T assumed that the Submarine Systems division's annual profit would remain constant and could be reinvested at an annual return of 15%, what would AT&T have considered to be the 20-year present value of its Submarine Systems division? (Assume a continuous stream.)

b. If prospective bidder A considered that the annual profits of this division would remain constant and could be reinvested at an annual return of 13%, what would bidder A consider to be the 20-year present value of AT&T's Submarine Systems? (Assume a continuous stream.)

c. If prospective bidder B considered that over a 20-year period, profits of the division would grow by 10% per year (after which it would be obsolete) and that profits could be reinvested at an annual return of 14%, what would bidder B consider to be the 20-year present value of AT&T's Submarine Systems? (Assume a continuous stream.)

16. **Buyout** On October 4, 1996, Tenet Healthcare Corporation, the second-largest hospital company in the United States at that time, announced that it would buy Ornda Healthcorp.
(Source: "Tenet to Acquire Ornda," *Wall Street Journal,* October 5, 1996.)

a. If Tenet Healthcare Corporation assumed that Ornda's annual revenue of $0.273 billion would increase by 10% per year and that the revenues could be continuously reinvested at an annual return of 13%, what would Tenet Healthcare Corporation consider to be the 15-year present value of Ornda Healthcorp at the time of the buyout?

b. If Ornda Healthcorp's forecast for its financial future was that its $0.273 billion annual revenue would remain constant and that revenues could be continuously reinvested at an annual return of 15%, what would Ornda Healthcorp consider its 15-year present value to be at the time of the buyout?

c. Tenet Healthcare Corporation bought Ornda Healthcorp for $1.82 billion in stock. If the sale price was the 15-year present value, did either of the companies have to compromise on what it believed to be the value of Ornda Healthcorp?

17. **Buyout** CSX Corporation, a railway company, announced in October of 1996 its intention to buy Conrail Inc. for $8.1 billion. The combined company, CSX-Conrail, would control 29,000 miles of track and have an annual revenue of $14 billion the first year after the merger, making it one of the largest railway companies in the country.
(Source: "Seeking Concessions from CSX-Conrail Is Seen as Most Likely Move by Norfolk," *Wall Street Journal,* October 5, 1996.)

a. If Conrail assumed that its $2 billion annual revenue would decrease by 5% each year for the next 10 years but that the annual revenue could be reinvested at an annual return of 20%, what would Conrail consider to be its 10-year present value at the time of CSX's offer? Is this more or less than the amount CSX offered?

b. CSX Corporation forecast that its Conrail acquisition would add $1.2 billion to its annual revenue the first year and that this added annual revenue would increase by 2% each year. Suppose CSX is able to reinvest that revenue at an annual return of 20%. What would CSX Corporation have considered to be the 10-year present value of the Conrail acquisition in October of 1996?

c. Why might CSX Corporation have forecast an increase in annual revenue when Conrail forecast a decrease?

18. **Buyout** Company A is attempting to negotiate a buyout of Company B. Company B accountants project an annual income of 2.8 million dollars per year. Accountants for Company A project that with Company B's assets, Company A could produce an income starting at 1.4 million dollars per year and

growing at a rate of 5% per year. The discount rate (the rate at which income can be reinvested) is 8% for both companies. Suppose that both companies consider their incomes over a 10-year period. Company A's top offer is equal to the present value of its projected income, and Company B's bottom price is equal to the present value of its projected income. Will the two companies come to an agreement for the buyout? Explain.

19. **Capital Value**   A company involved in videotape reproduction has just reported $1.2 million net income during its first year of operation. Projections are that net income will grow over the next 5 years at the rate of 6% per year. The *capital value* (present sales value) of the company has been set as its present value over the next 5 years. If the rate of return on reinvested income can be compounded continuously for the next 5 years at 12% per year, what is the capital value of this company?

20. **Population**   There were once more than 1 million elephants in West Africa. Now, however, the elephant population has dwindled to 19,000. Each year 17.8% of West Africa elephants die or are killed by hunters. At the same time, elephant births are decreasing by 13% per year.
(Source: Douglas Chawick, *The Fate of the Elephant.* Sierra Club Books, 1992.)

   a. How many of the current population of 19,000 elephants will still be alive 30 years from now?

   b. Considering that 47 elephants were born in the wild this year, write a function for the number of elephants that will be born *t* years from now and will still be alive 30 years from now.

   c. Estimate the elephant population of West Africa 30 years from now.

21. **Population**   In 1979 there were 12 million sooty terns (a bird) in the world. Assume that the percentage of terns that survive from year to year has

stayed constant at 83% and that approximately 2.04 million terns hatch each year.
(Source: Bryan Nelson, *Seabirds: Their Biology and Ecology.* New York: Hamlyn Publishing Group, 1979.)

   a. How many of the terns that were alive in 1979 are still alive?

   b. Write a function for the number of terns that hatched *t* years after 1979 and are still alive.

   c. Estimate the present population of sooty terns.

22. **Population**   From 1936 through 1957, a population of 15,000 muskrats in Iowa bred at a rate of 468 new muskrats per year and had a survival rate of 75%.
(Source: Paul L. Errington, *Muskrat Population.* Ames, IA: Iowa State University Press, 1963.)

   a. How many of the muskrats alive in 1936 were still alive in 1957?

   b. Write a function for the number of muskrats that were born *t* years after 1936 and were still alive in 1957.

   c. Estimate the muskrat population in 1957.

23. **Population**   There are approximately 200 thousand northern fur seals. Suppose the population is being renewed at a rate of $r(t) = 60 - 0.5t$ thousand seals per year and that the survival rate is 67%.
(Source: Delphine Haley, *Marine Mammals.* Seattle, WA: Pacific Search Press, 1978.)

   a. How many of the current population of 200 thousand seals will still be alive 50 years from now?

   b. Write a function for the number of seals that will be born *t* years from now and will still be alive 50 years from now.

   c. Estimate the northern fur seal population 50 years from now.

24. Explain, using related examples, the difference between a continuous income stream and a discrete income stream.

7.3 **Integrals in Economics**

This section uses improper integrals, which were discussed in Section 7.1

When you purchase an item in a store, you ordinarily have no control over the price that you pay. Your only choice is whether to buy or not to buy the item at the current price. In general, consumers hold to the view that price is a variable to which they can only respond. As the price per unit increases, consumers usually respond by purchasing (demanding) less. The typical relation between the price per unit (as input) and the quantity in demand (as output) is shown in Figure 7.6a.

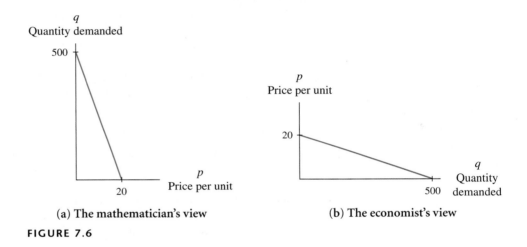

FIGURE 7.6

The traditional approach to graphing in economic theory is to put the price per unit along the vertical axis and the quantity in demand along the horizontal axis. (See Figure 7.6b.) In general, we choose to graph price per unit as input along the horizontal axis. This will help us understand price as input and visualize the definite integrals used later in this section. However, occasionally we will present both the mathematician's and the economist's graphical viewpoints.

Demand Curves

The graph relating quantity in demand q to price per unit p is called a **demand curve.** In economic theory, demand is actually a function that has several input variables, such as price per unit, consumers' ability to buy, consumers' need, and so on. The demand curve we consider here is a simplified version. We assume that all the possible input variables are constant except price. We denote this demand function as D with input p.

Even though the demand function is not a rate-of-change function, there are economic interpretations for the areas of certain regions lying beneath the demand curve. In order to interpret the area of these regions, you must understand how to interpret the information the demand curve represents.

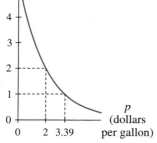

$D(p)$
(million
gallons)

FIGURE 7.7

For instance, suppose the graph in Figure 7.7 represents the weekly demand for regular unleaded gasoline in a California city. A point on the demand curve indicates the quantity that consumers will purchase at a given price. For instance, at $3.39 per gallon, consumers will purchase 1 million gallons of gas. At $2.00 per gallon, consumers will purchase 2 million gallons of gas.

Consumers' Willingness and Ability to Spend

Even though points on the demand curve tell us how much consumers will actually purchase at certain prices, consumers are willing and able to pay more than this amount for the quantity they purchase. For instance, consumers are willing and able to spend approximately $3.39 million for the first million gallons of regular unleaded gasoline, but they are willing and able to spend only approximately $2.00 million for the second million gallons. Thus, in total, consumers are willing and able to spend approximately $5.39 million for 2 million gallons of gas.

If the price of gas is $1.19 a gallon, consumers are willing and able to buy the third million gallons. That is, consumers are willing and able to spend approximately

(1 million gallons)($3.39 per gallon) + (1 million gallons)($2.00 per gallon)
+ (1 million gallons)($1.19 per gallon) = $6.58 million

for 3 million gallons of gas, even though in actuality they spend only

(3 million gallons)($1.19 per gallon) = $3.57 million

Consumers' willingness and ability to spend can be approximated graphically as the areas of stacked horizontal rectangles. The amount that consumers actually spend is depicted as the area of a single vertical rectangle. (See Figures 7.8a and b.)

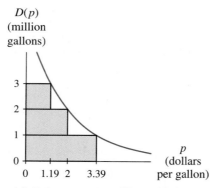

(a) Consumers are willing and able to spend about $6.58 million for 3 million gallons of gas.

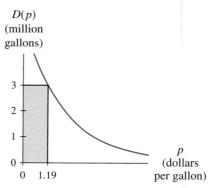

(b) Consumers pay only $3.57 million for 3 million gallons of gas.

FIGURE 7.8

You should have noticed that the amount that consumers are willing and able to spend for 3 million gallons was given as *approximately* $6.58 million. We can make this approximation better by considering smaller increments for price. If we were to approximate consumers' willingness and ability to spend using price increments of

$0.5 per gallon, $0.25 per gallon, $0.125 per gallon, etc., we would see that the areas of the stacked rectangles representing these approximations would become closer to being the true area depicted in Figure 7.9.

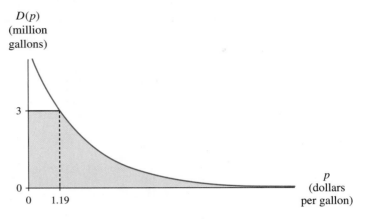

FIGURE 7.9

Thus consumers' willingness and ability to purchase 3 million gallons of gas can be visually represented by the sum of the area of the rectangle with width 1.19 under the horizontal line $D(p) = 3$ and the area under the demand curve from 1.19 to P, where P is the price above which consumers cannot and will not purchase any gas. We calculate the consumers' willingness and ability to spend as

$$3(1.19) + \int_{1.19}^{P} D(p)\,dp \text{ million dollars}$$

Suppose the demand for gas can be modeled by

$$D(p) = 5.43(0.607^p) \text{ million gallons}$$

where p dollars is the price per gallon. The only piece of information we still need is P, the price above which no gas will be purchased. You should notice that the demand function approaches 0 as p becomes large; however, it will never be exactly 0 for any p. Hence, we let P approach ∞. This is true for most demand functions in economics—some people will always want the product or service, regardless of the price. In this case, we consider the area under the demand curve as P becomes infinitely large. That is,

$$3(1.19) + \int_{1.19}^{\infty} 5.43(0.607^p)\,dp$$

$$= 3.57 + \lim_{P \to \infty} \int_{1.19}^{P} 5.43(0.607^p)\,dp$$

$$= 3.57 + \lim_{P \to \infty} \left(\frac{5.43(0.607^p)}{\ln 0.607} \right)\Bigg|_{1.19}^{P}$$

$$= 3.57 + \left(\lim_{P \to \infty} \frac{5.43(0.607^p)}{\ln 0.607} \right) - \frac{5.43(0.607^{1.19})}{\ln 0.607}$$

$$\approx 3.57 + 0 + 6.00$$

$$\approx \$9.57 \text{ million}$$

Thus consumers are willing and able to spend $9.57 million in order to purchase 3 million gallons of gas.

In general, we make the following definition:

Consumers' Willingness and Ability to Spend

For a continuous demand function $q = D(p)$, the maximum amount that consumers are willing and able to spend for a certain quantity q_0 of goods or services is the area of the shaded region in Figure 7.10.

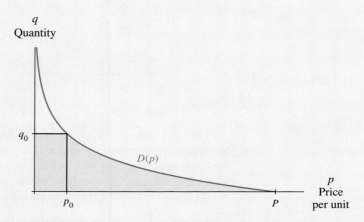

FIGURE 7.10

where p_0 is the market price at which q_0 units are in demand and P is the price above which consumers will purchase none of the goods or services. This area is calculated as

$$p_0 q_0 + \int_{p_0}^{P} D(p)\,dp$$

(Note that ∞ is used as the upper limit on the integral if the demand function approaches, but does not cross, the input axis.)

Because economists graph with the price on the vertical axis, in economics books the market price p_0 at which q_0 units are in demand as well as the price P above which consumers will purchase none of the goods or services both appear on the vertical axis whereas q_0 is on the horizontal axis. The area depicting the maximum amount that consumers are willing and able to spend appears (from the economists' viewpoint) as the shaded area in Figure 7.11. However, because price is the input regardless of graphical viewpoint, the area is calculated as

$$p_0 q_0 + \int_{p_0}^{P} D(p)\,dp$$

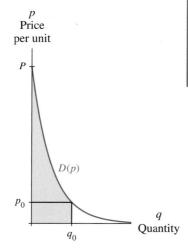

FIGURE 7.11

Consumers' Expenditure and Surplus

Now that we have considered what consumers are willing and able to spend for a certain quantity of a product, let us turn our attention to calculating what consumers actually spend for that quantity. We return to the discussion of gasoline demand.

As was previously mentioned, if the market price for gas is $1.19 per gallon, consumers will purchase 3 million gallons. The actual amount spent by consumers is (3 million gallons)($1.19 per gallon) = $3.57 million, even though they are willing and able to spend much more. This actual amount spent is (price)(quantity), which is the area of the rectangular region from the vertical axis to $p = 1.19$ with height 3 as shown in Figure 7.12. This amount is known as the **consumers' expenditure.** The amount that consumers are willing and able to spend but do not actually spend is known as the **consumers' surplus.**

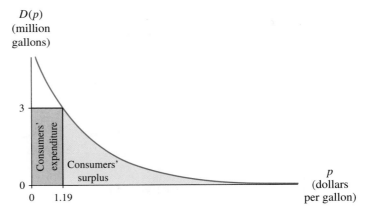

FIGURE 7.12

Earlier we found that consumers are willing and able to spend $9.57 million to purchase 3 million gallons of gas, so the consumers' surplus from buying 3 million gallons of gas at $1.19 per gallon is

$9.57 million $-$ $3.57 million = $6 million

Consumers' surplus can also be computed as the area between the demand function and the horizontal axis as

$$\text{Consumers' surplus} = \int_{1.19}^{\infty} 5.43(0.607^p)\,dp \approx \$6 \text{ million}$$

In general, we make the following definitions:

Consumers' Expenditure and Surplus

For a continuous demand function $q = D(p)$, the amount that consumers spend at a certain market price is called consumers' expenditure; it is represented by the rectangular area in Figure 7.13. Furthermore, the amount that consumers are willing and able to spend, but do not spend, for q_0 items at a market price p_0 is called consumers' surplus; it is represented by the area of the nonrectangular shaded region in Figure 7.13. The value P is the price above which consumers will purchase none of the goods or services.

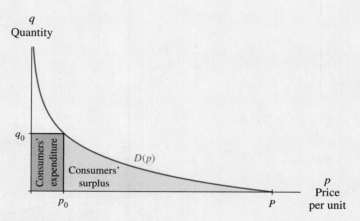

FIGURE 7.13

These areas are calculated as follows:

$$\text{Consumers' expenditure} = p_0 q_0$$

$$\text{Consumers' surplus} = \int_{p_0}^{P} D(p)\,dp$$

(Note that ∞ is used as the upper limit on the integral if the demand function approaches, but does not cross, the input axis.)

Again it is worth noting that because economists graph with the price on the vertical axis, graphs showing consumers' expenditure and surplus appear in economics books with the rectangle representing consumers' expenditure lying below the area representing consumers' surplus, as shown in Figure 7.14.

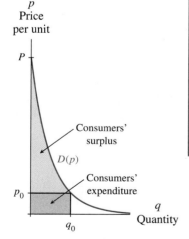

The economist's view
FIGURE 7.14

EXAMPLE 1   *Finding Areas Involving a Demand Curve*

7.3.1

Minivans   Suppose the demand for a certain model of minivan in the United States can be described as

$$D(p) = 14.12(0.933^p) - 0.25 \text{ million minivans}$$

when the market price is p thousand dollars per minivan.

a.   At what price per minivan will consumers purchase 2.5 million minivans?

b.   What is the consumers' expenditure when purchasing 2.5 million minivans?

c.   Does the model indicate a possible price above which consumers will purchase no minivans? If so, what is this price?

d.   When 2.5 million minivans are purchased, what is the consumers' surplus?

e.   What is the total amount consumers are willing and able to spend on 2.5 million minivans?

Solution

a.   We solve $D(p) = 2.5$ to find the market price at which consumers will purchase 2.5 million minivans. The equation

$$14.12(0.933^p) - 0.25 = 2.5$$

is satisfied when $p \approx 23.59033$. That is, at a market price p_0 of approximately $23,600 per minivan, consumers will purchase $q_0 = 2.5$ million minivans.

b.   When they purchase 2.5 million minivans, consumers' expenditure will be

$$p_0 q_0 \approx (23.59033 \text{ thousand dollars per minivan})(2.5 \text{ million minivans})$$
$$\approx \$59.0 \text{ billion}$$

c.   If the demand function approaches but does not cross the horizontal axis as price per unit increases without bound, then there is no price above which consumers will not purchase minivans. However, in this case, the demand function crosses the horizontal axis near $p = 58.16701$ (found by solving $D(p) = 0$). According to the model, the price above which consumers will purchase no minivans is approximately $p = \$58.2$ thousand per minivan.

d.   Consumers' surplus is the area of the region shaded in Figure 7.15, calculated as

$$\int_{p_0}^{P} D(p)\, dp \approx \int_{23.59033}^{58.16701} [14.12(0.933^p) - 0.25]\, dp$$
$$\approx 27.40482$$

To determine the appropriate units for consumers' surplus, remember that we are finding the area of a region whose width is measured in thousand dollars per minivan and whose height is measured in million minivans. Thus the units on consumers' surplus are (thousand dollars per minivan)(million minivans) which simplify to billion dollars.

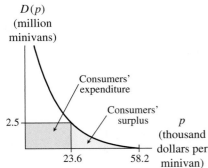

FIGURE 7.15

Therefore, we estimate the consumers' surplus when purchasing 2.5 million minivans to be $27.4 billion.

e. The amount that consumers are willing and able to spend for 2.5 million minivans is the combined area of the two shaded regions in Figure 7.15. This area is approximately 59.0 + 27.4 = $86.4 billion. ●

Supply Curves

We have seen that when prices go up, consumers usually respond by demanding less. However, manufacturers and producers respond to higher prices by supplying more. Thus a typical curve that relates the quantity supplied, S, to price per unit, p, is usually increasing and appears as shown in Figure 7.16.

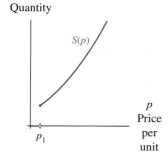

The graph that expresses the quantity supplied in terms of the price per unit is called a **supply curve.** You should note from Figure 7.16 that there is a price p_1 below which producers are not willing or able to supply any quantity of the product. The point $(p_1, S(p_1))$ is known in economics as the **shutdown point.** If the market price (and the corresponding quantity) fall below this point, producers will shut down their production.

The supply function has an interpretation very similar to that of the demand function. Suppose the quantity of regular unleaded gasoline that producers will supply is modeled as

FIGURE 7.16

$$S(p) = \begin{cases} 0 \text{ million gallons} & \text{when } p < 1 \\ 0.792p^2 - 0.433p + 0.314 \text{ million gallons} & \text{when } p \geq 1 \end{cases}$$

where the market price of gas is p dollars per gallon. This function is graphed in teal in Figure 7.17. At $1.24 a gallon, producers will supply 1 million gallons of gas. At $1.78 per gallon, producers will supply 2 million gallons, and if the price is $2.14 a gallon, producers will supply 3 million gallons.

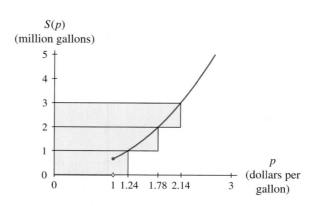

FIGURE 7.17

Producers' Willingness and Ability to Receive

We approximate the minimum that producers are willing and able to receive for 3 million gallons of gas by summing the amount they are willing and able to receive for the first million gallons, for the second million gallons, and for the third million gallons:

($1.24 per gallon)(first 1 million gallons) + ($1.78 per gallon)

(second 1 million gallons) + ($2.14 per gallon)(third 1 million gallons)

= $1,240,000 + $1,780,000 + $2,140,000 = $5,160,000

Thus the minimum that producers are willing and able to receive for 3 million gallons of gas is approximately $5,160,000. This amount can be thought of as the sum of the areas of the three stacked rectangles shaded in Figure 7.17.

As we use more intervals, the area of stacked rectangles comes closer to the true area of the region above the $q = S(p)$ curve and below the $q = 3$ line shown in Figure 7.18. Because a portion of $S(p)$ is zero, we find the total area by dividing the region into a rectangular region and the region below $q = 3$ and above $q = S(p)$ to the right of the shutdown point.

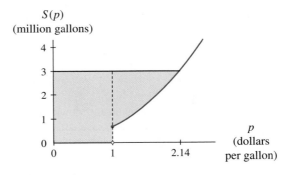

FIGURE 7.18

Therefore, the minimum amount that suppliers are willing and able to receive is calculated as

$$3(1) + \int_1^{2.14} [3 - S(p)]\,dp$$

$$= 3 + \int_1^{2.14} (-0.792p^2 + 0.433p + 2.686)\,dp$$

$$\approx \$4.5 \text{ million}$$

According to the supply model S, suppliers are willing and able to receive no less than $4.5 million for 3 million gallons of gas.

We make the following general definition:

Producers' Willingness and Ability to Receive

For a continuous or piecewise continuous supply function $q = S(p)$, the minimum amount that producers are willing and able to receive for a certain quantity q_0 of goods or services is the area of the shaded region in Figure 7.19

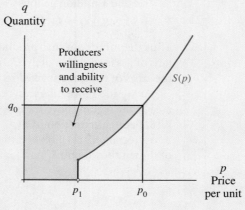

FIGURE 7.19

where p_0 is the market price at which q_0 units are supplied and p_1 is the shutdown price. This area is calculated as

$$p_1 q_0 + \int_{p_1}^{p_0} [q_0 - S(p)]\, dp$$

If there is no shutdown price, then $p_1 = 0$.

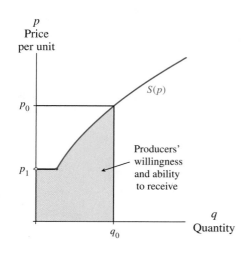

FIGURE 7.20

In the economist's viewpoint (with price graphed on the vertical axis), the minimum amount that producers are willing and able to receive for a certain quantity q_0 of goods or services is the shaded region shown in Figure 7.20. The market price p_0 at which q_0 units are supplied and the shutdown price p_1 are both graphed on the vertical axis. However, because price is the input and quantity is the output, the area is still calculated as $p_1 q_0 + \int_{p_1}^{p_0} [q_0 - S(p)]\, dp$.

Producers' Revenue and Surplus

The market price that will lead to the supply of 3 million gallons of gas is $p \approx \$2.14$ per gallon. The **producers' revenue** is (price)(quantity), which is the area of the rectangle shown in Figure 7.21: ($2.14 per gallon)(3 million gallons) $\approx \$6.4$ million. Producers will therefore receive

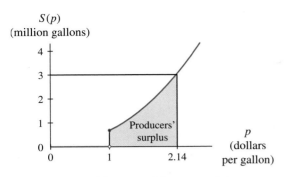

FIGURE 7.21 The area of the rectangle is producers' revenue.

$6.4 million − $4.5 million ≈ $1.9 million in excess of the minimum they are willing and able to receive. This excess amount is known as the **producers' surplus** and is the area of the shaded region in Figure 7.21.

We calculate the producers' surplus from the sale of 3 million gallons of gas at the market price of approximately $2.14 directly from the supply function as follows:

$$\int_1^{2.14} S(p)\,dp = \int_1^{2.14} (0.792p^2 - 0.433p + 0.314)\,dp$$

$$\approx \$1.9 \text{ million}$$

In general, we find the producers' total revenue and the producers' surplus as follows:

Producers' Revenue and Surplus

For a continuous or piecewise continuous supply function $q = S(p)$, the amount that producers receive at a certain market price is called the producers' revenue and is the area of the shaded rectangle in Figure 7.22a. Furthermore, the amount that producers receive above the minimum amount they are willing and able to receive for q_0 items at a market price p_0 is called the producers' surplus and is the area of the region between the supply function and the horizontal axis as shown in Figure 7.22b. The value p_1 is the price below which production shuts down. (If there is no shutdown price, then $p_1 = 0$.)

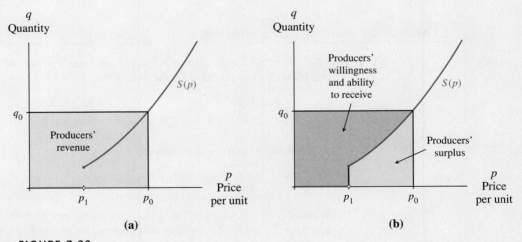

(a)   (b)

FIGURE 7.22

These areas are calculated as follows:

$$\text{Producers' revenue} = p_0 q_0$$

$$\text{Producers' surplus} = \int_{p_1}^{p_0} S(p)\,dp$$

Figure 7.23 is the economist's version of Figure 7.22. Because economists graph price vertically, producers' willingness and ability to receive and producers' surplus appear to be transposed as in Figure 7.23b. However, because the supply function S uses price per unit p as input, the calculations for producers' revenue and producers' surplus are the same as given in the preceding box.

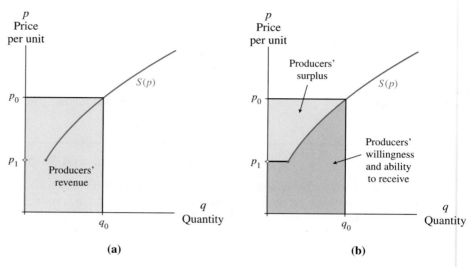

FIGURE 7.23

EXAMPLE 2   *Finding Areas Involving a Supply Curve*

Phones   Suppose the function for the average weekly supply of a certain brand of cellular phone can be modeled by the equation

$$S(p) = \begin{cases} 0 \text{ phones} & \text{when } p < 15 \\ 0.047p^2 + 9.38p + 150 \text{ phones} & \text{when } p \geq 15 \end{cases}$$

where p is the market price in dollars per phone.

a. How many phones (on average) will producers supply at a market price of $45.95?

b. What is the least amount that producers are willing and able to receive for the quantity of phones that corresponds to a market price of $45.95?

c. What is the producers' revenue when the market price is $45.95?

d. What is the producers' surplus when the market price is $45.95?

Solution

a. When the market price is $45.95, producers will supply an average of $S(45.95) \approx$ 680 phones each week.

b. The minimum amount that producers are willing and able to receive is the area of the labeled region in Figure 7.24 and is calculated as

$$(15)(680.247) + \int_{15}^{45.95} [680.247 - S(p)]\,dp$$
$$= 10{,}203.704 + (-0.0157p^3 - 4.69p^2 + 530.247p)\big|_{15}^{45.95}$$
$$\approx \$16{,}300.53$$

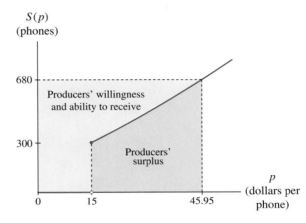

FIGURE 7.24

c. When the market price is $45.95, the producers' revenue is

(Quantity supplied at $45.95)($45.95 per phone)

$\approx$ (680.247 phones)($45.95 per phone)

$\approx \$31{,}257.35$

Graphically, the producers' revenue is the area of the rectangle in Figure 7.24.

d. When the market price is $45.95, the producers' surplus (see Figure 7.24) is

$$\int_{15}^{45.95} S(p)\,dp = (0.0157p^3 + 4.69p^2 + 150p)\Big|_{15}^{45.95} \approx \$14{,}956.82$$

Note that the producers' surplus plus the minimum amount that the producers are willing and able to receive is equal to the producers' revenue. ●

Social Gain

Consider the economic market for a particular item for which the demand and supply curves are shown in Figure 7.25. The point (p^*, q^*) where the demand curve and supply curve cross is called the **equilibrium point.** At the equilibrium price p^*, the quantity demanded by consumers coincides with the quantity supplied by producers. This quantity is q^*.

Economists consider that society is benefited whenever consumers and/or producers have surplus funds. When the market price of a product is the equilibrium price for that product, the total benefit to society is the consumers' surplus plus the producers' surplus. This amount is known as the **total social gain.**

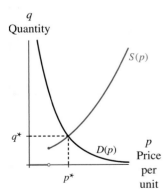

FIGURE 7.25 Market equilibrium (p^*, q^*) occurs when demand is equal to supply.

Market Equilibrium and Social Gain

Market equilibrium occurs when the supply of a product is equal to the demand for that product. If $q = D(p)$ is the demand function and $q = S(p)$ is the supply function, then the equilibrium point is the point (p^*, q^*), where p^* is the price that satisfies the equation $D(p) = S(p)$ and $q^* = D(p^*) = S(p^*)$.

The total social gain for a product is the sum of the producers' surplus and the consumers' surplus. When q^* units are produced and sold at a market price of p^*, the total social gain is the area of the entire shaded region in Figure 7.26. The value p_1 is the price below which production shuts down, and P is the price above which consumers will not purchase.

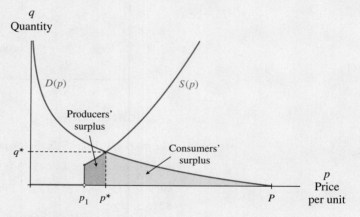

FIGURE 7.26

The combined area is calculated as

Total social gain = producers' surplus + consumers' surplus

$$= \int_{p_1}^{p^*} S(p)\, dp + \int_{p^*}^{P} D(p)\, dp$$

Once again we note that the economists' graphical viewpoint of Figure 7.26 differs from the mathematical graph in that economists represent the price (input) on the vertical axis and the quantity (output) on the horizontal axis. (See Figure 7.27.) The visual representation does not change the manner in which the calculations are performed in order to acquire the equilibrium point and total social gain.

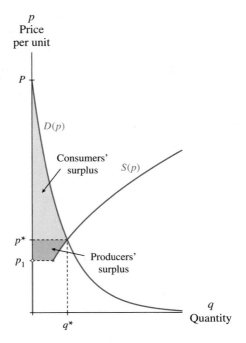

FIGURE 7.27

EXAMPLE 3   *Determining Market Equilibrium and Social Gain*

7.3.2

Gasoline The demand and supply functions for the gasoline example given near the beginning of this section are

$$\text{Demand} = D(p) = 5.43(0.607^p) \text{ million gallons}$$

and

$$\text{Supply} = S(p) = \begin{cases} 0 \text{ million gallons} & \text{when } p < 1 \\ 0.792p^2 - 0.433p + 0.314 \text{ gallons} & \text{when } p \geq 1 \end{cases}$$

where p is the market price in dollars per gallon.

a. Find the market equilibrium point for gasoline.

b. Find the total social gain when gasoline is sold at the market equilibrium price.

Solution

a. Solving

$$5.43(0.607^p) = 0.792p^2 - 0.433p + 0.314$$

for p yields $p^* \approx \$1.83$ per gallon. At this market price, $q^* \approx 2.2$ million gallons of gas will be purchased. [*Note:* q^* can be found as either $D(p^*)$ or $S(p^*)$.]

b. The total social gain at market equilibrium is the area of the shaded regions in Figure 7.28. We must find p_1, the shutdown price, and P, the price beyond which consumers will purchase no gasoline, before we can proceed.

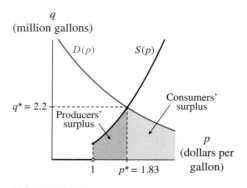

FIGURE 7.28

The shutdown price is given in the statement of the supply function as $p_1 = 1$. The demand function indicates that there is no price beyond which consumers will not purchase. Thus, $P \to \infty$.

Now we proceed with our calculation of total social gain:

$$\text{Total social gain} \approx \int_{1.00}^{1.8311} (0.792p^2 - 0.433p + 0.314)\, dp$$

$$+ \int_{1.8311}^{\infty} 5.43(0.607^p)\, dp$$

$$\approx 1.108 + 4.360 \approx \$5.5 \text{ million}$$

At the market equilibrium price of $1.83 per gallon, the total social gain is approximately $5.5 million. ●

7.3   Concept Inventory

O *Market price*

O *Demand curve*

O *Consumers' willingness and ability to spend*

O *Consumers' expenditure*

O *Consumers' surplus*

O *Supply curve*

O *Shutdown point*

O *Producers' revenue*

O *Producers' surplus*

O *Producers' willingness and ability to receive*

O *Market equilibrium*

O *Total social gain*

7.3   Activities

1. Give the economic name by which we call each of the following:

   a. The function relating the number of items the consumer will purchase at a certain price and the price per item

   b. The function relating the quantity of items the supplier of the items will sell at a certain price and the price per item

   c. The area of the region below the supply curve between the shutdown price and the market price

   d. The area of the region below the demand curve between the market price and the price above which consumers will cease to purchase

2. For each of the following amounts:

   i. Describe the region whose area gives the specified amount.

   ii. Illustrate that region by sketching an example.

   a. The maximum amount that consumers are willing and able to spend

   b. The minimum amount that producers are willing and able to receive

   c. The consumers' expenditure

   d. The total social gain at market equilibrium

3. Explain how to find each of the following:

   a. The price P above which consumers will purchase none of the goods or services

   b. The shutdown point

   c. The point of market equilibrium

4. **Demand**   The following two figures, drawn from the mathematician's and the economist's viewpoints, depict the same demand curve for a commodity. Consider p_0 to be the current market price of the commodity.

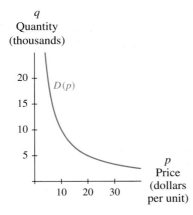

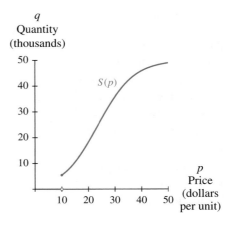

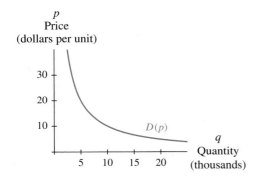

b. Sketch the supply curve from the economist's viewpoint. Include the shutdown price and the point (p_0, q_0) on the economist's graph.

c. Depict (by shading on each figure) the region representing producers' revenue.

6. Supply   The following two figures, drawn from the mathematician's and the economist's viewpoints, depict the same supply curve for a commodity. Consider $p_0 = \$3$ per unit to be the current market price of the commodity.

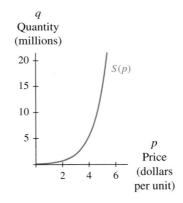

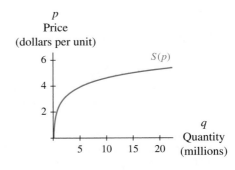

a. Use $p_0 = \$20$ per unit and the demand curve to estimate q_0 on the figure showing the mathematician's view. Transfer the point (p_0, q_0) to the figure showing the economist's view.

b. Depict (by shading on each figure) the regions representing consumers' expenditure and consumers' surplus.

c. Outline on each figure the region representing the amount that consumers are willing and able to spend.

d. Write a formula for the calculation of consumers' willingness and ability to spend.

5. Supply   The accompanying figure depicts a supply curve, drawn from the mathematician's viewpoint, for a commodity. Consider $p_0 = \$25$ per unit to be the current market price of the commodity and $p_1 = \$10$ per unit to be the shutdown price.

a. Use p_0 and the supply curve to estimate q_0 on the figure showing the mathematician's view.

a. Use p_0 and the supply curve to estimate q_0 on the figure showing the mathematician's view. Transfer the point (p_0, q_0) to and label the shutdown price on the figure showing the economist's view.

b. Depict (by shading on each figure) the regions representing producers' willingness and ability to receive and producers' surplus.

c. Outline on each figure the region representing producers' revenue.

d. Write a formula for producers' willingness and ability to receive.

e. Write a formula for producers' surplus.

7. **Social Gain**   The following two figures, drawn from the mathematician's and the economist's viewpoints, depict the same supply and demand curves for a product.

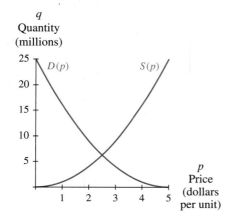

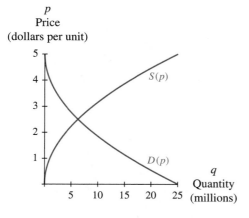

a. Mark on both figures the equilibrium point and label it (p^*, q^*). Estimate the values of and label p^* and q^* on the axes in both figures. Also esti-

mate the value of and label on both figures the shutdown price p_1 and the price P above which consumers will not purchase the product.

b. Depict (by shading on each figure) the regions representing consumers' surplus and producers' surplus.

c. Outline on each figure the region representing total social gain.

d. Write a formula for the calculation of total social gain.

8. **Demand**   The demand for wooden chairs can be modeled as

$$D(p) = -0.01p + 5.55 \text{ million chairs}$$

where p is the price (in dollars) of a chair.

a. According to the model, at what price will consumers no longer purchase chairs? Is this price guaranteed to be the highest price any consumer will pay for a wooden chair? Explain.

b. Find the quantity of wooden chairs that consumers will purchase if the market price is $99.95.

c. Determine the amount that consumers are willing and able to spend to purchase 3 million wooden chairs.

d. Find the consumers' surplus when consumers purchase 3 million wooden chairs.

9. **Demand**   The demand for ceiling fans can be modeled as

$$D(p) = 25.92(0.996^p) \text{ thousand ceiling fans}$$

where p is the price (in dollars) of a ceiling fan.

a. According to the model, is there a price above which consumers will no longer purchase fans? If so, what is it? If not, explain why not.

b. Find the amount that consumers are willing and able to spend to purchase 18 thousand ceiling fans.

c. Find the quantity of fans consumers will purchase if the market price is $100.

d. Find the consumers' surplus when the market price is $100.

10. **Demand**   The demand for a 12-ounce bottle of sparkling water is given in the accompanying table.

Price (dollars per bottle)	Demand (million bottles)
2.29	25
2.69	9
3.09	3
3.49	2
3.89	1
4.29	0.5

a. Find a model for demand as a function of price.

b. Does your model indicate a price above which consumers will purchase no bottles of water? If so, what is it? If not, explain.

c. Find the quantity of water that consumers will purchase if the market price is $2.59.

d. Find the amount that consumers are willing and able to spend to purchase the quantity you found in part c.

e. Find the consumers' surplus when the market price is $2.59.

11. **Demand** The average daily demand for a new type of kerosene lantern in a certain hardware store is as shown in the table.

Price (dollars per lantern)	Average quantity demanded (lanterns)
21.52	1
17.11	3
14.00	5
11.45	7
9.23	9
7.25	11

a. Find a model giving the average quantity demanded as a function of the price.

b. How much are consumers willing and able to spend each day for these lanterns if the market price is $12.34 per lantern?

c. Find the consumers' surplus if the equilibrium price for these lanterns is $12.34 per lantern.

12. **Supply** The willingness of saddle producers to supply saddles can be modeled by the following function:

$$S(p) = \begin{cases} 0 \text{ thousand saddles} & \text{if } p < 5 \\ 2.194(1.295^p) \text{ thousand saddles} & \text{if } p \geq 5 \end{cases}$$

when saddles are sold for p thousand dollars.

a. How many saddles will producers supply if the market price is $4000? $8000?

b. At what price will producers supply 10 thousand saddles?

c. Find the producers' revenue if the market price is $7500.

d. Find the producers' surplus if the market price is $7500.

13. **Supply** The willingness of answering machine producers to supply can be modeled by the following function:

$$S(p) = \begin{cases} 0 \text{ thousand answering machines} & \text{if } p < 20 \\ 0.024p^2 - 2p + 60 \text{ thousand} & \\ \text{answering machines} & \text{if } p \geq 20 \end{cases}$$

when answering machines are sold for p dollars.

a. How many answering machines will producers supply if the market price is $40? $150?

b. Find the producers' revenue and the producers' surplus if the market price is $99.95.

14. **Supply** The table shows the number of CDs that producers will supply at the given prices.

Price per CD (dollars)	CDs supplied (millions)
5.00	1
7.50	1.5
10.00	2
15.00	3
20.00	4
25.00	5

a. Find a model giving the quantity supplied as a function of the price per CD. *Note:* Producers will not supply CDs if the market price falls below $4.99.

b. How many CDs will producers supply if the market price is $15.98?

c. At what price will producers supply 2.3 million CDs?

d. Find the producers' revenue and producers' surplus if the market price is $19.99.

15. Supply The table shows the average number of prints of a famous painting that producers will supply at the given prices.

Price per print (hundred dollars)	Prints supplied (hundreds)
5	2
6	2.2
7	3
8	4.3
9	6.3
10	8.9

a. Find a model giving the quantity supplied as a function of the price per print. *Note:* Producers will not supply prints if the market price falls below $500.

b. At what price will producers supply 5 hundred prints?

c. Find the producers' revenue and producers' surplus if the market price is $630.

16. Market Equilibrium The daily demand for beef can be modeled by

$$D(p) = \frac{40.007}{1 + 0.033e^{0.35382p}} \text{ million pounds}$$

when the price for beef is p dollars per pound. Likewise, the supply for beef can be modeled by

$$S(p) = \begin{cases} 0 \text{ million pounds} & \text{if } p < 0.5 \\ \dfrac{51}{1 + 53.98e^{-0.3949p}} \text{ million pounds} & \text{if } p \geq 0.5 \end{cases}$$

when the price for beef is p dollars per pound.

a. How much beef is supplied when the price is $1.50 per pound? Will supply exceed demand at this quantity?

b. Find the point of market equilibrium.

17. Social Gain The average quantity of sculptures that consumers will demand can be modeled as $D(p) = -1.003p^2 - 20.689p + 850.375$ sculptures, and the average quantity that producers will supply can be modeled as

$$S(p) = \begin{cases} 0 \text{ sculptures} & \text{when } p < 4.5 \\ 0.256p^2 + 8.132p \\ + 250.097 \text{ sculptures} & \text{when } p \geq 4.5 \end{cases}$$

where the market price is p hundred dollars per sculpture.

a. How much are consumers willing and able to spend for 20 sculptures?

b. How many sculptures will producers supply at $500 per sculpture? Will supply exceed demand at this quantity?

c. Determine the total social gain when sculptures are sold at the equilibrium price.

18. Social Gain A florist constructs a table on the basis of sales data for roses.

Price of 1-dozen roses (dollars)	Dozens sold per week
10	190
15	145
20	110
25	86
30	65
35	52

a. Find a model for the quantity demanded.

b. Determine how much money consumers will be willing and able to spend for 80-dozen roses each week.

c. If the actual market price of the roses is $22 per dozen, find the consumers' surplus.

Suppose the suppliers of roses collect the data shown in the following table.

Price of 1-dozen roses (dollars)	Dozens supplied per week
20	200
18	150
14	100
11	80
8	60
5	50

d. Find an equation that models the supply data. Suppliers will supply no roses for prices below $5 per dozen.

e. What is the producers' surplus when the market price is $17 per dozen?

f. For what price will roses be sold at the equilibrium point?

g. What is the total social gain from the sale of roses at market equilibrium?

19. Social Gain The table gives both number of copies of a hard back science fiction novel in demand and the number supplied at certain prices.

Price (dollars per book)	Books demanded (thousands)	Books supplied (thousands)
20	214	120
23	186	130
25	170	140
28	150	160
30	138	190
32	128	210

a. Find an exponential model for demand given the price per book.

b. Find a model for supply given the price per book. *Note:* Producers are not willing to supply any books when the market price is less than $18.97.

c. At what price will market equilibrium occur? How many books will be supplied and demanded at this price?

d. Find the total social gain from the sale of a hard back science fiction novel at the market equilibrium price.

20. Social Gain The table shows both the number of a certain type of graphing calculator in demand and the number supplied at certain prices.

Price (dollars per calculator)	Calculators demanded (millions)	Calculators supplied (millions)
60	35	10
90	31	32
120	15	50
150	5	80
180	4	100
210	3	120

a. Find a model for demand given the price per calculator.

b. Find a model for supply given the price per calculator. *Note:* Producers are not willing to supply any of these graphing calculators when the market price is less than $47.50.

c. At what price will market equilibrium occur? How many calculators will be supplied and demanded at this price?

d. Find the producers' surplus at market equilibrium.

e. Estimate the consumers' surplus at market equilibrium.

f. Estimate the total social gain from the sale of this type of graphing calculator at the market equilibrium price.

SUMMARY

Improper Integrals

Improper integrals of the forms $\int_a^\infty f(x)\,dx$, $\int_{-\infty}^a f(x)\,dx$, and $\int_{-\infty}^\infty f(x)\,dx$ can be evaluated by substituting a constant for each infinity symbol, finding an antiderivative and evaluating it to determine an expression in terms of x and the constant(s), and then determining the limit of the resulting expression as the constant approaches infinity or negative infinity. If the limit does not exist, we say the integral diverges.

Streams in Business and Biology

An income stream is a flow of money into an interest-bearing account over a period of time. If the stream flows continuously into an account at a rate of $R(t)$ dollars per year and the account earns annual interest at the rate of $100r\%$ compounded continuously, then the future value of the account at the end of T years is given by

$$\text{Future value} = \int_0^T R(t)e^{r(T-t)}\,dt \text{ dollars}$$

The present value of an income stream is the amount that would have to be invested now in order for the account to grow to a given future value. The present value of a continuous income stream whose future value is given by the previous equation is

$$\text{Present value} = \int_0^T R(t)e^{-rt}\,dt \text{ dollars}$$

When the income stream comes into the account discretely, we determine future value by summing rather than integrating:

$$\text{Future value} = \sum_{d=0}^{D-1} R(d)\left(1 + \frac{r}{n}\right)^{D-d} \text{ dollars}$$

where n is the number of deposits made each year, $R(d)$ is the value of the dth deposit, $100r\%$ is the annual interest rate, and D is the total number of deposits made. Once the future value of a discrete income stream

is calculated, the present value is determined by solving for P in the formula

$$P\left(1 + \frac{r}{n}\right)^D = \text{future value}$$

Streams also have applications in biology and related fields. The future value (in b years) of a biological stream with initial population size P, survival rate s (in decimals), and renewal rate $r(t)$, where t is the number of years of the stream, is

$$\text{Future value} \approx Ps^b + \int_0^b r(t)s^{b-t}\,dt$$

Integrals in Economics

A demand curve and a supply curve for a commodity are determined by economic factors. The interaction between supply and demand usually determines the quantity of an item that is available. Areas of special interest that are determined as areas associated with supply and demand curves are consumers' expenditure, consumers' surplus, consumers' willingness and ability to spend, producers' willingness and ability to receive, producers' surplus, producers' revenue, and total social gain.

CONCEPT CHECK

Can you

O Evaluate improper integrals?
O Recognize that an improper integral diverges?
O Determine income flow rate functions?
O Calculate and interpret present and future values of discrete and continuous income streams?
O Find various quantities related to a demand function?
O Find various quantities related to a supply function?
O Find the market equilibrium point and total social gain?

To practice, try

Section 7.1	Activity 1
Section 7.1	Activity 7
Section 7.2	Activity 1
Section 7.2	Activities 7, 13
Section 7.3	Activity 9
Section 7.3	Activity 13
Section 7.3	Activity 17

REVIEW TEST

1. **Investment**   In preparing to start your own business (in 6 years), you plan to invest 10% of your salary each month in an account with a fixed rate of return of 5.3%. You currently make $3000 per month and expect your income to increase by $500 per year.

   a. Find a function for the yearly rate at which you will invest money in the account.

   b. If you start investing now, to what amount will your account grow in 6 years? (Consider a continuous stream.)

   c. How much would you have to invest now in one lump sum, instead of in a continuous stream, in order to build to the same 6-year future value?

2. **Investment**   A teacher is planning to retire in 8 years. To supplement her state retirement income, she plans to invest 7% of her salary each month until retirement in an annuity with a fixed rate of return of 5.2% compounded monthly. She currently makes $3100 per month and expects her income to increase, due to consulting work, by 0.4% per month.

   a. How much will be in the annuity at the end of 8 years?

   b. How much would she have to invest now, in one lump sum, to accumulate the same amount as the 8-year future value found in part *a*?

3. **Population**   Suppose a 1990 population of 10,000 foxes breeds at a rate of 500 pups per year and has a survival rate of 63%.

   a. Assuming that the survival and renewal rates remain constant, determine how many of the foxes alive in 1990 will still be alive in 2010.

   b. Write a function for the number of foxes that were born *t* years after 1990 and will still be alive in 2010.

   c. Estimate the fox population in the year 2010.

4. **Social Gain**   The average quantity of marble fountains that consumers will demand can be modeled as

   $$D(p) = -1.0p^2 - 20.6p + 900 \text{ fountains}$$

   and the average quantity that producers will supply can be modeled as

   $$S(p) = \begin{cases} 0 \text{ fountains} & \text{if } p < 2 \\ 0.3p^2 + 8.1p + 300 \text{ fountains} & \text{if } p \geq 2 \end{cases}$$

   when the market price is *p* hundred dollars per fountain.

   a. How much are consumers willing to spend for 30 fountains?

   b. How many fountains will producers supply at $1000 per fountain? Will supply exceed demand at this quantity?

   c. Determine the total social gain when fountains are sold at the equilibrium price.

Project 7.1 Arch Art

Setting

A popular historical site in Missouri is the Gateway Arch. Designed by Eero Saarinen, it is located on the original riverfront town site of St. Louis and symbolizes the city's role as gateway to the West. The stainless steel Gateway Arch (also called the St. Louis Arch) is 630 feet (192 meters) high and has an equal span.

In honor of the 200th anniversary of the Louisiana Purchase, which made St. Louis a part of the United States, the city has commissioned an artist to design a work of art at the Jefferson National Expansion Memorial National Historic Site. The artist plans to construct a hill beneath the Gateway Arch, located at the Historic Site, and hang strips of mylar from the arch to the hill so as to completely fill the space. (See the figure.) The artist has asked for your help in determining the amount of mylar needed.

Tasks

1. If the hill is to be 30 feet tall at its highest point, find an equation for the height of the cross section of the hill at its peak. Refer to the figure.

2. Estimate the height of the arch in at least ten different places. Use the estimated heights to construct a model for the height of the arch. (You need not consider only the models presented in this text.)

3. Estimate the area between the arch and the hill.

4. The artist plans to use strips of mylar 60 inches wide. What is the minimum number of yards of mylar that the artist will need to purchase?

5. Repeat Task 4 for strips 30 inches wide.

6. If the 30-inch strips cost half as much as the 60-inch strips, is there any cost benefit to using one width instead of the other? If so, which width? Explain.

Reporting

Write a memo telling the artist the minimum amount of mylar necessary. Explain how you came to your conclusions. Include your mathematical work as an attachment.

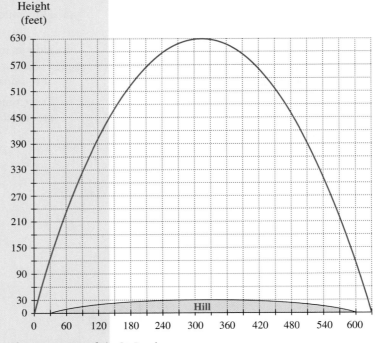

The Gateway Arch in St. Louis

Answers to Odd Activities

CHAPTER 1

Section 1.1

1. Input: weight of a letter Output: first-class domestic postage
   Input variable: w Output variable: $R(w)$
   Input units: ounces Output units: cents
   R is a function of w.

3. Input: day of the week Output: amount spent on lunch
   Input variable: m Output variable: $A(m)$
   Input units: none Output units: dollars
   A is not a function of m unless you always spend the same amount on lunch every Monday, Tuesday, and so on, or unless the input is the days in only 1 week.

5. This is a function.

7. This is not a function.

9. Graphs b and c represent functions.

11. **a.** $P(\text{Honolulu}) = 295$
    b. $P(\text{Providence, RI}) = 137.8$
    c. $P(\text{Portland, OR}) = 170.1$

13. **a.** In 1988 cotton exports had a value of $1,975,000,000.
    b. In 1992 cotton exports had a value of $1,999,000,000.

15. **a.**

Cost (dollars)

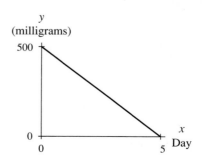

(This scatterplot belongs to 15a with Cost in dollars on the vertical axis labeled 18, 36, 54, 72, 90, 108, 126, 144 and CDs on the horizontal axis 1–10.)

b. $90
c. 2 CDs
d. 6 CDs
e. The average price of 3 CDs is $18 each.
   The average price of 6 CDs is $15 each.

17. **a.** Approximately $9000
    b. Approximately $340
    c. Approximately $105

d. The graph would pass through $(0, 0)$ but would lie below the graph in the activity because the same monthly payment would pay for a smaller loan amount.

19. **a.** 2.9%
    b. The cost-of-living increase was greatest in 2001 at 3.5%.
    c. 1998
    d. Benefits increased, but the percentage by which they increased decreased.

21. Weight (pounds)

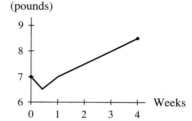

23. **a.** Approximately 5 inches
    b. For approximately 4 days
    c. Snow fell.
    d. The snow settled.
    e. The snow was deepest (4 feet 4 inches) around February 21.
    f. Warm temperatures probably caused the decline in late February.
    g. The most snow fell around February 18.

25. **a.**

x	0	1	2	3	4	5
y	500	400	300	200	100	0

b. $y = 500 - 100x$ milligrams after x days
c. x is between 0 and 5 days.
   y is between 0 and 500 milligrams.

y (milligrams)

d. The x-intercept is 5 days.
The y-intercept is 500 milligrams.
The y-intercept is the amount of the drug initially in the patient's body (500 milligrams). The x-intercept is the time when none of the drug remains (after 5 days).

e. y is always decreasing (for $0 \le x \le 5$).

f. After 3.5 days, approximately 150 milligrams of the drug remain. The exact value is $500 - 100(3.5) = 150$ mg.

g. There will be 60 milligrams of the drug after approximately 4.5 days. Solve $60 = 500 - 100x$ to find the exact value, $x = 4.4$ days.

27. $s = 5, t = 22$
$s = 10, t = 38$

29. $R(3) \approx 314.255$
$R(0) \approx 39.4$

31. $R(w) = 78.8$ when $w \approx 1.001$
$R(w) = 394$ when $w \approx 3.327$

33. $Q(x) = 515$ when $x = 10$
$Q(x) = 33.045$ when $x \approx 0.500$

35. Input is given.
$A(15) \approx 5{,}131{,}487.257$

37. Output is given.
When $y = 2.97, x \approx -3.203$

39. Answers will vary.

Section 1.2

1. a. $T(x) = 9088.859 + 1697.717 \ln x + 2424.764 + 915.025 \ln x = 11{,}513.623 + 2612.742 \ln x$ kidney and liver transplants, where x is the number of years since 1990

b. $T(5) \approx 15{,}719$ transplants

3. $N(t) = M(t) - W(t) = -0.01685t^2 + 3.26878t - 142.72712$ gallons or milk other than whole milk per person per year, where t is the number of years since 1900. $N(100) \approx 15.65$ gallons

5. a. $n(t) = h(t) - p(t) = \dfrac{100}{1 + 128.0427e^{-0.7211264t}}$
$- \dfrac{100}{1 + 913.7241e^{-0.607482t}}$ percent t years after 1924

b. $n(5) \approx 20\%$

7. $c(x) = n(x)p(x) = (-0.034x^3 + 1.331x^2 + 9.913x + 164.447)(-0.183x^2 + 2.891x + 20.215)$ cesarean-section deliveries x years after 1980

9. $R(y) = \dfrac{P(y)}{D(y)}$ percent in year y

11. $P(C(t)) = $ profit after t hours of production

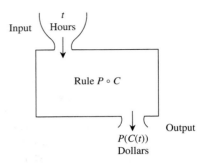

13. $P(C(t)) = $ average amount in tips t hours after 4 P.M.

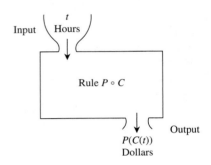

15. $f(t(p)) = 3e^{4p^2}$

17. $g(x(w)) = \sqrt{7(4e^w)^2 + 5(4e^w) - 2}$

19. $g(t(m)) = 3$

21. a. $S(85) = \$123.1$ million
$S(88) = \$159.4$ million
$S(89) = \$97.7$ million
$S(92) = \$53.3$ million

b.

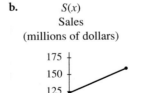
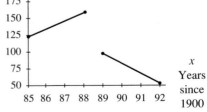

c. S is a function of x because each year corresponds to only one amount of sales. The fact that S is not defined between $x = 88$ and $x = 89$ does not prevent it from being a function.

23. a. The shipping charges encourage larger orders.

b.

Order amount	Shipping charge
$17.50	$3.50
$37.95	$6.83
$75.00	$11.25
$75.01	$9.00

c. $S(x) = \begin{cases} 0.2x \text{ dollars} & \text{when } 0 \le x \le 20 \\ 0.18x \text{ dollars} & \text{when } 20 < x \le 40 \\ 0.15x \text{ dollars} & \text{when } 40 < x \le 75 \\ 0.12x \text{ dollars} & \text{when } x > 75 \end{cases}$

where x is the order amount in dollars

d.

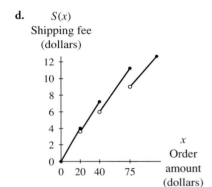

e. Answers will vary.

25.

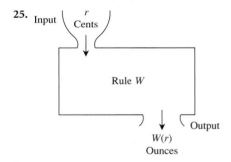

$W(r)$ is the weight in ounces of a first-class letter or parcel that costs r cents to mail. W is not an inverse function of r.

27.

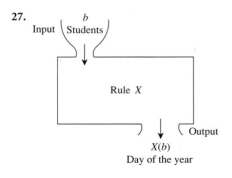

$X(b)$ is the day of the year corresponding to the birthday of b students. X is not an inverse function of b.

29. Reversing the inputs and outputs does not result in an inverse function.

31. Reversing the inputs and outputs does not result in an inverse function because the original table does not represent a function.

33. a. $m(2399) = 5$

   In May of 2001, 2399 complaints were received.

b. $C(1) + C(2) + C(3) = 4321$ complaints in the first quarter of the year.

c. C graphs as a diagonal line so it satisfies the Horizontal Line Test, and thus has an inverse function.

35. a.

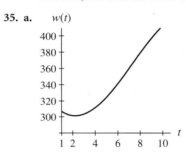

b. No, because the horizontal line at $w = 305$ crosses the graph of the function twice.

c. No, because it is not one-to-one.

37. Answers vary.

Section 1.3

1. a. 5   **b.** 0   **c.** 2.6

3. a. 3   **b.** 1   **c.** Does not exist

d. Does not exist $\lim\limits_{t \to \infty} m(t) \to -\infty$ **e.** 6

5. a. $y = -4$ is a horizontal asymptote on the left.

b. i. Yes   **ii.** Yes   **iii.** Yes   **iv.** No

7.

$x \to \dfrac{-1^-}{3}$	$\dfrac{x^3 + 6x}{3x + 1}$	$x \to \dfrac{-1^+}{3}$	$\dfrac{x^3 + 6x}{3x + 1}$
-0.35	42.858	-0.32	-48.819
-0.34	103.965	-0.332	-507.149
-0.334	1020.630	-0.3332	-5090.482
-0.3334	10,187.296	-0.33332	-50,923.814

$$\lim_{x \to \frac{-1}{3}^-} \frac{x^3 + 6x}{3x + 1} \to \infty; \quad \lim_{x \to \frac{-1}{3}^+} \frac{x^3 + 6x}{3x + 1} \to -\infty;$$

$$\lim_{x \to \frac{-1}{3}} \frac{x^3 + 6x}{3x + 1} \text{ does not exist.}$$

9.

$t \to \infty$	$\dfrac{14}{1+7e^{0.5t}}$
10	0.01346
20	$9.0799 \cdot 10^{-5}$
30	$6.1180 \cdot 10^{-7}$
40	$4.12231 \cdot 10^{-9}$
50	$2.7776 \cdot 10^{-11}$

$$\lim_{t \to \infty} \frac{14}{1 + 7e^{0.5t}} = 0$$

11. No. If a function is continuous, then the limit exists for all input values.

13. a. $A(n) = \left(1 + \dfrac{1}{n}\right)^n$ dollars

b–c.

Compounding	n	Amount
yearly	1	$2
semiannually	2	$2.25
quarterly	4	$2.44
monthly	12	$2.61
weekly	52	$2.69
daily	365	$2.71
every hour	8760	$2.72
every minute	525,600	$2.72
every second	31,536,000	$2.72

d. $2.72

e. $\displaystyle\lim_{n \to \infty} \left(1 + \dfrac{1}{n}\right)^n \approx \$2.72$

15. a. The concentration increases rapidly from 0 milligrams at time 0 hours to approximately 63 milligrams one-quarter hour later. It then decreases (at first rapidly, but then more slowly) until it is almost non-existent 6 hours after the initial dose is taken.

b. $\displaystyle\lim_{t \to \infty} 100\left(e^{-1.3t} - e^{-9.5t}\right) = 0$

c. Answers will vary.

17. No restriction

19. No restriction

21. a.

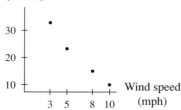

Rainfall (inches)

Wind speed (mph)

b. Because no vertical line passes through two points, the data represent a function.

c. Because the table does not contain an input of 6, we cannot read this information from the table.

d. Tables of data are always discrete representations.

23. a. The profit resulting from 25 calls in one day is $179.

b. Solving $P(C) = 180$ for c gives $c \approx 13.46$. Because the number of calls must be a positive integer, this input value does not have a meaningful interpretation in this context. The daily profit is never exactly $180. When 13 calls are made, the profit is slightly under $180, and when 14 calls are made, the profit is slightly over $180.

c.

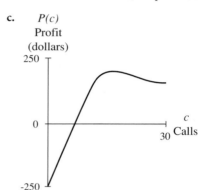

$P(c)$
Profit (dollars)

Calls

According to the graph, the maximum occurs when $c \approx 18.2$. Checking the integer input values on either side of 18.2, we find that $P(18) \approx \$201.85$ and $P(19) \approx \$201.37$. The maximum profit of $201.85 occurs when 18 calls are made.

25. a. Input: the average basic rate for a cable subscription
Output: the number of cable subscribers

b. Input units: dollars per month
Output units: million subscribers

c. No, because the input is an average that can be any real number.

d. $S(15) \approx 44.632$ million subscribers

e. Approximately 59.128 million dollars

27. a. Input: the number of hours after the market opened
Output: the price of Microsoft stock

b. Input units: hours
Output units: dollars per share

c. The model can be interpreted without restriction because the price changes in an approximately continuous manner throughout a trading day.

d. $P(3) \approx \$45.09$ per share

e. The model reaches an output of 47 for $h \approx 8.2$. The trading day is less than 8.2 hours long, so we conclude that the price was never $47 per share during this trading day.

29. a. $P(20) = 22.472$. Approximately 22.5% of 20-year-old workers have flex schedules.

  b. Input units: years (of age)

  Output units: percent

  c. No, age is a real number and those surveyed were at different ages in their age range at the time of survey.

31. Answers will vary.

Section 1.4

1. a. Slope $\approx \dfrac{-\$2.5 \text{ million}}{5 \text{ years}} = -\$0.5$ million per year

  The corporation's profit was declining by approximately a half a million dollars per year during the 5-year period.

  b. The rate of change is approximately $-\$0.5$ million per year.

  c. The vertical-axis intercept is approximately $2.5 million. This is the value of the corporation's profit in year zero. The horizontal-axis intercept is 5 years. This is the time when the corporation's profit is zero.

3. a. 382.5 donors per year

  b.

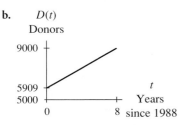

  c. The vertical-axis intercept is 5909 donors. This value is the number of donors in 1988, the starting year.

5. This question cannot be answered because the model does not include a description of the input variable.

7. a. Slope $= \dfrac{824.1 - 744.0}{1998 - 1997} = \$80.1$ million per year

  b. The revenue increased by $20.025 million each quarter of 1998.

  c.

Year	Revenue (millions of dollars)
1997	744.0
1998	824.1
1999	904.2
2000	984.3

  d. Revenue $= 744.0 + 80.1t$ million dollars, where t is the number of years after 1997

9. a. Rate of change $= \dfrac{\$97,500 - \$73,000}{2002 - 1990} \approx \$2042$ per year

  b. $\$97,500 + 3(\$2042) \approx \$103,600$

  c. The value was $75,000 in late 1991 and $100,000 in early 2004.

  d. $V(t) = 2041.667t + 73,000$ dollars t years after 1990

  $V(9) \approx \$91,400$

  The model assumes that the rate of increase of the market value remains constant.

11. a. $99.3 billion per year

  b. $1533.7 billion

  c. Answers will vary.

  d. Approximately 2005 ($t \approx 11.7$)

  e. Answers will vary.

13. a. 2000 dogs per hour or 48,000 dogs per day

  b. 3500 cats per hour or 84,000 cats per day

  c. $D(t) = 48,000t + 54,000,000$ dogs t days after the beginning of 1993

  d. $C(t) = 84,000t + 56,000,000$ cats t days after the beginning of 1993

  e. $D(365) + C(365) = 158,180,000$ dogs and cats; No

15. a. 78 million people per year

  b. $P(t) = 6 + 0.078t$ billion people t years after the beginning of 2000

  c. $t \approx 76.9$ years after the beginning of 2000, which corresponds to near the end of 2076. The article estimates that the world population will be 12 billion in 2050.

  d. The prediction in part c assumes that the world will grow at a constant rate of 78 million people per year between now and 2076. In making this prediction, the Census Bureau must have assumed that the growth rate will increase so that the 12 billion population will be reached sooner than our prediction based on the linear model.

17. a. $0.13 per year

  b. Answers may vary.

  c. Approximately $1.24 in 1998

  Approximately $1.90 in 2003

  d. 1998 interpolation, 2003 extrapolation

19. Extrapolating from a model is predicting values beyond the interval of the data, whereas interpolating is using a model to estimate what happened at some point within the interval of the data. Interpolation often yields a good approximation of what occurred. Extrapolation must be used with caution because many things that are not accounted for by the model may affect future events.

21. a. Answers may vary. Three possible models are

  $S1(x) = 499.3x - 976,088.3$ students in year x

  $S2(x) = 499.3x - 27,418.3$ students x years after 1900

  $S3(x) = 499.3x + 5036.2$ students x years after 1965

  b. Approximately 7533 students

c. The estimate from the models is 505 less than the actual enrollment. Answers may vary on whether the error is significant. For a school the size of the one in this activity, an error of 500 students could be significant, especially in housing and faculty loads.

d. These models should not be used to predict enrollment in the year 2000 because the data are too far removed from 2000 to be of any value in such a prediction.

23. a. $P(t) = 0.808t + 9.961$ dollars t years after 1981

b. The ticket prices rounded to the nearest dollar are

1984: $P(3) \approx \$12$ interpolation
1992: $P(11) \approx \$19$ interpolation
1999: $P(21) \approx \$27$ extrapolation

c. The model prediction is $1 more than the actual price. That's fairly accurate.

d. Answers will vary.

e. Answers will vary.

f. When a linear model is used to extrapolate, the underlying assumption is that the output will continue to increase at a constant rate. Often this may be a valid assumption for short-term extrapolation but not for long-term extrapolation.

g. Extrapolating from a model must always be done with caution. In order for the extrapolation to be accurate, the model must accurately describe the situation, and the future behavior of the output must match that of the model. Long-term extrapolations are always risky.

25. a. $F(y) = -0.152y + 19.514$ percent, where y is 81 for the 1981–82 school year, 82 for the 1982–83 school year, and so on

b. -0.152 percentage points per year

c. $F(93) \approx 5.4\%$

d. Solving $F(y) = 5$ gives $y \approx 95.7$. Thus the 1996–97 school year is the first year below the 5% level.

27. a. The scatter plot does reflect the statements about atmospheric release of CFCs.

b. Answers may vary. One possible model is

$$R(x) = \begin{cases} -15.37x + 1554.19 & \text{when } x \le 80 \\ \quad \text{million kilograms} \\ 9.165x - 412.5 & \text{when } 80 < x \le 88 \\ \quad \text{million kilograms} \\ -34.375x + 3413.283 & \text{when } x > 88 \\ \quad \text{million kilograms} \end{cases}$$

where x is the number of years after 1900. (Note that in computing the middle model, the 1980 data point was not included, but in computing the bottom model, the 1988 data point was included. This was done to improve the fit of the model. You will have to use your own discretion to determine whether or not to include the points at which the data is divided when finding a piecewise model.)

c.

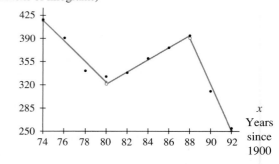

$R(x)$
CFC-12 release
(millions of kilograms)

d. i. What was the amount of CFCs released into the atmosphere in 1975? in 1995?
$C(75) \approx 401.4$ million kg, $C(95) \approx 147.7$ million kg

ii. At what rate was the release of CFCs declining between 1974 and 1980?
15.37 million kilograms per year between 1974 and 1980

iii. On the basis of data accumulated since 1987, in what year will there no longer be any CFCs released into the atmosphere?
This occurs when $x \approx 99.3$. This answer indicates that according to the model, there should have been no release of CFCs by early in 2000.

29. a. 1991 is the last year in which the data declined. In 1993 the data began to rise. We choose 1991 as the dividing point for a piecewise continuous model.

b. A model for the population of North Dakota x years after 1985 is

$$p(x) = \begin{cases} -7.35x + 676.3 & \text{when } 0 \le x < 6 \\ \quad \text{thousand people} \\ 2.122x + 619.730 & \text{when } 6 \le x \le 11 \\ \quad \text{thousand people} \end{cases}$$

c. The model estimates the 1997 population to be about 645,000 people. This is an overestimate of approximately 4000 people. This extrapolation is only one year beyond the data given.

31. Answers are located on the *Calculus Concepts* CD-ROM and the web site.

CHAPTER 1 REVIEW TEST

1. a. The number of acres increased, but by a smaller amount in 1992 than in 1991.

b. Between 1989 and 1991, the yearly gain of wetlands was increasing by approximately 33,000 acres per year.

2. a.

d
Day in
2000

Input

Rule T

$T(d)$
Tickets

Output

b. T is a function of d because for every possible day, there is only one associated number of tickets.

c. If the inputs and outputs are reversed, the result is not an inverse function because one number of tickets could have more than one day associated with it.

3. a. Rate of increase $= \dfrac{23\% - 18\%}{1995 - 1973} = \dfrac{5\%}{22 \text{ years}}$

≈ 0.23 percentage point per year

b. Percent in 2002

$= 23\% + (7 \text{ years}) \left(\dfrac{5}{22} \text{ percentage point per year}\right)$

$\approx 25\%$

This estimate is valid if the rate of change remains constant through 2002.

4. a. i. 7   **ii.** Does not exist   **iii.** 3

b. f is not continuous at $x = 1$ because $f(1)$ does not exist, and it is not continuous at $x = 2$ because the limit of f as x approaches 2 does not exist.

5. a. $C = 0.0342t + 11.39$ million square kilometers t years after 1900

b. C is continuous without restriction.

c. The amount of cropland increased by approximately 0.034 million square kilometers, or 34,000 square kilometers per year, between 1970 and 1990.

d. Answers will vary.

e. $C(95) = 14.64$ million square kilometers

CHAPTER 2

Section 2.1

1. $f(x) = 2(1.3^x)$ is the black graph. $f(x) = 2(0.7^x)$ is the teal graph.

3. $f(x) = 3(1.2^x)$ is the teal graph. $f(x) = 2(1.4^x)$ is the black graph.

5. $f(x)$ is increasing with a 5% change in output for every unit of input.

7. $y(x)$ is decreasing with a 13% change in output for every unit of input.

9. The number of bacteria declines by 39% each hour.

11. a. $I(t) = 4.81(1.0547^t)$ quadrillion Btu t years after 2005 through 2020

b. 2019

c. The projected petroleum product imports increase without bound as time increases.

13. a. $S(y) = 1.5(1.0746^y)$ dollars y years after 1997

b. $S(13) \approx \$3.82$

15. a. $W(t) = 3.3(0.9854^t)$ workers per beneficiary t years after 1996

b. $W(34) = 2$ workers per beneficiary. Fewer workers per beneficiary will mean that the social security program will have to find other means of supplementing payments rather than relying solely on withholdings from workers' wages.

17. a. $P(t) = 1.269(1.015646^t)$ billion people t years after 1900. The function is a good fit for the data. The 1960 data point is the one farthest from the function. The data points for 1974, 1987, and 1999 appear close to the graph.

b. $W(x) = 6 + 0.078x$ billion people x years after 1999

c. $W(t) = 6 + 0.078(t - 99)$ billion people t years after 1900

d. Population
(billions)

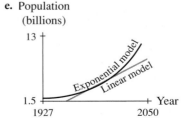

The linear model appears to be a better fit for the years 1960–1999; the exponential model is a better overall fit.

e. Population
(billions)

13

Exponential model
Linear model

1.5
1927   2050
Year

For years beyond 1999, the exponential model rises dramatically compared to the linear model. Both models should give reasonably close answers for populations between 1960 and 2000, but the exponential model should be used from 1900 through 1960. Which model will be more appropriate for years beyond 2000 depends on what happens to the world population growth rate in the future. See part f for a more detailed answer.

   f. The exponential model gives the population as approximately 6.0 billion in 2000 and 13.0 billion in 2050. The linear model gives 6.1 billion people in 2000 and 10.0 billion people in 2050. The linear model assumes that the growth rate (people per year) remains constant at the 1999 level. The exponential model assumes that the percentage growth rate remains constant at the level obtained by modeling the data for 1927 through 1999.

   g. The Census Bureau predicts a population of 12 billion in 2050. This is 1 billion less than the exponential model prediction and 2 billion more than the linear model prediction. The Census Bureau assumes that the percentage growth rate will decline but that the growth rate will increase.

19. a. $R(y) = 4.715(1.066857^y)$ million tons recycled, where y is the number of years since 1960

   b. Approximately 6.7% per year growth in the amount of MSW recycled

21. a. $F(x) = 345.957(0.942378^x)$ farms with milk cows where x is the number of years after 1980. This model indicates a $(1 - 0.942378)100\% \approx 5.8\%$ per year decline.

   b. $\lim\limits_{x \to \infty} 345.957(0.942378^x) = 0$. It is not reasonable that the number of farms with milk cows will approach zero in the future.

   c. Answers will vary. Possible reasons: The decline in farms with milk cows is probably indicative of an overall decline in farms because of the nature of the economy over the past decades. Specialization has led to fewer nondairy farms owning milk cows. The emphasis on decreasing fat and cholesterol in food may have produced a decrease in demand for dairy products.

23. Approximately 4960 years ago

25. a. Using the points $(0, 250)$ and $(0.5, 125)$, we obtain the model $P(h) = 250(0.25^h)$ mg of penicillin left after h hours.

   b. Solving $1 = 250(0.25^h)$ gives $h \approx 3.98$ hours. The second dose of penicillin should be taken approximately 4 hours after the first dose.

27. a. Approximately 11 years 1 month

   b. Approximately 8 years 8 months

   c. Approximately 10 years 3 months

29. Answers will vary.

Section 2.2

1. $f(x) = 2 \ln x$ is the teal graph. $f(x) = -2 \ln x$ is the black graph.

3. $f(x) = 2 \ln x$ is the teal graph. $f(x) = 4 \ln x$ is the black graph.

5. a. $L(d) = 158.574 - 42.877 \ln d$ ppm for soil that is d meters from the road

   b. $L(12) = 158.574 - 42.877 \ln 12 \approx 52$ ppm

   c. $E(d) = 123.238(0.932250^d)$ ppm for soil that is d meters from the road. The log model appears to be a better fit for small distances than the exponential model. Other than that, the two models are similar for distances between 5 and 20 meters. The log model will eventually be negative, whereas the exponential model will approach zero. The exponential end behavior seems to better describe the lead concentration as a function of distance from a road.

7. a. $B(t) = 8.435 - 0.639 \ln x$ percent for a maturity time of t years

   b. The model estimates 15-year bond rates as $B(15) \approx 6.70\%$, which is 0.3 percentage point less than the fund manager's estimate.

9. a. $S(x) = -38{,}217.374 + 23{,}245.372 \ln x$ dollars x years after 1970

   b. 1993: $S(23) \approx \$34{,}668$; 2000: $S(30) \approx \$40{,}845$. The 1993 estimate is probably the more accurate one because it is an interpolation, whereas the 2000 estimate is an extrapolation.

   c. Salaries are expected to have reached \$40,000 in 1999 $(x \approx 28.9)$.

11. a. The data exhibit concave-down behavior and show no indication of a horizontal asymptote in either the positive or negative input directions.

   b. $C(x) = 5.005 + 2.002 \ln x$ pounds per person per year when the yearly family income is $10{,}000x$ dollars.

   c. $C(3.5) \approx 7.5$ pounds per person per year

13. a. $S(x) = -350{,}193.616 + 78{,}843.360 \ln x$ million dollars x years after 1990. The graph of this equation appears linear when graphed on the scatter plot.

   b. $C(t) = 6721.530 + 2213.132 \ln t$ million dollars t years after 1992. The graph of this equation is a better fit for the data than the equation in part a.

   c. $A(y) = 7734.238 + 1633.996 \ln y$ million dollars t years since 1993 plus 0.5. This equation fits the data on the left slightly better than the one in part b.

d. All three models will increase infinitely as the input increases. The data show a decline from 1996 to 1997. A quadratic model may be a better model for these data.

e. A log model must have inputs greater than zero.

15. a. $pH(x) = -9.792 \cdot 10^{-5} - 0.434 \ln x$, where x is the H_3O^+ concentration in moles per liter

b. $pH(1.585 \cdot 10^{-3}) = -9.792 \cdot 10^{-5} - 0.434 \ln (1.585 \cdot 10^{-3})$
≈ 2.8

c. Approximately $1.0 \cdot 10^{-5}$ mole per liter

d. Beer is acidic with a pH of approximately 4.5.

17. a. $15, 44.632 million subscribers
$30, 70.417 million subscribers
$45, 85.501 million subscribers

b. $R(s) = 4.519(1.027246^s)$ dollars per month when s million people subscribe to cable TV

19. a. Yes, air pressure can be considered to be a function of altitude.

b. $P(a) = 35.54 (0.9542^a)$ inches of mercury, where a is the altitude in thousands of feet

21. Answers will vary.

Section 2.3

1. Exponential or logarithmic

3. Logarithmic

5. None. The scatter plot exhibits a change in concavity, so it is not linear, exponential, or logarithmic. It does not level off, so it is not logistic. It is possible to model this data with a piecewise continuous model using three linear pieces.

7. Increasing; limiting value is 100; horizontal asymptotes are $y = 0$ and $y = 100$.

9. Decreasing; limiting value is 39.2; horizontal asymptotes are $y = 0$ and $y = 39.2$.

11. a. $C(t) = \dfrac{37.195}{1 + 21.374e^{-0.182968t}}$ countries t years after 1840. The model is a good fit.

b.

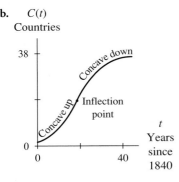

$C(t)$
Countries

13. a. A plow sulky is a horse-drawn plow with a seat so that the person plowing can ride instead of walk. It was a precursor to the tractor.

b. $P(t) = \dfrac{2591.299}{1 + 16.848e^{-0.194833t}}$ patents t years after 1871

c. Patents begin with one innovative idea and grow almost exponentially as more and more improvements and variations are patented. Eventually, however, the patent market becomes saturated, and the number of new patents dwindles to none as newer ideas and products are invented.

15. a. The data are concave down from January through April and concave up from April through June. This is not the concavity exhibited by a logistic model.

b. The entire data set does appear to be logistic.

c. $P(t) = \dfrac{42,183.911}{1 + 21,484.253e^{-1.248911t}}$ polio cases t months after December 1948
The model appears to be a good fit.

d. The model is a poor fit for the January through June data.

17. a. The limiting value appears to be approximately $2U/100\mu L$. The inflection point occurs at approximately 9 minutes. (Answers may vary.)

b. $A(m) = \dfrac{1.937}{1 + 29.064e^{-0.421110m}}$ U/100μL after m minutes. The limiting value is about 1.94 U/100μL.

c. Approximately 0.74 U/100μL

19. a. $P(x) = \dfrac{11.742}{1 + 154.546e^{-0.025538x}}$ billion people x years after 1800. The equation appears to be a good fit for the later (1960–2071) data but a poor fit for the early (1800–1960) data.

b. According to the model, the world population will level off at 11.7 billion. This is probably not an accurate prediction of future world population.

c. The model will probably be a poor estimate of the 1850 population because it appears to be a poor fit for the years between 1800 and 1960. It is a good fit around 1990, however, so it will probably give a good estimate of population in that year.

21. Answers are located on the *Calculus Concepts* CD-ROM and web site.

23. a. The data appear to be concave up from 1990 through 1994 and concave down from 1996 through 2000. The concavity changes between 1994 and 1996. Either a logistic or a cubic function could be used to model these data.

b. $D(t) = \dfrac{552.278}{1 + 2.773e^{-0.306931t}}$ billion dollars of debt t

years after 1990. This equation is not a particularly good fit for the data. It does not display the change in concavity described in part a.

c. $C(t) = \dfrac{311.126}{1 + 51.770e^{-0.866231t}}$ billion dollars over \$165 billion t years after 1990. This equation is an excellent fit for the data.

d. $F(t) = \dfrac{311.126}{1 + 51.770e^{-0.866231t}} + 165$ billion dollars t years after 1990

e. $y = 165$ and $y \approx 476.126$. The function in part b has asymptotes $y = 0$ and $y \approx 552.278$.

25. a. A scatter plot shows an inflection point with leveling-off behavior at both ends.

b. $P(t) = \dfrac{235.641}{1 + 9.637e^{-0.172050t}}$ thousand people above 2,700,000 t years after 1980. This equation is not a good fit for the data. It does not exhibit the same inflection point or the same horizontal asymptotes as the data.

c. $I(t) = \dfrac{97.500}{1 + 1405.903e^{-0.591268t}}$ thousand people above 2,760,000 t years after 1980. This equation is an excellent fit for the data.

d. $N(t) = \dfrac{97.500}{1 + 1405.903e^{-0.591268t}} + 2760$ thousand people t years after 1980. The horizontal asymptotes are $y = 2760$ and $y = 2857.5$.

27. a. $b(x) = 15.501(0.972249^x)$ percent when the mother gained x pounds. This function does not give a good fit.

b. $B(x) = 78.196(0.863408^x)$ percent when the mother gained x pounds; Yes

c. $L(x) = 78.196(0.863408^x) + 5$ percent when the mother gained x pounds; $L(14) \approx 15\%$. About 15% of babies born to women who gained 14 pounds had low birthweight.

29. a. $c(t) = 2.053(1.022247^t)$ million children t years after 1970. The graph of this equation is not nearly so curved as the data suggest.

b. Answers will vary. One possible model is $C(t) = 0.086(1.119883^t) + 2.1$ million children t years after 1970.

31. Answers will vary.

Section 2.4

1. Concave up, decreasing from $x = 0.75$ to $x = 3$, increasing from $x = 3$ to $x = 4$

3. Concave up, decreasing from $x = 13.5$ to $x = 18$, increasing from $x = 18$ to $x = 22.5$

5. Concave down, always decreasing

7. a.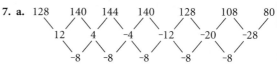

Second differences are constant, so the data are quadratic.

b. After 3.5 seconds the height is 44 feet. After 4 seconds the height is 0 feet.

c. $H(s) = -16s^2 + 32s + 128$ feet after s seconds

d. $H(s) = 0$ when $s = -2$ and $s = 4$
   The missile hits the water after 4 seconds.

9. a. $J(12) \approx 3729$ jobs   b., c. Answers will vary.

11. a. Because the data are evenly spaced and the second differences are constant, the data are perfectly quadratic.

b. 26.5 years of age

c. $A(x) = 0.0035x^2 - 0.405x + 32$ years of age x years after 1900

d. $A(100) = 26.5$ years of age

e., f. Answers will vary.

13. a. The data do not appear to be concave up or concave down. A linear model is $B(x) = 0.002x + 1.880$ dollars to make x ball bearings.

b. Overhead is \$1.88.

c. $B(5000) \approx \$13.64$

d. $B(5100) - B(5000) \approx \$0.24$
   This answer could also be found by multiplying the slope of the model by 100. This value is called marginal cost.

e. $C(u) = 0.002(500u) + 1.880$ dollars to make u cases of ball bearings

15. a. $V(t) = 0.092t^2 + 0.720t + 149.554$ pounds per person t years after 1980

b. The model appears to be a good fit.

c. $V(21) \approx 205.4$ pounds per person
   Because this estimate is only 1 year beyond the known data, it is probably a good estimate.

d. According to the model, the consumption will exceed 225 pounds per person in 2005.

e. Answers will vary.

17. a. $Q(x) = 0.057x^2 - 3.986x + 69.429$ thousands of tons of lead x years after 1940
   $E(x) = 211.196(0.821575^x)$ thousands of tons of lead x years after 1940
   The exponential equation is a poor fit. The quadratic equation appears to be a good fit.

b. The data suggest that lead usage is approaching zero as time increases. The exponential function approaches zero as time increases.

c. The quadratic model is the more appropriate one. $Q(15) \approx 22.5$ thousands of tons of lead

19.

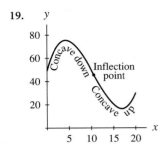

21.

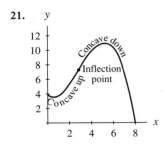

23.

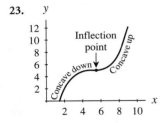

25. a. The scatter plot indicates an inflection point.

b. $A(t) = 0.427t^3 - 5.286t^2 + 22.827t + 3.014$ million dollars t years after 1990

c. $A(3) \approx \$35$ million; $A(9) \approx \$92$ million
The 1993 estimate is more likely to be accurate because it is an interpolation rather than an extrapolation.

d. The 1993 estimate exceeded the actual amount by $1 million. The 1999 estimate is $7 million short of the actual amount. These figures confirm the statements in part *c*.

27. a. The scatter plot suggests an inflection point.

b. $G(x) = (5.051 \cdot 10^{-5})x^3 - 0.007x^2 + 0.085x + 105.027$ males per 100 females x years after 1900. The graph of G rises beyond 2000. Answers vary on whether the gender ratio will rise as indicated by the model.

c. Possible answers include: Health issues in the early 1900s probably contributed to a greater death rate among women, who often died in childbirth. The west-ern expansion and gold and silver rush days probably attracted more male immigrants than female ones. Male deaths from World War II through the Vietnam Conflict contributed to the declining ratio.

29. a. The number of females and males is approximately equal for 40-year-olds.

b. Using "under age" as 0 and "100 and over" as 100 for modeling, but not prediction purposes, the model is $C(a) = (-9.590 \cdot 10^{-5})a^3 + (8.421 \cdot 10^{-4})a^2 + 0.037a + 104.601$ males per 100 females for an age of a years.
$L(a) = \dfrac{104.3}{1 + 9.817 \cdot 10^{-4}e^{0.081848a}}$ males per 100 females for an age of a years
The logistic equation fits the data better than the cubic equation, especially for ages above 60.

c. Among 86-year-olds there are approximately twice as many women as men. This implies that men die younger than women.

31. a. The behavior of the data changes abruptly in 1999, resulting in a sharp point. None of the five models we have studied will reflect that behavior.

b. $E(t) = \begin{cases} 4.500t^3 - 51.321t^2 + 233.440t - 90.262 \\ \qquad \text{million dollars} \qquad\qquad \text{when } t < 9 \\[2mm] -317.5t^2 + 5782.5t - 25,198 \\ \qquad \text{million dollars} \qquad\qquad \text{when } t \geq 9 \end{cases}$
where t is the number of years since 1990.

c. $E(1) \approx \$96$ million
$E(12) \approx -\$1.528$ billion

d. Answers will vary.

e. The model indicates a rapid decline of Gap Inc.'s earnings.

f. Answers will vary.

33. Answers will vary.

Section 2.5

1. Because the data are concave up and display a minimum, a quadratic is the only appropriate model.

3. The data appear to be essentially linear. Any concavity is probably not obvious enough to warrant the use of a more complex model.

5. Because of the inflection point, a cubic or logistic model would be an appropriate choice. The choice would depend on the desired behavior of the model outside the range of the data.

7. a. Linear data have constant first differences and lie in a line.

b. Quadratic data have constant second differences and are either concave up or concave down.

c. Exponential data have constant percentage differences and are concave up, approaching the horizontal axis.

9. a. The scatter plot is decreasing and concave up indicating that a quadratic or log function could be used to model the data. Note that the initial decline is probably not steep enough for a good exponential function fit.

b. $W(t) = 0.0053t^2 - 1.148t + 118.972$ seconds and $S(t) = 172.311 - 25.273 \ln t$ seconds where t is the number of years after 1900. The quadratic function better follows the curvature of the data.

c. $T(x) = 92.114 - 8.838 \ln x$ seconds where x is the number of years after 1946. The equation provides a good fit to the data.

d. Both functions provide good fits to the data, but the quadratic function better fits the first three data points than does the log function. It is not likely that the winning times will reach a minimum and increase in future years. Therefore, the log function better describes the end behavior in context.

11. a. The best model choices are quadratic and log. We show both models.
Quadratic: $Q(t) = -31.286t^2 + 1292.171t + 3321.286$ million dollars worth of CDs shipped t years after 1990
Log: $L(t) = -29,492.053 + 14,221.876 \ln t$ million dollars t years after 1980

b. $Q(11) \approx \$13,750$ million; $L(21) \approx \$13,807$ million
$Q(15) \approx \$15,665$ million; $L(25) \approx \$16,286$ million

13. a. The scatter plot is decreasing. It is concave up to the left of approximately 50 thousand miles and concave up to the right. The presence of an inflection point and the end behavior (leveling off) indicate that a logistic equation may fit the data.
$$R(m) = \frac{8147.558}{1 + 0.084e^{0.051302m}} + 12,000 \text{ dollars when}$$
the Jeep has accumulated m thousand miles

b. $R(52) \approx \$15,676$

15. a. $E(y) = -0.009y^3 + 0.203y^2 + 0.126y + 21.618$ billion dollars y years after 1990

b. The data appear to have an inflection point near 1998 and the context does not suggest that the expenditure will level off past 2001.

17. a. The scatter plot is concave down from 1980 through 1987. From 1987 through 1997, the scatter plot appears to be logistic.

b. The data exhibit two changes in concavity: one abrupt change around 1987 and a smoother change between 1991 and 1993. This concavity description does not

match that of a cubic model that has one concavity change.

c. A cubic model is an extremely poor fit to the data.

d. Answers will vary. Once possible model is
$$P(t) = \begin{cases} -2.1t^2 - 6.3t + 2914 \\ \quad \text{thousand people} \qquad \text{when } 0 \le t \le 7 \\ \dfrac{97.5}{1 + 1405.903e^{-0.591268t}} + 2760 \\ \quad \text{thousand people} \qquad \text{when } 7 < t \le 17 \end{cases}$$
where t is the number of years since 1980.

e. Answers will vary.

19. It is important to reduce the magnitude of the input values for exponential and logistic models in order to avoid numerical computation error.

21. Answers will vary.

23. Answers found on the *Calculus Concepts* CD-ROM and web site.

CHAPTER 2 REVIEW TEST

1. a. $C(x) = 2.2(1.021431^x)$ million children x years after 1970

b. Approximately 2.1% per year

c. Solving $C(x) = 5$ gives $x \approx 38.7$, which corresponds to the fall of 2009.

d. Answers will vary.

2. a. The data are increasing and concave up, indicating that either an exponential or a quadratic model is appropriate. The data suggest a limiting value on the left, so an exponential model may be the better choice.

b. $P(t) = 0.975 (1.540440^t)$ dollars t years after 1989

c. Answers will vary.

3. a. The scatter plot is essentially concave up and then concave down. A cubic model appears to be appropriate.

b. $F(t) = -0.049t^3 + 1.560t^2 - 13.485t + 110.175$ degrees Fahrenheit t hours after midnight August 27

c. $F(17.5) \approx 91°F$

d. Solving $F(t) = 90$ gives three solutions: $t \approx 1.88$, 12.37, and 17.82. The first solution lies outside the time frame of the data. The other two solutions correspond to 12:22 P.M. and 5:49 P.M.

4. a. The statement is false because the input data are not evenly spaced.

b. The scatter plot suggests a concave-down shape. This shape could be modeled by a quadratic or log function. It is also possible to use the right side of a logistic function to model the data.

c. $Q(t) = (-4.988 \cdot 10^{-4})t^2 + 0.0191t - 7.995$ billion people t years after 1900. The graph of this equation is an excellent fit for the data.

$$L(t) = \frac{10.764}{1 + 20.429e^{-0.032771t}}$$ billion people t years after 1900. The graph of this equation is also an excellent fit for the data.

$G(t) = -27.890 + 7.408 \ln t$ billion people t years after 1900. The graph of this equation is a reasonably good fit for the data, although not so good a fit as the other two models. Shifting the input data to the left might produce a better fit.

d. $\lim_{t \to \infty} Q(t) \to -\infty$

$\lim_{t \to \infty} L(t) \approx 10.8$ billion people

$\lim_{t \to \infty} G(t) \to \infty$

Because the United Nations study suggested that the population will stabilize, the logistic model is the most appropriate one even though the data do not suggest an inflection point.

CHAPTER 3

Section 3.1

1. The stock price rose an average of 46 cents per day during the 5-day period.
3. The company lost an average of $8333.33 per month during the past 3 months.
5. Unemployment has risen an average of 1.3 percentage points per year in the past 3 years. *Note:* Whenever you are writing a change for a function whose output is a percentage, the correct label (unit of measure) is *percentage points*. The same is true for a rate of change. The phrase *percentage points per year* indicates that the effect is additive, whereas the phrase *percent per year* indicates a multiplicative effect.
7. The ACT composite average for females increased by 0.4 point between 1990 and 2002. This change represents an approximate 2% increase. The average female score increased by an average of 0.03 point per year between 1990 and 2002.
9. The number of Internet users in China grew by 11.1 million between 1997 and 2000. This growth represented a 1233% increase. The number of Internet users increased at an average rate of 3.7 million users per year between 1997 and 2000.

11. **a.** Slope of secant line $= \dfrac{6229.09 - 6228.98}{1996 - 1982}$

$= \dfrac{0.11 \text{ feet}}{14 \text{ years}} \approx 0.008$ foot per year

b. In the 14-year period from 1982 through 1996, the lake level rose an average of 0.008 foot per year.

c. The lake level dropped below the natural rim because of drought conditions in the early 1990s but rose again to normal elevation by 1996. The average rate of change tells us that the level of the lake in 1996 was close to the 1982 level. Although the average rate of change is nearly zero, the graph shows that the lake level changed dramatically during the 14-year period.

13. **a.** The balance increased by $1908.80 - $1489.55 = $419.25, or 28.1%.

b. Average rate of change $= \dfrac{\$419.25}{4 \text{ years}} \approx \$104.81/\text{year}$

From the end of year 1 through the end of year 5, the balance increased at an average rate of $104.81 per year.

c. You could estimate the amount in the middle of the fourth year, but doing so might not be as accurate as using a model to find the amount.

d. $A(t) = 1400(1.063962^t)$ dollars after t years

Amount in middle of year 4 $= A(3.5) \approx \$1739.28$

Amount at end of year 4 $= A(4) \approx \$1794.04$

Average rate of change $=$

$\dfrac{\$1794.04 - \$1739.28}{\frac{1}{2} \text{ year}} = \$109.52$ per year

15. **a.** Average rate of change ≈ -0.81 year per year

b. Average rate of change between ages 10 and 20 $= -0.96$ year per year

Average rate of change between ages 20 and 30 $= -0.89$ year per year

The magnitude of the average rate of change between ages 10 and 20 is greater than the average rate of change between ages 20 and 30.

17. **a.** $p(55) - p(40) \approx 31.70 - 21.45 \approx 10.3$ million people

Percentage change $\approx 48\%$

b. $\dfrac{p(85) - p(83)}{85 - 83} \approx \dfrac{3.52 \text{ million people}}{2 \text{ years}}$

≈ 1.8 million people per year

19. **a.** Between 0 and 2 seconds, the average rate of change is 0 feet per second because the secant line is horizontal.

b. Slope ≈ -50 feet/second

c. $-50 \dfrac{\text{feet}}{\text{second}} \cdot \dfrac{3600 \text{ seconds}}{\text{hour}} \cdot \dfrac{1 \text{ mile}}{5280 \text{ feet}} \approx -34$ mph

21. a.

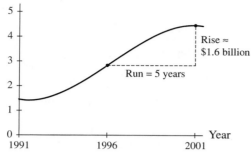

Sales
(billions of dollars)

Kelly's sales of services increased by approximately $320 million per year on average between 1996 and 2001.

b. Approximately a 42% increase

23. 501%

25. a. i. 3   **b. i.** 85.7%
      ii. 3   **ii.** 46.2%
      iii. 3   **iii.** 31.6%

c. The average rate of change of any linear function over any interval will be constant because the slope of a line (and, therefore, of any secant line) is constant. The percentage change is not constant.

27. a.

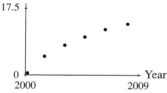

Cost
(billions of dollars)

The data suggest a log model: concave down with no sign of leveling off.

b. $C(t) = 0.689 + 7.258 \ln t$ billion dollars t years after 1999

c. $C(6) - C(1) \approx \$13$ billion

d. $\dfrac{C(6) - C(1)}{6 - 1} \approx \$2.6$ billion per year

e. $\dfrac{C(6) - C(1)}{C(1)} \cdot 100\% \approx 1888\%$

29. a. Solve $\dfrac{152.6 - x}{x} = 0.197$ for x to obtain $x \approx 127.5$ births per 100,000 in 1995.

Solve $\dfrac{152.6 - x}{x} = 3.124$ for x to obtain $x \approx 37.0$ births per 100,000 in 1980.

b. $M(y) = \dfrac{176.984}{1 + 968.131e^{-0.291929y}} + 29$ births per 100,000 y years after 1970

c. $M(25) \approx 135.9$ births per 100,000 in 1995
$M(10) \approx 32.3$ births per 100,000 in 1980
Both estimates are interpolations.

d. The increased use of fertility drugs is the primary factor in the rise of the multiple-birth rate. Women having children later in life also contributes to this rise.

31. Answers will vary.

Section 3.2

1. a. A continuous graph or model is defined for all possible input values on an interval. A continuous model with discrete interpretation has meaning for only certain input values on an interval. A continuous graph can be drawn without lifting the pencil from the paper. A discrete graph is a scatter plot. A continuous model or graph can be used to find average or instantaneous rates of change. Discrete data or a scatter plot can be used to find average rates of change.

b. An average rate of change is a slope between two points. An instantaneous rate of change is the slope at a single point on a graph.

c. A secant line connects two points on a graph. A tangent line touches the graph at a point and is tilted the same way the graph is tilted at that point.

3. Average rates of change are slopes of secant lines. Instantaneous rates of change are slopes of tangent lines.

5. Average speed $= \dfrac{19 - 0 \text{ miles}}{17 \text{ minutes}} \cdot \dfrac{60 \text{ minutes}}{\text{hour}} \approx 67.1$ mph

7. a. The slope is positive at A, negative at B and E, and zero at C and D.

b. The graph is steeper at point B than at point A.

9. a.   **b.**

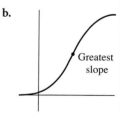

11. The lines at A and C are not tangent lines.

13.

15. a, b. *A*: concave down; tangent line lies above the curve
B: inflection point; tangent line lies below the curve on the left, above the curve on the right
C: inflection point; tangent line lies above the curve on the left, below the curve on the right
D: concave up; tangent line lies below the curve

c.

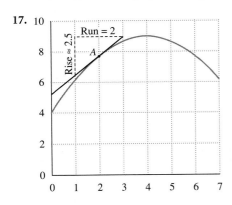

d. *A*, *D*: positive slope; *C*: negative slope (inflection point); *B*: zero slope (inflection point)

17.

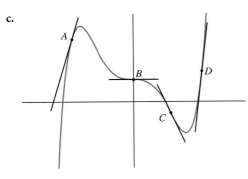

Slope $\approx \dfrac{2.5}{2} = 1.25$ (Answers may vary.)

19.

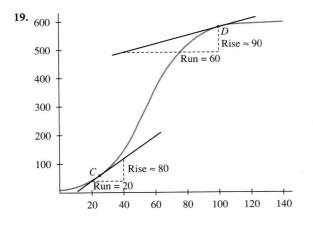

$$\text{Slope at } C \approx \frac{80}{20} = 4 \qquad \text{Slope at } D \approx \frac{90}{60} = 1.5$$

(Answers may vary.)

21. a. Million subscribers per year

b. The number of subscribers was growing at a rate of 23.1 million per year in 2000.

c. 23.1 million subscribers per year

d. 23.1 million subscribers per year

23. a, b. *A*: 1.3 mm per day per °C; *B*: 5.9 mm per day per °C; *C*: -4.2 mm per day per °C

c. The growth rate is increasing by 5.9 mm per day per °C.

d. The slope of the tangent line at 32°C is -4.2 mm per day per °C.

e. At 17°C, the instantaneous rate of change is 1.3 mm per day per °C.

25. a. The slope at the solstices is zero.

Declination of sun

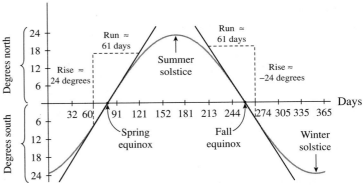

b. The steepest points on the graph are those where the graph crosses the horizontal axis. The slopes are estimated as

$$\frac{24 \text{ degrees}}{61 \text{ days}} \approx 0.4 \text{ degree per day and}$$

$$\frac{-24 \text{ degrees}}{61 \text{ days}} \approx -0.4 \text{ degree per day}$$

A negative slope indicates that the sun is moving from north to south.

c. The points identified in part *b* correspond to the spring and fall equinoxes.

27. a, b, c. Any line tangent to *p* is the graph itself. The slope of any tangent line will be approximately 2370 thousand people per year.

d. The slope of the graph at every point will be 2370 thousand people per year.

e. The instantaneous rate of change is 2370 thousand people per year.

29.

Employees

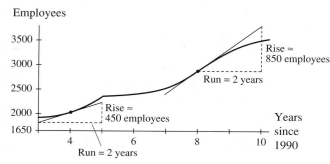

a. Slope in 1994 ≈ $\dfrac{450 \text{ employees}}{2 \text{ years}}$

= 225 employees per year

b. The slope of the graph at 1995 does not exist because the graph has a sharp point at 1995. Not enough information is given to estimate the rate of change of the underlying situation in 1995.

c. Slope in 1998 ≈ $\dfrac{850 \text{ employees}}{2 \text{ years}}$

= 425 employees per year

31. a. $\dfrac{31 - 47}{2000 - 1998} = \dfrac{-16 \text{ million}}{2 \text{ years}}$

= -8 million subscribers per year

b.

Analog subscribers (millions)

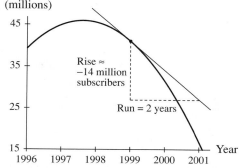

Slope ≈ $\dfrac{-14 \text{ million subscribers}}{2 \text{ years}}$

= -7 million subscribers per year

The two estimates differ by 1 million.

33. Drawing a tangent line at a break point on a piecewise continuous function is possible only when the function is continuous and smooth at the break point. Consider zooming in on the graph of *f* around the break point.

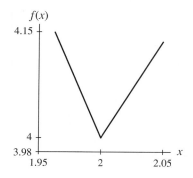

There is a sharp point on the graph of *f* at *x* = 2, so the limiting position of secant lines does not exist there.

Zooming in on the graph of *g* around *x* = 3, we see that this function is continuous and smooth at that break point.

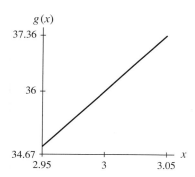

Thus the limiting position of secant lines exists at *x* = 3 and a tangent line can be drawn there.

Section 3.3

1. a. Miles per hour
   b. Speed or velocity
3. a. The number of words per minute cannot be negative.
   b. Words per minute per week
   c. The student's typing speed could actually be getting worse, which would mean that $W'(t)$ is negative.
5. a. When the ticket price is $65, the airline's weekly profit is $15,000.
   b. When the ticket price is $65, the airline's weekly profit is increasing by $1500 per dollar of ticket price. Raising the ticket price a little will increase profit.

c. When the ticket price is $90, the profit is declining by $2000 per dollar of ticket price. An increase in price will decrease profit.

7. $t(x)$

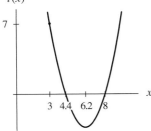

9. a. At the beginning of the diet, you weigh 167 pounds.
b. After 12 weeks of dieting, your weight is 142 pounds.
c. After 1 week of dieting, your weight is decreasing by 2 pounds per week.
d. After 9 weeks of dieting, your weight is decreasing by 1 pound per week.
e. After 12 weeks of dieting, your weight is neither increasing nor decreasing.
f. After 15 weeks of dieting you are gaining weight at a rate of a fourth of a pound per week.

g. $W(t)$
Weight
(pounds)

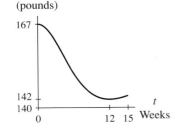

11. We know that the following points are on the graph: (1940, 4), (1970, 12), (2000, 33), and (1980, 18).

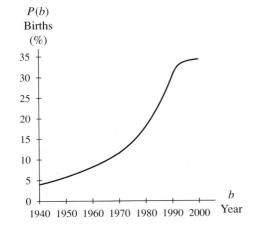

13. a. It is possible for profit to be negative if costs are more than revenue.
b. It is possible for the derivative to be negative if profit declines as more shirts are sold (because the price is so low, the revenue for that shirt is less than the cost associated with the shirt).
c. If $P'(200) = -1.5$, the fraternity's profit is declining. Profit may still be positive (which means that the fraternity is making money), but the negative rate of change indicates that it is not making the most profit possible.

15. a. Years per percentage point
b. As the rate of return increases, the time it takes the investment to double decreases.
c. i. When the interest rate is 9%, it takes 7.7 years for the investment to double.
ii. When the interest rate is 5%, the doubling time is decreasing by 2.77 years per percentage point. A 1-percentage-point increase in the rate will decrease the doubling time by approximately 2.8 years.
iii. When the interest rate is 12%, the doubling time is decreasing by 0.48 year per percentage point. A one-percentage-point increase in the rate of return will result in a decrease in doubling time of approximately half a year.

17.

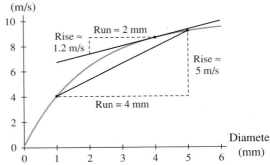

a. The slope of the secant line gives the average rate of change. Between 1 mm and 5 mm, the terminal speed of a raindrop increases an average of about 1.2 m/s per mm.
b. The slope of the tangent line gives the instantaneous rate of change of the terminal speed of a 4 mm raindrop.
c. Slope $\approx \dfrac{1.2 \text{ m/s}}{2 \text{ mm}} = 0.6$ m/s per mm

A 4-mm-raindrop's terminal speed is increasing by approximately 0.6 m/s per mm.

d. By sketching a tangent line at 2 mm and estimating its slope, we find that the terminal speed of a 2-mm raindrop is increasing by approximately 1.8 m/s per mm.

e. Percentage rate of charge

$$\approx \frac{1.8 \text{ m/s per mm}}{6.4 \text{ m/s}} \cdot 100\% \approx 28\% \text{ per mm}$$

The terminal speed of a 2-mm raindrop is increasing by about 28% per mm as the diameter increases.

19. $G(t)$

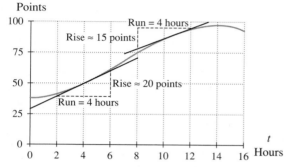

a, b. Slope at 4 hours $\approx \dfrac{20 \text{ points}}{4 \text{ hours}} = 5$ points per hour

Slope at 11 hours $\approx \dfrac{15 \text{ points}}{4 \text{ hours}} = 3.75$ points per hour

(Answers will vary.)

c. Average rate of change $\approx \dfrac{36 \text{ points}}{6 \text{ hours}}$

$$= 6 \text{ points per hour}$$

d. Percentage rate of change $\approx \dfrac{5 \text{ points/hour}}{50 \text{ points}} \cdot 100\%$

$$= 10\% \text{ per hour}$$

When you have studied for 4 hours, the number of points you will make on your test is increasing by about 10% per hour.

e. $G(4.6) \approx 0.6(5) + 50 = 53$ points.

21. a. Rate of change in 1970 ≈ 0.85 death per year; 7.7% per year

Rate of change in 1980 ≈ 1.1 deaths per year; 5.2% per year

b. Answers will vary.

23. Rate of change in 1995

$$\approx \frac{158.0 \text{ thousand deaths} - 78.2 \text{ thousand deaths}}{4 \text{ years}}$$

$$\approx 20 \text{ thousand deaths per year}$$

In 1995 the number of alcohol-related deaths was increasing by approximately 20,000 deaths per year.

25. Answers will vary.
27. Answers will vary.

Section 3.4

1. a.

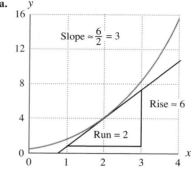

b.

Input of close point on left	Secant line slope	Input of close point on right	Secant line slope
1.9	2.67868	2.1	2.87094
1.99	2.76300	2.01	2.78222
1.999	2.77163	2.001	2.77355
1.9999	2.77249	2.0001	2.77268
1.99999	2.77258	2.00001	2.77260
Limit ≈ 2.77		Limit ≈ 2.77	

The slope of the line tangent to $y = 2^x$ at $x = 2$ is approximately 2.77.

3. a, b.

Input of close point on left	Secant line slope	Input of close point on right	Secant line slope
1.9	19.21	2.1	20.1
1.99	19.9201	2.01	20.01
1.999	19.99200	2.001	20.001
1.9999	19.99920	2.0001	20.0001
1.99999	19.99992	2.00001	20.00001
Limit = 20		Limit = 20	

c. Although the slopes of the two pieces are the same on either side of $x = 2$, the graph is not continuous at $x = 2$ because the pieces do not join at $x = 2$. The graph of f is not smooth at $x = 2$ because it is not continuous at that input.

d. Because the graph is not continuous at $x = 2$, it is not possible to sketch a tangent line at $x = 2$.

5. a.

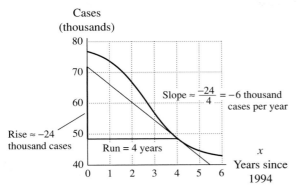

Cases (thousands)

Slope $\approx \dfrac{-24}{4} = -6$ thousand cases per year

Rise ≈ -24 thousand cases

Run = 4 years

x
Years since 1994

b.

Input of close point on left	Secant line slope	Input of close point on right	Secant line slope
3.9	-6.2828	4.1	-5.8463
3.99	-6.0849	4.01	-6.0412
3.999	-6.0652	4.001	-6.0609
3.9999	-6.0633	4.0001	-6.0628
3.99999	-6.0631	4.00001	-6.0630
Limit ≈ -6.06		Limit ≈ -6.06	

The slope of the line tangent to the graph at $x = 4$ is approximately -6.06 thousand cases per year.

c. In 1998 the number of AIDS cases diagnosed was decreasing by approximately 6.06 thousand cases per year.

7. a.

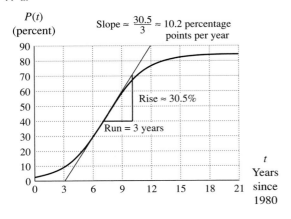

$P(t)$ (percent)

Slope $\approx \dfrac{30.5}{3} \approx 10.2$ percentage points per year

Rise ≈ 30.5%

Run = 3 years

t
Years since 1980

b.

Input of close point on left	Secant line slope	Input of close point on right	Secant line slope
6.9	10.15555	7.1	10.18681
6.99	10.17156	7.01	10.17468
6.999	10.17298	7.001	10.17330
6.9999	10.17312	7.0001	10.17315
6.99999	10.17314	7.00001	10.17314
Limit ≈ 10.17		Limit ≈ 10.17	

The line tangent to the graph of P at $t = 7$ has a slope of approximately 10.17 percentage points per year.

c. In 1987 the percentage of households with VCRs was increasing by approximately 10.2 percentage points per year.

d. The tangent line approximation is fast but can be inaccurate. The numerical approximation is somewhat tedious but usually accurate.

9. a.

Input of close point on left	Secant line slope	Input of close point on right	Secant line slope
9.9	7.37148	10.1	7.50528
9.99	7.43131	10.01	7.44469
9.999	7.43733	10.001	7.43867
9.9999	7.43793	10.0001	7.43806
9.99999	7.43799	10.00001	7.43800
Limit ≈ 7.44		Limit ≈ 7.47	

The derivative (slope of the tangent line) of S at $t = 10$ is approximately \$7.44 billion per year.

b. In 2000 the annual total U.S. factory sales of consumer electronics goods to dealers was increasing by approximately \$7.44 billion per year.

11. a. $\dfrac{32.9 - 43.1}{2000 - 1998} = \dfrac{-10.2 \text{ violent crimes per 1000 persons}}{2 \text{ years}}$

$= -5.1$ violent crimes per 1000 persons against males per year

b. $V(t) = -4.57t + 55.386$ violent crimes per 1000 persons against males t years after 1995

Slope $= -4.57$ violent crimes per 1000 persons against males per year

The symmetric difference quotient is the average rate of change between 1998 and 2000. The slope of the linear equation can be thought of as an average of the rates of change for all years between 1995 and 2001. The answer to part *a* is the more accurate answer.

13. a. $\dfrac{318.7 - 294.1}{1999 - 1997} = \dfrac{24.6 \text{ points}}{2 \text{ years}}$

$= 12.3$ index points per year

b. $L(x) = 65.411 + 144.812 \ln x$ points x years after 1985

$Q(x) = -0.498x^2 + 25.586x + 58.985$ points x years after 1985

c. Log model estimate: 11.1 index points per year

Quadratic model estimate: 12.6 index points per year

d. The symmetric difference quotient is faster than the numerical estimation and gives an answer comparable to that obtained from the quadratic model.

15. a. Approximately $1.18 per 100 pounds per month

b. $1.67 per 100 pounds per month

c. The derivative of p does not exist at $m = 5$.

d. Because we do not have data available, we can use the model to estimate cattle prices in August and October of 1998 and find the symmetric difference quotient using those two points:

$\dfrac{p(6) - p(4)}{6 - 4} = \dfrac{78.918 - 78.4706}{6 - 4}$

$= \dfrac{\$0.4474}{2 \text{ months}}$

$\approx \$0.22$ per 100 pounds per month

(There are other valid methods for estimating this rate of change.)

17. a. $C(P(x)) = \dfrac{1.02^x}{1.5786}$ U.S. dollars when x mountain bikes are sold

b. $P(400) \approx \$2755$ Canadian

$D[P(400)] \approx \$1745$ U.S.

c. To find $\dfrac{dC}{dx}$ when $x = 400$, we numerically investigate the secant line slopes of $C \circ P$ as x approaches 400.

Input of close point on left	Secant line slope	Input of close point on right	Secant line slope
399.9	34.5214	400.1	34.5899
399.99	34.5522	400.01	34.5590
399.999	34.5553	400.001	34.5560
399.9999	34.55564	400.0001	34.55571
399.99999	34.55568	400.00001	34.55568
Limit $\approx$ 34.56		Limit $\approx$ 34.56	

When 400 mountain bikes are sold, the profit (in U.S. dollars) is increasing at a rate of approximately $34.56 per mountain bike sold.

19. Answers are located on the *Calculus Concepts* CD-ROM and web site.

21. Answers will vary.

Section 3.5

1. a. $T(13) = 67.946$ seconds

b. $T(13 + h) = 0.181(13 + h)^2 - 8.463(13 + h) + 147.376$

$= 0.181h^2 - 3.757h + 67.946$

c. $\dfrac{T(13 + h) - T(13)}{13 + h - 13} = \dfrac{0.181h^2 - 3.757h}{h}$

d. $\lim\limits_{h \to 0} \dfrac{0.181h^2 - 3.757h}{h} = \lim\limits_{h \to 0} (0.181h - 3.757)$

$= -3.757$ seconds per year of age

The swim time for a 13-year-old is decreasing by 3.757 seconds per year of age. This tells us that as a 13-year-old athlete gets older, the athlete's swim time improves.

3. a. $c(8) = 307.41$

b. $c(8 + h) = -0.498(8 + h)^2 + 20.603(8 + h) + 174.458$

$= -0.498h^2 + 12.635h + 307.41$

c. $\dfrac{c(8 + h) - c(5)}{8 + h - 8} = \dfrac{-0.498h^2 + 12.635h}{h}$

d. $\lim\limits_{h \to 0} \dfrac{-0.498h^2 + 12.635h}{h} = \lim\limits_{h \to 0} (-0.498h + 12.635)$

$= 12.635$ index points per year

e. The consumer price index for college tuition was increasing by 12.635 index points per year in 1998.

f. Activity 13 of Section 3.4 used a symmetric difference quotient to estimate this rate of change as 12.3 points per year and used the numerical method with a quadratic model to estimate it as 12.6 points per year. Because we are dealing with a model that is not an exact fit to the data, none of these estimates is exact. Assuming that you both have the data and can find a good model for the data, then

    i. Because the algebraic method uses a rounded model, the numerical method with an unrounded model will probably produce the most accurate answer as long as the model is an excellent fit to the data.

    ii. Use the data and a symmetric difference quotient to obtain a quick, rough estimate.

    iii. Because both the numerical method and the algebraic method are somewhat time-consuming, use the data and a symmetric difference quotient to obtain a fairly good estimate without taking much time.

5.   **i.** $g(4) = -89$

    ii. $g(4 + h) = -6(4 + h)^2 + 7 = -6h^2 - 48h - 89$

    iii. $\dfrac{c(4 + h) - c(4)}{4 + h - 4} = \dfrac{-6h^2 - 48h}{h}$

    iv. $\lim\limits_{h \to 0} \dfrac{-6h^2 - 48h}{h} = \lim\limits_{h \to 0} (-6h - 48) = -48$

$\dfrac{dg}{dt} = -48$ when $t = 4$

7. a.   i. $H(t) = -16t^2 + 100$

ii. $H(t + h) = -16(t + h)^2 + 100$

$$= -16h^2 - 32th - 16t^2 + 100$$

iii. $\dfrac{H(t + h) - H(t)}{t + h - t} = \dfrac{-16h^2 - 32th}{h}$

iv. $\lim\limits_{h \to 0} \dfrac{-16h^2 - 32th}{h} = \lim\limits_{h \to 0} (-16h - 32t) = -32t$

$\dfrac{dH}{dt} = -32t$ feet per second t seconds after the object is dropped

b. After 1 second, the velocity of the object was $-32(1) = -32$ feet per second; that is, the object was falling at a rate of 32 feet per second.

9. a. $D(a) = -0.045a^2 + 1.774a - 16.064$ million drivers age a years

b. $D(a + h) = -0.045(a + h)^2 + 1.774(a + h) - 16.064$

$$= -0.045a^2 - 0.09ah - 0.045h^2 + 1.774a + 1.774h - 16.064$$

$\dfrac{D(a + h) - D(a)}{a + h - a} = \dfrac{-0.09ah - 0.045h^2 + 1.774h}{h}$

$\lim\limits_{h \to 0} \dfrac{-0.09ah - 0.045h^2 + 1.774h}{h}$

$= \lim\limits_{h \to 0} (-0.09a - 0.045h + 1.774) = -0.09a + 1.774$

Thus $D'(a) = -0.09a + 1.774$ million drivers per year of age.

c. $D'(20) = -0.026$ million drivers per year of age. For 20-year-olds, the number of licensed drivers is decreasing by 26,000 drivers per year of age. This tells us that there are fewer 21-year-olds who are licensed drivers than there are 20-year-olds.

d. Approximately -1.7% per year. The number of licensed drivers was falling by approximately 1.7% per year of age when the driver was 20 years old in 1997.

11.   i. $f(x) = 3x - 2$

ii. $f(x + h) = 3(x + h) - 2 = 3x + 3h - 2$

iii. $\dfrac{f(x + h) - f(x)}{x + h - x} = \dfrac{3h}{h}$

iv. $\lim\limits_{h \to 0} \dfrac{3h}{h} = \lim\limits_{h \to 0} 3 = 3$. Therefore, $\dfrac{df}{dx} = 3$.

13.   i. $f(x) = 3x^2$

ii. $f(x + h) = 3(x + h)^2 = 3x^2 + 6xh + 3h^2$

iii. $\dfrac{f(x + h) - f(x)}{x + h - x} = \dfrac{6xh + 3h^2}{h}$

iv. $\lim\limits_{h \to 0} \dfrac{6xh + 3h^2}{h} = \lim\limits_{h \to 0} (6x + 3h) = 6x$

Therefore, $f'(x) = 6x$.

15.   i. $f(x) = x^3$

ii. $f(x + h) = (x + h)^3 = x^3 + 3x^2h + 3xh^2 + h^3$

iii. $\dfrac{f(x + h) - f(x)}{x + h - x} = \dfrac{3x^2h + 3xh^2 + h^3}{h}$

iv. $\lim\limits_{h \to 0} \dfrac{3x^2h + 3xh^2 + h^3}{h} = \lim\limits_{h \to 0} (3x^2 + 3xh + h^2) = 3x^2$

Therefore, $f'(x) = 3x^2$.

17. Answers will vary.

CHAPTER 3 REVIEW TEST

1. a. i. A, B, C **ii.** E **iii.** D

b. The graph is steeper at B than it is at A, C, or D.

c. Below: C, D, E; above: A; At B: above to the left of B, below to the right of B

d.

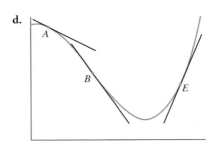

e.   i. Feet per second per second. This is acceleration.

ii, iii. The roller coaster was slowing down until D, when it began speeding up.

iv. The roller coaster's speed was slowest at D.

v. The roller coaster was slowing down most rapidly at B.

2. a. $\dfrac{1200 - 510}{2002 - 1996} = 115$ million cards per year. Between 1996 and 2002, the number of Visa cards increased at an average rate of 115 million cards per year.

b. $\dfrac{1200 - 1000}{1000} \cdot 100\% = 20\%$. Between 2000 and 2002, the number of Visa cards increased by about 20%.

3. a, b.

$E(x)$
Employees

[graph showing employees from 1992 to 2001, y-axis from 0 to 40,000; tangent lines marked "Run = 1 year", "Rise ≈ 6600", "Run = 4 years", "Rise ≈ 6000 employees"; x-axis labeled "Year" with 1992, 1995, 1998, 2001]

The slope of the secant line gives the average rate of change between 1993 and 1997. The slope of the tangent line gives the instantaneous rate of change in 1998.

c. In 1998 the number of employees was increasing by approximately 6600 employees per year. If we estimate the number of employees in 1998 as 16,000, then this rate represents an increase of approximately 41% per year.

d. Between 1993 and 1997, the number of Dell employees increased at an average rate of 1500 employees per year.

4. a. i. An average 22-year-old athlete can swim 100 meters in 49 seconds.

ii. The swim time for a 22-year-old is declining by half a second per year of age.

b. A negative rate of change indicates that an average swimmer's time improves as age increases.

5. a. $M(t) = 0.975(1.543534^t)$ dollars per share t years after 1989

b.

Input of close point on left	Secant line slope	Input of close point on right	Secant line slope
9.9	31.8027	10.1	33.2136
9.99	32.4275	10.01	32.5685
9.999	32.4909	10.001	32.5050
9.9999	32.4972	10.0001	32.4986
9.99999	32.4978	10.00001	32.4980
Limit ≈ 32.50		Limit ≈ 32.50	

c. In 1999 the yearly high stock price for Microsoft was increasing by approximately $32.50 per share per year.

d. $M(t) = 0.975(\ln 1.543534)(1.543534^t)$ dollars per share per year t years after 1989

$M(10) \approx \$32.50$ per share per year

6. (1) Begin with a point: $(x, 7x + 3)$

(2) Choose a close point: $(x + h, 7(x + h) + 3)$

(3) Find a formula for the slope between the two points. Simplify completely.

$$\text{Slope} = \frac{7x + 7h + 3 - (7x + 3)}{x + h - x} = \frac{7h}{h}$$

(4) Find the limit of the slope as $h \to 0$: $\lim_{h \to 0} \frac{7h}{h} = 7$

CHAPTER 4

Section 4.1

1. The slopes are negative to the left of A and positive to the right of A. The slope is zero at A.

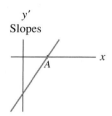

3. The slopes are positive everywhere, near zero to the left of $x = 0$, and increasingly positive to the right of $x = 0$.

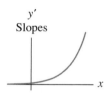

5. The slope is zero everywhere.

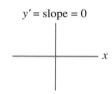

7. The slopes are negative everywhere. The magnitude is large close to $x = 0$ and is near zero to the far right.

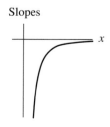

9. The slopes are negative to the left and right of *A*. The slope appears to be zero at *A*.

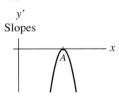

11. a.

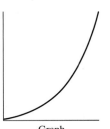

Graph   Slope graph

b.

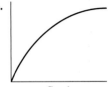

Graph   Slope graph

13. a. Table values may vary.

Year	1991	1993	1997	1999	2001
Slope	−6.6	−6.2	−2.5	1.0	5.4

b. Rate of change
of average bill
(dollars per year)

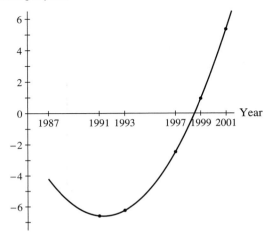

15. a. (Table values may vary.)

Year	1985	1990	1995	1997	2000
Slope	7.8	46.8	79.2	55.2	21.4

b. Rate of change
(cases per year)

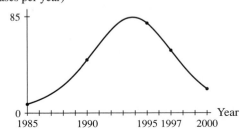

17. a. The average rate of change during the year (found by estimating the slope of the secant line drawn from September to May) is approximately 14 members per month. (Answers will vary.)

b. By estimating the slopes of tangent lines, we obtain the following. (Answers will vary.)

Month	Slope (members per month)
Sept	98
Nov	−9
Feb	30
Apr	11

c. Members
per month

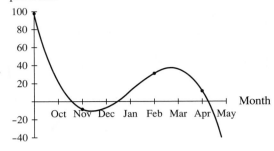

d. Membership was growing most rapidly around March. This point on the membership graph is an inflection point.

e. The average of change is not useful in sketching an instantaneous rate-of-change graph.

19.

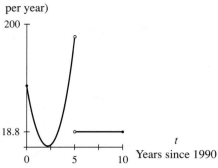

j'(t)
(inmates
per year)

21. a. Profit is increasing on average by approximately $600 per car.

b.

Number of cars	Slope (dollars per car)
20	0
40	160
60	750
80	10
100	−1200

Answers will vary.

c.

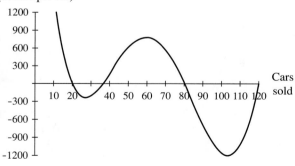

Average monthly profit (dollars per car)

d. The average monthly profit is increasing most rapidly for about 60 cars sold and is decreasing most rapidly when about 100 cars are sold. The corresponding points on the graph are inflection points.

e. Average rates of change are not useful in graphing instantaneous rates of change.

23. The derivative does not exist at $x = 0$, $x = 3$, and $x = 4$ because the graph is not continuous at those inputs.

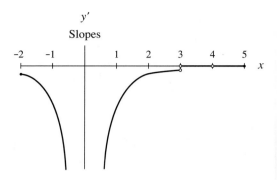

25. The derivative does not exist at $x = 2$ and $x = 3$ because the slopes from the right and left are different at those inputs.

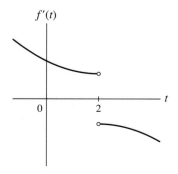

27. Answers will vary.

29. Answers will vary.

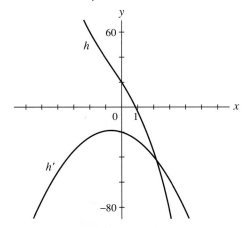

31. a. $p'(m)$

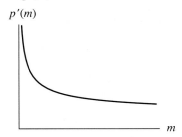

b. i. $p(m) = m + \sqrt{m}$

   ii. $p(m + h) = m + h + \sqrt{m + h}$

   iii. $\dfrac{p(m + h) - p(m)}{m + h - m} = \dfrac{h + \sqrt{m + h} - \sqrt{m}}{h}$

$$= \left(\dfrac{h + \sqrt{m + h} - \sqrt{m}}{h}\right)\left(\dfrac{\sqrt{m + h} + \sqrt{m}}{\sqrt{m + h} + \sqrt{m}}\right)$$

$$= \dfrac{h(\sqrt{m + h} + \sqrt{m}) + (m + h) - m}{h(\sqrt{m + h} + \sqrt{m})}$$

$$= \dfrac{h(\sqrt{m + h} + \sqrt{m}) + h}{h(\sqrt{m + h} + \sqrt{m})}$$

   iv. $\displaystyle\lim_{h \to 0} \dfrac{h(\sqrt{m + h} + \sqrt{m}) + h}{h(\sqrt{m + h} + \sqrt{m})}$

$$= \lim_{h \to 0} \dfrac{(\sqrt{m + h} + \sqrt{m}) + 1}{\sqrt{m + h} + \sqrt{m}}$$

$$= \dfrac{2\sqrt{m} + 1}{2\sqrt{m}} = 1 + \dfrac{1}{2\sqrt{m}}$$

$\dfrac{dp}{dm} = 1 + \dfrac{1}{2\sqrt{m}}$. The graph of $\dfrac{dp}{dm}$ is the same as the one in part *a*.

33. Answers will vary.

Section 4.2

1.

3.

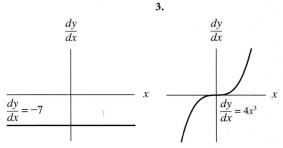

5.

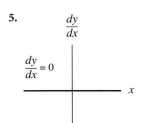

7. $\dfrac{dy}{dx} = 14x - 12$

9. $\dfrac{dy}{dx} = 15x^2 + 6x - 2$

11. $\dfrac{dy}{dx} = 18x^{-3} = \dfrac{18}{x^3}$

13. $h'(x) = -0.049\left(\dfrac{1}{2}x^{-1/2}\right) = \dfrac{-0.0245}{\sqrt{x}}$

15. a. $A'(t) = 0.1333$ dollars per year t years after 1990

   b. $A(10) \approx \$1.50$

   c. $A'(9) = \$0.1333$ per year. The transaction fee in 1999 was increasing by approximately \$0.13 per year.

17. a. $P'(t) = 15.48$ thousand people per year t years after 1950

   b. $P(20) = 795$ thousand people

   c. $P'(40) = 15.48$ thousand people per year

19. $T'(x) = -1.6t + 2°F$ per hour t hours after noon

   a. $T'(1.5) = -0.4°F$ per hour

   $T'(-5) = 10°F$ per hour

   b. $T'(-5) = 10°F$ per hour

   c. $T'(0) = 2°F$ per hour

   d. $T'(4) = -4.4°F$ per hour

21. a. $N(-5) \approx 33.4$ million senior citizens in 1995

   $N(30) \approx 70.7$ million senior citizens in 2030

b. $N'(-5) \approx 0.015$ million senior citizens per year in 1995
$N'(30) \approx 2.1$ million senior citizens per year in 2030

c. $\dfrac{N'(30)}{N(30)} \cdot 100\% \approx 3\%$ per year in 2030

d. Solve $70.68 = 0.201P$ for P to obtain $P \approx 351.6$ million senior citizens in 2030.

23. a, b. In 1970 the number of births was falling at a rate of approximately 207 births per year. In 1995 the number of births was rising at a rate of about 322 births per year.

25. a. $R(x) = -3.68x^3 + 47.958x^2 - 80.759x + 166.98$ billion dollars of revenue when \$$x$ billion is spent on advertising

b. $R'(x) = -11.039x^2 + 95.916x - 80.759$ billion dollars of revenue per billion advertising dollars when \$$x$ billion is spent on advertising

c. $R'(5) \approx \$122.8$ billion of revenue per billion advertising dollars

d. $R(5) \approx \$502.2$ billion

e. $\dfrac{R'(4.6)}{R(4.6)} \cdot 100\% \approx 25.3\%$ per billion advertising dollars

27. a. $P(x) = 175 - 0.0146x^2 + 0.7823x - 46.9125 - \dfrac{49.6032}{x}$ dollars when x windows are produced each hour

b. $P'(x) = -0.0292x + 0.7823 + \dfrac{49.6032}{x^2}$ dollars per window when x windows are produced each hour

c. $P(80) \approx \$96.61$

d. $P'(80) \approx -\$1.55$ per window
When 80 windows are produced each hour, the profit from the sale of one window is decreasing by about \$1.55 per window produced.

29. a.
$$f'(x) = \begin{cases} 44.408x - 108.431 \text{ million lb per year} \\ \qquad\qquad\qquad \text{when } 0 \le x < 10 \\ 171.244x - 2253.951 \text{ million lb per year} \\ \qquad\qquad\qquad \text{when } 10 < x \le 20 \end{cases}$$
x years after 1970

b. In 1970: $f'(0) = -108.431$ million pounds per year
In 1980: $f'(10)$ does not exist. The rate of change of the amount of fish produced cannot be calculated directly from this model. One of the estimation techniques discussed in Example 3 should be used.
In 1990: $f'(20) = 1170.93$ million pounds per year

c. $\dfrac{f'(20)}{f(20)} \cdot 100\% \approx 16.9\%$ per year

31. Answers will vary.

Section 4.3

1.
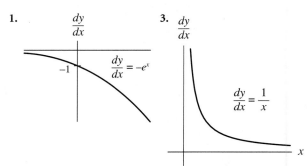
$\dfrac{dy}{dx}$ $\dfrac{dy}{dx} = -e^x$

3. $\dfrac{dy}{dx}$ $\dfrac{dy}{dx} = \dfrac{1}{x}$ x

5.
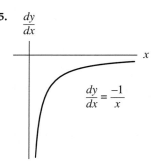
$\dfrac{dy}{dx}$ x $\dfrac{dy}{dx} = \dfrac{-1}{x}$

7. $h'(x) = 14x + \dfrac{13}{x}$

9. $\dfrac{dg}{dx} = 17\,(\ln 4.962)(4.962^x)$

11. $f'(x) = 100{,}000 \ln\!\left(1 + \dfrac{0.05}{12}\right)^{12}\!\left(1 + \dfrac{0.05}{12}\right)^{12x} \approx$
$4989.61218\!\left(1 + \dfrac{0.05}{12}\right)^{12x} \approx 4989.61218(1.05116^x)$

13. $\dfrac{dy}{dx} = \dfrac{-4.2}{x} + 3.3(\ln 2.9)(2.9^x)$

15. a. $h = \ln 2.5 \approx 0.916$ hours

b. $\dfrac{dV}{dh} = e^h$ quarts per hour

c. $V(0.4) \approx 1.49$ quarts per hour ≈ 0.025 quart per minute
$V(0.7) \approx 2.01$ quarts per hour ≈ 0.034 quart per minute
$V\!\left(\dfrac{55}{60}\right) \approx 2.5$ quarts per hour ≈ 0.042 quart per minute

17. a. $\dfrac{dy}{dx} = (\ln e)e^x = (1)e^x = e^x$

b. $\dfrac{dy}{dx} = (\ln e^k)e^{kx} = k(\ln e)e^{kx} = k(1)e^{kx} = ke^{kx}$

19. a. $w(7) \approx 25.64$ grams; $w'(7) \approx 1.05$ grams per week

b. $\dfrac{w(9) - w(4)}{9 - 4} \approx \dfrac{5.977 \text{ grams}}{5 \text{ weeks}} \approx 1.195$ grams per week on average

c. The older the mouse is, the more slowly it gains weight because the rate-of-change formula $w'(x) = \frac{7.37}{x}$ has the age of the mouse in the denominator.

21. a. $\frac{318.7 - 294.1}{1999 - 1997} = 12.3$ index points per year

  b. $CPI(x) = -351.521 + 227.777 \ln x$, where x is the number of years since 1980

  c. $CPI'(18) = \frac{227.777}{18} \approx 12.7$ index points per year

23. a. $M(t) = -0.216t^3 + 7.383t^2 + 661.969t + 19,839.889$ dollars t years after 1947

  b. $M'(t) = -0.649t^3 + 14.765t^2 + 661.969$ dollars per year t years after 1947

  c.

Year	t	$M'(t)$	$\frac{M'(t)}{M(t)} \cdot 100$
1972	25	626	1.7% per year
1980	33	443	1.1% per year
1984	37	320	0.7% per year
1992	45	12.9	0.3% per year
1996	49	-172	-0.4% per year

  d. There seems to be no relationship between re-election and the rate of change in median family income. This may be an indication that the model does not provide sufficient information to answer this question accurately because it models the 50-year trend in median family income rather than the median income close to each election year.

25. a. $A'(r) = 10{,}000e^{10r}$ dollars per 100 percentage points when the interest rate is $100r\%$

  b. \$20,137.53 per 100 percentage points or \$201.38 per percentage point. The rate of change is large because it approximates by how much the value will increase when r changes by 1. Because the interest rate is input in decimals, an increase of 1 in r corresponds to a change in the interest rate of 100 percentage points.

  c. $A'(r) = 100e^{0.1r}$ dollars per percentage point when the interest rate is $r\%$

  d. \$201.38 per percentage point

27. a.
$$p'(t) = \begin{cases} -23.73t^2 + 241.92t + 193.92 \text{ people/year} \\ \qquad\qquad \text{when } 0.7 \le t < 13 \\ 45{,}544\,(\ln 0.8474)(0.8474^t) \text{ people/year} \\ \qquad\qquad \text{when } 13 < t \le 55 \end{cases}$$
where t is the number of years since the beginning of 1860

  b. $p'(10) \approx 240$ people per year
$p'(13)$ does not exist.
$p'(40) \approx -10$ people per year
The rate of change of the population in 1873 could be estimated by
  i. Calculating the model estimate for the populations in 1872 and 1874: 5954 and 4484
  ii. Finding the average rate of change (symmetric difference quotient) for those years:
$$\frac{4484 - 5954}{1874 - 1872} \approx -735 \text{ people per year}$$
(There are other valid methods of estimating this rate of change.)

29. Answers will vary.

Section 4.4

1. a. $f(x(2)) = f(6) = 140$

  b. $\dfrac{df}{dx} = -27$

  c. $\dfrac{dx}{dt} = 1.3$

  d. $\dfrac{df}{dt} = (-27)(1.3) = -35.1$

3. The value of the investor's gold is increasing at a rate of (0.2 troy ounce per day)(\$323.10 per troy ounce) = \$64.62 per day.

5. a. $R(476) = \$10{,}000$ Canadian
On November 25, 2002, sales were 476 units, producing revenue of \$10,000 Canadian.

  b. $C(10{,}000) = \$6334.7$ U.S.
On November 25, 2002, 10,000 Canadian dollars were worth \$6334.7 U.S.

  c. $\dfrac{dR}{dx} = \$2.6$ Canadian per unit
Revenue was increasing by 2.6 Canadian dollars per unit sold.

  d. $\dfrac{dC}{dr} = \$0.6335$ U.S. per Canadian dollar
The exchange rate was \$0.6335 U.S. per Canadian dollar

  e. $\dfrac{dC}{dx} = (2.6$ Canadian dollar per unit)(\$0.6335 U.S. per Canadian dollar) $\approx \$1.65$ U.S. per unit
On November 25, 2002, revenue was increasing at a rate of \$1.65 U.S. per unit sold.

7. a. $p(10) \approx 12.009$ thousand people
In 2010 the city had a population of approximately 12,000 people.

b. $g(p(10)) \approx g(12.009) \approx 22$ garbage trucks
In 2010 the city owned 22 garbage trucks.

c. $p'(t) = \dfrac{31.2e^{-0.02t}}{(1 + 12e^{-0.02t})^2}$ thousand people per year

$p'(10) \approx 0.22$ thousand people per year
In 2010, the population was increasing at a rate of approximately 220 people per year.

d. $g'(p) = 2 - 0.003p^2$ trucks per thousand people
$g'(12.009) \approx 1.6$ trucks per thousand people
In 2010, when the population was about 12,000, the number of garbage trucks needed by the city was increasing by 1.6 trucks per thousand people.

e. When $t = 10$,

$$\frac{dg}{dt} = \frac{dg}{dp} \cdot \frac{dp}{dt}$$

$$\approx \frac{1.6 \text{ trucks}}{\text{thousand people}} \cdot \frac{0.22 \text{ thousand people}}{\text{year}}$$

$$\approx 0.34 \text{ trucks per year}$$

In 2010, the number of trucks needed by the city was increasing at a rate of 0.34 truck per year, or 1 truck every 3 years.

f. See interpretations in parts *a* through *e*.

9. $c(x(t)) = 3(4 - 6t)^2 - 2$

$\dfrac{dc}{dt} = 6(4 - 6t)(-6) = -144 + 216t$

11. $h(p(t)) = \dfrac{4}{1 + 3e^{-0.5t}}$

$\dfrac{dh}{dt} = \dfrac{4(3)(0.5)e^{-0.5t}}{(1 + 3e^{-0.5t})^2} = \dfrac{6e^{-0.5t}}{(1 + 3e^{-0.5t})^2}$

13. $k(t(x)) = 4.3(\ln x)^3 - 2(\ln x)^2 + 4\ln x - 12$

$\dfrac{dk}{dx} = \dfrac{1.29(\ln x)^2}{x} - \dfrac{4\ln x}{x} + \dfrac{4}{x}$

15. $p(t(k)) = 7.9(1.046^{14k^3 - 12k^2})$

$\dfrac{dp}{dk} = 7.9(\ln 1.046)(1.046^{14k^3 - 12k^2})(42k^2 - 24k)$

17. Inside function: $u = 3.2x + 5.7$
Outside function: $f = u^5$

$\dfrac{df}{dx} = 5(3.2x + 5.7)^4(3.2)$

19. Inside function: $u = x^2 - 3x$
Outside function: $f = u^{\frac{1}{2}}$

$\dfrac{df}{dx} = \frac{1}{2}(x^2 - 3x)^{-\frac{1}{2}}(2x - 3)$

21. Inside function: $u = 35x$
Outside function: $f = \ln u$

$\dfrac{df}{dx} = \dfrac{1}{35x}(35) = \dfrac{1}{x}$

23. Inside function: $u = 16x^2 + 37x$
Outside function: $f = \ln u$

$\dfrac{df}{dx} = \dfrac{1}{16x^2 + 37x}(32x + 37)$

25. Inside function: $u = 0.695x$
Outside function: $f = 72.378e^u$

$\dfrac{df}{dx} = 72.378e^{0.695x}(0.695)$

27. Inside function: $u = 0.0856x$
Outside function: $f = 1 + 58.32e^u$

$\dfrac{df}{dx} = 58.32e^{0.0856x}(0.0856)$

29. Inside function: $u = 4x + 7$
Outside function: $f = 350u^{-1}$

$\dfrac{df}{dx} = -350(4x + 7)^{-2}(4)$

31. Inside function:

$u = 1 + 8.976e^{-1.243x} \begin{cases} \text{inside:} & w = -1.243x \\ \text{outside:} & u = 1 + 8.976e^w \end{cases}$

Outside function: $f = 3706.5u^{-1} + 89070$

$\dfrac{df}{dx} = -3706.5(1 + 8.976e^{-1.243x})^{-2}(8.976e^{-1.243x})(-1.243)$

$= \dfrac{(3706.5)(8.976)(1.243)e^{-1.243x}}{(1 + 8.976e^{-1.243x})^2}$

33. Inside function: $u = \sqrt{x} - 3x$
Outside function: $f = 3^u$

$\dfrac{df}{dx} = 3^{\sqrt{x} - 3x}(\ln 3)\left(\dfrac{1}{2\sqrt{x}} - 3\right)$

35. Inside function: $u = 2^x$
Outside function: $f = \ln u$

$\dfrac{df}{dx} = \dfrac{1}{2^x}(2^x)(\ln 2) = \ln 2$

37. Inside function:

$u = 1 + Ae^{-Bx} \begin{cases} \text{inside:} & w = -Bx \\ \text{outside:} & u = 1 + Ae^w \end{cases}$

Outside function: $f = Lu^{-1}$

$\dfrac{df}{dx} = -L(1 + Ae^{-Bx})^{-2}[Ae^{-Bx}(-B)] = \dfrac{LABe^{-Bx}}{(1 + Ae^{-Bx})^2}$

39. a. $A(t) = 1500e^{0.04t}$ dollars after t years
b. $A'(t) = 1500(0.04)e^{0.04t} = 60e^{0.04t}$ dollars per year after t years
c. $A'(1) \approx \$62.45$ per year
$A'(2) \approx \$65.00$ per year
d. The rates of change in part *c* tell you approximately how much interest your account will earn during the second and third years. The actual amounts are $A(2) - A(1) \approx \$63.71$ during the second year and $A(3) - A(2) \approx \$66.31$ during the third year.

41. a. $R'(q) = 0.0314(0.62285)e^{0.62285q}$ million dollars per quarter q quarters after the beginning of 1998

b.

Quarter ending	June 1998	June 1999	June 2000
$R(q)$ (millions of dollars)	3.0	4.2	18.8
$R'(q)$ (millions of dollars per quarter)		0.07	0.82	9.92
$\dfrac{R'(q)}{R(q)} \cdot 100\%$ (% per quarter)		2.3	19.5	52.7

43. a. A logistic model is probably a better model because of the leveling-off behavior, although neither model should be used to extrapolate beyond one 24-hour day.

b. $C(t) = \dfrac{1342.077}{1 + 36.797e^{-0.258856t}}$ calls t hours after 5 A.M.

c. $C'(t) = \dfrac{(1342.077)(36.797)(0.258856)e^{-0.258856t}}{(1 + 36.797e^{-0.258856t})^2}$ calls per hour t hours after 5 A.M.

d. Noon: $C'(7) \approx 42$ calls per hour

10 P.M.: $C'(17) \approx 74$ calls per hour

Midnight: $C'(19) \approx 58$ calls per hour

4 A.M.: $C'(23) \approx 28$ calls per hour

e. The rates of change give approximate hourly calls. This information could be used to determine how many dispatchers would be needed each hour.

45. a. $c'(4) \approx 1575$ deaths per week

b. $\dfrac{c'(4)}{c(4)} \cdot 100\% \approx 108\%$ increase per week

c. $c'(8) \approx 25{,}331$ deaths per week

$\dfrac{c'(8)}{c(8)} \cdot 100\% \approx 48\%$ increase per week

d. Although the rate of change is larger, it represents a smaller proportion of the total number of deaths that had occurred at that time.

47. a. The data are essentially concave up, which indicates that a quadratic or exponential model may be appropriate. Looking at the second differences and percentage differences indicates that a quadratic model is the better choice.

b. $u(x) = 177.356x^2 - 342.240x + 5914.964$ units per week x years after 1990

c. $C(u(x)) = 3250.23 + 74.95 \ln(177.356x^2 - 342.240x + 5914.964)$ dollars per week x years after 1990

d. $\dfrac{dC}{dx} = \left(\dfrac{74.95}{177.356x^2 - 342.240x + 5914.964}\right) \cdot$ $(354.712x - 342.240)$ dollars per week per year x years after 1990

e.

Year	2002	2003	2004	2005
x	12	13	14	15
$C(x)$ ($/week)	4015.95	4026.40	4036.31	4045.72
$C'(x)$ ($/w/y)	10.73	10.18	9.66	9.17

f, g. Although a graph of C may not appear ever to decrease, a graph of $\dfrac{dC}{dx}$ is negative between $x = 0$ and $x = 0.965$, indicating that cost was decreasing from the end of 1990 to (almost) the end of 1991. A close-up view of the graph of C between $x = 0$ and $x = 2$ confirms this.

49. Composite functions are formed by making the output of one function (the inside) the input of another function (the outside). It is imperative that the output of the inside and the input of the outside agree in the quantity that they measure as well as in the units of measurement.

Section 4.5

1. $h'(2) = f(2)g'(2) + f'(2)g(2) = 6(3) + (-1.5)(4) = 12$

3. a. i. In 2007 there will be 75,000 households in the city.

ii. In 2007 the number of households will be declining at a rate of 1200 per year.

iii. In 2007, 52% of households will own a computer.

iv. In 2007, the percentage of households in the city with a computer will be increasing by 5 percentage points per year.

b. Input: the number of years since 2005

Output: the number of households with computers

c. $N(2) = h(2)c(2) = (75{,}000)(0.52) = 39{,}000$ households with computers

$N'(2) = h(2)c'(2) + h'(2)c(2) = (75{,}000)(0.05) + (-1200)(0.52) = 3126$ households per year

In 2007 there will be 39,000 households with computers in the city, and that number will be increasing at a rate of 3126 households per year.

5. a. i. $S(10) \approx \$15.24$; $S'(10) \approx -\$0.02$ per week

After 10 weeks, 1 share is worth $15.24, and the value is declining by $0.02 per week.

    ii. $N(10) = 125$ shares; $N'(10) = 5$ shares per week

       After 10 weeks, the investor owns 125 shares and is buying 5 shares per week.

    iii. $V(10) = S(10)N(10) \approx \$1905$; $V'(10) = S(10)$

       $N'(10) + S'(10)N(10) \approx \$73.50$ per week

       After 10 weeks, the investor's stock is worth approximately \$1905, and the value is increasing at a rate of \$73.50 per week.

  b. $V'(x) = \left(15 + \dfrac{2.6}{x+1}\right)(0.5x) + (100 + 0.25x^2) \cdot$

    $\dfrac{-2.6}{(x+1)^2}$ dollars per week after x weeks

7. $(500 \text{ acres})(5 \text{ bushels/acre/year}) +$
$(130 \text{ bushels/acre})(50 \text{ acres/year}) = 9000 \text{ bushels/year}$

9. a. $(17{,}000)(0.48) = 8160$ voters

  b. $(8160)(0.57) \approx 4651$ votes for candidate A

  c. $17{,}000\,(0.48)(-0.03) + 17{,}000(0.57)(0.07) \approx 434$
    votes for candidate A per week

11. $f'(x) = (3x^2 + 15x + 7)(96x^2) + (6x + 15)(32x^3 + 49)$

13. $f'(x) = (12.8893x^2 + 3.7885x + 1.2548) \cdot$
$[29.685(\ln 1.7584)(1.7584^x)] + (25.7786x + 3.7885) \cdot$
$[29.685(1.7584^x)]$

15. $f'(x) = (5.7x^2 + 3.5x + 2.9)^3 \cdot$
$[-2(3.8x^2 + 5.2x + 7)^{-3}(7.6x + 5.2)] +$
$[3(5.7x^2 + 3.5x + 2.9)^2(11.4x + 3.5)](3.8x^2 + 5.2x + 7)^{-2}$

17. $f'(x) = 12.624(14.831^x)(-2x^{-3}) + 12.624(\ln 14.831) \cdot$
$(14.831^x)(x^{-2})$

19. $f'(x) = (79.32x)\left(\dfrac{1984.32(7.68)(0.859347)e^{-0.859347x}}{(1 + 7.68e^{-0.859347x})^2}\right) +$
$79.32\left(\dfrac{1984.32}{1 + 7.68e^{-0.859347x}} + 1568\right)$

21. $f'(x) = -430(0.62^x)[6.42 + 3.3(1.46^x)]^{-2}[3.3(\ln 1.46)$
$(1.46^x)] + [6.42 + 3.3(1.46^x)]^{-1}(430 \ln 0.62)(0.62^x)$

23. $f'(x) = 4x\left[\dfrac{1}{2}(3x+2)^{-1/2}(3)\right] + 4\sqrt{3x+2}$

    $= \dfrac{6x}{\sqrt{3x+2}} + 4\sqrt{3x+2}$

25. $f'(x) = -14{,}000x(1 + 12.6e^{-0.73x})^{-2}(12.6(-0.73)e^{-0.73x}) +$
$(1 + 12.6e^{-0.73x})^{-1}(14{,}000)$

    $= \dfrac{14{,}000(12.6)(0.73)xe^{-0.73x}}{(1 + 12.6e^{-0.73x})^2} + \dfrac{14{,}000}{1 + 12.6e^{-0.73x}}$

27. a. Price $= 0.049m + 1.144$ dollars m months after December

    Quantity sold $= -0.946m^2 + 0.244m + 279.911$ units sold m months after December

b. $R(m) = (0.049m + 1.144)(-0.946m^2 + 0.244m + 279.911)$ dollars of revenue m months after December

c. $R(8) \approx \$340.05$; $R(9) \approx \$325.78$

d. Because the revenue in September is less than that in August, the rate of change in August in probably negative.

e. $R'(m) = (0.049m + 1.144)(-1.893m + 0.244) + 0.049(-0.946m^2 + 0.244m + 279.911)$ dollars of revenue per month m months after December

f. $R'(2) \approx \$9.17$ per month
   $R'(8) \approx -\$12.04$ per month
   $R'(9) \approx -\$16.55$ per month

29. a. $P(t) = 0.023t^2 + 1.764t + 205.895$ million people t years after 1970

  b. $F(t) = (0.023t^2 + 1.764t + 205.895) \cdot$
$$\left[\dfrac{6.53(0.941163)^t + 22}{100}\right]$$
million people t years after 1970

  c. $F'(t) = (0.023t^2 + 1.764t + 205.895) \cdot$
$$\left(\dfrac{6.53}{100}(\ln 0.941163)(0.941163^t)\right) + (0.046t + 1.764) \cdot$$
$$\left(\dfrac{6.53}{100}(0.941163^t) + 22\right)$$
million people per year t years after 1970

  d. 1990: $F'(20) \approx 0.35$ million people per year
     1995: $F'(25) \approx 0.46$ million people per year
     2000: $F'(30) \approx 0.55$ million people per year

31. a. $m(x) = 0.209x + 8.208$ million men 65 years or older x years after 1970

    $p(x) = \dfrac{0.0223x^2 - 1.085x + 20.07}{100}$ percentage (expressed as a decimal) of men 65 or older below poverty level x years after 1970

  b. $n(x) = m(x)p(x)$
$$= (0.209x + 8.208)\left[\dfrac{0.0223x^2 - 1.085x + 20.07}{100}\right]$$
million men 65 or older below poverty level

  c. 1990: $n'(20) \approx -0.009$ million men per year
     2000: $n'(30) \approx 0.05$ million men per year

33. a. $E(x) = \dfrac{0.73(1.2912^x) + 8}{100}(-0.026x^2 - 3.842x + 538.868)$ women receiving epidurals at the Arizona hospital x years after 1980

    $E'(x) = \dfrac{0.73(1.2912^x) + 8}{100}(-0.052x - 3.842) +$
$\dfrac{0.73(\ln 1.2912)(1.2912^x)}{100}(-0.026x^2 - 3.842x + 538.868)$ women per year x years after 1980

b. Increasing by $p'(17) \approx 14.4$ percentage points per year

c. Decreasing by approximately 5 births per year ($b'(17) \approx -4.7$)

d. Increasing by $E'(17) \approx 64$ women per year

e. Profit $= \$57 \cdot E(17) \approx \$17,043$ (using a value of 299 for the number of births) or $\$17,071$ (using an unrounded number of births)

35. a. $E(x) = -151.516x^3 + 2060.988x^2 - 8819.062x + 195,291.201$ students enrolled x years after the 1980–81 school year

$D(x) = -14.271x^3 + 213.882x^2 - 1393.655x + 11,697.292$ students dropping out x years after the 1980–81 school year

b. $P(x) = \dfrac{D(x)}{E(x)} \cdot 100$ percent x years after the 1980–81 school year

c. $P'(x) = D(x)(-1[E(x)]^{-2}E'(x)) + D'(x)[E(x)]^{-1}$ percentage points per year x years after the 1980–81 school year

d.

x	$P'(x)$ (percentage points per year)	x	$P'(x)$ (percentage points per year)
0	-0.44	5	-0.19
1	-0.38	6	-0.19
2	-0.32	7	-0.22
3	-0.26	8	-0.29
4	-0.21	9	-0.41

In the 1980–81 school year, the rate of change was most negative with a value of -0.44 percentage points per year. This is the most rapid decline during this time period. The rate of change was least negative in the 1985–86 school year with a value of -0.187 percentage points per year.

e. Negative rates of change indicate that high school attrition in South Carolina was improving during the 1980s.

37. The inputs must correspond in order for the result of the multiplication to be meaningful.

CHAPTER 4 REVIEW TEST

1. a. $x \approx 0.8$

b. Positive slope: $0.8 < x < 2$
Negative slope: $-3 < x < 0, 0 < x < 0.8$

c. $x = 0$

d. $f'(-2) \approx -4, f'(1) \approx 1.1$

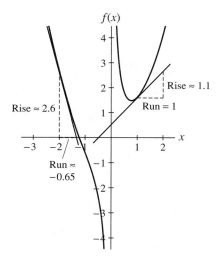

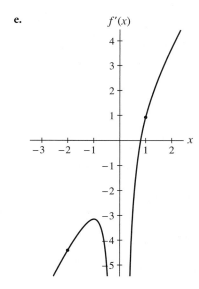

e.

2. a. $\dfrac{dD}{dt} = \dfrac{8.101(214.8)(0.797)e^{-0.797}}{(1 + 214.8e^{-0.797})^2}$ pounds per person per year t years after 1980

b. $D'(10) \approx 0.4$ pound per person per year. In 1990 the average annual per capita consumption of turkey in the United States was increasing by 0.4 pound per person per year.

c. $D'(21) \approx 0.00007$ pound per person per year. There was essentially no growth in the per capita consumption of turkey in 2001.

3. a. $\dfrac{6890 - 4865}{2000 - 1996} = 506.3$ billion per year

b.

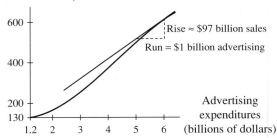

A(t)
(billions of dollars)

c. $A'(18) \approx \$496.4$ billion per year

4. a. Rewrite $P(t)$ as $P(t) = 100N(t)[A(t)]^{-1}$. Use the Product Rule to find the derivative.

$$P'(t) = 100N(t)\{-1[A(t)]^{-2}A'(t)\} + 100N'(t)[A(t)]^{-1}$$

percentage points per year t years after 1980

b. Input units: years
Output units: percentage points per year

CHAPTER 5

Section 5.1

1. $32\% - (4 \text{ percentage points per hour})\left(\frac{1}{3} \text{ hour}\right) \approx 30.7\%$

3. $f(3.5) \approx f(3) + f'(3)(0.5)$
$= 17 + (4.6)(0.5) = 19.3$

5. a. Increasing production from 500 to 501 units will increase cost by approximately \$17.

b. If sales increase from 150 to 151 units, then profit will increase by approximately \$4.75.

7. A marginal profit of -\$4 per shirt means that the fraternity is currently losing \$4 for each additional shirt sold. The fraternity should consider selling fewer shirts or increasing the sales price.

9.   Premium
(dollars)

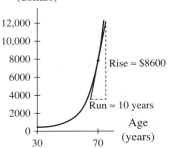

Slope of tangent line $\approx \$860$ per year of age
Annual premium for 70-year-old $\approx \$7850$
Premium for 72-year-old $\approx \$7850 + 2(\$860) = \$9570$
(Answers will vary.)

11. a.

Revenue
(billions of dollars)

Slope of tangent line ≈ 97 billion dollars of sales per billion advertising dollars
Revenue when \$6 billion is spent on advertising $\approx \$614$ billion
Revenue when 6.5 billion is spent on advertising $\approx 614 + \frac{1}{2}(97) = \$662.5$ billion
(Answers will vary.)

b. $R(6.5) \approx \$658$ billion

c. Answers will vary.

13. a. In 2000 the population of South Carolina was increasing by 53.6 thousand people per year.

b. Between 2000 and 2003, the population increased by approximately 160.8 thousand people.

c. By finding the slope of the tangent line at 2000 and multiplying by 3, we determine the change in the tangent line from 2000 through 2003 and use that change to estimate the change in the population function.

15. a. The population was growing at a rate of about 2.53 million people per year in 1998.

b. Between 1998 and 1999, the population of Mexico increased by approximately 2.53 million people.

17. a. In 1998 the amount was increasing by 1.15 million pieces per year.

b. We would expect an increase of approximately 1.15 million pieces between 1998 and 1999.

c. $p(24) - p(23) \approx 1.3$ million pieces

d. $101.9 - 100.4 = 1.5$ million pieces

e. As long as the data in part d were correctly reported, the answer to part d is the most accurate one.

19. a. $R(x) = (-7.032 \cdot 10^{-4})x^2 + 1.666x + 47.130$ dollars when x hot dogs are sold

b. Cost: $c(x) = 0.5x$ dollars per hot dog

Profit: $p(x) = R(x) - c(x) = (-7.032 \cdot 10^{-4})x^2 + 1.166x + 47.130$ dollars per hot dog

c.

x (hot dogs)	R'(x) (dollars per hot dog)	c'(x) (dollars per hot dog)	p'(x) (dollars per hot dog)
200	1.38	0.50	0.88
800	0.54	0.50	0.04
1100	0.12	0.50	-0.38
1400	-0.30	0.50	-0.80

If the number of hot dogs sold increases from 200 to 201, the revenue increases by approximately $1.38 and the profit increases by approximately $0.88. If the number increases from 800 to 801, the revenue increases by $0.54, but the profit sees almost no increase (4 cents). If the number increases from 1100 to 1101, the increase in revenue is only about 12 cents. Because this marginal revenue is less than the marginal cost at a sales level of 1100, the result of the sales increase from 1100 to 1101 is a decrease of $0.38 in profit. If the number of hot dogs increases from 1400 to 1401, revenue declines by approximately 30 cents and profit declines by approximately 80 cents.

d.   Dollars

The marginal values in part c are the slopes of the graphs shown in the figure above. For example, at $x =$ 800, the slope of the revenue graph is $0.54 per hot dog, the slope of the cost graph is $0.50 per hot dog, and the slope of the profit graph is $0.04 per hot dog. We see from the graph that maximum profit is realized when about 800 hot dogs are sold.

Revenue is greatest near $x = 1100$, so the marginal revenue there is small. However, once costs are factored in, the profit is actually declining at this sales level. This is illustrated by the graph.

21. a. United States: $A(t) = 0.109t^3 - 1.555t^2 + 10.927t + 100.320$; Canada: $C(t) = 0.150t^3 - 2.171t^2 + 15.814t + 99.650$; Peru: $P(t) = 85.112(2.01325^t)$; Brazil: $B(t) = 73.430(2.61594^t)$. For all models t is the number of years since 1980.

b, c.

	U.S.	Canada	Peru	Brazil
Rate of change in 1987 (CPI points per year)	5.2	7.5	7984	59,193
1988 CPI estimate	143	163	19,134*	136,451*

*Based on 1987 data rather than $P(7)$ and $B(7)$ because the models differ significantly from the data.

23. a. $P(A) = -0.158A^3 + 5.235A^2 - 23.056A + 154.884$ thousand dollars of profit when A thousand dollars is spent on advertising

b. When $10,000 is spent on advertising, profit is increasing by $34.3 thousand per thousand advertising dollars. If advertising is increased from $10,000 to $11,000, the car dealership can expect an approximate monthly increase in profit of $34,300.

c. When $18,000 is spent on advertising, revenue is increasing by $12.0 thousand per thousand advertising dollars. If advertising is increased from $18,000 to $19,000, the sporting goods company can expect an approximate monthly increase in profit of $12,000.

25. a. $A(t) = 300\left(1 + \dfrac{0.065}{12}\right)^{12t}$ dollars after t years

b. $A(t) \approx 300(1.066972^t)$ dollars after t years

c. $A(2) \approx \$341.53$

d. $A'(2) \approx \$22.14$ per year

e. $\dfrac{1}{4}A'(2) \approx \$5.53$

27. Answers will vary.

29. Answers will vary.

Section 5.2

1. Quadratic, cubic, product, quotient, composite and piecewise functions could have relative maxima or minima.

3. *y*

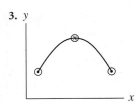

The derivative is zero at the absolute maximum point.

5. *y*

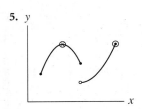

The derivative is zero at the absolute maximum point marked with an *X*.

7. *y*

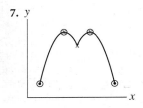

The derivative is zero at both absolute maximum points. The derivative does not exist at the relative minimum point.

9. Answers will vary. One such graph is $y = x^3$.

11. a. All statements are true.

b. $f'(2)$ does not exist because f is not continuous at $x = 2$.

c. $f'(x)$ is less than zero for $x < 2$ because the graph to the left of $x = 2$ is decreasing.

d. $f'(2)$ does not exist because f is not smooth at $x = 2$.

13. Answers will vary. One possibility is

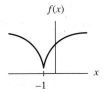

$f(x)$

15. Answers will vary. One possibility is

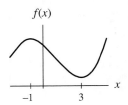

$f(x)$

17. a. The relative maximum value is approximately 19.888, which occurs at $x \approx 3.633$. The relative minimum value is approximately 11.779, which occurs at $x \approx 11.034$.

b. The absolute maximum and minimum are the relative maximum and minimum found in part *a*.

c.

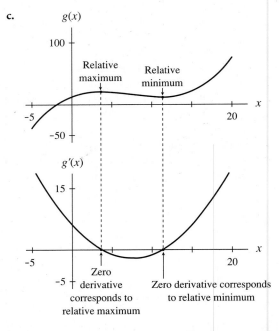

19. a. The greatest percentage of eggs hatching (95.6%) occurs at 9.4°C.

b. 9.4°C ≈ 49°F

21. a. $C(0) \approx 123$ cfs and $C(24) \approx 140$ cfs

b. The highest flow rate is about 387.6 cfs; it occurs when $h \approx 8.9$ hours. The lowest flow rate is about 121.3 cfs; it occurs when $h \approx 0.4$ hour.

23. a. $S(x) = 0.181x^2 - 8.463x + 147.376$ seconds at age x years.

b. The model gives a minimum time of 48.5 seconds occurring at 23.4 years.

c. The minimum time in the table is 49 seconds, which occurs at 24 years of age.

25. a. An exponential model for the data is $R(p) = 316.765(0.949^p)$ dozen roses when the price per dozen is p dollars.

b. $E(p) = 316.765p(0.949^p)$ dollars spent on roses each week when the price per dozen is p dollars

c. A price of $19.16 maximizes consumer expenditure.

d. A price of $25.16 maximizes profit.

e. Marginal values are with respect to the number of units sold or produced. In this activity, the input is price, so derivatives are with respect to price and are not marginals.

27. a. $G(t) = 0.008t^3 - 0.347t^2 + 6.108t + 79.690$ million tons of garbage taken to a landfill t years after 1975

b. $G'(t) = 0.025t^2 - 0.693t + 6.108$ million tons of garbage per year t years after 1975

c. In 2005, the amount of garbage was increasing by 8.1 million tons per year.

d.

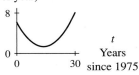

Because the derivative graph exists for all input and never crosses the horizontal axis, $G(t)$ has no relative maxima.

29. a. $A(p) = 568.074(0.965582^p)$ tickets sold on average when the price is p dollars
$A(p) = 0.15p^2 - 16.007p + 543.286$ tickets sold on average when the price is p dollars

  The exponential model probably better reflects the probable attendance if the price is raised beyond $35 because attendance is likely to continue to decline. (The quadratic model will begin to increase around $53.)

b. $R(p) = 568.074p(0.965582^p)$ dollars of revenue when the ticket price is p dollars

c. (28.55, 5966.86) is the maximum point on the revenue graph. This corresponds to a ticket price of $28.55, which results in revenue of approximately $5967. The resulting average attendance is approximately 209.

31. The absolute minimum is approximately -6.3. There is no absolute maximum because $\lim\limits_{x \to \pm\infty} y \to \infty$.

33. Answer located on the *Calculus Concepts* CD-ROM and web site.

Section 5.3

1. a. Visual estimates of the inflection points are (1982, 25) and (2018, 25).

b. The input values of the inflection points are the years in which the rate of crude oil production is estimated to be increasing and decreasing most rapidly. We estimate that the rate of production was increasing most rapidly in 1982, when production was approximately 25 billion barrels per year, and that it will be decreasing most rapidly in 2018, when production is estimated to be approximately 25 billion barrels per year.

3. a. Derivative
b. Function
c. Second derivative

5. a. Second derivative
b. Derivative
c. Function

7. a.

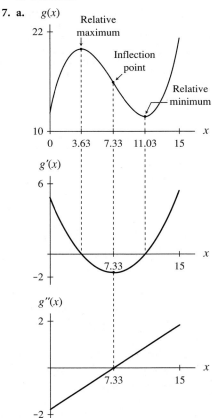

b. The inflection point on the graph of g is approximately (7.333, 15.834). This is a point of most rapid decline.

9. a. The inflection point is approximately (1.838, 22.5). After about 1.8 hours of study (1 hour and 50 minutes), the percentage of new material being retained is increasing most rapidly. At that time, approximately 22.5% of the material has been retained.

b. The answer agrees with the one given in the discussion at the end of the section.

11. a.

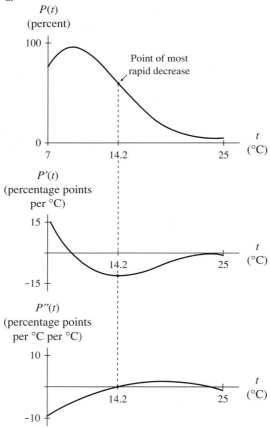

b. Because the graph of P'' crosses the t-axis twice, there are two inflection points. These are approximately (14.2, 59.4) and (23.6, 5.8). The point of most rapid decrease is (14.2, 59.4). (The other inflection point is a point of least rapid decrease.) The most rapid decrease occurs at 14.2°C, when 59.4% of eggs hatch. At this temperature, the percentage of eggs hatching is declining by 11.1 percentage points per °C. A small increase in temperature will result in a relatively large increase in the percentage of eggs not hatching.

13. a.

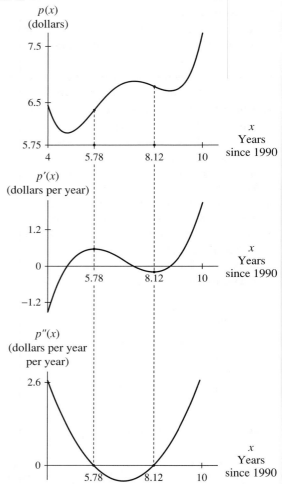

The maximum and minimum points of the graph of p' and the x-intercepts of the graph of p'' correspond to the inflection points of the graph of p.

b. The x-intercepts of p'' are $x \approx 5.78$ and $x \approx 8.12$. These are the inputs of the inflection points of the graph of p. This function can be considered essentially continuous without restriction if you consider that it models the average price on a daily basis. If you assume it models a yearly average price, then it must be interpreted discretely. We are not given enough information to know for certain, so either one or the other assumption needs to be made.

In the continuous without restriction case, we conclude that the price was declining most rapidly in January of 1999 ($x \approx 8.12$) at a rate of -$0.19 per year and increasing most rapidly in September of 1996

($x \approx 5.78$) at a rate of $0.57 per year. In the case of discrete interpretation, we compare $p'(8)$ and $p'(9)$ and conclude that the price was declining most rapidly in 1998 at a rate of -$0.19 per year. We compare $p'(5)$ with $p'(6)$ and conclude that the price was increasing most rapidly in 1996 at a rate of $0.55 per year.

c. The price was declining most rapidly at the end of 1994 at a rate of -$1.45 per year and increasing most rapidly at the end of 2000 at a rate of $2.10 per year.

d. The price was declining most rapidly in 1998 at a rate of -$0.19 per year and increasing most rapidly in 1996 at a rate of $0.55 per year.

15. a. (0.418, 9740.089) is a relative minimum point, and (13.121, 20,242.033) is a relative maximum point on the cubic model.

b. The inflection point is approximately (6.8, 14,991.1).

c. **i.** The inflection point occurs between 1981 and 1982, shortly after the team won the National Championship. This is when the number of donors was increasing most rapidly.

ii. The relative maximum occurred around the same time that a new coach was hired. After this time, the number of donors declined.

17. a. The greatest rate occurs at $h \approx 3.733$, or approximately 3 hours and 44 minutes after she began working.

b. Her employer may wish to give her a break after 4 hours to prevent a decline in her productivity.

19. a. $H(w) = \dfrac{10{,}111.102}{1 + 1153.222e^{-0.727966w}}$ total labor hours after w weeks

b. $H'(w) = \dfrac{10{,}111.102(1153.222)(0.727966)e^{-0.727966w}}{(1 + 1153.222e^{-0.727966w})^2}$

labor hours per week after w weeks

c.

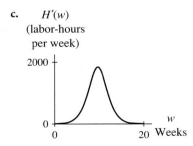

$H'(w)$
(labor-hours per week)

The derivative gives the manager information about the number of labor hours spent each week.

d. The maximum point is about (9.685, 1840.134). In the tenth week the most labor hours are needed. That number is $H'(10) \approx 1816$ labor hours.

e. The point of most rapid increase on the graph of H' is (7.876, 1226.756). This occurs approximately 8 weeks into the job, and the number of labor hours per week is increasing by approximately $H''(8) \approx 513$ labor hours per week per week.

f. The point of most rapid decrease on the graph of H' is (11.494, 1226.756). This occurs approximately 12 weeks into the job, and the number of labor hours per week is changing by about $H''(12) = -486$ labor hours per week per week.

g. By solving the equation $H'''(w) = 0$, we can find the input values that correspond to a maximum or minimum point on the graph of H'', which corresponds to inflection points on the graph of H', the weekly labor hour curve.

h. The second job should begin about 4 weeks into the first job.

21. a. Between 1980 and 1985, the average rate of change was smallest at 1 million tons per year.

b. $g(t) = 0.008t^3 - 0.347t^2 + 6.108t + 79.690$ million tons t years after 1970

c. $g''(t) = 0.051t - 0.693$ million tons per year per year t years after 1970

d. Solving $g''(t) = 0$ gives $t \approx 13.684$ and $g(13.684) \approx 120$ million tons of garbage.

e.

$g'(t)$
(millions of tons per year)

10

1

0   13.68   30 t Years since 1970

$g''(t)$
(millions of tons per year per year)

1

13.68   30 t Years since 1970

-1

f. The year with the smallest rate of change is 1984, with $g(14) \approx 120$ million tons of garbage, increasing at a rate of $g'(14) \approx 1.4$ million tons per year.

23. a. The first differences are greatest between 6 and 10 minutes, indicating the most rapid increase in activity.

   b. $A(m) = \dfrac{1.930}{1 + 31.720e^{-0.439118m}}$ U/100μL

   m minutes after the mixture reaches 95°C; The inflection point is (7.872, 0.965). After approximately 7.9 minutes, the activity was approximately 0.97 U/100μL and was increasing most rapidly at a rate of approximately 0.212 U/100μL/min.

   Because the graph of g'' crosses the t-axis at 13.68, we know that input corresponds to an inflection point of the graph of g. Because $g'(t)$ is a minimum at that same value, we know that it corresponds to a point of slowest increase on the graph of g.

25. a. $L'(x) = \dfrac{-251.3(0.1376)(0.2854)e^{0.2854x}}{(1 + 0.1376e^{0.2854x})^2}$ million tons

   per year x years after 1970

   b. 1970: decreasing by 7.63 million tons per year
   1995: decreasing by 0.41 million tons per year

   c. Emissions were declining most rapidly in 1977 at a rate of 17.9 million tons per year. At that time, yearly emissions were 124.7 million tons.

27. The graph of f is always concave up. A concave-up parabola fits this description.

29. a. The graph is concave up to the left of $x = 2$ and concave down to the right of $x = 2$.

   b.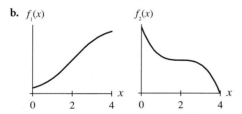

31. Cubic and logistic models have inflection points. In some cases product, quotient, and composite functions may have inflection points.

33. Answer located on the *Calculus Concepts* CD-ROM and web site.

Section 5.4

1. The equation for surface area in terms of diameter d is $S = \pi dh + \pi\left(\dfrac{d}{2}\right)^2 + \pi\left(\dfrac{d}{2} + \dfrac{9}{8}\right)^2$, and the volume equation is $V = \pi\left(\dfrac{d}{2}\right)^2 h = 808.5$. Solving the volume

equation for h in terms of d gives $h = \dfrac{808.5}{\pi\left(\dfrac{d}{2}\right)^2} = \dfrac{3234}{\pi d^2}$.

Substituting this expression into the surface area equation results in the equation we seek to optimize:

$$S = \dfrac{3234}{d} + \pi\left(\dfrac{d}{2}\right)^2 + \pi\left(\dfrac{d}{2} + \dfrac{9}{8}\right)^2$$

Setting the derivative equal to zero and solving for d gives $d \approx 9.73568$, which is twice the optimal radius found in Example 3.

3. 450 square feet, occurring when one side is 30 feet and each of the other two sides is 15 feet.

5. a. A corner cut of approximately 1.47 inches will result in the maximum volume of 52.5 cubic inches of sand.

   b. Corner cuts of approximately 1.12 inches and 1.86 inches will result in boxes with volume 50 cubic inches.

   c. Answers will vary.

7. a. 42 cases

   b. $\dfrac{42}{x}$ orders rounded up to the nearest integer

   c. $\dfrac{42}{x}$($12 per order) $= \dfrac{504}{x}$ dollars to order

   d. $\left(\dfrac{x}{2}\text{ cases to store}\right)$($4 per case to store) $= 2x$ dollars to store

   e. $C(x) = \dfrac{504}{x} + 2x$ dollars when x cases are ordered

   f. $C(x)$ is a minimum when $x \approx 15.9$ cases. The manager must order 3 times a year. Because the order sizes are not all the same (16 cases ordered 3 times a year are more cases than the cafeteria needs), we cannot just substitute the optimal value of x into the cost equation to obtain the total cost. There are two most cost-effective ways to order: 15, 15, and 12 cases, or 16, 16, and 10 cases (assuming that the storage fees can be calculated as $1.33 per case for 4 months). The total cost associated with each option is $63.93.

9. A radius of approximately 1.2 inches and a height of approximately 4.8 inches will minimize the cost.

11. a. The chain link side should be approximately 6.6 feet, and the cinder block sides should be approximately 18.2 feet.

   b. The chain link side should be approximately 5.7 feet, and the cinder block sides should be approximately 21 feet.

   c. The answer to part *a* does not change. The answer to part *b* becomes: The chain link side should be 6 feet and the cinder block sides should be 20 feet.

13. a. $R(n) = 350 + 2n(44 - n) + 35n$ dollars when n students go on the trip

b. $S(n) = 350 + 2n(44 - n)$ dollars when n students go on the trip

c. The bus company's revenue is maximized when 31 students go on the trip and is minimized when 10 students go on the trip.

d. The sorority pays the most when 22 students go and pays the least when 44 students go.

15. a. $m(s) = -0.0015s^2 + 0.1043s + 3.5997$ mpg at a speed of s mph

b. i. $\frac{400}{s}$ hours

   ii. $\frac{6200}{s}$ dollars

   iii. $\frac{400}{m(s)}$ gallons, where $m(s)$ is the output of the model in part a

   iv. $\frac{460}{m(s)}$ dollars (assuming a price of $1.15 per gallon)

   v. $C(s) = \frac{6200}{s} + \frac{460}{m(s)}$ dollars

c. 60.3 miles per hour

d. 700-mile trip: 1200-mile trip:

   i. $\frac{700}{s}$ hours   **i.** $\frac{1200}{s}$ hours

   ii. $\frac{10,850}{s}$ dollars   **ii.** $\frac{18,600}{s}$ dollars

   iii. $\frac{700}{m(s)}$ gallons   **iii.** $\frac{1200}{m(s)}$ gallons

   iv. $\frac{805}{m(s)}$ dollars   **iv.** $\frac{1380}{m(s)}$ dollars

   (assuming a price of $1.15 per gallon)

   v. 700-mile trip: $C(s) = \frac{10,850}{s} + \frac{805}{m(s)}$ dollars

   1200-mile trip: $C(s) = \frac{18,600}{s} + \frac{1380}{m(s)}$ dollars

   Optimal speed: Optimal speed:
   60.3 miles per hour   60.3 miles per hour
   The optimal speed remains constant. It does not depend on trip length.

e. Price of gas: Optimal speed for 400-mile trip:
   $1.35/gallon   58.8 mph
   $1.55/gallon   57.5 mph
   $1.75/gallon   56.4 mph
   As the gas price increases, the optimal speed decreases.

f. Wages of driver: Optimal speed for 400-mile trip:
   $17.50/hour   61.4 mph
   $20.50/hour   62.9 mph
   $25.50/hour   64.9 mph
   As the wages increase, the optimal speed increases.

17. Answers will vary.

Section 5.5

1. $\frac{df}{dt} = 3\frac{dx}{dt}$ **3.** $\frac{dk}{dy} = 12x\frac{dx}{dy}$ **5.** $\frac{dg}{dt} = 3e^{3x}\frac{dx}{dt}$

7. $\frac{df}{dt} = 62(\ln 1.02)(1.02^x)\frac{dx}{dt}$

9. $\frac{dh}{dy} = 6\frac{da}{dy} + 6\ln a\frac{da}{dy} = 6(1 + \ln a)\frac{da}{dy}$

11. $\frac{ds}{dt} = \frac{\pi r h}{\sqrt{r^2 + h^2}}\frac{dh}{dt}$

13. $0 = \frac{\pi r}{\sqrt{r^2 + h^2}}\left(r\frac{dr}{dt} + h\frac{dh}{dt}\right) + \pi\sqrt{r^2 + h^2}\frac{dr}{dt}$

15. a. Approximately 52.4 gallons per day

b. $\frac{dw}{dt} \approx 0.4323$; The amount of water transpired is increasing by approximately 0.4323 gallon per day per year. In other words, in 1 year, the tree will be transpiring about 0.4 gallon more each day than it currently is transpiring.

17. a. $B = \frac{45}{0.00064516h^2}$ points

b. $\frac{dB}{dt} = \frac{-90}{0.00064516h^3}\frac{dh}{dt}$

c. $\frac{dB}{dt} \approx -0.2789$ point per year

19. a. Approximately 0.0014 cubic foot per year

b. Approximately 0.0347 cubic foot per year

21. a. $L = \left(\frac{M}{48.10352\,K^{0.4}}\right)^{5/3} = \left(\frac{M}{48.10352}\right)^{5/3}K^{-2/3}$

b. $\frac{dL}{dt} = \left(\frac{M}{48.10352}\right)^{5/3}\left(\frac{-2}{3}K^{-5/3}\right)\frac{dk}{dt}$

c. The number of worker hours should be decreasing by approximately 57 worker hours per year.

23. The balloon is approximately 1529.7 feet from the observer, and that distance is increasing by approximately 1.96 feet per second.

25. The runner is approximately 67.08 feet from home plate, and that distance is decreasing by about 9.84 feet per second.

27. a. Approximately 4188.79 cubic centimeters

b. By about -167.6 cubic centimeters per minute

29. Approximately 5.305 centimeters per second

31. Begin by solving for h: $h = \frac{V}{\pi r^2} = \frac{V}{\pi}r^{-2}$

   Differentiate with respect to t (V is constant):

   $\frac{dh}{dt} = \frac{V}{\pi}(-2r^{-3})\frac{dr}{dt}$

   Substitute $\pi r^2 h$ for V: $\frac{dh}{dt} = \frac{\pi r^2 h}{\pi}(-2r^{-3})\frac{dr}{dt}$

Simplify: $\dfrac{dh}{dt} = \dfrac{-2h}{r}\dfrac{dr}{dt}$

Rewrite: $\dfrac{dr}{dt} = \dfrac{r}{-2h}\dfrac{dh}{dt}$

33. Answers will vary.

CHAPTER 5 REVIEW TEST

1. a. T has a relative maximum point at (0.682, 143.098) and a relative minimum point at (3.160, 120.687). These points can be determined by finding the values of x between 0 and 6 at which the graph of T' crosses the x-axis. (There is also a relative maximum to the right of $x = 6$.)

  b. T has two inflection points: (1.762, 132.939) and (5.143, 149.067). These points can be determined by finding the values of x between 0 and 6 at which the graph of T'' crosses the x-axis. These are also the points at which T' has a relative maximum and relative minimum.

  c.

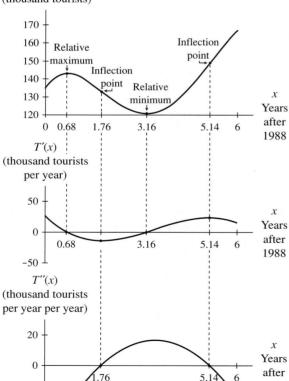

d. The number of tourists was greatest in 1994 at 166.8 thousand tourists. The number was least in 1991 at 120.9 thousand.

e. The number of tourists was increasing most rapidly in 1993 at a rate of 23.1 thousand tourists per year. The number of tourists was decreasing most rapidly in 1990 at a rate of 13.3 thousand tourists per year.

2. a. $(4.5 \text{ thousand people per year})\left(\dfrac{1}{4}\text{ year}\right) = 1.125$ thousand people

  b. $202 + \dfrac{1}{2}(4.5) = 204.25$ thousand people

3. The optimal distance is $x \approx 2.89$ miles. The total cost is $1,642,527.

4.

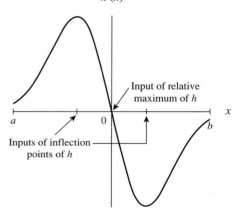

$h'(x)$

5. $\dfrac{ds}{dt} = 0.000013(2wv)\dfrac{dv}{dt}$
  $= -31.2$ feet per second
  The length of the skid marks is decreasing by 31.2 feet per second.

CHAPTER 6

Section 6.1

1. a. Thousand bacteria per hour
  b. Hours
  c. Thousand bacteria
  d. Thousand bacteria
  e. Thousand bacteria

3. a. The area would represent how much farther a car going 60 mph would require to stop than a car going 40 mph.

  b. i. The heights are in feet per mile per hour, and the widths are in miles per hour.
    ii. The area is in feet.

5. a. On the horizontal axis, mark integer values of x between 0 and 8. Construct a rectangle with width from $x = 0$ to $x = 1$ and height $f(1)$. Because the width of the rectangle is 1, the area will be the same as the height. Repeat the rectangle constructions between each pair of consecutive integer input values. Note that for the rectangles that lie below the horizontal axis, the heights are the absolute values of the function values. Also note that the fourth rectangle has height 0. Sum the areas (heights) of the rectangles to obtain the area estimate.

b. Repeat part *a*, except that the height of each rectangle is determined by the function value corresponding to the left side of the interval. In the case of the first rectangle, the height is $f(0)$. When we use left rectangles, the fifth rectangle has height 0.

7. Divide the interval from a to b into four equal subintervals. Determine the midpoint of each subinterval, and substitute into the function to find the heights of the rectangles. Multiply each height by the width of the subintervals, and add the four resulting areas.

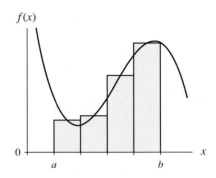

9. Answers will vary.

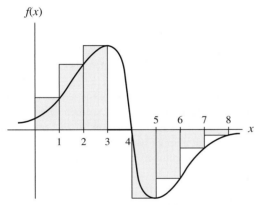

Activity 5 part *a*: Right rectangles

11. a.   Velocity (feet per minute)

b. 2420 feet

c. The robot traveled 2420 feet during the $3\frac{1}{2}$-minute experiment.

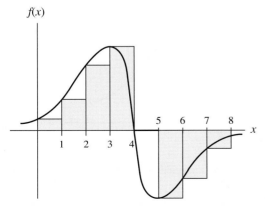

Activity 5 part *b*: Left rectangles

13. a.   Speed (mph)

Distance traveled $\approx \dfrac{115}{6} \approx 19.2$ miles

b.

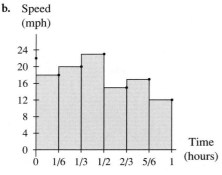

Speed (mph)

Time (hours)

$$\text{Distance traveled} \approx \frac{105}{6} = 17.5 \text{ miles}$$

15. a.

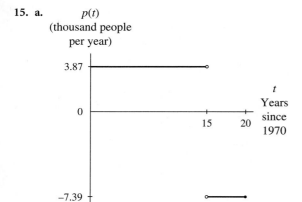

$p(t)$
(thousand people per year)

t
Years since 1970

b. The population of North Dakota grew by about 58.1 thousand people between 1970 and 1985.

c. The population of North Dakota declined by about 37.0 thousand people between 1985 and 1990.

d. The population was $58.1 - 37.0 = 21.1$ thousand people greater in 1990 than it was in 1970.

e. You would need the population in some year between 1970 and 1990.

17. a.

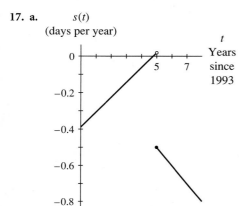

$s(t)$
(days per year)

t
Years since 1993

b. The rate-of-change graph is negative for approximately one quarter of 1998. However, these data should be discretely interpreted. The average hospital stay decreased between 1993 and 2000.

c. Area above ≈ 0.0024 day

d. Area below ≈ 1.127 days

e. The average hospital stay decreased by approximately 1.12 days between 1993 and 2000. We cannot determine the average stay in 2000 without knowing the average length of stay for some year between 1993 and 2000.

19. a.

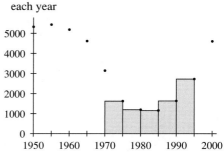

Births each year

From the end of 1970 through the end of 1995, there were 57,505 live births to U.S. women 45 years of age and older. This approximation to the actual value uses the number of births at the end of each 5-year period as the number of births each year during that period.

b. $B(x) = 0.279x^3 - 16.189x^2 + 101.463x + 5383.594$ births x years after 1950

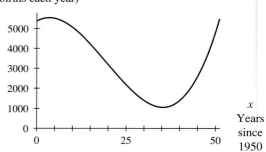

$B(x)$
(births each year)

x
Years since 1950

c. Using 36 rectangles, we estimate the number of births to be 83,888.

d. To find the exact total, we would need exact data for all years from 1964 through 2000.

21. a. To convert the air speeds to miles per second, multiply by 1.15 mph per knot and divide by 3600 seconds per hour.

b. $S(t) = -(2.108 \cdot 10^{-5})t^2 + (1.554 \cdot 10^{-3})t - 9.880 \cdot 10^{-5}$ miles per second t seconds after the plane began to taxi

c. Using 18 midpoint rectangles, the area is about 0.209 mile. Answers may vary.

d. It took approximately 0.2 mile of runway for the Cessna to taxi for takeoff (assuming no headwind).

23. a.

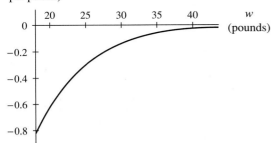

$P(w)$
(percentage points per pound)

b. The percentage of low-birthweight babies was declining as the mother's weight gain increased from 18 to 43 pounds.

c. The signed area is approximately -5.3. When the mother gained between 18 and 43 pounds during pregnancy, the percentage of low-birthweight babies decreased by approximately 5.3 percentage points.

d. We cannot answer this question with the information given.

25. a. Declining positive rate-of-change data indicate that life expectancies were increasing at a slower and slower rate.

b. $E(t) = (4.199 \cdot 10^{-4})t^2 - 0.022t + 0.359$ years per year t years after 1970

c. Life expectancy for women increased approximately 5.8 years from 1970 through 2010.

27. a. $A \approx 43.799$ days after Sept. 30, 1995
$B \approx 273.382$ days after Sept. 30, 1995

b. In about 44 days after Sept. 30, 1995, the level of the lake fell by approximately 0.398 foot.

c. Between about 44 and 273 days after Sept. 30, 1995, the lake level rose by about 3.250 feet.

d. The lake level was approximately $3.250 - 0.398 \approx 2.853$ feet higher 273 days after Sept. 30, 1995.

29. a.

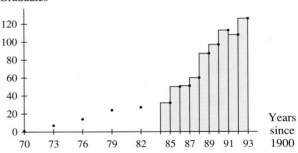

Graduates

There were 724 graduates from the beginning of 1985 through the end of 1993. This is exact as long as the data were correctly reported.

b. $g(t) = 0.016t^3 - 0.233t^2 + 2.669t + 1.633$ graduates t years after 1970

c, d. Summing values of $g(t)$ for $t = 0$ through $t = 23$ gives a total of 971 graduates which is close to the actual number of 987.

31. Answers will vary.

33. Answers are located on the *Calculus Concepts* CD-ROM and web site.

Section 6.2

1. a. Thousand people
b. Thousand people
c. Thousand people

3. a. This is the change in the number of organisms when the temperature increases from 25°C to 35°C.

b. If the graph of A is always positive, this is the change in the number of organisms when the temperature increases from 30°C to 40°C.

5. a, b. Between 0 and 300 boxes and between 400 and 600 boxes

c. NA   **d.** 300   **e.** 400
f. 350   **g.** dollars   **h.** less

7. a. 17.91

b. Between 3 and 11 weeks of age, the mouse gained 17.91 grams.

c. 21.91 grams

9. a. The yield from the oil field during the first 5 years is approximately 11.29 thousand barrels.

b. The yield from the oil field during the first 10 years is approximately 12.48 thousand barrels.

c. $\int_0^5 r(t)\,dt$ and $\int_0^{10} r(t)\,dt$

d. The first 5 years account for approximately 90.5% of the first 10 years' production.

11. a. $A \approx 17.3$ seconds

b. From 0 to 17.3 seconds, the car's speed increased by approximately 174.7 feet per second (or 119.1 mph).

c. From 17.3 to 35 seconds, the car's speed decreased by approximately 94.9 feet per second (or 64.7 mph).

d. The car's speed after 35 seconds was approximately 79.8 feet per second (or 54.4 mph) faster than it was at 0 seconds.

13. a. The heights will be in meters per second, and the widths will be in microseconds.

b. The area units will be in millimeters.

c. $V(m) = -(1.589 \cdot 10^{-6})m^2 + 0.001m + 0.137$ millimeters per microsecond after m microseconds

d. The crack traveled approximately 10.2 millimeters.

e. $\int_0^{60} V(m)\,dm$

15. a. For $n = 5$, labor-hours ≈ 9859.

For $n = 10$, labor-hours $\approx 10{,}097$.

For $n = 20$, labor-hours $\approx 10{,}100$.

b. i

17. a. $\int_0^{720} c(m)\,dm \approx 2602$ customers

b. During the 12-hour sale, approximately 2602 customers entered the store.

19. a. Blood pressure rises when the rate of change is positive. In the table, this is from around 2 A.M. to almost 2 P.M. Blood pressure falls when the rate of change is negative, from around 2 P.M. to almost 2 A.M.

b. In the table, the greatest rate of change occurs at 8 A.M., and the most negative rate of change occurs at 8 P.M.

c. $B(t) = 0.030t^2 - 0.718t + 3.067$ mm Hg per hour, where t is the number of hours since 8 A.M.

d. The model is zero at $t \approx 5.59$ hours and at $t \approx 18.13$ hours. These are the times when the blood pressure is highest and lowest, respectively.

e. From 8 A.M. to 8 P.M., diastolic blood pressure rose by about 2.54 mm Hg.

f. $\int_0^{12} B(t)\,dt$

g. Not enough information to answer.

21. Answers are located on the *Calculus Concepts* CD-ROM and web site.

23. Answers will vary.

Section 6.3

1. a.

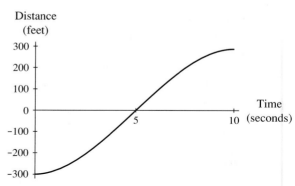

b. $D(x) = \int_5^x f(t)\,dt$

c. The accumulation function gives the distance traveled between 5 seconds and x seconds. For times before 5 seconds, the accumulation function is the negative of the distance traveled, because we are looking back in time.

3. a. The area between days 0 and 18 represents how much the price of the technology stock declined ($15.40 per share) during the first 18 trading days of 2003.

b. The area between days 18 and 47 represents how much the price of the technology stock rose ($55.80 per share) between days 18 and 47.

c. $40.40 more

d. $11.10 less

e.

x	$\int_0^x r(t)\,dt$	x	$\int_0^x r(t)\,dt$
0	0	35	15.0
8	-7.1	47	40.4
18	-15.4	55	29.3

f.

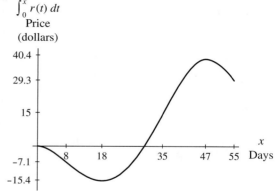

g. $156.30

5. a. No

b. The peak corresponds to the time when the number of subscribers was increasing most rapidly.

c. The number of new subscribers t days after the end of the twentieth week.

d. 280 subscribers

e.

Week	t (days)	Area	Week	t (days)	Area
4	28	350	28	196	8820
8	56	924	36	252	10,430
12	84	1960	44	308	10,920
16	112	3500	52	364	11,060
20	140	5390			

f.

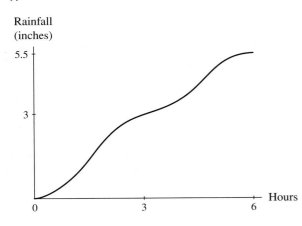

$$\int_{140}^{t} n(x)\,dx$$

Subscribers

7.

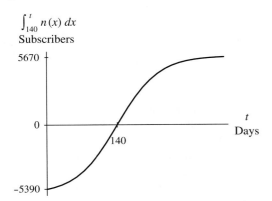

Rainfall (inches)

9. a.

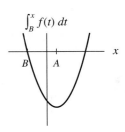

$$\int_{A}^{x} f(t)\,dt$$

b.

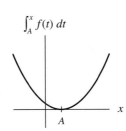

$$\int_{B}^{x} f(t)\,dt$$

11. a.

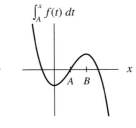

$$\int_{0}^{x} f(t)\,dt$$

b.

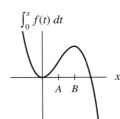

$$\int_{A}^{x} f(t)\,dt$$

c.

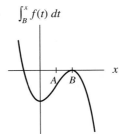

$$\int_{B}^{x} f(t)\,dt$$

13.

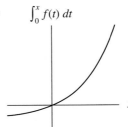

$$\int_{0}^{x} f(t)\,dt$$

15. Derivative graph: b; accumulation graph: f

17. Derivative graph: f; accumulation graph: e

19. Derivative: left table; accumulation function: right table

21. a. $\dfrac{\text{Million dollars of revenue}}{\text{Thousand advertising dollars}}$

b.

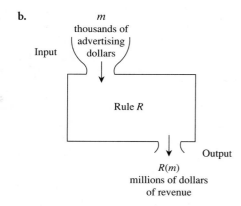

Input — *m* thousands of advertising dollars

Rule *R*

Output — *R(m)* millions of dollars of revenue

c. When *m* thousand dollars are being spent on advertising, the annual revenue is $R(m)$ million dollars.

23. a. $\dfrac{\text{Milligrams per liter}}{\text{Hour}}$

b.

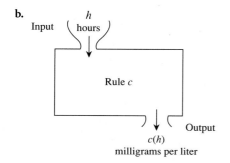

Input — *h* hours

Rule *c*

Output — *c(h)* milligrams per liter

c. The concentration of a drug in the bloodstream is $c(h)$ milligrams per liter h hours after the drug is given.

25. Answers will vary.

Section 6.4

1. a. b   **b.** a   **c.** c   **3. a.** c   **b.** a   **c.** b

5. Answers will vary.

7. Answers will vary.

9. $\displaystyle\int 19.436(1.07^x)\,dx = \dfrac{19.436(1.07^x)}{\ln 1.07} + C$

11. $\displaystyle\int [6e^x + 4(2^x)]\,dx = 6e^x + \dfrac{4(2^x)}{\ln 2} + C$

13. $\displaystyle\int (10^x + 4x^{\frac{1}{2}} + 8)\,dx = \dfrac{10^x}{\ln 10} + 4\left(\dfrac{2}{3}\right)x^{\frac{3}{2}} + 8x + C$

15. $S(m) = \dfrac{6250(0.92985^m)}{\ln 0.92985} + C$ CDs m months after the beginning of the year

17. $C(x) = 0.7925 \ln |x| + \dfrac{0.3292(0.009324^x)}{\ln 0.009324} + K$ dollars per unit when x units are produced

19. $F(t) = \dfrac{1}{3} t^3 + t^2 - 20$

21. $F(z) = \dfrac{-1}{z} + e^z + \left(\dfrac{3}{2} - e^2\right)$

23. $w(t) = 7.372 \ln t + (26 - 7.372 \ln 7)$
$\approx 7.372 \ln t + 11.65475$ grams at age $(t + 2)$ weeks

25. $G(t) = \dfrac{1.667 \cdot 10^{-4}}{3} t^3 - \dfrac{0.01472}{2} t^2 - 0.103t + 119.015$
males per 100 females t years after 1900

27. a. $P(x) = \dfrac{891.6}{\ln 1.5}(1.5^x) - 2196.887$ thousand cellular phone subscribers x years after 1988

b. $P(6) - P(2) \approx 20{,}099.8$ thousand subscribers ≈ 20 million subscribers

29. a, b. Velocity: $v(t) = -32t$ feet/second
Distance: $s(t) = -16t^2 + 540$ feet
where t is the number of seconds after the penny was dropped

c. Solving for t in $s(t) = 0$, we obtain $t \approx \pm 5.8$ seconds. The penny will hit the ground approximately 5.8 seconds after it was dropped.

d. $v(5.809) = -32(5.809) \approx -185.9$ feet/second
$$= \left(\dfrac{-185.9 \text{ feet}}{1 \text{ second}}\right)\left(\dfrac{3600 \text{ seconds}}{1 \text{ hour}}\right)\left(\dfrac{1 \text{ mile}}{5280 \text{ feet}}\right)$$
$$= \dfrac{-126.75 \text{ miles}}{1 \text{ hour}}, \text{ or } -126.75 \text{ mph}$$

31. a. The impact velocity is -64.99 feet/second, or -44.31 mph.

b. Air resistance probably accounts for the difference.

33. a. $N(x) = 593 \ln |x| + 138x - 748.397$ employees x years after 1996

b. The function in part *a* applies from 1997 $(x = 1)$ through 2002 $(x = 6)$.

c. There are two ways to estimate the number of employees the company hired. If we consider the function to be continuous with discrete interpretation, then the number can be calculated from the function n by summing the yearly totals:
$$n(1) + n(2) + n(3) + n(4) + n(5) + n(6) \approx 2281$$
We can also estimate the total number of employees hired between 1997 and 2002 as
$$\int_1^6 n(x)\,dx = N(6) - N(1) \approx 1753$$
This estimate treats the function as continuous without restriction and is, therefore, probably less accurate than the first estimate. If any employees were fired or quit between 1997 and 2002, neither estimate would represent the number of employees at the end of 2002.

Section 6.5

1. c   **3.** c   **5.** b   **7.** a

9. a.

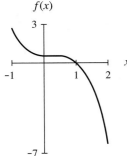

b. Area = $\int_{-1}^{1.0544} f(x)\,dx - \int_{1.0544}^{2} f(x)\,dx \approx 3.822$; No

c. −0.615

11. a.

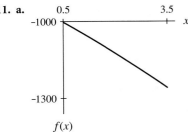

b. Area = $-\int_{0.5}^{3.5} f(x)\,dx \approx 3378.735$; No

c. −3378.735

13. $\int_{5}^{15} P(x)\,dx \approx 2305.357$

Between 1985 and 1995, the number of international calls billed in the United States increased by 2305.4 million calls.

15. $\int_{0}^{5} r(x)\,dx \approx 195.639$

The corporation's revenue increased by 195.6 million dollars between 1987 and 1992.

17. a. $\int_{0}^{70} s(t)\,dt = 7.938$

In the 70 days after April 1, the snow pack increased by 7.938 equivalent cm of water.

b. $\int_{72}^{76} s(t)\,dt = -22.768$

Between 72 and 76 days after April 1, the snow pack decreased by 22.768 equivalent centimeters of water.

c. It is not possible to find $\int_{0}^{76} s(t)\,dt$ because $s(t)$ is not defined between $t = 70$ and $t = 72$.

19. a. $\int_{0}^{0.8955} T(h)\,dh \approx 1.48$

The temperature rose 1.48°F between 8 A.M. and 8:54 A.M.

b. $\int_{0.8955}^{1.75} T(h)\,dh \approx -1.61$

After rising 1.48°F, the temperature then fell 1.61°F during the field trip.

c. No, the highest temperature reached was $71 + 1.48 = 72.48$°F.

21. a. An exponential model for the data is $f(x) = 0.161(1.076186^{x})$ trillion cubic feet per year x years after 1900.

b. From 1940 through 1960, 138.3 trillion cubic feet of natural gas was produced.

c. $\int_{40}^{60} f(x)\,dx$

23. a. A quadratic model for the data is

$C'(x) = (7.714 \cdot 10^{-5})x^2 - 0.047x + 8.940$ dollars per CD

when x CDs are produced each hour.

b. $C(x) = \dfrac{7.714 \cdot 10^{-5}}{3}x^3 - \dfrac{0.047}{2}x^2 + 8.940x$
$\quad - 143.893$ dollars per CD

when x CDs are produced each hour.

c. $\int_{200}^{300} C'(x)\,dx \approx 196.14$ dollars

When production is increased from 200 to 300 CDs per hour, cost increases by about $196.14.

25. a, b.

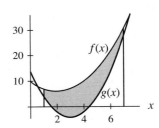

c. 54

27. a, c.

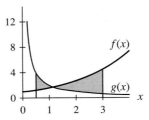

b. $x \approx 1.134$

d. $\int_{0.5}^{3} [f(x) - g(x)]\,dx \approx 2.812$

e. Area ≈ 4.172

29. a. When the amount invested in capital increases from $1500 to $5500, profit increases by approximately $1.33 thousand.

b. Area $= \int_{1.5}^{5.5} [r'(x) - c'(x)]\,dx = 13.29$

31. a. The population of the country grew by 3690 people in January.

b. The population declined by 9720 people between the beginning of February and the beginning of May.

c. −6030 people

d. Because the graphs intersect, the area of R_1 represents an increase in population, and the area of R_2 represents a decrease. The net change is the difference Area of R_1 − area of R_2.

The total area is the sum:

Area of R_1 + area of R_2.

33. a. Before fitting models to the data, add the point (0,0), and convert the data from miles per hour to feet per second by multiplying each speed by $\left(\frac{5280\ \text{feet}}{1\ \text{mile}}\right)$ $\left(\frac{1\ \text{hour}}{3600\ \text{seconds}}\right)$. The speed of the Supra after t seconds can be modeled as

$s(t) = -0.702t^2 + 20.278t + 2.440$ feet per second

The speed of the Carrera after t seconds can be modeled as

$c(t) = -0.643t^2 + 18.963t + 5.252$ feet per second

b. Approximately 17.96 feet

c. Approximately 18.04 feet

35. a.

Rate of change of revenue
(billions of dollars per year)

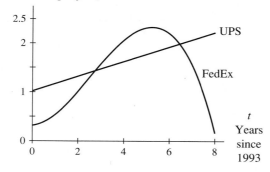

FedEx: $F(t) = -0.026t^3 + 0.198t^2 + 0.06t + 0.317$
billion dollars per year t years after 1993

UPS: $U(t) = 0.15t + 1.022$ billion dollars per year t years after 1993

b. Area of region on left $\approx$ $1.28 billion
Area of region in middle $\approx$ $1.20 billion

Area of region on right $\approx$ $1.59 billion
Between the beginning of 1993 and late 1996 ($t \approx 2.8$), UPS's accumulated revenue exceeded that of FedEx by approximately $1.28 billion. Between late 1996 and the spring of 2000 ($t \approx 6.3$), FedEx's accumulated revenue exceeded that of UPS by approximately $1.2 billion. From the spring of 2000 until the end of 2001, UPS's accumulated revenue exceeded that of FedEx by about $1.59 billion.

c. $\int_0^8 [F(t) - U(t)]\,dt \approx$ $1.68 billion. This value is the net amount by which UPS's accumulated revenue exceeded that of FedEx between 1993 and 2001.

37. a. $f(x) = \begin{cases} 0 \text{ tons per year} & \text{when } 0 \le x < 5 \\ \dfrac{557.960}{1 + 91.202e^{-0.318025x}} \\ \quad\text{tons per year} & \text{when } x \ge 5 \end{cases}$

x years after 1990

b, c.

Carbon production
and absorption
(tons per year)

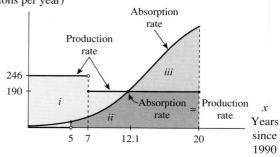

d. i. 2054.9 tons

ii. 2137.1 tons

iii. 1269.0 tons

e. The factory produces 4192 tons, and the trees absorb 3406 tons. This does not comply with the federal regulation.

39. Answers are located on the *Calculus Concepts* on CD-ROM and web site.

41. Answers will vary.

Section 6.6

1. a. $V(t) = -1.664t^3 + 5.867t^2 + 1.640t + 60.164$ mph
t hours after 4 P.M.

b. 68.99 mph

c. 72.23 mph

3. a.
$$R(y) = \begin{cases} \dfrac{0.719}{1 + 0.005e^{0.865563y}} + 0.62 \\ \quad \text{dollars per minute} \quad \text{when } 2 \le y < 10 \\ -0.045y + 1.092 \\ \quad \text{dollars per minute} \quad \text{when } 10 \le y \le 20 \end{cases}$$

where y is the number of years since 1980

b. $0.99 per minute

c. $0.67 per minute

5. a. 87.8 million people

b. Solving population $\approx$ 87.8 for t gives $t \approx 94.1$ years since 1900. This corresponds to early 1995.

c. 2.29 million people per year

7. a. -100.6 yearly accidents per year

b. 2810.5 yearly accidents

c.

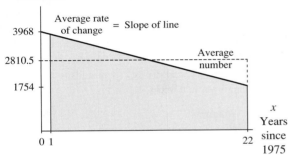

9. a. 2.28 feet per second squared

b. 129.7 feet per second

c. 4540.7 feet

d. 4540.7 feet

e.

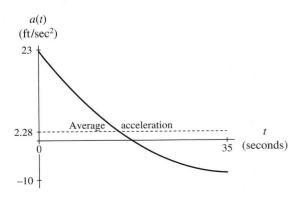

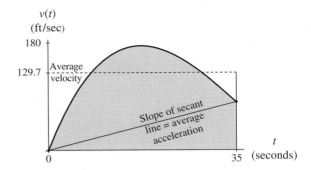

Area of shaded region = distance traveled = area of rectangle with height 129.7 ft/sec and width 35 seconds.

11. a. $V(t) = 1.033t + 138.413$ meters per second t microseconds after the experiment began

b. 174.58 meters per second

13. a. $B(t) = 0.030t^2 - 0.718t + 3.067$ mm Hg per hour t hours after 8 A.M.

b. 0.21 mm Hg per hour

c. 93.4 mm Hg

15. -0.32 seconds per year

17. a. The average of the data is highest between 10 A.M. and 6 P.M.

b. $C(x) = \begin{cases} 4.75x + 2.833 \text{ ppm} & \text{when } 0 \le x < 4 \\ 0.536x^2 - 7.871x \\ \quad + 45.200 \text{ ppm} & \text{when } 4 \le x \le 12 \\ -5.5x + 93.667 \text{ ppm} & \text{when } 12 < x \le 16 \end{cases}$

x hours after 6 A.M. Average concentration between 10 A.M. and 6 P.M. is 19.4 ppm.

c. Average $\approx$ 16.9 ppm.

d. Severe pollution warning.

19. Consider the two graphs of a function f shown below, where A is the average value of f from a to b and k is an arbitrary constant.

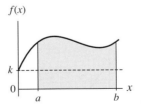

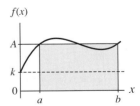

We know that the areas of the two shaded regions are equal. If we remove from each graph the rectangular region with height k and width $b - a$, the areas of the resulting regions are still equal, because we have removed the same area from each. See the following graphs.

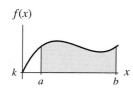

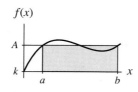

It is true for the graphs shown in this section with vertical axis shown from k rather than from zero that the area of the region between the function and $y = k$ from a to b is the same as the area of the rectangle with height equal to the average value minus k and width equal to $b - a$.

21. Answers are located on the *Calculus Concepts* CD-ROM and web site.

Section 6.7

1. $\int 2e^{2x}dx = e^{2x} + C$

3. Not possible using the techniques discussed in this text.

5. $\int (1 + e^x)^2\,dx = \int (1 + 2e^x + e^{2x})\,dx$

$$= x + 2e^x + \frac{e^{2x}}{2} + C$$

7. Not possible using the techniques discussed in this text.

9. Approximately 2.5452

11. Exactly $\dfrac{(\ln 5)^2}{2} - \dfrac{(\ln 2)^2}{2}$

13. Approximately 3.6609

15. Exactly $\ln 17 - \ln 10$

17. Approximately 8.7595

19. Exactly $1\frac{1}{12}$

CHAPTER 6 REVIEW TEST

1. a.
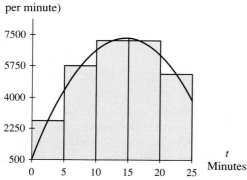

b. 139,237.5 cubic feet

c. In the first 25 minutes that oil was flowing into the tank, approximately 139,238 cubic feet of oil flowed in.

2. a. A quadratic model for the data is

$S(t) = -1.643t^2 + 16.157t + 0.2$ miles per hour

t hours after midnight

b.

n	Sum
5	127.131
10	126.869
20	126.803
40	126.786
80	126.782
160	126.781
Limit ≈ 126.8 miles	

3. a. The area beneath the horizontal axis represents the amount of weight that the person lost during the diet.

b. The area above the axis represents the amount of weight that the person regained between weeks 20 and 30.

c. The person's weight was 11.3 pounds less at 30 weeks than it was at 0 weeks.

d.
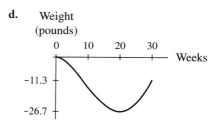

e. The graph in part d is the change in the person's weight as a function of the number of weeks after the beginning of dieting.

4. a. $R(t) = 10\left(\dfrac{-3.2}{3}t^3 + \dfrac{93.3}{2}t^2 + 50.7t\right) + 5000$ cubic feet

after t minutes

b. $R(10) - R(0) \approx 41{,}053.3$ ft³

c. Solving for t in the equation $R(t) = 150{,}000$, we find that the tank will be full after approximately 28 minutes.

5. a. $2598.60

b. At the end of the third quarter of the third year, the $10,000 had increased by $2598.60 so that the total value of the investment was $12,598.60.

c. $944.94 per year

6. $\int_{79}^{88} [m(t) - w(t)]\, dt \approx \$123$ thousand

Between the beginning of 1980 and the end of 1988, a man earning the average full-time wage would have earned approximately \$123,000 more than a woman earning the average full-time wage.

CHAPTER 7

Section 7.1

1. 15

3. 0.3

5. −0.02

7. Diverges

9. Diverges

11. Diverges

13. a. Approximately 0.0002 milligram

Approximately 0.0015 milligram

b. $\int_0^\infty r(t)\,dt = \lim_{N\to\infty} \dfrac{-1.55(0.9999999845^t)\cdot 10^{-6}}{\ln(0.9999999845)}\Big|_0^N =$

$-99.99999923 \approx -100$ milligrams

15. a. $p_0 \approx \$28.04$

b. $C = qp_0 + \int_{p_0}^\infty D(p)\,dp$

$\approx 150(28.04) + \int_{28.04}^\infty 499.589(0.958^p)\,dp$

$\approx \$7702$ million

Consumers are willing and able to spend about \$7.7 million for 150,000 books.

17. $\int_0^\infty 0.1e^{-0.1x}\,dx = \lim_{N\to\infty}\int_0^N 0.1e^{-0.1x}\,dx$

$\lim_{N\to\infty}\left(-e^{-0.1x}\big|_0^N\right) = \lim_{N\to\infty}\left[-e^{-0.1N} - (-e^0)\right]$

$\lim_{N\to\infty}\left(-e^{-0.1N}\right) + \lim_{N\to\infty}\left(e^0\right) = 0 + 1 = 1$

Section 7.2

1. a. i. $R(m) = 0.2\left(\dfrac{\$47{,}000}{12}\right) = \$783.33$ per month

ii. $R(m) = 0.2\left(\dfrac{47{,}000}{12} + 100m\right) = 783.33 + 20m$
dollars a month after m months

iii. $R(m) = 0.2\left(\dfrac{47{,}000}{12}(1.005^m)\right) = 783.33(1.005^m)$
dollars a month after m months

b. i. \$53,493.40

ii. \$92,082.72

iii. \$61,818.49

The first option is the only one that will not result in the amount needed for the down payment.

3. \$11.2 billion

5. a. \$51.5 billion

b. \$27.8 billion

7. a. $\int_0^6 6000e^{0.0634(6-x)}\,dx \approx \$43{,}804.70$

b. $\sum_{m=0}^{71} 500\left(1 + \dfrac{0.0634}{12}\right)^{(72-m)} \approx \$43{,}896.84$

c. Answers may vary.

9. a. $r(q) = 82.1(1.05^q)(0.15)$ million dollars per quarter q quarters after the third quarter of 2002

b. $R(q) = 82.1(1.05^q)(0.15)(1.09)^{16-q}$ million dollars per quarter for money invested q quarters after the third quarter of 2002

c. If the investment begins with the fourth quarter 2002 profits, then the initial investment is based on a profit of $(82.1)(1.05) = \$86.205$ million. Thus we calculate

$\sum_{q=0}^{15} 86.205(0.15)(1.05^q)(1.09)^{16-q} \approx \$629.8$ million

11. a. \$79.87 thousand   **b.** \$58.29 thousand

13. a. \$28,324.60   **b.** \$28,445.37   **c.** Answers will vary.

15. a. \$5.4 billion   **b.** \$6.1 billion   **c.** \$11.2 billion

17. a. \$7.3 billion   **b.** \$5.6 billion   **c.** Answers will vary.

19. \$5.2 million

21. Answers given are based on the end of 2003.

a. 0.14 million terns

b. $T(t) = 2.04(0.83)^{24-t}$ million terns born t years after 1979

c. 10.96 million terns

23. a. None

b. $S(t) = (60 - 0.5t)(0.67)^{50-t}$ thousand seals born t years from now

c. 90.5 thousand seals

Section 7.3

1. a. The demand function

b. The supply function

c. The producers' surplus

d. The consumers' surplus

3. a. To find the price P above which consumers will purchase none of the goods or services, either find the smallest positive value for which the demand function is zero, $D(p) = 0$, or, if $D(p)$ is never exactly zero but approaches zero as p increases without bound, then let $P \to \infty$.

b. The supply function, S, is often a piecewise continuous function with the first piece being the 0 function. The value p at which $S(p)$ is no longer 0 is the shutdown price. The shutdown point is $(p_1, S(p_1))$. When S is a continuous function the shutdown point is $(0, S(0))$.

c. The market equilibrium price, p_0, can be found as the solution to $S(p) = D(p)$. That is, it is the price at which demand is equal to supply. The equilibrium point is the point $(p_0, D(p_0)) = (p_0, S(p_0))$.

5. a. $q_0 \approx 27.5$ thousand units

b, c.

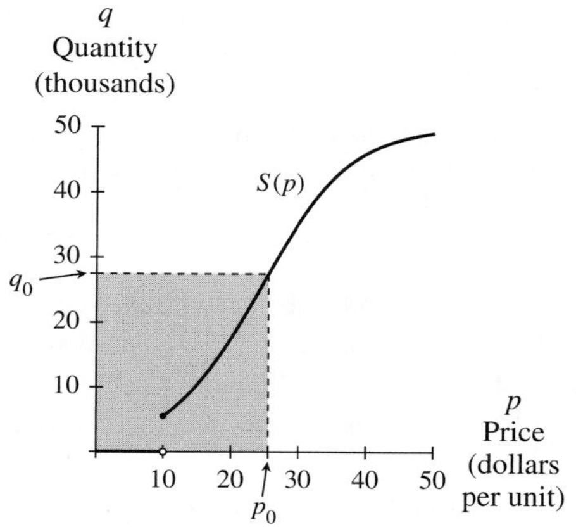

(a) Mathematician's viewpoint

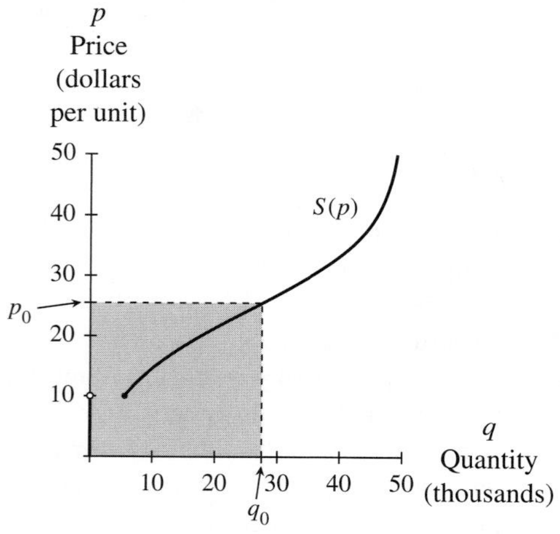

(b) Economist's viewpoint

7. a.

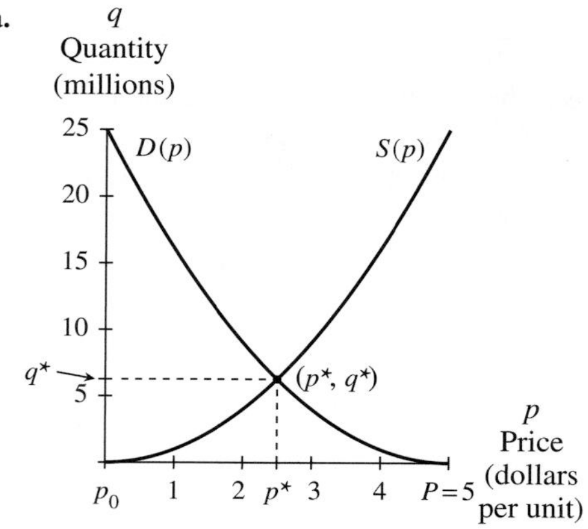

(a) Mathematician's viewpoint

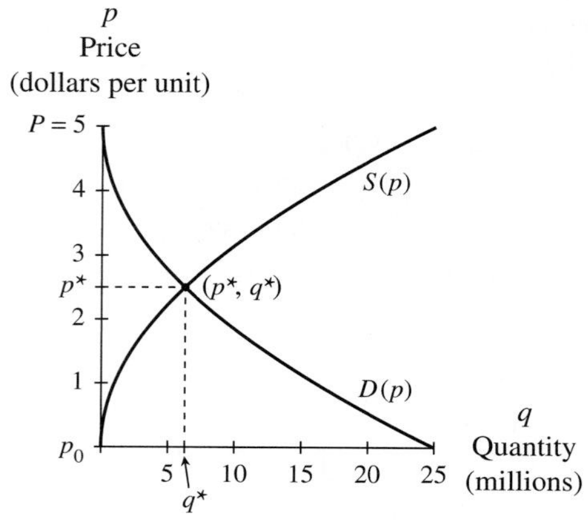

(b) Economist's viewpoint

$p^\star \approx 2.5$ dollars per unit, $q^\star \approx 6.25$ million units, $p_1 = 0$ dollars per unit, $P \approx 5$ dollars per unit

b, c.

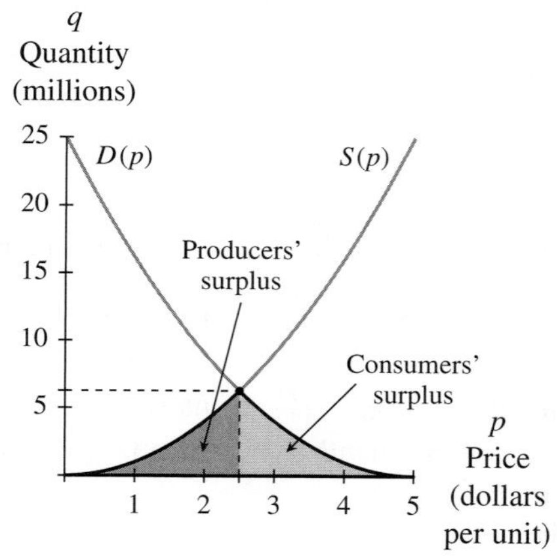

(c) Mathematician's viewpoint

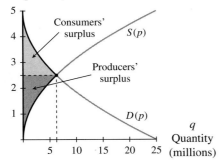

p
Price
(dollars per unit)

(d) Economist's viewpoint

d. $TSG = \int_0^{2.5} S(p)\,dp + \int_{2.5}^5 D(p)\,dp$ million dollars

9. a. D is an exponential demand function and so does not have a finite value p at which $D(p) = 0$. Thus the equation does not indicate a price above which consumers will purchase none of the goods or services.

b. $6128.6 thousand

c. 17.4 thousand fans

d. $4331.5 thousand

11. a. $D(p) = 0.025p^2 - 1.421p + 19.983$ lanterns when the market price is $p per lantern

b. $109.18

c. $31.89

13. a. 18.4 thousand answering machines; 300 thousand answering machines

b. $9981 thousand; $3131 thousand

15. a. $S(p) = \begin{cases} 0 \text{ hundred prints} & \text{when } p < 5 \\ 0.300p^2 - 3.126p + \\ 10.143 \text{ hundred prints} & \text{when } p \ge 5 \end{cases}$

where p hundred dollars is the price of a print

b. $837.12

c. Producers' revenue = $148.5 thousand
Producers' surplus = $27.3 thousand

17. a. $408.3 hundred

b. 297 sculptures; No

c. $4542.2 hundred

19. a. $D(p) = 499.589(0.958086^p)$ thousand books when the market price is $p per book

b.

$S(p) = \begin{cases} 0 \text{ thousand books} & \text{when } p < 18.97 \\ 0.532p^2 - 20.060p + \\ 309.025 \text{ thousand books} & \text{when } p \ge 18.97 \end{cases}$

where $p is the price of a book

c. Approximately $27.15; 156.2 thousand books

d. $4728.6 thousand

CHAPTER 7 REVIEW TEST

1. a. $R(t) = (0.1)(3000 \cdot 12 + 500t) = 3600 + 50t$ dollars per year after t years

b. $29,064

c. $17,664

2. a. $35,143.80

b. $18,277.65

3. a. Approximately 1 fox

b. $f(t) = 500(0.63)^{20-t}$ foxes born t years after 1990 that will still be alive in 2010

c. Approximately 1083 foxes

4. a. Approximately $635.4 hundred

b. 411 fountains. No, because $D(10) = 594$, supply is smaller than demand at this point.

c. Approximately $6236.0 hundred

Index of Applications

General Applications

Subject Index

polynomial models, 117–129
  cubic, 124–129
  quadratic, 118–124
  selection of, 139
  shifts built into, 136
population
  as biological stream, 485–486
  as continuous function, 42–43
Power Rule
  for antiderivatives, 419
  for derivatives, 248–249, 254
present value
  of continuous stream, 480–481
  of discrete stream, 483
price, 13
  demand curves and, 491–498, 503–506
  equilibrium, 503–506
  shutdown point of, 498, 499, 500
  supply curves and, 498–506
principal, 25
problem-solving strategy, for optimization, 341–342
producers' revenue, 500–503
producers' surplus, 500–506
producers' willingness and ability to receive, 499–500, 502–503
product functions, 21
  antiderivatives of, 461–462
  derivatives of, 280–285
Product Rule, 280–285
profit, 13
  marginal, 303–305
Pythagorean Theorem, 6

Q

quadratic functions, 118, 122
quadratic models, 118–124
  cubic models distinguished from, 127
  exponential models distinguished from, 122–124, 141–142
  finding from data, 118–121
  graphs of, 118, 121, 122
  selection of, 137, 139, 140, 141–142
  shifts built into, 136
Quotient Rule, 285

R

range, 7
rate
  of biological streams, 485–486
  of continuous income streams, 477–478

  of discrete income streams, 482–483
rate of change, 191–193. *See also* derivatives
  approximating change with, 299–305
  average, 158–166, 191, 450–455
  of future value, 479
  instantaneous, 173–183, 191, 198
  of linear function, 52, 54, 55
  percentage, 199–201
  of quadratic function, 118
  recovering function from, 409–410
  units of, 409–410
rate-of-change formulas, 256, 267
  for composite functions, 271–276
  for constant function, 248, 254
  for constant multipliers, 249–250, 254
  for differences, 250–253, 254
  for exponential functions, 261–264, 266
  for linear function, 248
  for logarithmic functions, 264–266
  piecewise functions and, 254–256
  for powers, 248–249, 254
  for product functions, 280–285
  for quotient functions, 285
  for sums, 250–253, 254
rate-of-change graphs, 234–242
  accumulated change and, 368–375, 378
recovering a function, 409–410
rectangles
  representing accumulated change, 371–378, 386–390, 394
regression
  linear, 57
related rates, 353–358
relative extrema, 311–319
  definition of, 311–312
  endpoints and, 313
  existence of, 315–319
  smooth functions and, 318–319
  steps for finding, 321
relative maxima or minima. *See* relative extrema
renewal functions, 485–486
results of change, 368–370. *See also* accumulated change
  approximation of, 300–303
revenue, 13
  marginal, 303–305
  producers', 500–503
right limits, 37–39
right triangles, 6
right-rectangle approximation, 373–375
  with count data, 377–378
rise, 51, 52–53
rounding, 60–62

Rule of 72, 231
run, 51, 52–53

S

saturation level, 103
scatter plots, 42, 57
  model selection and, 137–140
  slope graph for, 240–241
secant lines, 161
  finding change from, 161–162
  finding slope from, 206–208, 217–223
  symmetric difference quotient and, 182–183
  tangent lines and, 177–178, 181, 206–208
second derivatives
  concavity and, 333–334
  inflection points and, 327–334
  nonexistent, 333
  notation for, 329
second differences, 118, 119, 140, 141–142
sharp point, 174, 181, 210
  relative extremum at, 318–319
shifting data, 136–137. *See also* aligning data
  exponential, 107–108, 111, 136, 137
  logarithmic, 92, 95–96, 136–137
  logistic, 107–111, 136, 137
  polynomial, 136–137
shutdown point, 498, 499, 500
Simple Power Rule, 248–249
  for antiderivatives, 419
  for derivatives, 248–249, 254
slope. *See also* derivatives
  as average rate of change, 161–162
  finding by algebraic method, 217–223
  finding by numerical method, 206–212
  of graph, 175–176
  at inflection point, 327
  as instantaneous rate of change, 173–177, 191
  of line, 51–53, 54
  of piecewise continuous function, 210–212
  tangent and (*see* slopes of tangent lines)
  undefined, 241–242
  units of, 55
  zero, 312
slope graphs, 234–242
  of accumulation functions, 415–416
  maxima and minima of, 316